BIOARCHAEOLOGY IN THE CARIBBEAN

Bioarchaeology in the Caribbean assembles leading and emerging scholars in Caribbean bioarchaeology, offering an overview of current research in genomic analyses, deathways, demography and health, diet and population mobility, and research ethics.

Chapters emphasize the importance of culture in human adaptation and behavior at both population and individual levels. The first volume to focus solely on Caribbean bioarchaeology, this book is a landmark in this rapidly advancing area of scholarship, providing insight into current research methods and theoretical debates. The Caribbean region has a long and diverse history, and the chapters reflect this, discussing Indigenous, African, and European colonial populations, temporally spanning the "Archaic" period, the Early and Late "Ceramic" periods, the time of first European contact, and the Colonial period.

Bioarchaeology in the Caribbean will appeal to undergraduates, postgraduates, and researchers in bioarchaeology and Caribbean bioarchaeology and archaeology, in particular, as well as local stakeholders in the Caribbean (museum and archaeology professionals).

Darlene Weston is an Associate Professor of Biological Anthropology at the University of British Columbia in Vancouver, Canada.

Yadira Chinique de Armas is an Associate Professor of Biological Anthropology at the Department of Anthropology, University of Winnipeg, Canada.

BIOARCHAEOLOGY IN THE CARIBBEAN

Edited by
Darlene Weston and Yadira Chinique de Armas

Routledge
Taylor & Francis Group
LONDON AND NEW YORK

Designed cover image: Photograph taken by Joao Martínez and edited by
Luis Manuel Viera Sanfiel

First published 2025
by Routledge
4 Park Square, Milton Park, Abingdon, Oxon, OX14 4RN

and by Routledge
605 Third Avenue, New York, NY 10158

Routledge is an imprint of the Taylor & Francis Group, an informal business

British Library Cataloguing-in-Publication Data
A catalogue record for this book is available from the British Library

Library of Congress Cataloging-in-Publication Data
Names: Weston, Darlene, editor. | Chinique de Armas, Yadira, editor.
Title: Bioarchaeology in the Caribbean / edited by Darlene Weston and
Yadira Chinique de Armas.
Description: Abingdon, Oxon; New York, NY: Routledge, [2025] |
Collection of essays by Kathrin Nägele and 26 others. |
Includes bibliographical references and index.
Identifiers: LCCN 2024058807 (print) | LCCN 2024058808 (ebook) |
ISBN 9780367001155 (hbk) | ISBN 9781032193878 (pbk) |
ISBN 9780429444449 (ebk)
Subjects: LCSH: Human remains (Archaeology)—Caribbean Area—Methodology. |
Caribbean Area—Antiquities. | Genomics—Caribbean Area. | Slavery—
Caribbean Area—History. Classification: LCC F1619 .B56 2025 (print) |
LCC F1619 (ebook) | DDC 972.9—dc23/eng/20250203
LC record available at https://lccn.loc.gov/2024058807
LC ebook record available at https://lccn.loc.gov/2024058808

ISBN: 9780367001155 (hbk)
ISBN: 9781032193878 (pbk)
ISBN: 9780429444449 (ebk)

DOI: 10.4324/9780429444449

Typeset in Times New Roman
by codeMantra

CONTENTS

FIGURES

TABLES

CONTRIBUTORS

Jada Benn Torres (Ph.D., University of New Mexico) is an Associate Professor of Anthropology and Director of the Laboratory of Genetic Anthropology and Bio-cultural Studies in the Department of Anthropology at Vanderbilt University in Nashville, Tennessee. Her research in the Anglophone Caribbean explores genetic ancestry and population history of African and Indigenous Caribbean peoples.

Scott Burnett is a Professor of Anthropology at Eckerd College with research interests in skeletal biology, bioarchaeology, and dental anthropology. Much of his work examines worldwide developmental skeletal variation to better understand humans, past and present. His dental research also includes studies of morphology, wear, methodological biases, and intentional dental modification. Dr. Burnett has served as President of the Dental Anthropology Association and as Chair of the Comparative Cultures Collegium at Eckerd College. He co-edited the volume *A World View of Bioculturally Modified Teeth* (University Press of Florida) and has published over two dozen book chapters and articles.

Yadira Chinique de Armas is an Associate Professor in Bioanthropology at the Department of Anthropology, University of Winnipeg, with expertise in human skeletal biology, stable isotope analysis and starch analysis. Her research focuses on Caribbean human-environmental interactions, with emphasis on their subsistence strategies and mobility. Her current research deals with the biocultural diversity, adaptations and interactions of circum-Caribbean indigenous groups, the biological and cultural determinants of breastfeeding and weaning practices in ancient human populations and the origins of agriculture in the Antilles and Nicaragua. Currently, she is the leader of three multidisciplinary archaeological projects in Cuba and Nicaragua and the coordinator of the Research Program '*Cuba Aborigen*'.

Patrice Courtaud is a research engineer at the CNRS in the laboratory Préhistoire à l'Actuel: Culture, Environnement, Anthropologie (PACEA) and Director of the Fédération des Sciences Archéologiques de Bordeaux (FSAB) at the University of Bordeaux (F). He conducts research on the funerary practices, and more generally in archaeothanaology, of ancient populations. He is particularly interested in the cemeteries of the colonial period in the overseas French territories. In different areas, he leads excavations on sepulchral sites of other periods from Mesolithic to historical times to document the diversity of funerary behaviors.

Marie-France Deguilloux is a palaeogeneticist and head of the palaeogenetic platform of UMR PACEA (University of Bordeaux), which hosts expert anthropologists and archaeologists. This position has allowed her to develop collaborations to reconstruct ancient human population dynamics. More recently, the Bordeaux team has specifically investigated the origin and interaction of ancient communities from both mainland French territory and the Antilles. At the regional level, ancient DNA data have revealed a correlation between cultural and genetic exchanges, while at the local level, it has facilitated the comparison of funerary practices and documented the social organization of ancient groups (matrimonial systems, lineage systems, etc.).

Harper Dine is a Ph.D. candidate in Anthropology at Brown University, where she studies Maya archaeology and paleoethnobotany, focusing on ancient Maya agricultural strategies and foodways. Her interests span broader themes of food security and political economy. She received her BA from the University of Miami, where she trained in the analysis of dietary stable isotopes. She has participated in fieldwork in Yucatán, Chiapas, Puerto Rico, and Florida.

Scott M. Fitzpatrick is a Professor in the Department of Anthropology and Associate Director for Research and Collections at the Museum of Natural and Cultural History, University of Oregon. He is an archaeologist who specializes in the prehistory and historical ecology of island and coastal regions, particularly the Pacific and Caribbean. Much of his research has focused on colonization events, seafaring strategies, adaptations to smaller islands, exchange systems, and human impacts on ancient environments. Dr. Fitzpatrick is the founding co-editor of *The Journal of Island and Coastal Archaeology* and has published several books and more than 140 journal articles and book chapters.

Naila García Méndez is a Ph.D. student in Biomedical Sciences at the University of Talca in Chile. Her current research interests focus on population genetics and its relationship to complex diseases. During her undergraduate and graduate studies, she has gained experience in human osteoarchaeology and in the application of geometric morphometrics to the study of human skeletal remains.

Ulises M. González Herrera is a Postdoctoral Fellow at the University of Winnipeg and an Assistant Professor and Researcher at the Department of Archeology, Cuban Institute of Anthropology of the Agency of Social and Humanistic Sciences of the Ministry of Science, Technology and Environment (CITMA). His research interest is oriented towards ethnohistorical studies on Hispanic narrative sources from the 15th and 16th centuries, spatial archeology and functional analysis of grinding tools. He is currently head of the research project *Human landscape, environmental changes and socio-ecological resilience in Guacanayabo: The Playa del Mango archaeological site, an exemplary case study.*

John Krigbaum is a Professor of Anthropology at the University of Florida. He received his Anthropology Ph.D. from New York University in 2001, where he focused his studies on biological anthropology and bioarchaeology and conducted dissertation research on the prehistory and isotope ecology of tropical Southeast Asia. Dr. Krigbaum directs the Bone Chemistry Lab, where he leads collaborative training and research initiatives using a variety of isotope systems and tissues recovered from diverse settings. In addition to more than 70 journal articles and book chapters, he co-edited the volume *Isotope Research in Zooarchaeology: Methods, Applications, and Advances* (University Press of Florida).

Jason E. Laffoon (Ph.D., Leiden University, 2012) is currently Assistant Professor and Department Chair of the Department of Archaeological Science, Faculty of Archaeology, Leiden University. His main research interests focus on the study of patterns of mobility, migration, diet, and exchange. Jason has conducted archaeological fieldwork in Europe, North America, South America, and the Caribbean, and is currently involved in numerous research projects applying archaeological science approaches to address fundamental questions about past human lifeways.

Fanny Mendisco is an engineer within the paleogenetic platform of the PACEA laboratory (UMR 5199, University of Bordeaux). Her research focuses on the study of the history of Holocene human populations through paleogenetic and paleogenomic analyses. Mendisco's research includes a study of the evolutionary history of Amerindian populations in the Lesser Antilles, with the aim of characterizing not only the presumed migrations but also the modalities of interaction (biological, cultural) between the different groups involved in these processes.

Dany Morales Valdés is a Senior Researcher with a Master's in Anthropology, specializing in Physical Anthropology (University of Havana, 2009). He has been working in the Archaeology Department of the Cuban Institute of Anthropology since 2005. He has participated in research on Indigenous archaeology in Cuba, with a focus on physical anthropology, rock art, and the broader heritage of Indigenous archaeology. He is an active member of the Speleological Society of Cuba since 2004 and the Cuban Group for Rock Art Research (GCIAR).

Luz Veronica Muñoz Guevara is an Adjunct Professor in the Department of Anthropology and Sociology at the University of Puerto Rico Recinto Río Piedras. She has experience in the bioarchaeology of Caribbean Indigenous populations with emphasis on their physical traits, paleopathologies and mortuary practices.

Reginald Murphy (Ph.D., University of Calgary) is an archaeologist and has been conducting excavations and research in the Eastern Caribbean for the past 22 years. He works in a variety of capacities in cultural resources management and as the secretary general for the UNESCO National Commission for Antigua and Barbuda. Murphy is affiliated with the University of Nebraska-Lincoln, School of Global Integrative Studies and Anthropology, and the Graduate Centre, City University New York. He works in close collaboration with colleagues on numerous sites, including the Royal Naval Hospital Cemetery, the Galleon Beach Cemetery, the Indian Creek pre-Columbian site, and most recently mapping the early impacts of climate change on Antigua and Barbuda.

Kathrin Nägele (Ph.D., Max Planck Institute for the Science of Human History, 2021) is currently a postdoctoral researcher at the Max Planck Institute for Evolutionary Anthropology in Leipzig, Germany. During her doctoral studies, she gained expertise in retrieving DNA from regions with poor DNA preservation and applied her experience in tropical areas, focusing on the Caribbean and the Pacific.

Maria Nieves-Colón (Ph.D., Arizona State University, 2017) is Assistant Professor of Anthropology at the University of Minnesota Twin Cities. Her research uses genomics to investigate human population history in the Caribbean and Latin America. A recent publication is "Anthropological genetic insights on Caribbean population history" in *Evolutionary Anthropology* (2022).

Marie-Hélène Pemonge is an engineer at the CNRS (National Center for Scientific Research) and specializes in paleogenetic studies. She is the operational manager of the paleogenetic platform of PACEA (UMR 5199, University of Bordeaux). The platform's high-level equipment allows reliable paleogenetic/paleogenomic analyses of human, animal, and plant remains covering a wide chronological period. The topics of research carried out are numerous, including documentation of dynamics of human migrations, study of funeral groups, domestication, etc.

Ali Pourmand (Ph.D., Tulane University, 2006) is an Associate Professor in the Rosenstiel School of Marine, Atmospheric, and Earth Science at the University of Miami. His research focuses on applications of isotope geochemistry and extraction chromatography in paleoclimatology, paleoceanography, geoarchaeology, and cosmochemistry.

William J. Pestle is Director of Education at the Kelsey Museum of Archaeology at the University of Michigan. He is a bioarchaeologist who specializes in the study of the ancient peoples of Latin America and the Caribbean, and he has active research projects in Puerto Rico and the Atacama Desert of northern Chile. His academic background includes a BA from the University of Michigan, an MSc from the University of Bradford (UK), and a PhD from the University of Illinois at Chicago. His research efforts have been supported by the National Science Foundation and National Endowment for the Humanities, among others.

Mirjana Roksandic is a Professor in the Department of Anthropology and coordinator of the interdisciplinary program in Bioanthropology at the University of Winnipeg. Her main research topics include Pleistocene hominin evolution in Europe, particularly the Eastern Mediterranean, and mortuary rituals among sedentary and semi-sedentary archaeological hunter-gatherers. She is interested in when, how, and why humans have moved across continents and vast expanses of lands and seas in the context of adaptation to climatic fluctuations and social pressures. Prof. Roksandic has active projects in Serbia, Cuba, Nicaragua, and Tanzania funded by both NSERC and SSHRC grants, and is the recipient of the Nellie McClung Foundation Manitoba Women Trailblazer Award.

Thomas Romon is archaeologist at INRAP (Institut National de Recherches Archéologiques Préventives) in Guadeloupe and is attached to the laboratory Préhistoire à l'Actuel: Culture, Environnement, Anthropologie (PACEA) at the University of Bordeaux (F). For 20 years, he has worked on the funerary practices of pre-Columbian populations and of the cemeteries of the colonial period in the Lesser Antilles.

Hannes Schroeder (DPhil, Oxford) is an Associate Professor, Globe Institute, Section for Molecular Ecology at the University of Copenhagen. His research focuses on the use of ancient genomics to explore past population histories and the spread and evolution of infectious diseases. He has led several large research projects and his research has been published in *Nature*, *Science* and *PNAS*. He was co-developer of the Ancient Genome Atlas and heads the ERC Consolidator Project AlpGen.

Kristrina A. Shuler (Ph.D., Southern Illinois University, Carbondale, US, 2005) is an Associate Professor of Anthropology at Auburn University. She is a biological anthropologist who utilizes archaeological, historical, and contemporary data to explore the complex entanglements of health and labor in food production systems from the Caribbean, Latin America, and the eastern U.S.

Jessica Stone is a National Science Foundation SBE postdoctoral fellow in the Department of Anthropology at the University of Minnesota Twin Cities. She is a bioarchaeologist who integrates ancient DNA, isotopes, osteology, and archaeological methods in her study of the process of initial human settlement and subsequent adaptation to island and coastal environments in Micronesia and the Caribbean. Her research has been published in several journal articles and book chapters, and she co-edited the volume *The Archaeology of Island Colonization: Global Approaches to Initial Human Settlement* (University Press of Florida).

Tamara L. Varney (Ph.D., University of Calgary) is Professor of Biological Anthropology in the Department of Anthropology at Lakehead University in Thunder Bay, Ontario, Canada. She has conducted bioarchaeological research on Antigua for 24 years, focused primarily on 18th- and 19th-century cemetery sites. Her expertise is in extrapolating human behavior via stable isotope and trace element analyses.

Christopher K. Waters (Ph.D., Syracuse University) is the Manager of the National Parks Authority Heritage Department in Antigua and Barbuda and the Focal Point for the Antigua Naval Dockyard and Related Archaeological Sites UNESCO World Heritage Site.

Darlene Weston (Ph.D., University College London) is an Associate Professor of Biological Anthropology at the University of British Columbia in Vancouver, Canada. As a human osteologist, paleopathologist, and mortuary archaeologist, she investigates how European contact influences the health, demography, and diet of Caribbean Amerindian populations. She also works with First Nations communities in Canada, facilitating the repatriation of ancestors from museum repositories and collaboratively creating osteobiographies before reburial.

INTRODUCTION

The Study of Archaeological Human Remains in the Antilles: The Development of the Bioarchaeological Approach

Darlene Weston and Yadira Chinique de Armas

Introduction

The study of archaeological human remains in the Caribbean archipelago has undergone a conceptual transformation in the last 60 years. From being a descriptive discipline focused on craniology and craniometry, bioarchaeology in the region today embraces the best spirit of a multidisciplinary science that aims to understand the biological aspects of the human skeleton in their cultural and environmental contexts. This new approach had its roots in a long and strong tradition in osteology initiated at the end of the nineteenth century in the Dominican Republic (Llenas, 1821), Cuba (Poey, 1865; Montane, 1874, 1877, 1885, 1887), Bahamas (Brooks, 1888), and Jamaica (Flower, 1891, 1895; Duerden, 1895, 1897; Haddon, 1897) and further developed by local researchers during the second half of the twenty century, with Cuban scholars at the forefront (*e.g.*, Royo, 1943; Rivero de La Calle, 1949, 1960; Herrera Fritot, 1964, 1965).[1] Although this period was characterized by academic interest in the study of the human skull, including intentional cranial modification, their work contributed significantly to the advancement of methodology in measuring and documenting various aspects of the entire skeleton (Crespo Torres et al., 2013).

A major change in regional osteological practice occurred in what Goodwin (1979) described as the *modern period* of Caribbean osteology (1965–1979). During those years, osteological analyses were influenced by the theoretical perspectives of *cultural ecology* that recognized the intertwined relationship between humans and their environment (Meggers, 1954; Steward, 1955). Regional osteological studies during this time reflected the interest of researchers in pathological conditions such as caries, treponematosis, osteomyelitis, and their causalities (e.g., Bullen, 1964; Torres and Rivero de La Calle, 1972; Luna Calderon, 1976

DOI: 10.4324/9780429444449-1

1977a, 1977b), with physical anthropologists Adelaide Kendall Bullen (Unites States, 1908–1987), Manuel Rivero de La Calle (Cuba, 1926–2001), and Fernando Luna Calderón (Dominican Republic, 1945–2005) as prominent researchers in the field. With the new potential of osteology to contribute to the understanding of human adaptations, local archaeologists increased their interest in burial sites, mortuary practices, and the integral study of the human skeleton (Luna Calderón, 1973, 1977; Harris, 1971). During the next 20 years, skeletal biology consolidated its place in the archaeological practice of the Greater (e.g., Luna Calderon, 1993; Crespo Torres, 1994, 1998; Rodríguez Lopéz, 1999) and the Lesser Antilles (e.g., Versteeg et al., 1990; Versteeg and Schinkel, 1992), with the biological anthropologist, Edwin Crespo Torres (Puerto Rico, 1961–2019), substantially contributing to the field from Puerto Rico. These years witnessed new developments in dental anthropology, including studies on the use of teeth as tools (Rivero de La Calle and Toribio, 1984; Crespo Torres, 1994), the reporting of new wear patterns (Crespo Torres, 1994), and dental morphological analysis (Crespo Torres, 1994; Coppa et al., 1995).

By the end of the 90s, new theoretical and methodological approaches from the field of bioarchaeology had been consolidated in the region, bringing a multidisciplinary approach to the study of burial sites (e.g., Righter, 2002; Rodríguez Suárez et al., 2006; Curet and Stringer, 2010; Valcárcel Rojas et al., 2011). Notably, studies aiming to understand mortuary behaviour became a common practice (e.g., Curet and Oliver, 1998; Siegel, 1999; Keegan, 2009; Crespo Torres, 2010; Hoogland et al., 2010; Valcárcel Rojas et al., 2011; Van den Bel and Romon, 2010), with an emerging interest in archaeothanatology (e.g., Hoogland and Hofman, 2013). In this context, osteochemical analysis of human remains experienced tremendous growth, opening up new questions in Caribbean archaeology on peopling of the islands, identity, population movements, diversity, and effects of colonization and slavery. Isotope analysis, for instance, became a standard practice, contributing to the understanding of Indigenous dietary traditions (e.g., Keegan and DeNiro, 1988; Stokes, 1998; Pestle, 2010; 2013; Chinique de Armas et al., 2015, 2022a; Laffoon et al., 2016, 2020; Pestle et al., 2023; Schulting et al., 2021), including breastfeeding and weaning (Chinique de Armas et al., 2017, 2022b; Chinique de Armas and Pestle, 2018), mobility patterns (e.g., Laffoon et al., 2013, 2017, 2018; Chinique de Armas et al., 2025), and the notion of diversity in lifeways. More recently, the development of genomics overcame the limitations of the pioneer mitochondrial DNA analyses (e.g., Lalueza Fox et al., 2003), showing great potential to contribute to the debate on population dynamics, including population origins and interactions (e.g., Nieves Colón, 2020; Nägele et al., 2020).

This volume constitutes the first interregional effort dedicated to visualizing bioarchaeological research in the Caribbean. Here we present a diverse set of papers representing the theoretical and methodological approaches currently used to understand diversity, adaptation, and people's lifestyles. May this volume serve to encourage the integration of bioarchaeological research at the regional level and promote debate on urgent issues in the region such as the decolonization of precolonial Antillean history, slavery, and ethics in the management and study of human remains.

Organization of the Volume

This book is divided into four sections. The first three chapters explore the complex biological roots of circum-Caribbean populations through analysis of their DNA. The first chapter, by Nägele and Nieves-Colón, introduces DNA methods and sampling strategies unique to Caribbean contexts, summarizes major findings and new avenues of research related to settlement patterns, exchange and kinship, and the continuity of pre-contact ancestry among current populations, and proposes a future path focusing on ethical and sustainable research. The second chapter, by Mendisco et al., analyses genetic data from archaeological and present-day populations to investigate the origins of the Indigenous communities of the Guadeloupe Archipelago, recognizing the impact of Europeans and the trans-Atlantic slave trade and emphasizing the importance of using both ancient and modern DNA to elucidate evolutionary histories. The third chapter in this section, by Benn Torres and Schroeder, focuses on the demographic histories of populations of African descent in the Caribbean. Using case studies across the Americas and recognizing the ethical implications and obligations of their research, they demonstrate how genetics can be used to understand how the diversity of experiences across the region has resulted in the formation of distinct Caribbean communities. These three papers are exemplary of the advanced scientific methods that are currently being applied to research on both modern and past Caribbean populations, while always considering and being cognizant of the ethical implications of their work.

The second section of the book includes six chapters, ordered chronologically, that explore the lifeways, demography, and health of Caribbean populations via osteological analyses. In this section's first chapter, Stone et al. recount their analyses of pre-contact skeletal remains from Grand Bay, Carriacou, in the Southern Grenadines – the first analytic work in this region from this time period. Next, Muñoz Guevara reviews the funerary customs of Pre-Columbian populations from Puerto Rico, highlighting the work of the late Edwin Crespo-Torres and providing a detailed analysis of individuals from the La Gallera site. Following is a chapter by Weston, who provides a bio-cultural analysis of population demography and health relating to the skeletal population from Juan Dolio, Dominican Republic – a population that spanned both pre- and post-colonial time periods and had to navigate a world of colonial domination. Next, García Méndez and colleagues present their results from the Cuban site of Playa del Mango. This site offers an unparalleled opportunity to explore the physical characteristics and lifeways of Indigenous people from Banwaroid tradition, an early Antillean group that inhabited eastern Cuba. Courtaud and Romon's chapter describes the analysis of slave cemeteries in the French West Indies, highlighting the characteristics that distinguish them from

European ones, particularly their marginalization. Lastly, within the framework of decolonizing archaeology, Shuler offers a synthesis and critical reappraisal of research at the Newton Plantation Burial Ground in Barbados, highlighting new research by BIPOC scholars who are offering new perspectives. All these papers illustrate how osteological research in the Caribbean has moved on from the basic recording of physical data to offer more sophisticated and nuanced analyses of lifeways and deathways that are bio-cultural and contextual in nature.

The third section of the book includes four chapters that explore Caribbean diet and population mobility using stable isotopes and trace element analyses. The section starts with Chinique de Armas and Laffoon's research on breastfeeding and weaning practices among circum-Caribbean populations. Highlighting the role of individual variability, they see a diversity of feeding practices in the region and few distinct patterns. In their second paper, Laffoon and Chinique de Armas summarize previous research on paleodiet and paleomobility in the Caribbean, emphasizing the need to look at both in combination as well as the potential and limitations of using multiple isotopic proxies. These themes are featured as they present results from the site of Cueva del Perico I, Cuba. Next, Pestle et al. use stable isotopes, Bayesian mixture modelling, radiocarbon dating, and strontium analysis to investigate dietary practices in a Later Ceramic Age site from Puerto Rico. They reveal variability that correlates with an increase in community cultural differentiation after AD 600. To conclude this section, Varney et al. describe their analysis of individuals of European and African descent buried in the Royal British Naval Hospital Cemetery in Antigua c. AD 1800. Using stable isotopes and trace elements, they provide greater insight into individual identities, particularly pertaining to ancestry, social status, and health. Like the papers focusing on genetics, the papers in this section demonstrate how the application of new and evolving scientific techniques can be used to glean previously unimaginable information from the remains of past populations, providing insight into dietary practices, population movements, and aspects of individual identities.

The volume concludes with a chapter by Roksandic, who provides a personal account of her experiences navigating the ethics of research in bioarchaeology. Emphasizing that ethics are historically bound, ever-changing, and situation-specific, Roksandic discusses archaeology as a colonial science, the issues surrounding international collaborations, working with sensitive materials, and disciplinary imbalances. The need to consider the ethical implications of one's work is a thread that weaves its way through the chapters of this book – an important advance in bioarchaeological research in the Caribbean, where concerns of descent communities were, until recently, largely overlooked when planning, implementing, and disseminating research.

Note

1 Historical accounts of the development of Caribbean osteological studies can be found in Goodwin (1979), Crespo Torres et al. (2013) and Crespo (2019).

References

Brooks, W. K. 1888. On the Lucayan Indians. *Memoirs of the National Academy of Sciences*, 10: pp. 215–223.

Bullen, A. K. 1964. Skeletal remains and dental pathology of the Savanne Suazey site, Grenada, West Indies. In *The Archaeology of Granada*. Bullen, R. P. (Ed), pp. 13–17. Contribution of the Florida State Museum, Social Sciences 11. Gainesville, FL: University of Florida Press.

Chinique de Armas, Y., and Pestle, W. 2018. Assessing the association between subsistence strategies and the timing of weaning among indigenous archaeological populations of the Caribbean. *International Journal of Osteoarchaeology*, 28 (5): pp. 492–509. https://doi.org/10.1002/oa.2695.

Chinique de Armas, Y., Buhay, W. M., Rodríguez, R., Bestel, S, Smith, D., Mowat, S. D., and Roksandic, M. 2015. Starch analysis and isotopic evidence of consumption of cultigens among fisher–gatherers in Cuba: The archaeological site of Canímar Abajo, Matanzas. *Journal of Archaeological Science*, 58: pp. 121–132. https://dx.doi.org/10.1016/j.jas.2015.03.003.

Chinique de Armas, Y., González, U. M., Reyes, I., Buhay, W. M., Skelton, S., Rodríguez, R., and Laffoon, J. 2022a. Multiproxy paleodietary reconstruction using stable isotopes and starch analysis: The case of the archaeological site of Playa del Mango, Granma, Cuba. *Journal of Archaeological Science,* (46): pp. 1–18. https://doi.org/10.1016/j.jasrep.2022.103671.

Chinique de Armas, Y., Mavridou, A. M., Garcell, J., Hanson, K., Laffoon, J. 2022b. Tracking breastfeeding and weaning practices in ancient populations by combining carbon, nitrogen and oxygen stable isotopes from multiple non adult tissues. *PloS One*, 17(2): pp. e0262435. https://doi.org/10.1371/journal.pone.0262435.

Chinique de Armas, Y., Roksandic, M., Nikitović, D., Rodríguez, R., Smith, D., García, D., Buhay, W. M. 2017. Isotopic reconstruction of the weaning process in the archaeological population of Canímar Abajo, Cuba: A Bayesian probability mixing model approach. *PLoS One*, 12 (5): pp. 1–23. https://dx.doi.org/10.1371/journal.pone.0176065.

Coppa, A., Chiarelli, B., Cucina, A., Luna Calderón, F., Mancinelli, D. 1995. Dental anthropology and paleodemography of the pre-Columbian populations of Hispaniola from the 3rd millennium BC to the Spanish conquest. *Human Evolution,* 10 (2): pp. 153–167.

Crespo-Torres, E. 1994. *Dental Analysis of Human Burials Recovered from Punta Candelero: A Prehistoric Site on the Southeast Coast of Puerto Rico*. Master's thesis, Arizona: Arizona State University.

Crespo-Torres, E. 1998. Análisis osteológico del hombre de Puerto Ferro. In *Tiempo, Población y Sociedad. Homenaje al Maestro Arturo Romano*. Jaen, M. T., López, S. A., Márquez L., and Hernández, O. P. (Eds.), pp. 331–342. Mexico, D. F.: INAH.

Crespo-Torres, E. 2010. Ancient bones tell stories: Osteobiography of human remains from Tibes. In Tibes. People, Power, *and Ritual at the Center of the Cosmos*. Curet, A. L., and Stringer, L. M. (Eds.) pp. 191–208. Tuscaloosa: The University of Alabama Press.

Crespo-Torres, E., Mickleburgh, H. L., and Valcárcel, R. R. 2013. The study of pre-Columbian human remains in the Caribbean archipelago. From descriptive osteology to a bioarchaeological approach. In *Oxford Handbook of Caribbean Archaeology*. Keegan, W. F., Hoffman, C., and Rodriguez, R. (Eds.), pp. 436–451. Oxford: Oxford University Press. https://doi.org/10.1093/oxfordhb/9780195392302.013.0122.

Curet, L. A., and Oliver, J. R. 1998. Mortuary practices, social development, and ideology in pre-Columbian Puerto Rico. *Latin American Antiquity*, 9(3): pp. 217–239.

Curet, L. A, and Stringer, L. M. 2010. *Tibes: People, Power and Ritual at the Center of the Cosmos (Caribbean Archaeology and Ethnohistory).* Tuscaloosa: University of Alabama.

Duerdon, J. E. 1895. Discovery of aboriginal Indian remains in Jamaica. *Nature*, 52(1338): pp. 173–174. https://doi.org/10.1038/052173d0.

Duerdon, J. E. 1897. Aboriginal Indian remains in Jamaica. *Journal of Institute of Jamaica*, 2(4): pp. 21–22.

Flower, W. H. 1891. Exhibition of two skulls from a Cave in Jamaica. *Journal of the Anthropological Institute of Great Britain and Ireland*, 20: pp. 110–112. https://doi.org/10.2307/2842230.

Flower, W. H. 1895. On recently discovered remains of the aboriginal inhabitants of Jamaica. *Nature*, 52(1355): pp. 607–608. https://doi.org/10.1038/052607a0.

Goodwin, C. R. 1979. The history and development of osteology in the Caribbean area. *Revista/Review Interamericana*, 8: pp. 463–494.

Haddon, A. C. 1897. Note on the craniology of the aborigines of Jamaica. *Journal of the Institute of Jamaica*, 2(4): pp. 23–24.

Harris, P. O. B. 1971. Preliminary report on Banwari Trace: A pre-ceramic site in Trinidad. In *Proceedings of the Fourth International Congress for the Study of the Pre-Colombian Culture of the Lesser Antilles*, pp. 115–125. Saint Lucia.

Herrera-Fritot, R. 1964. *Craneotrigonometría: Tratado práctico de geometría craneana.* Havana: Department of Anthropology, National Science Commission of the Republic of Cuba.

Herrera-Fritot, R. 1965. *Nueva técnica para calcular la capacidad craneana.* Havana: Academia de Ciencias de Cuba.

Hoogland M.L.P., and Hofman C.L. 2013. From Corpse thaphonomy to mortuary behavior in the Caribbean. In: *The Oxford Handbook of Caribbean Archaeology*. Keegan, W. F., Hofman, C. L., and Ramos, R. R. (Eds.), pp. 452-469. New York: Oxford University Press.

Keegan, W. F. 2009. Central Plaza Burials in Saladoid Puerto Rico: An Alternative Perspective. *Latin American Antiquity,* 20 (2): 375–385. https://doi.org/10.1017/S1045663500002686

Keegan, W. F., and DeNiro, M. J. 1988. Stable carbon-and nitrogen-isotope ratios of bone collagen used to study coral-reef and terrestrial components of prehistoric Bahamian diet. *American Antiquity*, 53 (2): pp. 320–336. https://doi.org/10.2307/281022.

Laffoon, J. E. 2013. Paleomobility research in Caribbean contexts: New Perspectives from isotope analysis. In *The Oxford Handbook of Caribbean Archaeology*. Keegan, W. F., Hofman, C., and Rodríguez, R. (Eds.), pp. 418–435. Oxford: Oxford University Press.

Laffoon, J. E., Espersen, R., and Mickleburgh, H. L. 2018. The life history of an enslaved African: Multiple isotope evidence for forced childhood migration from Africa to the Caribbean and associated dietary change. *Archaeometry*, 60 (2): pp. 350–365. https://doi.org/10.1111/arcm.12354.

Laffoon, J. E., Hoogland, M. L., Davies, G. R., and Hofman, C. L. 2016. Human dietary assessment in the Pre-colonial Lesser Antilles: New stable isotope evidence from Lavoutte, Saint Lucia. *Journal of Archaeological Science*, 5: pp. 168–180. https://doi.org/10.1016/j.jasrep.2015.11.020.

Laffoon, J. E., Rojas, R., Weston, D., Hoogland, M. L., Davies, G. R., and Hofman, C. L. 2020. Diverse and dynamic dietary patterns in early colonial Cuba: New insights from multiple isotope analyses. *Latin American Antiquity*, 31 (1): pp. 103–121. https://doi.org/10.1017/laq.2019.103.

Laffoon, J. E., Sonnemann, T. F., Shafie, T., Hofman, C. L., Brandes, U., and Davies, G. R. 2017. Investigating human geographic origins using dual-isotope ($^{87}Sr/^{86}Sr$, $\delta^{18}O$) assignment approaches. *PLoS One*, 12 (2): e0172562. https://doi.org/10.1371/journal.pone.0172562.

Lalueza-Fox, C., Calderón, F. L., Calafell, F., Morera, B., and Bertranpetit, J. 2003. Mitochondrial DNA from pre-Columbian Ciboneys from Cuba and the prehistoric colonization of the Caribbean. *American Journal of Physical Anthropology*, 121(2): pp. 97–108.

Llenas, A. 1821. *Découverte d'un Crane d'Indien Ciguayo à Saint-Domingue*. Nantes: Imprimeur de la Societe Academique.

Luna, C. F. 1973. Estudio comparativo y preliminar de dos cementerios neo-Indios: La Cucama y La Unión. *Boletín del Museo del Hombre Dominicano*, 3: pp. 11–47.

Luna, C. F. 1976. *Atlas de Paleopatología Ósea*. Santo Domingo: Universidad Central del Este.

Luna, C. F. 1977a. Primeras evidencias de sífilis en las Antillas precolombinas. Volume CCXLIII, *Cuadernos del Centro Domincano de Investigaciones Antropológicas (CENDIA)*, 2. Santo Domingo, Dominican Republic: Universidad Autonoma de Santo Domingo.

Luna, C. F. 1977b. Enterramientos humanos y patologías de Cueva Berna. In *Arqueología de Cueva de Berna.* Veloz-Maggiolo, M., Ortega, E., Nadal, J., Luna-Calderon, F., and Rimoli, R. O. (Eds.), pp. 27–32. San Pedro de Macoris: Universidad Central del Este.

Luna, C. F. 1993. Paleopatología de los grupos Taínos de la Española. In *La Cultura Taína*. pp. 165–174. Madrid: Biblioteca del V Centenario.

Meggers, B. J. 1954. Environmental limitation on the development of culture. *American Anthropologist*, 56: pp. 801–824. https://doi.org/10.1525/aa.1954.56.5.02a00060.

Montane, L. 1874. *Étude anatomique du crane chez les microcéphales, accompagnée de six planches lithographies par Format.* Paris: Librairie J. B. Bailliere.

Montane, L. 1877. El cráneo desde el punto de vista antropológico. *Anales de la Academia de Ciencias*, 14: pp. 96–98.

Montane, L. 1885. Un caribe cubano. Estudio craneológico. *Revista Cubana*, 2: pp. 97–114.

Montane, L. 1887. *Un cráneo Caribe.* Havana: Imprenta de Soler, Álvarez y Cia.

Nägele, K., Posth, C., Orbegozo, M., Chinique de Armas, Y., Hernández Godoy, S. T., González, U. M., Nieves-Colón, M. A., Sandoval-Velasco, M., Mylopotamitaki, D., Radzeviciute, R., Laffoon, J., Pestle W. J., Ramos-Madrigal J., Lamnidis T. C., Schaffer, W. C., Carr, R. S., Day, J. S., Arredondo, C., Rangel, A., Martínez-Fuentes, A. J., Crespo-Torres, E., Roksandic, I., Stone A. C., Lalueza-Fox, C., Hoogland, M., Roksandic, M., Hofman, C. L., Krause, J., and Schroeder H. 2020. Genomic insights into the early peopling of the Caribbean. *Science,* 369(6502): pp. 456–460. https://doi.org/10.1126/science.aba8697.

Nieves-Colón, M. A., Pestle W. P., Reynolds A. W, Llamas B., de la Fuente, C., Fowler K., Skerry K. M., Crespo-Torres E., Bustamante C. D, and Stone A. C. 2020. Ancient DNA reconstructs the genetic legacies of precontact Puerto Rico communities. *Molecular Biology and Evolution*, 37 (3): pp. 611–626. https://doi.org/10.1093/molbev/msz267.

Pestle, W. J. 2010. Diet and society in prehistoric Puerto Rico an isotopic approach. *PhD. dissertation*. Chicago: University of Illinois.

Pestle, W. J. 2013. Stable isotope analysis of Paleodiet in the Caribbean. In *The Oxford Handbook of Caribbean Archaeology*. Keegan W. F., Hofman, C. L., and Rodríguez, R. (Eds.), pp. 407–417. Oxford: Oxford University Press.

Pestle, W. J., Pérez, E. M., and Koski-Karell, D. 2023. Reconsidering the lives of the earliest Puerto Ricans: Mortuary archaeology and bioarchaeology of the Ortiz site. *PLoS One*, 18 (4): p. e0284291. https://doi.org/10.1371/journal.pone.0284291.

Poey, F. 1865. Cráneo de un Indio Caribe. In *Repertorio físico-natural de la isla de Cuba I*. Poey, F. (Ed.), pp.150–158. Havana: Imprenta del Gobierno y Capitania.

Righter, E. 2002 (Ed.). *The Tutu Archaeological Village Site: A Multidisciplinary Case Study in Human Adaptation*. London: Routledge. https://doi.org/10.4324/9780203165843.

Rivero de la Calle, M. 1949. *Deformación craneana en los aborígenes de Cuba: Estudio comparativo*. Ph.D. dissertation. Escuela de Ciencias, University of Havana.

Rivero de la Calle, M. 1960. Deformación craneana en los aborígenes de Cuba. Estudio comparativo. *Actes du VIe Congrès International des Sciences Anthropologiques et Ethnologique*, 1: pp. 251–260.

Rivero de la Calle, M., and Toribio Suarez, L. 1984. *Estudio odontométrico de la dentición permanente en los aborígenes de Cuba. Paper presented at X Congreso Nacional de Estomatología*, Havana.

Rodríguez Lopéz, M. 1999. Excavaciones en "Maruca", una comunidad Arcaica del sur de Puerto Rico. *Proceedings of the Congress of the International Asociation for Caribbean Archaeology*, 17: pp. 166–180. New Providence, Bahamas.

Rodríguez Suárez, R., Arredondo, C., Rangel, A., Hernández Godoy, S., de Lara, O., González, U. M., Martínez, J. G., and Pereira, O. 2006. *5000 años de ocupación prehispánica en Canímar Abajo*. Cuba: Matanzas, 30/6/2011. Available at https://cubaarqueologica.org/document/anto6_rodriguez.pdf.

Royo, G. F. 1943. Ensayo sobre cránea cubana precolombina. *Actas y Documentos del Primer Congreso Histórico y Municipal Interamericano*: pp. 23–28. Cuba: Havana.

Schulting, R. J., Snoeck, C., Pouncett, J., Brock, F., Ramsey, C. B., Higham, T., Devièse, T., Delancy, K., Pateman, M., Keegan, W., and Ostapkowicz, J. 2021. Six centuries of adaptation to a challenging island environment: AMS 14C dating and stable isotopic analysis of pre-Columbian human remains from the Bahamian archipelago reveal dietary trends. *Quaternary Science Reviews*, 254: pp. 106780. https://doi.org/10.1016/j.quascirev.2020.106780.

Siegel, P. E. 1999. Contested places and places of contest: The evolution of social power and ceremonial space in prehistoric Puerto Rico. *Latin American Antiquity,* 10 (3): pp. 209–238.

Steward, J. H. 1955. The concept and method of cultural ecology. In *Theory of Culture Change*. Steward, J. H. (Ed.), pp. 30–42. Champaign Urbana: University of Illinois Press.

Stokes, A. V. 1998. *A biogeographic Survey of Prehistoric Human diet in the West Indies Using Stable Isotopes*. PhD. dissertation. Florida: University of Florida.

Torres, V. P., and Rivero de la Calle, M. 1972. Paleopatología de los aborígenes de Cuba. *Serie Espeleológica y Carsológica*, 32: 1–28. Havana: Academia de Ciencias de Cuba.

Valcárcel Rojas, R., Weston, D. A., Laffoon, J. E., Mickleburgh, H. L., and Duijvenbode, A. 2011. El Chorro de Maita: A diverse approach to a context of diversity. In *Leiden in the Caribbean IV: From Prehistory to Ethnography in the Circum-Caribbean*. Hofman C. L., Duijvenbode, A. (Eds.), pp. 225–252. Leiden: Sidestone Press.

Van den Bel, M., and Romon, T. 2010. A Troumassoid site at Trois-Rivières, Guadeloupe FWI funerary practices and house patterns at La Pointe de Grande Anse. *Journal of Caribbean Archaeology,* 9: pp. 1–17.

Versteeg, A. H., and Schinkel, K. 1992. *The Archaeology of St. Eustatius: The Golden Rock Site.* Amsterdan: St. Eustatius Historical Foundation, The Foundation for Scientific Research in the Caribbean Region.

Versteeg, A. H., Tacoma, J., and Van de Velde, P. 1990. *Archaeological Investigations on Aruba: The Malmok Cemetery* (No. 2). Oranjestad: Archaeological Museum Aruba.

Genomic Analyses

1

ANCIENT DNA ANALYSIS AND ITS APPLICATIONS TO PRE-CONTACT CARIBBEAN ARCHAEOLOGY

Kathrin Nägele and Maria Nieves-Colón

Introduction

Paleoanthropologists and archaeologists investigate questions of ancient human history and shape our understanding of human evolution through the study of fossil and material remains. The young field of archaeogenetics utilizes genetic analysis of the remains of human ancestors to contribute to the reconstruction of population histories and past events. Especially in the Caribbean, where the European invasion had a dramatic effect on language, culture, and genetics, archaeological evidence is the most important tool to reconstruct the pre-contact history and diversity of the region. Adding archaeogenetics to the toolbox can help to investigate open questions such as the origins of the first settlers of the Caribbean, interactions between different groups, and – on a site-specific level – infer kinship structures and patterns of regional change.

Ancient DNA and the Human Genome

Ancient DNA (aDNA) from humans and their pathogenic or commensal microorganisms can be obtained from a variety of biological substrates, including mineralized tissues such as bone and teeth (Hagelberg et al. 1989; Campos et al. 2012), calcified dental biofilms (Warinner et al. 2014, 2015), preserved or mummified soft tissues (Pääbo 1989; Schuenemann et al. 2017), hair (Rasmussen et al. 2010), coprolites, and cesspit sediments (Hagan et al. 2020; Sabin et al. 2020). It can also be extracted from animal, insect, and plant remains such as feathers (Hartnup et al. 2011), egg or mollusk shells (Oskam et al. 2010; Sullivan et al. 2021), cobs (Vallebueno-Estrada et al. 2016), seeds (Ramos-Madrigal et al. 2019), and herbarium specimens (Kistler et al. 2020). Recent research has also shown that aDNA can preserve cultural artifacts that come in contact with bodily fluids or tissues, such as

DOI: 10.4324/9780429444449-3

clay pipes (Schablitsky et al. 2019), parchment (Teasdale et al. 2015), vaccination equipment (Duggan et al. 2020), chewed plant materials (LeBlanc et al. 2007; Jensen et al. 2019), and even food storage and drinking vessels (Bro-Jørgensen et al. 2018; Hansson and Foley 2008). Lastly, aDNA can be recovered from sediments and ice cores and thus be used to reconstruct the biodiversity and species composition of ancient ecosystems (Dommain et al. 2020; Slon et al. 2017).

Depending on the research questions asked and the organism of interest, aDNA studies may target different genomic markers. In humans, most aDNA research historically focused on the mitochondrial genome (mtDNA). MtDNA is the circular, haploid genome found within the mitochondria of eukaryotic cells. Several characteristics make mtDNA an excellent locus for aDNA research. First, because each cell contains between 100 and 10,000 mitochondria, mtDNA is present at high copy numbers (Giles et al. 1980) and is more likely to be preserved in ancient or degraded tissues. Second, mtDNA is maternally inherited and does not recombine. Therefore, mtDNA diversity patterns are informative of female-mediated genealogies, ancestries, and population histories. Third, mtDNA molecules are short, only 16,569 base pairs (bp) in length. This short length makes mtDNA easier to sequence and analyze than the much larger nuclear genome, which stretches for approximately 3 billion bp. Fourth, mtDNA accumulates mutation at a high rate relative to the rest of the human genome, and therefore, it provides information at relatively recent evolutionary timescales. Lastly, mtDNA lineages, also known as haplogroups, have been well studied and characterized in many modern human populations. This large body of previous research facilitates comparative analyses (Bisso-Machado and Fagundes 2021). However, a major limitation of mtDNA is that, as a single locus, it only provides information about one line of female ancestors out of the many ancestors that populate an individual's genealogy (Mathieson and Scally 2020). To gain a more complete picture of genetic diversity over time, it is necessary to investigate the nuclear genome.

The human nuclear genome consists of 22 autosomes and a pair of sex chromosomes (X or Y). It is diploid, passed on by both parents and subject to recombination every generation. An exception to this pattern is the non-recombining portion of the Y chromosome (also known as the NRY region), which is passed on directly from fathers to chromosomal male offspring. The main drawback of the nuclear genome for aDNA studies is that it is present at only two copies per eukaryotic cell and therefore more subject to degradation over time. However, because the nuclear genome has many independent loci, it provides information about all of an individual's genetic ancestors (Mathieson and Scally 2020) and therefore has much more statistical power and deeper resolution for reconstructing past events than mtDNA (Stoneking and Krause 2011).

Ancient DNA Preservation and Damage

The first challenge of aDNA research is recovering sufficient DNA molecules. aDNA is abundant only in meager copy numbers, while environmental contaminants such as bacterial, viral, fungal, plant, and contemporary human DNA dominate the extract (Huson et al. 2007; Kazarina et al. 2019; Mann et al. 2018). To reduce contamination, several measures can be taken. During excavation, **the use of gloves when retrieving elements, as well as foregoing a washing of the bones, can decrease contamination by handlers** or cross-contamination with other elements excavated (Llamas et al. 2017). Before sampling, decontamination procedures such as the use of separate pre- and post-amplification laboratory facilities, UV-radiation of elements, and the use of negative controls (Pääbo 1989; Pääbo et al. 2004) can also increase aDNA yields. Post-sequencing, bioinformatic tools for statistical analysis allow quantification of endogenous DNA and authentication through the investigation of DNA damage (Fellows Yates et al. 2021; Peltzer et al. 2016).

DNA damage arises because DNA molecules have limited chemical stability. Their fragmentation starts right after death once cellular repair mechanisms stop working (Lindahl 1993) and the body's enzymes expedite the decomposition process (Darzynkiewicz et al. 1997). During the decomposition of soft tissues, DNA is further fragmented and destabilized through microbial digestive processes (Eglinton and Logan 1991; Lindahl 1993). If water is present in the depositional environment, hydrolytic reactions alter the chemistry of the DNA molecule. Deamination leads to the loss of DNA bases, causing strand breaks and driving further fragmentation (Dabney et al. 2013; Lindahl and Andersson 1972). As a result, aDNA molecules have a very short fragment length of, on average, 40 bp (Krause et al. 2010). Another type of damage typical for aDNA is depurination, where the base cytosine is chemically altered to uracil, a base in vivo only observed in RNA (Dabney et al. 2013). This chemical alteration can lead to misincorporations that artificially duplicate the number of molecules and ultimately lead to sequencing errors. As a result, the sequences contain artificially high amounts of cytosine and guanine bases, particularly at the end of reads (Dabney et al. 2013). It is possible to repair this damage enzymatically in the lab (Rohland et al. 2015), or it can be removed bioinformatically before data analysis. While DNA damage complicates retrieval and analysis and can result in systematic errors if not taken into account, the detection of these typical damage patterns can authenticate the presence of true aDNA molecules.

The rate at which DNA damage and fragmentation occur varies and depends on several factors. While time is an essential factor (Adler et al. 2011; Allentoft et al. 2012), other environmental conditions such as soil acidity, mineral content, and macro- and microclimate (Allentoft et al. 2012) can significantly influence the degradation process, complicating the predictability of DNA preservation. In tropical environments, the two most apparent factors are humidity and temperature, both aiding hydrolytic reactions and driving the fragmentation and damage of DNA molecules. *Regions with high and fluctuating temperatures generally show the worst rates of aDNA preservation* (Kistler et al. 2017), while in regions with stable cold and dry conditions, aDNA as old as 700,000 years (Orlando et al. 2013), or possibly older (van der Valk et al. 2021), can be recovered. Another factor is the microclimate and environmental context from which remains are recovered. With more stable conditions, *cave sites usually show better preservation compared to open sites* (Mendisco et al. 2014). It has also recently become apparent that the *presence of carbon-rich minerals is beneficial for aDNA preservation* (Slon et al. 2022), while acidic soils have a detrimental effect. For example, human skeletal remains excavated from archaeological sites in the limestone-rich Canimar River Delta in Cuba, dating to between 1500 and 3000 years before the present (BP), show very good aDNA preservation. In contrast, skeletal remains excavated from open-air sites in the Cauto River Delta in Granma, Cuba, dating to 1500–1200 BP, show poor aDNA preservation (Nägele et al. 2020).

Sampling and Laboratory Methods

Although aDNA can be retrieved from a variety of substrates, comparisons of the DNA yields of different anatomical elements have shown that the dense bone surrounding the inner ear, in the petrous portion of the temporal bone, preserves DNA molecules better than other tissues (Hansen et al. 2017; Parker et al. 2020; Pinhasi et al. 2015). In the challenging conditions posed by tropical climates, the **petrous part of the temporal bone is currently the preferred element for aDNA sampling**. However, methods for sampling petrous portions are highly destructive and can result in extensive damage to the cranium, especially if the bones are in fragile conditions (Pinhasi et al. 2015; Sirak et al. 2017). Thus, for a variety of reasons, including preservation concerns, the absence of these elements due to degradation, ethical concerns (Charlton et al. 2019; Fox and Hawks 2019), or the need to preserve the skeletal morphology for osteological study (Ponce de León et al. 2018), destructive sampling of the petrous portion is not always feasible and alternative substrates must also be considered (Parker et al. 2020). While preserving DNA molecules less effectively than the petrous portion, teeth have shown to be a good source of aDNA, even in tropical climates (Fernandes et al. 2021; Nägele et al. 2020; Nieves-Colón et al. 2020; Ramos Madrigal et al. forthcoming), as the enamel encapsulates the DNA and protects it from environmental factors. Finally, dental calculus has also been successfully sampled as a source of human (Nieves-Colón et al. 2020; Ziesemer et al. 2019) and microbial DNA (Ziesemer et al. 2015) in ancient samples from tropical regions.

After the proper tissues or substrates have been selected, the standard aDNA laboratory workflow (Figure 1.1) begins with sampling and DNA extraction. To minimize the impact of destructive sampling, aDNA extraction protocols attempt to isolate the most amount of DNA from the smallest amount of material input (e.g., 50–100 mg of bone or tooth powder) (Dabney et al. 2013; Rohland and Hofreiter 2007). Minimally invasive approaches (Bolnick et al. 2012; Harney et al. 2021) and methods that combine sampling for several types of molecular analyses, such as aDNA and radiocarbon dating (Korlević et al. 2018), have also been developed. Despite these efforts, most aDNA extraction protocols still carry a significant risk of destruction for irreplaceable ancient samples. Thus, before embarking on any aDNA research project, it is necessary to carefully consider the potential benefits of the information that aDNA may provide versus the need for preserving rare, valuable, or culturally significant remains (Austin et al. 2019; Fox and Hawks 2019; Nieves-Colón et al. 2021).

After sampling and pulverization, the DNA has to be extracted from the retrieved bone powder. Standard aDNA extraction protocols isolate and purify DNA from debris and other cellular components (Dabney et al. 2013; Glocke and Meyer 2017), seeking to preserve short DNA molecules that are likely to contain the bulk of authentic aDNA fragments – referred to as "endogenous DNA". However, aDNA extraction procedures cannot distinguish between ancient and modern contaminant DNA, and at the end of this first stage, extracts are likely to contain a mix of ancient and contaminant DNA molecules (Carpenter et al. 2013). After extraction, aDNA extracts are converted into double- or single-stranded libraries by ligating synthetic adapters of known sequences to the ends of all the molecules contained in the extract (Gansauge et al. 2017; Meyer and Kircher 2010). The adapters contain the binding sites necessary for amplifying and later sequencing the libraries on next-generation sequencing (NGS) platforms. One benefit of this approach is that once prepared, aDNA libraries are theoretically an "immortal" repository of the target individual's DNA, as they can be continuously amplified and used as input for several different downstream processes (Orlando et al. 2021).

Although aDNA can be sequenced through a variety of methods, including Sanger sequencing (Sanger et al. 1977), today most sequencing is performed using NGS, specifically Illumina's sequencing by synthesis platform. This technology is particularly well suited for aDNA because it can handle short sequence reads, has a low error rate, and can produce massive amounts of data at relatively low costs by sequencing all DNA fragments present in a library in parallel (Goodwin et al. 2016). This approach of non-specific or non-targeted sequencing is known as shotgun sequencing. In aDNA workflows, shotgun sequencing is often used as an initial screening step to assess levels of endogenous DNA preservation, examine overall damage patterns, and even screen for the presence of specific taxa, such as pathogens or commensal microbes (Orlando et al. 2021). However, in most cases, especially in tropical regions, the fraction of endogenous DNA present in the library is too small for shotgun sequencing to be an effective strategy for reaching the necessary genomic coverage and depth for performing robust data analyses (Carpenter et al. 2013). To increase the amount of endogenous aDNA, it is often necessary to use target enrichment methods.

Target enrichment methods selectively capture endogenous aDNA molecules from the mix of DNA fragments present in a library. In addition to removing contaminant DNA molecules, enrichment increases the amount of endogenous aDNA from any target of interest that may be present in the DNA library, such as, for example, ancient human or pathogen DNA (Andrades Valtueña et al. 2017; Bos et al. 2016). Target enrichment works through the hybridization of the target DNA to

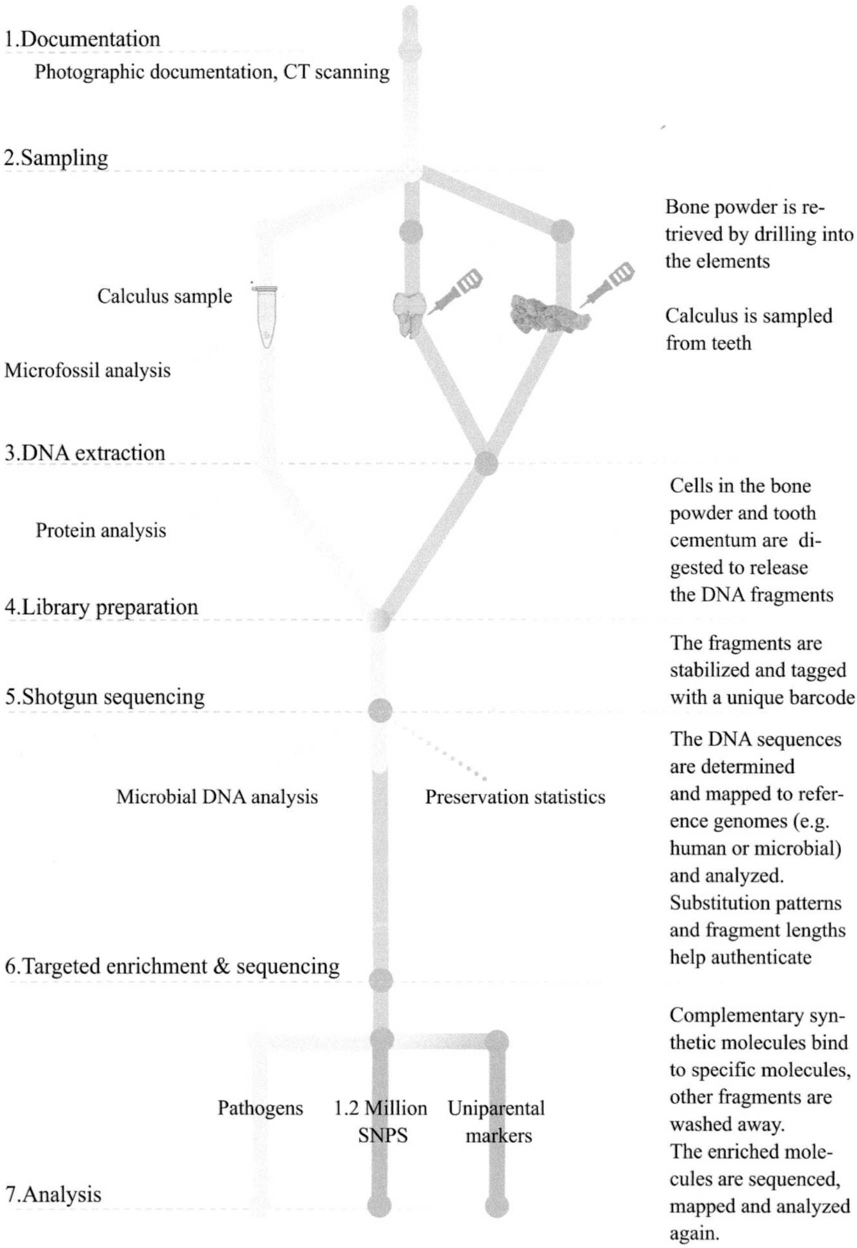

FIGURE 1.1 Schematic overview of the workflow in an aDNA laboratory from receiving an archaeological or skeletal element to data analysis.

complementary probes produced from synthetic DNA strands, DNA amplicons, or RNA (Carpenter et al. 2013; Enk et al. 2014; González Fortes and Paijmans 2019). In human aDNA studies, commonly used aDNA enrichment assays target thousands or even millions of genome-wide variants, the whole mtDNA genome, or portions of the nuclear genome (Ávila-Arcos et al. 2011; Carpenter et al. 2013; Fu et al. 2013; Maricic et al. 2010). The target DNA binds to the probes while other molecules can be washed away. The targeted molecules are then amplified and sequenced. Sequencing molecules deriving only from the targeted organism results in more analyzable data at a much lower cost.

However, one limitation of enrichment approaches is that they may introduce bias in the resulting dataset (Seguin-Orlando et al. 2015). While shotgun sequencing provides non-specific information about the entire genome, enrichment focuses on genomic regions previously ascertained and selected to be most informative about intra- or inter-specific variation (Orlando et al. 2021). This may create challenges for downstream analysis, especially when probes are generated based on a subset of present-day genetic diversity but seek to capture the variation in an ancient or extinct population. For example, due to a history of serial bottlenecks and founder effects, Indigenous populations of the Americas have lower genetic variation than other world populations (Bergström et al. 2020). Current capture assays are optimized to distinguish between worldwide populations and contain only few markers that can distinguish between closely related Indigenous American populations, much less between ancient groups in the Caribbean, leading to little power in investigating intra-regional variation. Additionally, aDNA sequences enriched with one assay may not be easily comparable to datasets produced with another due to poor overlap of selected genomic regions. Despite these challenges, the use of target enrichment combined with NGS is currently the best approach for obtaining genomic scale data from degraded, ancient sources.

Ancient DNA Analysis

Once sequenced, aDNA data can be queried to answer a variety of questions about past populations at both macro and micro scales. At the macro level, aDNA studies can investigate and compare patterns of within and between group diversity. Such comparisons are often used in large-scale paleogenomic studies, which may include tens to hundreds of ancient genomes from different time periods and geographic sites (Mathieson et al. 2018; McColl et al. 2018; Narasimhan et al. 2019). Such studies seek to leverage the temporal dimension of aDNA datasets to reconstruct the demographic histories of large territorial regions, such as continents (Posth et al. 2018; Raghavan et al. 2015), sub-continents (Haak et al. 2015; Lazaridis et al. 2014; Olalde et al. 2018; Skoglund et al. 2012), and island systems (Fehren-Schmitz et al. 2017; Fernandes et al. 2021; Matisoo-Smith 2015; Nägele et al. 2020; Posth et al. 2018; Pugach et al. 2021). Common questions addressed in these studies include tracing the origins of ancient populations (Slatkin and

Racimo 2016), reconstructing historical settlement or migration patterns (Olalde et al. 2018; Perri et al. 2021; Rivollat et al. 2020), investigating large-scale interaction networks (Feuerborn et al. 2021; Lord et al. 2018), animal dispersals (Rabinow 2020; Woods et al. 2021), domestication (Lord et al. 2018; Kimura et al. 2016; Perri et al. 2021), and epidemic transmission patterns (Duchêne et al. 2020; Kerner et al. 2021; Kocher 2021).

At the micro-level, aDNA studies focus on research questions pertaining to local sites or communities. These studies often compare patterns of genetic diversity within and between specific burial populations, archaeological sites, or individuals with the goal of investigating local or regional scale population dynamics. Micro-level aDNA studies may leverage aDNA data to answer questions about ancient social structures or relatedness networks (Amorim et al. 2018; Cassidy et al. 2020; Stone and Stoneking 1993), examine intra- and inter-site mobility (Bongers et al. 2020), reconstruct ancient residence patterns (Žegarac et al. 2021), diets (Warinner et al. 2015), activities or behaviors (Racimo et al. 2020), By sampling coprolites and dental calculus, it is also possible to examine changes in communities of human commensals over time (Arriola et al. 2020; Sabin et al. 2020; West et al. 2017). This microbiome can be informative of larger lifestyle changes (Obregon-Tito et al. 2015), but until now has not been shown to be informative enough to infer human population history (Fellows Yates et al. 2021). Lastly, aDNA studies can determine the presence of pathogenic organisms (Barquera and Krause 2020; Guzmán-Solís et al. 2021) and reconstruct human disease adaptation through the characterization of ancient immunogenetic systems (Barquera and Krause 2020).

A range of statistical approaches and computational tools are available to analyze aDNA datasets. Selecting the appropriate tool depends on the data quality, the overall research goals, and the specific hypotheses to be tested. Studies seeking to characterize population history often use unsupervised analytical methods – which do not require assigning group labels *a priori* – to obtain a first approximation of genetic diversity patterns in ancient populations (Stoneking and Krause 2011). Commonly used unsupervised approaches include multivariate and clustering methods. Multivariate methods, such as principal components analysis (PCA) (Figure 1.2a) or multidimensional scaling, visually represent patterns of genetic distance between individuals in a two-dimensional graph (Patterson et al. 2006). Clustering approaches partition genetic diversity into pre-defined clusters, and individuals are placed into clusters based on shared similarities in allele frequencies (Alexander et al. 2009; Rosenberg and Jakobsson 2008). These classifications can then be used to infer patterns of shared ancestry and admixture (Lawson et al. 2018). These methods provide a first approximation of population structure and are often used to formulate hypotheses based on the distance or similarity of the population of interest to other groups. However, unsupervised and multivariate approaches do not allow for formal reconstructions of demographic processes. Thus, they must be followed up with additional analyses such as formal statistical tests, phylogenetic reconstruction, or genetic modeling.

Formal tests, such as those implemented in f-3 statistics, can detect instances of extended mixing between populations by comparing **allele frequencies** to determine how much **genetic drift** populations share relative to an **outgroup** (Patterson et al. 2012). Tests such as those implemented in the f4-statistics framework can examine differential affinities, identify and quantify instances of **admixture**, and test for genetic continuity over time (Durand et al. 2011; Patterson et al. 2012). Other approaches use patterns of allelic diversity (diversity estimator π) or measure distances between individuals or groups (Fst) (Holsinger and Weir 2009). Measures of genetic distance can be used as input for constructing phylogenetic trees or phylogeographic networks (Figure 1.2b,c). Trees and networks are visual representations of the branching patterns between sequences (Bandelt et al. 1995, 1999; Mardulyn 2012). They are used to investigate historical and evolutionary relationships between lineages, populations, or taxa. Dated aDNA sequences add a temporal dimension to phylogenetic reconstruction by calibrating or anchoring the tree to a specific timeframe. This calibration permits the direct estimation of the timing of sequence divergence and changes in mutation rates over time (Molak et al. 2012; Rieux and Balloux 2016). These analyses can be used to infer patterns of sequence diversification or even speciation (Chang et al. 2017). Recently developed admixture graph methods extend the power of classical phylogenetic approaches by explicitly considering admixture (Pickrell and Pritchard 2012; Reich et al. 2010). However, one major drawback of these methods is that modeling more complex scenarios of population interactions, perhaps involving unknown populations, is not currently possible.

Complex demographic processes such as gene flow and admixture can also be detected and dated using model-based approaches. These analyses compare patterns of genetic diversity simulated under different scenarios to observed patterns collected from real datasets (Leonardi et al. 2017). In aDNA studies, models can be used to test the interpretations derived from first-pass exploratory analyses such as PCA or to test scenarios predicted by other lines of evidence such as archaeological or paleontological data (Nägele et al. 2020). At a smaller scale, other types of analyses can reconstruct patterns of genetic relatedness among ancient individuals. For example, genomic segments that are shared between individuals due to identity-by-descent inheritance can quantify genetic relatedness and allow reconstructions of kinship (Monroy Kuhn et al. 2018; Vai et al. 2020). Especially useful for this type of analysis are runs of homozygosity (ROH), long continuous segments of homozygous genotypes, transmitted from parents to offspring without recombination (Ringbauer et al. 2021). By estimating patterns of shared alleles and reconstructing ROH segments in ancient individuals, kinship tests can infer patterns of inbreeding and reconstruct ancient family pedigrees (Cassidy et al. 2020; Fernandes et al. 2020).

Lastly, molecular sexing approaches can determine sex chromosome karyotypes in ancient individuals. In humans, most females carry XX and most males carry XY karyograms – although variations of this inheritance pattern exist. Molecular

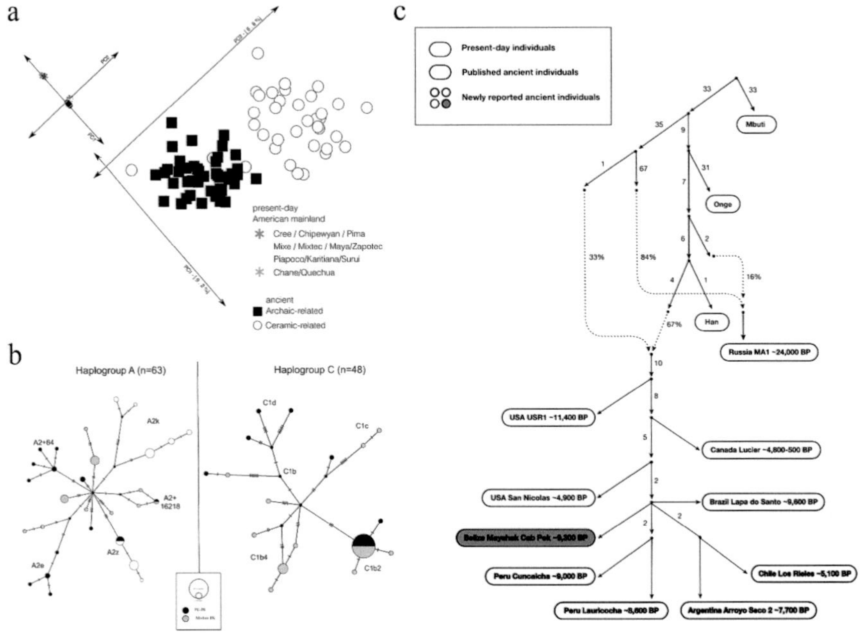

FIGURE 1.2 Example figures. PCA (a), adapted from Nägele et al. (2020), visualizing genetic distances between individuals. Figure showing differences between individuals from Archaic Age contexts (squares) and Ceramic Age contexts (circles); closer clustering suggests higher genetic similarity. Phylogenetic Network (b), Median-joining network for mitochondrial haplogroups A and C, including pre-contact and present-day Puerto Rican genomes published in Nieves-Colón et al. (2020). Admixture tree model (c) published in Posth et al. (2018), based on differential affinities between population groups, showing the branching pattern suggestive of the peopling of the Americas and a star-like radiation of all South American sequences included in the model.

sexing methods rely on detecting unique genetic elements in the X or Y chromosomes (Stone et al. 1996) or comparing the number of sequence reads aligning to the X or Y chromosomes versus the autosomes (Skoglund et al. 2013; Buonasera et al. 2020). Importantly, while these analyses may be informative about chromosomal sex arrangements, they are not necessarily indicative of gender identification or lived gender experiences in the past. Such inferences necessitate other types of supporting evidence (Moilanen et al. 2021; Haas et al. 2020).

Although all of these analytical methods are powerful, they also have inherent challenges and limitations. The millions of markers in genomic data are too complex to be easily processed, visualized, and communicated, and require a reduction of complexity. This reduction introduces subjective simplifications that may not represent the full complexity of the data. Most aDNA analyses (and especially modeling approaches) are sensitive to underlying assumptions, the selection of reference populations, and the choice of parameters or significance thresholds, all of which may impact data interpretation. Additionally, ancient sequences must be combined with sequences derived from present-day populations to carry out comparative analyses or trace changes in genetic diversity over time. But present-day populations are not always good proxies for ancient genetic variation (Pickrell and Reich 2014), especially in regions of the world that have experienced large-scale changes or recent population turnovers, such as the Caribbean. Analyses can also be affected by the characteristics of degraded aDNA sequences, such as low genomic coverage or patterns of extensive damage (Leonardi et al. 2017; Orlando et al. 2021). Researchers must be careful to take these characteristics into account when carrying out analyses and drawing inferences from the results (Callaway 2016; Gallego Llorente et al. 2015). Lastly, another major challenge is that analyses conducted with large-scale paleogenomics datasets are computationally intensive. Access to high-performance computational resources and terabytes of data storage are needed. Additionally, training in computer programming, bioinformatics, and familiarity with the management of computational pipelines and workflows (Fellows Yates et al. 2021; Peltzer et al. 2016) are now essential skills for any aDNA researcher.

Caribbean Human Population History: Archaeological Evidence

Archaeological research shows that the Caribbean was settled 7000–5000 years before the present, during a period denominated as the Archaic Age. The first occupational sites are found at the extreme ends of the archipelago. Trinidad, where the site of Banwari Trace is located, has the oldest evidence of human occupation (Tankersley et al. 2018). This island was – at the time – still connected to the mainland of Venezuela, facilitating human settlement compared to the other islands, which remained disconnected from each other throughout the Holocene. Apart from the Banwari Trace, the oldest dates for human occupation are in the northern Caribbean islands, such as Cuba, Puerto Rico, St. Martin, and Anguilla (Napolitano et al. 2019). Stylistic comparisons of recovered artifacts at these sites imply connections to North-, Central-, and South American sites (Coe 1957; Cruxent and Rouse 1961; Febles 1991; Rouse 1960, 1994; Wilson et al. 1998), leaving open questions about the origins and settlement routes people took in the Archaic Age.

The diverse archaeological materials recovered from the islands dating between the initial settlement and the European invasion (Bérard 2013; Chinique de Armas et al. 2016; Febles and Rives 1991; Hofman 2013) indicate biological and cultural diversity existed throughout the archipelago. Particularly after the arrival of people practicing larger-scale agriculture and crafting ornate ceramics, distinct from those found during the Archaic Age. Their arrival around 2500 BP marked the beginning of the Ceramic Age in the Caribbean. Archaeological, linguistic, and genetic evidence all agree that the likely origin of Ceramic Age populations was in the Orinoco River delta region of northeastern South America (Cruxent and Rouse 1961; Granberry and Vescelius 2004; Keegan 1994; Schroeder et al. 2018). But the route through which the islands were settled during the Ceramic Age remains debated. The Stepping Stone model assumes a dispersal from South to North, using the Lesser Antilles as stepping stones (Rouse 1992). In contrast, the Southward Expansion model (Fitzpatrick et al. 2010), based mostly on the available radiocarbon dates for the first occupations of each island (Napolitano et al. 2019), proposes a leap from the South American mainland directly to Puerto Rico before expanding south into the Lesser Antilles and west and north to Hispaniola, Cuba, and the Bahamas. Archaeologists have also suggested that multiple dispersals may have occurred during the Ceramic Age. For instance, the La Hueca material culture (Chanlatte Baik and Narganes-Storde 2005), first identified on Vieques Island off the eastern coast of Puerto Rico, exhibits a distinct bird shape in its decorations and ornaments, reminiscent of the Andean condor, suggesting connections to Andean cultures. The emergence of new ceramic styles across the Caribbean throughout the Late Ceramic Age has also been thought to be the result of additional migrations from the mainland (Rouse 1992), although other researchers suggest it may be the result of local diversification (Keegan and Hofman 2017).

During the European invasion starting in the late fifteenth century, the accompanying chroniclers and other colonial writers produced descriptions of the Caribbeans' Indigenous populations that were fragmented, inaccurate, or heavily biased by the colonial perspective (Anderson-Córdova 2017; Sued-Badillo 2020b). Early Spanish accounts reported three Indigenous groups were present on the islands: the "noble Taino" of the Greater Antilles, the "enigmatic Guanahatabey" in the westernmost tip of Cuba, and the "fierce Caribs" in the Lesser Antilles. Although these groupings undoubtedly obscure the cultural diversity in the islands and do not reflect the self-identification of these groups, they have continued to be used by anthropologists, archaeologists, and historians into the present day (Curet 2006).

In fact, a major debated question in Caribbean archaeology concerns the origins and existence of the "Carib" in the southern Lesser Antilles. Much of what is known about these populations stems from accounts made by chroniclers (Kim 2013; Sued-Badillo 2007). Partly due to European misinterpretations of Indigenous perspectives (Keegan 1996) and partly due to the success of Indigenous Lesser Antillean communities in resisting colonialism (Beckles 1992), the "Carib" were described by chroniclers as fierce, warlike, and practicing cannibalism (for an in depth discussion of the lack of evidence for cannibalism in the archaeological record of the Caribbean, please see Moore 1973; Myers 1984; Reid 2009). Religious missionaries who lived in the Lesser Antilles during the seventeenth century speculated about the origins of the Carib, proposing that they had arrived to the Antilles in a separate, later migration sometime during the thirteenth century (De Navarrete 1837; Grunberg 2011; Roux 2011).

Debates about potential distinct Carib origins are ongoing, with some researchers attempting to reconcile the archaeological evidence with the records of the colonizers. Currently, two major models of "Carib" origins exist (Keegan and Hofman 2017). The first model suggests that the "Island Carib" were connected to Carib speakers from the South American mainland and entered the Antilles in a separate migration from other Antillean groups, just before the European invasion. The second model proposes that the "Island Carib" arose locally during the Late Ceramic period, spoke an Arawakan language (Davis and Goodwin 1990), and were primarily connected to Antillean Indigenous communities. Distinguishing between these models is complicated by disagreements on how to identify pre-contact Carib sites. The distinctive Cayo pottery style, first identified in St. Vincent (Boomert 1986), as well as the Suazey ceramic complex (Bullen 1964; Bullen and Bullen 1970, 1972), have been interpreted as pre-colonial evidence of the Island Caribs. The discovery of several colonial-period Carib sites in Dominica, St. Vincent, and other parts of the Lesser Antilles (Keegan and Hofman 2017; Lenik 2012) has further complicated the question of Carib origins. Cayo pottery found at these early colonial sites bear strong similarities to the Koriabo ceramics from the Guianas, suggesting some connections to the mainland, but also have a mix of stylistic elements from ceramic traditions in the Greater Antilles (Boomert 2004, 2011). Thus, archaeological approaches have, as of yet, been unable to fully resolve the "Island Carib Problem" (Allaire 1977), and the existence of pre-contact "Caribs" as an ethnic entity continues to be debated. However, the present-day Indigenous peoples of the Lesser Antilles self-identify as "Kalinago" or "Carib" and trace their ancestry to the historic Island Caribs (Hofman and Hoogland 2012; Hofman et al. 2014; Honychurch 1997; Twinn 2006).

Caribbean Human Population History: Ancient DNA Evidence

Reconstructions of the history and diversity of the ancient Caribbean have largely relied on historical records and archaeological materials. Together with these other lines of evidence, archaeogenetics has recently provided novel insights into the origins, diversity, and social relations of pre-contact Indigenous Caribbean communities. Before the combination of targeted sampling, recovery of short molecules, and targeted enrichment, archaeogenetics of the Caribbean was mostly limited to analyses of small fragments of the mtDNA genome (Nieves-Colón, 2022). But since the publication of the first nuclear genome from a pre-contact individual excavated in the Bahamas (Schroeder et al. 2018), the number of nuclear genomes has increased tremendously (Figure 1.3) and provided novel insights into the population history of the Caribbean (Fernandes et al. 2021; Nägele et al. 2020; Nieves-Colón et al. 2020). As of this writing, whole ancient genomes are available for 267 individuals from various islands (Figure 1.3) (Fernandes et al. 2021; Schroeder et al. 2018; Nägele et al. 2020; Nieves-Colón et al. 2020).

Analysis of these genomes has demonstrated that Archaic Age and Ceramic Age settlers were genetically different and had different affinities to other Indigenous American populations. This finding suggests that the genetic differentiation seen between Ceramic and Archaic populations is the result of different origins and not solely the effect of **genetic drift**. The ancient genomic data also support

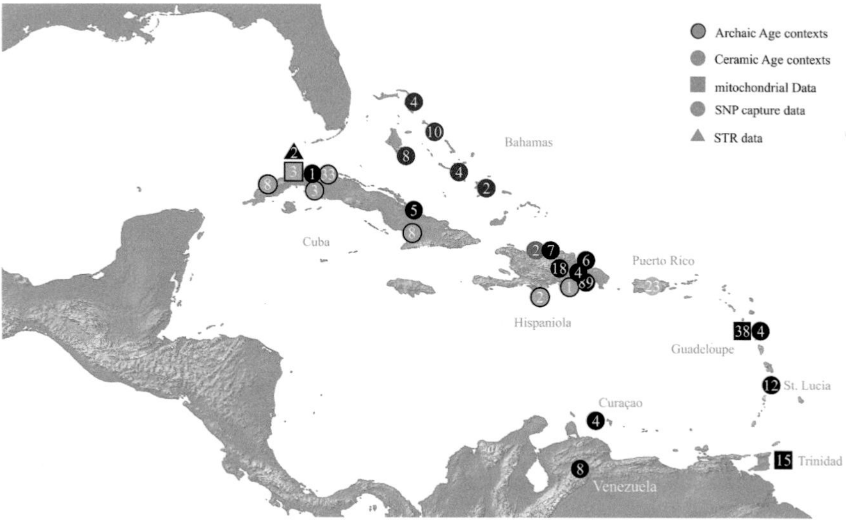

FIGURE 1.3 Map of the Caribbean showing currently available ancient whole genomic data (circles, Fernandes et al. 2020; Schroeder et al. 2018; Nägele et al. 2020; Nieves-Colón et al. 2020; mitochondrial data is available for all individuals), additional mitochondrial sequences (squares, Lalueza-Fox et al. 2001, 2003; Mendisco et al. 2015; Nieves-Colón et al. 2020), and STRs data (triangle, Lleonart et al. 1999). Individuals from an Archaic Age context are shown with a dark outline. Numbers in the circles represent the number of individuals available from the respective region/island.

archaeological hypotheses that point to the origin of Ceramic Age settlers in north-eastern South America, finding their closest genetic relations among modern Arawakan speakers such as the Indigenous Palikur of Guiana and Brazil, as well as Piapoco speakers from northeastern Colombia. However, the analysis of aDNA recovered from eight ancient individuals from the site of Las Locas in Venezuela (dated ~2300 BP) found that the Ceramic Age population of this region shows more genetic links to Indigenous populations of the Panama Isthmus region than to those of the ancient Caribbean. This suggests that despite the location of this site in a purported source region, the origins of Caribbean Ceramic Age settlers likely lie further east (Fernandes et al. 2021). Interestingly, these studies also found that the ancient inhabitants of Curaçao were genetically a mixture of Ceramic Age populations dispersing to the Lesser Antilles and pre-contact populations from Venezuela.

Genomic studies have investigated whether patterns of genetic differentiation and diversification (usually represented as branching patterns on a phylogenetic tree, see Figure 1.2b) between Ceramic Age Caribbean populations mirror patterns of initial island settlement. Nägele et al. (2020) and Fernandes et al. (2020) used a modeling approach to test the predictions of archaeological dispersal route models against the genetic branching patterns of Ceramic Age populations from sites in the Lesser

Antilles. Both found the genetic data are a better fit to the Stepping Stone hypothesis than to the Southward Expansion model. However, the oldest individuals included in this analysis post-date the first occurrence of decorated ceramics in the Caribbean by more than 1,000 years (Cruxent and Rouse 1958). Future analyses, including individuals from older sites or from currently unsampled locations, might change this result.

Ancient DNA has also shed light on local patterns of inter-island genetic exchange and kinship in the Caribbean. By examining ROH and identity-by-descent patterns among precontact individuals, Fernandes et al. (2020) identified 19 pairs of cousins buried across the islands of Hispaniola and the Bahamas, as well as two close relatives buried in different sites of Hispaniola (Fernandes et al. 2021). This finding supports archaeological reconstructions that indicate that expansive inter-island connections and exchange networks existed throughout the Caribbean during the Ceramic Age (Laffoon 2013; LeFebvre et al. 2018; Hofman et al. 2021).

There have also been attempts to use aDNA to address a highly controversial topic in Caribbean archaeology: the question of the "Carib" expansion. Analyses of ancient mtDNA (Mendisco et al. 2015) and genomic datasets (Fernandes et al. 2021; Nägele et al. 2020) from individuals excavated from contexts previously discussed as "Carib" in the Lesser Antilles (characterized as part of the Suazey ceramic complex; Bullen 1964; Bullen and Bullen 1970; Bullen and Bullen 1972), find some population structure between ancient communities but no evidence for a separate "Carib migration" from Central or South America or a connection to the Bahamas as previously proposed based on cranial morphology (Ross et al. 2020). To settle the question of whether or not the "Caribs" as a genetic entity – possibly with origins on the South American mainland – existed, human skeletal remains clearly associated with pre-contact "Carib" contexts (both Cayo pottery style and Suazey ceramic complex) need to be analyzed. Given the very poor preservation of skeletal remains found to date on St. Vincent and Dominica (Hofman et al. 2020), it is unlikely ancient genomics will put this discussion to rest.

aDNA has contributed to our understanding of the first settlement of the Caribbean by characterizing the genetic diversity of Archaic Age settlers and investigating the population dynamics of the Archaic to Ceramic Age transition. Surprisingly, although archaeological evidence suggests interactions occurred between Archaic and Ceramic Age populations (Hofman et al. 2014), very little genetic mixture has been identified. Out of all individuals analyzed by Fernandes et al. (2020) and Nägele et al. (2020), only three showed genetic contribution from both groups: Two in Haiti (Diale 1, 750–1200 BP, ~18% Archaic Age contribution) and one in the Dominican Republic (La Caleta, 350–1300 BP, ~12% Archaic Age contribution). This finding contrasts with patterns seen in other parts of the world, including other island systems, where the encounter of two groups usually leaves descendants with genetic mixtures (Lipson et al. 2018; Oliveira et al. 2022; Posth et al. 2018). However, a major caveat is that the number of individuals available from the Archaic period and from the regions where this encounter most likely happened is still small. Future targeted sampling efforts in the Greater Antilles, where interactions between the two groups likely occurred, might provide new evidence for admixture. But, for now, the question remains as to why there is so little mixture. Genetic analyses alone cannot give insights into possible cultural barriers or reconstruct the events that might have defined the Archaic-Ceramic transition. Additionally, the population history of Caribbean Archaic Age communities prior to the arrival of the Ceramic Age settlers is poorly understood.

Unlike the Ceramic Age migrations, archaeological, linguistic, and genetic evidence have not been able to conclusively identify the geographic source region for the Archaic Age dispersals in the American mainland (Allaire 1999; Fernandes et al. 2021; Fitzpatrick et al. 2021; Nägele et al. 2020; Rodríguez Ramos et al. 2013). Genomic analysis of individuals excavated from Archaic Age contexts from Cuba (Nägele et al. 2020) and Hispaniola (Fernandes et al. 2021) indicates that even though a small degree of population structure differentiated them, these groups had overall low levels of genetic variability, indicating small population sizes. These analyses also found that while Archaic Age Caribbean groups have an Indigenous American origin, their specific sub-continental origins within the continental Americas are unclear. In fact, comparisons with present-day populations and all ancient individuals sequenced to date from the Circum-Caribbean and mainland Americas could not identify which population is genetically the most closely related to Archaic Age settlers. All tested Indigenous groups from South and Central America were found to be equally related to ancient Caribbean individuals from Archaic Age contexts, although Indigenous groups from North America, such as the Chipewyan, Cree, and Ojibwe, were found to be the most genetically distant.

Two aDNA studies analyzing genomes from Archaic Age contexts disagree in their results. Modeling Archaic Age ancestries relative to other ancient genomes from the Americas, Fernandes et al. (2021) claim a single origin either in South or Central America. In contrast, Nägele et al. (2020) find the best fit to be a mixture of South American ancestry with a contribution from a very ancient lineage dating to the earliest peopling of the Americas, for which the only currently available representatives are ancient genomes from the California Channel Islands (Scheib et al. 2018). However, serial reductions of population size during the initial settlement of the islands (bottlenecks) led to a decrease in genetic variability in ancient Caribbean populations (Benn Torres et al. 2015; Schroeder et al. 2018; Vilar et al. 2014). Therefore, only a small number of genetic markers vary sufficiently between island populations and can be included in the analysis, reducing the statistical power of the models. Both studies agree that future modeling could be significantly improved by increasing the quality of the available aDNA data and adding more sequences from other ancient Circum-Caribbean populations. Currently, a lack of Archaic Age genomes from the Lesser Antilles and Puerto Rico leaves the picture of ancient genomic diversity incomplete. Thus, the geographic origins and number of migrations that brought Archaic Age peoples to the Caribbean remain unclear. Whether the differences in material culture (Febles and Rives 1991) and consumption patterns (Chinique de Armas et al. 2016) visible in the Archaic Age archaeological record of islands such as Cuba correspond with different patterns of genetic diversity remains to be investigated. For now, these differences are extremely difficult to dissect given the limitations of modeling methods, a lack of comparative ancient genomes from the mainland Circum-Caribbean, and the small number of Archaic Age individuals available for genetic sampling.

Transformations in population structure, in the Caribbean as well as in other parts of the world, have often been accompanied, if not caused by, diseases. Ancient genomic material can contain the pathogens from which ancient individuals suffered. Although detection is extremely difficult (Bos et al. 2015; Nelson et al. 2020), it has higher odds of success when targeting bones altered by the disease (Bos et al. 2019). aDNA has given insights into the spread and evolution of pathogens such as *Yersinia pestis* (Spyrou et al. 2018) and shown that tuberculosis was a common disease in the pre-contact Americas (Nelson et al. 2020). Most recently, the first viral aDNA was retrieved from Archaic and Ceramic Age individuals from the Canimar River basin in Cuba (Kocher 2021). The two individuals, dating to between 1600 and 1300 BP were infected with the Hepatitis B virus (HBV), which is transmitted sexually or congenitally and affects the liver (Liaw and Chu 2009). The viral HBV lineages carried by these individuals belong to a branch that was introduced to the Americas with some of its earliest peoples. The lineage present in Cuba 1600 years ago split from the mainland lineages around 4000 years ago, providing a snapshot of the earliest occupation through the lens of virus evolution.

The European invasion starting in the late fifteenth century completely transformed the linguistic, cultural, and genetic landscape of the Caribbean. Although they did not disappear entirely, Indigenous populations experienced a strong decline due to mass displacement, warfare, enslavement, and exposure to new epidemic diseases (Anderson-Córdova 2017; Cook 2002). In response, the colonizers imported enslaved laborers from across the mainland Americas and sub-Saharan Africa, the latter through the Maafa (trans-Atlantic slave trade)[1] (Ani 1994; Palmer 2003; Rivera-Pagán 2003). Although Indigenous and African peoples resisted colonization through multiple means, including flight, rebellion, and marronage (Stone 2013), their communities and lifeways were forever transformed by colonization (Ulloa Hung and de Ruiter 2011).

Genetic studies have documented the impact of colonial and postcolonial migrations on the genetic diversity of present-day Caribbean populations, many of whom carry admixture from Indigenous, European, and African ancestors. These admixture proportions are often highly sex-biased and differ between, and sometimes among, island populations (Adhikari et al. 2016; Fortes-Lima et al. 2018). For instance, mtDNA studies have shown high frequencies of Native American mitochondrial haplogroups are found in present-day populations from the Greater Antilles (Bryc et al. 2010; Bukhari et al. 2017; Marcheco-Teruel et al. 2014; Martínez-Cruzado et al. 2005; Mendizabal et al. 2008; Vilar et al. 2014), and some groups in the Lesser Antilles (Benn Torres et al. 2015; Schurr et al. 2016). In contrast, other populations have higher proportions of Sub-Saharan African ancestry in mtDNA loci (Benn Torres et al. 2007, 2019; Bukhari et al. 2017; Madrilejo et al. 2015; Mendisco et al. 2019; Newman et al. 2013). In most Caribbean populations, European or African paternal ancestries are significantly enriched, while proportions of Native American ancestry on the Y-Chromosome, and across the whole genome are generally low (Benn Torres et al. 2007; Bukhari et al. 2017; Bryc et al. 2010; Fortes-Lima et al. 2018; Mendisco et al. 2019; Moreno-Estrada et al. 2013; Via et al. 2011; Vilar et al. 2014). To some extent, the Native American ancestry of Caribbean populations may have been shaped by the post-contact displacement of Indigenous peoples during the *encomienda* period, including the forced importation of enslaved groups from across the Americas (Anderson-Córdova 2017; Valcárcel Rojas and Ulloa Hung 2018). However, mtDNA (Nieves-Colón et al. 2020) and whole genomic analyses have demonstrated there is genetic continuity between Ceramic Age (but interestingly not Archaic Age) Indigenous groups and present-day populations in Puerto Rico and Cuba (Fernandes et al. 2021; Schroeder et al. 2018). Moreover, several of these mtDNA lineages have genomic signatures indicative of local evolution and diversification on the islands (Benn Torres et al. 2015; Martinez-Cruzado 2010; Nieves-Colón et al. 2020; Vilar et al. 2014). Lastly, self-identified Indigenous groups or communities of Indigenous descent and heritage exist throughout the Antilles (Benn Torres 2014; Higman 2021).

Ethics and Sustainability in Ancient DNA Research

aDNA analyses have become an important tool in the study of Caribbean archaeology and history. However, the methods, geography, and implications of ancient genomics findings for present-day Caribbean communities call for several ethical considerations. An important drawback of aDNA research is that it is inherently destructive. Although laboratory methods are continuously optimized and improved to require less input material (Bolnick et al. 2012; Harney et al. 2021; Sirak et al. 2017), the process of sampling can damage or destroy entire skeletal elements (Austin et al. 2019; Charlton et al. 2019; Fox and Hawks 2019; Hansen et al. 2017; McDonough et al. 2018). Human tissues and skeletal remains are the remains of ancestors who may have living descendants and are part of the cultural heritage of the countries in which they were found or excavated. For these reasons, it is paramount to obtain permission from the appropriate institutions and stakeholder communities, regardless of when the element was collected and where it is currently curated (Prendergast and Sawchuk 2018). It is of the utmost importance to also carefully assess whether the element or artifact should be sampled now – starting with the least invasive method – or whether research should be postponed until better methods with higher chances of a positive outcome are established (Sirak and Sedig 2019).

The long history of colonialism in the Caribbean also presents particular challenges for aDNA research, especially when it comes to defining and assigning names to genetic populations, lineages, or clusters. The Caribbean is a contextually diverse region, but this is only poorly reflected in the terms aDNA studies have used to describe ancient populations. Using terms established to describe archaeological cultures or phenomena (e.g., Taino, Saladoid, Guanahatabey) implies a connection between material culture and genetics that does not necessarily exist (Eisenmann et al. 2018). For instance, a widely applied naming scheme in aDNA studies combines the geographic location of a sampled individual with archaeological information. In spaces like the Caribbean, where contexts are diverse and overlapping geographically and temporally, this approach can be challenging and communication between archaeologists and geneticists can become cumbersome. However, the use of colonial exonyms is problematic as well, as detailed above in the discussion of the term "Carib". Using terms by which groups self-identify today to describe ancient communities is similarly problematic, even if people today derive some elements of their identity and shared heritage from those ancient cultures. At the same time, identity cannot be inferred from genetic ancestry (Elliott and Brodwin 2002), and in the absence of written records from Indigenous perspectives, we cannot know how ancient Caribbean peoples self-identified. As several Caribbeanist scholars have pointed out (Curet 2006, 2014; Deagan and Cruxent 2008; Sued-Badillo 2020a; Yaremko 2009), the names and terms coined by colonial writers largely reflect colonial European understandings of culture, race, and human difference. Rather than considering such categorizations as objective descriptions of the islands' cultural landscapes, these terms must be employed critically and their inherent biases taken into account when considering whether to use them to describe archaeological or genetic groups. The development of a novel, unified terminology that reconciles the intricacies of multiple disciplines and reflects ongoing efforts to decolonize Caribbean archaeology (González-Tennant 2014; Kulstad González 2020; Pagán-Jiménez and Rodríguez Ramos 2008; Rodríguez Ramos 2010; Rivera-Collazo 2011) can only be achieved through cross-disciplinary conversations. Until then, providing terminology sections defining the terms applied and justifying their use can be a sensible approach.

Many of the ethical pitfalls in working in the Caribbean can be avoided by a culture of equitable and sustainable collaboration. In the past, European and North-American scientists have perpetuated colonial practices, exploiting the cultural heritage and biology of the islands and further disenfranchising local communities and scholars. This behavior has affected genetic research and its ability to answer open questions directly. Unfortunately, some of these practices are still ongoing in aDNA. Many have used museum specimens curated outside the original communities' sphere of influence and without their consent. As aDNA is used to reconstruct human history, it impacts local identities and can be used and abused in political agendas and discussions about land rights. Involving participating descendant communities (Sankatsing Nava and Hofman 2018) and local scholars in the design of studies, supporting locally-led research efforts by facilitating capacity building, ensuring knowledge transfer by training Caribbean students, and communicating research results to the local communities is imperative to drive a culture of new relational ethics.

Conclusion

At both macro- and micro-scales, aDNA studies are strengthened by the integration of multiple sources of evidence about the past. Paleontological, paleoenvironmental, archaeological, historical, or ethnographic datasets, as well as input from descendant communities, can properly contextualize and increase the resolution of aDNA findings, as well as generate new avenues for future inquiry.

GLOSSARY

Admixture In genetics, admixture describes the mixture of two previously diverged genetic lineages.

Allele Frequency Relative frequency of a gene variant in a population.

Bottleneck A genetic bottleneck is a strong reduction of the genetic variation of a population as the result of a decline in population size.

Founder effect Loss of genetic diversity as a result of the establishment of new populations from a subgroup of a larger population.

Genetic drift Random change of the frequency of a gene variant in a certain population. Founder effects and bottlenecks are examples of genetic drift.

Outgroup In phylogenetics, a more distantly related group that serves as a reference to the ingroup, whose evolutionary relationships are to be investigated.

PCA Principal component analysis (PCA) is a tool for exploratory data analysis that reduces the dimensionality of the data. Usually, the first two to three components are used to visualize the dataset, as they represent the most variation.

Population "Population" has various definitions in different disciplines, taking into account shared language, culture, or identity. In the most basic sense, a population is defined as a group sharing the same geographical distribution. In the more specific sense of population genetics, populations are also specified by individuals who have shared genetic variants.

Population structure A population is structured when there are systematic differences in the allele frequencies between subpopulations within a larger group.

STR – short tandem repeats Also known as "microsatellites", are short tracts of repeated sequence motifs, generally 1–6 or more base pairs long. They can be repeated between 5 and 50 times. In population genetics, short tandem repeats (STRs) are used to determine genetic relatedness between individuals (e.g. kinship) and populations.

Note

1 M. Ani (1994) applied the term "Maafa"– Swahili for "disaster" – to describe the abduction and enslavement of African people. The term is preferred over the euphemistic "slave trade" as it better reflects the dehumanizing experiences African people suffered during and after the colonial period (Carroll and Jamison 2011).

References

Adhikari, Kaustubh, Javier Mendoza-Revilla, Juan Camilo Chacón-Duque, Macarena Fuentes-Guajardo, and Andrés. Ruiz-Linares. 2016. "Admixture in Latin America." *Current Opinion in Genetics & Development* 41: 106–114. https://doi.org/10.1016/j.gde.2016.09.003

Adler, Christina J., Wolfgang Haak, Denise Donlon, Alan Cooper, and Genographic Consortium. 2011. "Survival and recovery of DNA from ancient teeth and bones." *Journal of Archaeological Science* 38 (5): 956–964. https://doi.org/10.1016/j.jas.2010.11.010

Alexander, David H., John Novembre, and Kenneth Lange. 2009. "Fast model-based estimation of ancestry in unrelated individuals." *Genome Research* 19 (9): 1655–1664. https://doi.org/10.1101/gr.094052.109

Allaire, Louis. 1977. *Later Prehistory in Martinique and the Island Caribs: Problems in Ethnic Identity* [Unpublished doctoral dissertation]. Yale University.

Allaire, Louis. 1999. "Archaeology of the Caribbean region." In Salomon, F., and Schwartz, S.B. (Eds.), *The Cambridge History of the Native Peoples of the Americas: South America.* Cambridge University Press, 668–733.

Allentoft, Morten E., Matthew Collins, David Harker, James Haile, Charlotte L. Oskam, Marie L. Hale, Paula F. Campos, Jose A. Samaniego, M. Thomas P. Gilbert, Eske Willerslev, Guojie Zhang, R. Paul Scofield, Richard N. Holdaway, and Michael Bunce. 2012. "The half-life of DNA in bone: Measuring decay kinetics in 158 dated fossils." *Proceedings of the Royal Society B: Biological Sciences* 279 (1748): 4724–4733. https://doi.org/10.1098/rspb.2012.1745

Amorim, C.E.G., S. Vai, C. Posth, A. Modi, I. Koncz, S. Hakenbeck, M.C. La Rocca, B. Mende, D. Bobo, W. Pohl, L.P. Baricco, E. Bedini, P. Francalacci, C. Giostra, T. Vida, D. Winger, U. von Freeden, S. Ghirotto, M. Lari, ... and K.R. Veeramah. 2018. "Understanding 6th-century barbarian social organization and migration through paleogenomics." *Nature Communications* 9 (1): 1–11. https://doi.org/10.1038/s41467-018-06024-4.

Anderson-Córdova, K.F. 2017. *Surviving Spanish conquest: Indian Fight, Flight, and Cultural Transformation in Hispaniola and Puerto Rico.* Tuscaloosa, AL: University of Alabama Press.

Andrades Valtueña, A., A. Mittnik, F.M. Key, W. Haak, R. Allmäe, A. Belinskij, M. Daubaras, M. Feldman, R. Jankauskas, I. Janković, K. Massy, M. Novak, S. Pfrengle, S. Reinhold, M. laus, M.A. Spyrou, A. Szécsényi-Nagy, M. Tõrv, S. Hansen, ... and J. Krause. 2017. "The Stone Age plague and its persistence in Eurasia." *Current Biology* 27 (23): 3683–3691. e8. https://doi.org/10.1016/j.cub.2017.10.025

Ani, Marimba. 1994. *Yurugu: An African–Centred Critique of European Cultural Thought and Behaviour*. Trenton, NJ: Africa World Press.

Arriola, Luis A., Alan Cooper, and Laura S. Weyrich. 2020. "Palaeomicrobiology: Application of ancient DNA sequencing to better understand bacterial genome evolution and adaptation." *Frontiers in Ecology and Evolution* 8: 40. https://doi.org/10.3389/fevo.2020.00040.

Austin, Rita M., Sabrina B. Sholts, LaShanda Williams, Logan Kistler, and Courtney A. Hofman. 2019. "Opinion: To curate the molecular past, museums need a carefully considered set of best practices." *Proceedings of the National Academy of Sciences USA* 116 (5): 1471–1474. https://doi.org/10.1073/pnas.1822038116.

Ávila-Arcos, María C., Enrico Cappellini, J. Alberto Romero-Navarro, Nathan Wales, J. Victor Moreno-Mayar, Morten Rasmussen, Sarah L. Fordyce, Rafael Montiel, Jean.-Philippe Vielle-Calzada, Eske Willerslev, and M. Thomas P. Gilbert 2011. "Application and comparison of large-scale solution-based DNA capture-enrichment methods on ancient DNA." *Scientific Reports* 1 (1): 1–5. https://doi.org/10.1038/srep00074.

Bandelt, Hans -J., Peter Forster, Bryan C. Sykes, and Martin B. Richards. 1995. "Mitochondrial portraits of human populations using median networks." *Genetics* 141 (2): 743–753.

Bandelt, Hans-Jurgen, Peter Forster, and Arne Röhl. 1999. "Median-joining networks for inferring intraspecific phylogenies." *Molecular Biology and Evolution* 16 (1): 37–48.

Barquera, Rodrigo, and Johannes Krause. 2020. "An ancient view on host pathogen interaction across time and space." *Current Opinion in Immunology* 65: 65–69.

Beckles, Hilary Mc D. 1992. "Kalinago (Carib) resistance to European colonisation of the Caribbean." *Caribbean Quarterly* 38 (2–3): 1–124.

Benn Torres, Jada. 2014. "Prospecting the past: Genetic perspectives on the extinction and survival of indigenous peoples of the Caribbean." *New Genetics and Society* 33 (1): 21–41.

Benn Torres, J., Rick A. Kittles, and Anne C. Stone. 2007. "Mitochondrial and Y chromosome diversity in the English-speaking Caribbean." *Annals of Human Genetics* 71 (6): 782–790.

Benn Torres, Jada, Victoria Martucci, Melinda C. Aldrich, Miguel G. Vilar, Taryn MacKinney, Muhammad Tariq, Jill B. Gaieski, Ricardo Bharath Hernandez, Zoila E. Browne, and Marlon Stevenson. 2019. "Analysis of biogeographic ancestry reveals complex genetic histories for indigenous communities of St. Vincent and Trinidad." *American Journal of Physical Anthropology* 169 (3): 482–497.

Benn Torres, Jada, Miguel G. Vilar, Gabriel A. Torres, Jill B. Gaieski, Ricardo Bharath Hernandez, Zoila E. Browne, Marlon Stevenson, Wendell Walters, Theodore G. Schurr, and Genographic Consortium. 2015. "Genetic diversity in the Lesser Antilles and its implications for the settlement of the Caribbean Basin." *PLoS One* 10 (10): e0139192.

Bérard, Benoît. 2013. "The Saladoid." In *The Oxford Handbook of Caribbean Archaeology*. Oxford: Oxford University Press, 184–197.

Bergström, Anders, Shane A. McCarthy, Ruoyun Hui, Mohamed A. Almarri, Qasim Ayub, Petr Danecek, Yuan Chen, Sabine Felkel, Pille Hallast, and Jack Kamm. 2020. "Insights into human genetic variation and population history from 929 diverse genomes." *Science* 367 (6484): eaay5012.

Bisso-Machado, Rafael, and Nelson J.R. Fagundes. 2021. "Uniparental genetic markers in Native Americans: A summary of all available data from ancient and contemporary populations." *American Journal of Physical Anthropology* 176 (3): 445–458.

Bolnick, Deborah A., Holly M. Bonine, Jaime Mata-Míguez, Brian M. Kemp, Meradeth H. Snow, and Steven A. LeBlanc. 2012. "Nondestructive sampling of human skeletal remains yields ancient nuclear and mitochondrial DNA." *American Journal of Physical Anthropology* 147 (2): 293–300.

Bongers, Jacob L., Nathan Nakatsuka, Colleen O'Shea, Thomas K. Harper, Henry Tantaleán, Charles Stanish, and Lars Fehren-Schmitz. 2020. "Integration of ancient DNA with transdisciplinary dataset finds strong support for Inca resettlement in the south Peruvian coast." *Proceedings of the National Academy of Sciences* 117 (31): 18359–18368.

Boomert, Arie. 1986. "The Cayo complex of St. Vincent: Ethnohistorical and archaeological aspects of the Island Carib problem." *Antropológica* 66: 3–68.

Boomert, Arie. 2004. "Koriabo and the polychrome tradition: The late-prehistoric era between the Orinoco and Amazon mouths." In Delpeuch, A., and Hofman, C. (Eds.), *Late Ceramic Age Societies in the Eastern Caribbean.* Oxford: BAR International Series. Pp. 251–268.

Boomert, Arie. 2011. "From Cayo to Kalinago." In Hofman, C., and van Duijvenbode, A. (Eds.), *Communities in Contact. Essay in Archaeology, Ethnohistory and Ethnography*, pp. 291–306. Leiden: Sidestone Press.

Bos, Kirsten I., Alexander Herbig, Jason Sahl, Nicholas Waglechner, Mathieu Fourment, Stephen A. Forrest, Jennifer Klunk, Verena J. Schuenemann, Debi Poinar, and Melanie Kuch. 2016. "Eighteenth century Yersinia pestis genomes reveal the long-term persistence of an historical plague focus." *Elife* 5: e12994.

Bos, Kirsten I., Günter Jäger, Verena J. Schuenemann, Åshild J. Vågene, Maria A. Spyrou, Alexander Herbig, Kay Nieselt, and Johannes Krause. 2015. "Parallel detection of ancient pathogens via array-based DNA capture." *Philosophical Transactions of the Royal Society B: Biological Sciences* 370 (1660): 20130375.

Bos, Kirsten I., Denise Kühnert, Alexander Herbig, Luis Roger Esquivel-Gomez, Aida Andrades Valtueña, Rodrigo Barquera, Karen Giffin, Aditya Kumar Lankapalli, Elizabeth A. Nelson, and Susanna Sabin. 2019. "Paleomicrobiology: Diagnosis and evolution of ancient pathogens." *Annual Review of Microbiology* 73: 639–666.

Bro-Jørgensen, Maiken Hemme, Christian Carøe, Filipe G. Vieira, Sofia Nestor, Ann Hallström, Kristian M. Gregersen, Vivian Etting, M. Thomas P. Gilbert, and Mikkel-Holger S. Sinding. 2018. "Ancient DNA analysis of Scandinavian medieval drinking horns and the horn of the last aurochs bull." *Journal of Archaeological Science* 99: 47–54.

Bryc, Katarzyna, Adam Auton, Matthew R. Nelson, Jorge R. Oksenberg, Stephen L. Hauser, Scott Williams, Alain Froment, Jean-Marie Bodo, Charles Wambebe, and Sarah A. Tishkoff. 2010. "Genome-wide patterns of population structure and admixture in West Africans and African Americans." *Proceedings of the National Academy of Sciences* 107 (2): 786–791.

Bukhari, Areej, Javier Rodriguez Luis, Miguel A. Alfonso-Sanchez, Ralph Garcia-Bertrand, and Rene J. Herrera. 2017. "Taino and African maternal heritage in the Greater Antilles." *Gene* 637: 33–40.

Bullen, Adelaide K., and Ripley P. Bullen. 1970. "'The Lavoutte Site, St. Lucia: A Carib ceremonial centre." In *Proceedings of the 3rd International Congress for the Study of Pre-Columbian Cultures of the Lesser Antilles*. Grenada: Grenada National Museum, pp. 61–86.

Bullen, Ripley P. 1964. *The Archaeology of Grenada*. West Indies: University of Florida.

Bullen, Ripley P., and Adelaide K. Bullen. 1972. *Archaeological Investigations on St. Vincent and the Grenadines*. West Indies: Bryant Foundation.

Buonasera, Tammy, Jelmer Eerkens, Alida de Flamingh, Laurel Engbring, Julia Yip, Hongjie Li, Randall Haas, Diane DiGiuseppe, Dave Grant, and Michelle Salemi. 2020. "A comparison of proteomic, genomic, and osteological methods of archaeological sex estimation." *Scientific Reports* 10 (1): 1–15.

Callaway, Ewen. 2016. "Error found in study of first ancient African genome." *Nature*. https://doi.org/10.1038/nature.2016.19258

Campos, Paula F., Oliver E. Craig, Gordon Turner-Walker, Elizabeth Peacock, Eske Willerslev, and M. Thomas P. Gilbert. 2012. "DNA in ancient bone–where is it located and how should we extract it?" *Annals of Anatomy-Anatomischer Anzeiger* 194 (1): 7–16.

Carpenter, Meredith L., Jason D. Buenrostro, Cristina Valdiosera, Hannes Schroeder, Morten E Allentoft, Martin Sikora, Morten Rasmussen, Simon Gravel, Sonia Guillén, and Georgi Nekhrizov. 2013. "Pulling out the 1%: Whole-genome capture for the targeted enrichment of ancient DNA sequencing libraries." *The American Journal of Human Genetics* 93 (5): 852–864.

Carroll, Karanja Keita, and DeReef F. Jamison. 2011. "African-centered psychology, education and the liberation of African minds: Notes on the psycho-cultural justification for reparations." *Race, Gender & Class* 18 (1–2): 52–72.

Cassidy, Lara M., Ros Ó. Maoldúin, Thomas Kador, Ann Lynch, Carleton Jones, Peter C Woodman, Eileen Murphy, Greer Ramsey, Marion Dowd, and Alice Noonan. 2020. "A dynastic elite in monumental Neolithic society." *Nature* 582 (7812): 384–388.

Chang, Dan, Michael Knapp, Jacob Enk, Sebastian Lippold, Martin Kircher, Adrian Lister, Ross DE MacPhee, Christopher Widga, Paul Czechowski, and Robert Sommer. 2017. "The evolutionary and phylogeographic history of woolly mammoths: A comprehensive mitogenomic analysis." *Scientific Reports* 7 (1): 1–10.

Chanlatte Baik, Luis, and Yvonne Narganes-Sorde. 2005. *Cultura La Hueca*, 1st ed. San Juan, Puerto Rico: Museo de Historia, Antropología y Arte. Universidad de Puerto Rico, Recinto de Río Piedras.

Charlton, Sophy, Thomas Booth, and Ian Barnes. 2019. "The problem with petrous? A consideration of the potential biases in the utilization of pars petrosa for ancient DNA analysis." *World Archaeology* 51 (4): 574–585.

Chinique de Armas, Yadira, Mirjana Roksandic, Roberto Rodriguez Suarez, David G. Smith, and William M. Buhay. 2016. "Isotopic evidence of variation in subsistance strategies and food consumption patterns among 'Fischer-Gatherer' populations of Western Cuba." In *Cuban Archaeology in the Caribbean*, edited by Ivan Roksandic. University of Florida Press, pp. 125–146.

Coe, William R. 1957. "A distinctive artifact common to Haiti and Central America." *American Antiquity* 22 (3): 280–282.

Cook, Noble David. 2002. "Sickness, starvation, and death in early Hispaniola." *Journal of Interdisciplinary History* 32 (3): 349–386.

Cruxent, Jose M., and Iriving Rouse. 1958. "59, An archaeological chronology of Venezuela, 2 vols., Washington, Pan American Union." *Social Science Monography* 6: 604.

Cruxent, Jose M., and Irving Rouse. 1961. "An archaeological chronology of Venezuela." *American Antiquity* 26 (3): 2.

Curet, L. Antonio. 2006. "Las crónicas en la arqueología de Puerto Rico y del Caribe." *Caribbean Studies* 34 (1): 163–199.

Curet, L. Antonio. 2014. "The Taino: Phenomena, concepts, and terms." *Ethnohistory* 61 (3): 467–495. https://doi.org/10.1215/00141801-2681759.

Dabney, J., M. Knapp, I. Glocke, M.T. Gansauge, A. Weihmann, B. Nickel, C. Valdiosera, N. Garcia, S. Paabo, J.L. Arsuaga, and M. Meyer. 2013. "Complete mitochondrial genome sequence of a Middle Pleistocene cave bear reconstructed from ultrashort DNA fragments." *Proceedings of the National Academy of Sciences of the United States of America* 110 (39): 15758–15763. https://doi.org/10.1073/pnas.1314445110.

Darzynkiewicz, Zbigniew, Gloria Juan, Xun Li, Wojciech Gorczyca, Tomoyuki Murakami, and Frank Traganos. 1997. "Cytometry in cell necrobiology: Analysis of apoptosis and accidental cell death (necrosis)." *Cytometry: The Journal of the International Society for Analytical Cytology* 27 (1): 1–20.

Davis, Dave D., and R. Christopher Goodwin. 1990. "Island Carib origins: Evidence and nonevidence." *American Antiquity 55*: 37–48.

Deagan, Kathleen, and José María Cruxent. 2008. *Columbus's Outpost among the Taínos.* Yale University Press.

De Navarrete, Martin Fernández. 1837. *Colección de las viajes y descubrimentos que hicieron por mar los españoles desde fines del siglo XV…*, 5.

Dommain, René, Morgan Andama, Molly M. McDonough, Natalia A. Prado, Tobias Goldhammer, Richard Potts, Jesús E. Maldonado, John Bosco Nkurunungi, and Michael G. Campana. 2020. "The challenges of reconstructing tropical biodiversity with sedimentary ancient DNA: A 2200-year-long metagenomic record from Bwindi impenetrable forest, Uganda." *Frontiers in Ecology and Evolution* 8: 218. https://doi.org/10.3389/fevo.2020.00218

Duchêne, Sebastián, Simon YW Ho, Ann G. Carmichael, Edward C. Holmes, and Hendrik Poinar. 2020. "The recovery, interpretation and use of ancient pathogen genomes." *Current Biology* 30 (19): R1215–R1231.

Duggan, Ana T., Jennifer Klunk, Ashleigh F. Porter, Anna N. Dhody, Robert Hicks, Geoffrey L. Smith, Margaret Humphreys, Andrea M. McCollum, Whitni B. Davidson, and Kimberly Wilkins. 2020. "The origins and genomic diversity of American Civil War Era smallpox vaccine strains." *Genome Biology* 21 (1): 1–11.

Durand, Eric Y., Nick Patterson, David Reich, and Montgomery Slatkin. 2011. "Testing for ancient admixture between closely related populations." *Molecular Biology and Evolution* 28 (8): 2239–2252.

Eglinton, Geoffrey, and Graham A. Logan. 1991. "Molecular preservation." *Philosophical Transactions of the Royal Society of London. Series B: Biological Sciences* 333 (1268): 315–328.

Eisenmann, Stefanie, Eszter Bánffy, Peter van Dommelen, Kerstin P. Hofmann, Joseph Maran, Iosif Lazaridis, Alissa Mittnik, Michael McCormick, Johannes Krause, and David Reich. 2018. "Reconciling material cultures in archaeology with genetic data: The nomenclature of clusters emerging from archaeogenomic analysis." *Scientific Reports* 8 (1): 1–12.

Elliott, Carl, and Paul Brodwin. 2002. "Identity and genetic ancestry tracing." *BMJ* 325 (7378): 1469–1471.

Enk, Jacob M., Alison M. Devault, Melanie Kuch, Yusuf E. Murgha, Jean-Marie Rouillard, and Hendrik N. Poinar. 2014. "Ancient whole genome enrichment using baits built from modern DNA." *Molecular Biology and Evolution* 31 (5): 1292–1294.

Febles, Jorge. 1991. "Estudio comparativo de las industrias de la piedra tallada de Aguas Verdes (Baracoa) y Playitas (Matanzas): Probable relación de estas industrias con otras del SE de los Estados Unidos."In Rodriguez, M.A. (Ed.), *Arqueologia de Cuba y de otras dreas Antillanas*. La Habana: Editorial Academia, 312–371.

Febles, Jorge, and Alexis Rives. 1991. "Cluster analysis: Un experimento aplicado a la industria de la piedra tallada del protoarcaico de Cuba." In Rodriguez, M.A. (Ed.), *Arqueología de Cuba y de otras áreas antillanas*. La Habana: Editorial Academia, 115–124.

Fehren-Schmitz, Lars, Kelly M. Harkins, and Catrine L. Jarman. 2017. "Paleogenomic investigations of human remains from Rapa Nui." *American Journal of Physical Anthropology* 162 (S64): 178

Fellows Yates, James A., Irina M. Velsko, Franziska Aron, Cosimo Posth, Courtney A. Hofman, Rita M. Austin, Cody E. Parker, Allison E. Mann, Kathrin Nägele, and Kathryn Weedman Arthur. 2021. "The evolution and changing ecology of the African hominid oral microbiome." *Proceedings of the National Academy of Sciences* 118 (20): e2021655118. https://doi.org/10.1073/pnas.2021655118

Fellows Yates, James A., Thiseas C. Lamnidis, Maxime Borry, Aida Andrades Valtueña, Zandra Fagernäs, Stephen Clayton, Maxime U. Garcia, Judith Neukamm, and Alexander Peltzer. 2021. "Reproducible, portable, and efficient ancient genome reconstruction with NF-core/eager." *PeerJ* 9: e10947.

Fernandes, Daniel M., Alissa Mittnik, Iñigo Olalde, Iosif Lazaridis, Olivia Cheronet, Nadin Rohland, Swapan Mallick, Rebecca Bernardos, Nasreen Broomandkhoshbacht, and Jens Carlsson. 2020. "The spread of steppe and Iranian-related ancestry in the islands of the western Mediterranean." *Nature Ecology & Evolution* 4 (3): 334–345.

Fernandes, Daniel M., Kendra A. Sirak, Harald Ringbauer, Jakob Sedig, Nadin Rohland, Olivia Cheronet, Matthew Mah, Swapan Mallick, Iñigo Olalde, and Brendan J. Culleton. 2021. "A genetic history of the pre-contact Caribbean." *Nature* 590 (7844): 103–110.

Feuerborn, Tatiana R., Alberto Carmagnini, Robert J Losey, Tatiana Nomokonova, Arthur Askeyev, Igor Askeyev, Oleg Askeyev, Ekaterina E. Antipina, Martin Appelt, and Olga P. Bachura. 2021. "Modern Siberian dog ancestry was shaped by several thousand years of Eurasian-wide trade and human dispersal." *Proceedings of the National Academy of Sciences* 118 (39): e2100338118. https://doi.org/10.1073/pnas.2100338118

Fitzpatrick, Scott M., Matthew F. Napolitano, and Jessica H. Stone. 2021. "What is the most parsimonious explanation for where Pre-Columbian Caribbean peoples originated?" In Napolitano, M.F., Stone, J.H., and DiNapoli, R.J. (Eds.), *The Archaeology of Island Colonization*. Gainesville, FL: University Press of Florida, pp. 231–264.

Fitzpatrick, Scott M., Michiel Kappers, and Christina M. Giovas. 2010. "The southward route hypothesis: Examining Carriacou's chronological position in Antillean prehistory." In: Fitzpatrick, S.M., and Ross, A.H. (Eds.), *Island Shores Distant Pasts: Archaeological and Biological Approaches to the PreColumbian Settlement of the Caribbean*. Gainesville: University Press of Florida, pp. 163–176.

Fortes-Lima, Cesar, Jonas Bybjerg-Grauholm, Lilia Caridad Marin-Padrón, Enrique Javier Gomez-Cabezas, Marie Bækvad-Hansen, Christine Søholm Hansen, Phuong Le, David Michael Hougaard, Paul Verdu, and Ole Mors. 2018. "Exploring Cuba's population structure and demographic history using genome-wide data." *Scientific Reports* 8 (1): 1–13.

Fox, Keolu, and John Hawks. 2019. *Use Ancient Remains More Wisely*. Nature Publishing Group.

Fu, Qiaomei, Alissa Mittnik, Philip L.F. Johnson, Kristen Bos, Martina Lari, Ruth Bollongino, Chengkai Sun, Liane Giemsch, Ralf Schmitz, Joachim Burger, Anna Maria Ronchitelli, Fabio Martini, Renata G. Cremonesi, Jiří Svoboda, Peter Bauer, David Caramelli, Sergi Castellano, David Reich, Svante. Pääbo, and Johannes Krause. 2013. "A revised timescale for human evolution based on ancient mitochondrial genomes." *Current Biology* 23 (7): 553–559. https://doi.org/10.1016/j.cub.2013.02.044.

Gallego Llorente, Marcos, Eppie Jones, Anders Eriksson, Veronika Siska, K.W. Arthur, J.W. Arthur, M.C. Curtis, Jay Stock, M. Coltorti, and P. Pieruccini. 2015. "Ancient Ethiopian genome reveals extensive Eurasian admixture throughout the African continent." *Science* 350 (6262): 820–822.

Gansauge, Marie-Theres, Tobias Gerber, Isabelle Glocke, Petra Korlević, Laurin Lippik, Sarah Nagel, Lara Maria Riehl, Anna Schmidt, and Matthias Meyer. 2017. "Single-stranded DNA library preparation from highly degraded DNA usingT4DNA ligase." *Nucleic Acids Research* 45 (10): e79. https://doi.org/10.1093/nar/gkx033.

Giles, Richard E., Hugues Blanc, Howard M. Cann, and Douglas C. Wallace. 1980. "Maternal inheritance of human mitochondrial DNA." *Proceedings of the National Academy of Sciences* 77 (11): 6715–6719.

Glocke, Isabelle, and Matthias Meyer. 2017. "Extending the spectrum of DNA sequences retrieved from ancient bones and teeth." *Genome Research* 27 (7): 1230–1237.

González Fortes, Gloria, and Johanna L.A. Paijmans. 2019. "Whole-genome capture of ancient DNA using homemade baits." *Methods in Molecular Biology*, 1963: 93–105.

González-Tennant, Edward. 2014. "The "color" of heritage: Decolonizing collaborative archaeology in the Caribbean." *Journal of African Diaspora Archaeology and Heritage* 3 (1): 26–50.

Goodwin, Sara, John D. McPherson, and W. Richard McCombie. 2016. "Coming of age: Ten years of next-generation sequencing technologies." *Nature Reviews Genetics* 17 (6): 333–351.

Granberry, Julian, and Gary Vescelius. 2004. *Languages of the Pre-Columbian Antilles*. Tuscaloosa, AL: University of Alabama Press.

Grunberg, Bernard. 2011. "An ethnohistorical approach of the carib through written sources." In *Communities in Contact: Essays in Archaeology, Ethnohistory & Ethnography of the Amerindian Circum-Caribbean*. 327.

Guzmán-Solís, Axel A., Viridiana Villa-Islas, Miriam J. Bravo-López, Marcela Sandoval-Velasco, Julie K. Wesp, Jorge A. Gómez-Valdés, María de la Luz Moreno-Cabrera,

Alejandro Meraz, Gabriela Solís-Pichardo, and Peter Schaaf. 2021. "Ancient viral genomes reveal introduction of human pathogenic viruses into Mexico during the transatlantic slave trade." *Elife* 10: e68612.

Haak, Wolfgang, Iosif Lazaridis, Nick Patterson, Nadin Rohland, Swapan Mallick, Bastien Llamas, Guido Brandt, Susanne Nordenfelt, Eadaoin Harney, Kristin Stewardson, Qiaomei Fu, Alissa Mittnik, Eszter Bánffy, Christos Economou, Michael Francken, Susanne Friederich, Rafael Garrido Pena, Fredrik Hallgren, Valery Khartanovich, Aleksandr Khokhlov, Michael Kunst, Pavel Kuznetsov, Harald Meller, Oleg Mochalov, Vayacheslav Moiseyev, Nicole Nicklisch, Sandra L. Pichler, Roberto Risch, Manuel A. Rojo Guerra, Christina Roth, Anna Szécsényi-Nagy, Joachim Wahl, Matthias Meyer, Johannes Krause, Dorcas Brown, David Anthony, Alan Cooper, Kurt Werner Alt, and David Reich. 2015. "Massive migration from the steppe was a source for Indo-European languages in Europe." *Nature* 522 (7555): 207–211. https://doi.org/10.1038/nature14317.

Haas, Randall, James Watson, Tammy Buonasera, John Southon, Jennifer C. Chen, Sarah Noe, Kevin Smith, Carlos Viviano Llave, Jelmer Eerkens, and Glendon Parker. 2020. "Female hunters of the early Americas." *Science Advances* 6 (45): eabd0310.

Hagan, Richard W., Courtney A. Hofman, Alexander Hübner, Karl Reinhard, Stephanie Schnorr, Cecil M. Lewis Jr, Krithivasan Sankaranarayanan, and Christina G. Warinner. 2020. "Comparison of extraction methods for recovering ancient microbial DNA from paleofeces." *American Journal of Physical Anthropology* 171 (2): 275–284.

Hagelberg, Erika, Bryan Sykes, and Robert Hedges. 1989. "Ancient bone DNA amplified." *Nature* 342 (6249): 485–485.

Hansen, Henrik B., Peter B. Damgaard, Ashot Margaryan, Jesper Stenderup, Niels Lynnerup, Eske Willerslev, and Morten E. Allentoft. 2017. "Comparing ancient DNA preservation in petrous bone and tooth cementum." *PloS One* 12 (1): e0170940.

Hansson, Maria C., and Brendan P. Foley. 2008. "Ancient DNA fragments inside Classical Greek amphoras reveal cargo of 2400-year-old shipwreck." *Journal of Archaeological Science* 35 (5): 1169–1176.

Harney, Éadaoin, Olivia Cheronet, Daniel M. Fernandes, Kendra Sirak, Matthew Mah, Rebecca Bernardos, Nicole Adamski, Nasreen Broomandkhoshbacht, Kimberly Callan, and Ann Marie Lawson. 2021. "A minimally destructive protocol for DNA extraction from ancient teeth." *Genome Research* 31 (3): 472–483.

Hartnup, Katie, Leon Huynen, Rangi Te Kanawa, Lara D. Shepherd, Craig D. Millar, and David M. Lambert. 2011. "Ancient DNA recovers the origins of Māori feather cloaks." *Molecular Biology and Evolution* 28 (10): 2741–2750.

Higman, B.W. 2021. "Indigenous history and identity in the Caribbean." In *The Routledge Companion to Global Indigenous History*. Routledge, 500–523.

Hofman, Corinne L. 2013. "The Post-Saladoid in the Lesser Antilles (AD 600/800–1492)." In W. F. Keegan, C.L. Hofman, and R.R. Ramos (Eds.), *The Oxford Handbook of Caribbean Archaeology.* Oxford: Oxford University Press, 205–220.

Hofman, Corinne L., Angus Mol, R. Rodríguez Ramos, and Sebastiaan Knippenberg. 2014. "Networks set in stone: Archaic-ceramic interaction in the early precolonial northeastern Caribbean." In Berard, B., and Losier, C. (Eds.), *Archéologie Caraïbe*. Leiden: Sidestone Press, 119–132.

Hofman, Corinne L., Lewis Borck, Jason E. Laffoon, Emma R. Slayton, Rebecca B. Scott, Thomas W. Breukel, Catarina Guzzo Falci, Maroussia Favre, and Menno L.P. Hoogland. 2021. "Island networks: Transformations of inter-community social relationships in the Lesser Antilles at the advent of European colonialism." *The Journal of Island and Coastal Archaeology* 16 (2–4): 290–316.

Hofman, Corinne L., and Menno L.P. Hoogland. 2012. "Caribbean encounters: Rescue excavations at the early colonial Island Carib site of Argyle, St. Vincent." *Analecta Praehistorica Leidensia* 43 (44): 63–76.

Hofman, Corinne L., Menno L.P. Hoogland, Katarina Jacobson, Sébastien Manem, and Arie Boomert. 2020. "Cayo in the Lesser Antilles: A network of peoples, places and practices in the late 15th to Early 17th Century." *Ko iabO* 33.

Holsinger, Kent E., and Bruce S. Weir. 2009. "Genetics in geographically structured populations: Defining, estimating and interpreting F ST." *Nature Reviews Genetics* 10 (9): 639–650.

Honychurch, Lennox. 1997. *Carib to Creole: A History of Contact and Culture Exchange*. D.Phil Dissertation. University of Oxford

Huson, Daniel H., Alexander F. Auch, Ji Qi, and Stephan C. Schuster. 2007. "MEGAN analysis of metagenomic data." *Genome Research* 17 (3): 377–386.

Jensen, Theis Z.T., Jonas Niemann, Katrine Højholt Iversen, Anna K. Fotakis, Shyam Gopalakrishnan, Åshild J. Vågene, Mikkel Winther Pedersen, Mikkel-Holger S. Sinding, Martin R. Ellegaard, and Morten E. Allentoft. 2019. "A 5700 year-old human genome and oral microbiome from chewed birch pitch." *Nature Communications* 10 (1): 1–10.

Kazarina, Alisa, Guntis Gerhards, Elina Petersone-Gordina, Janis Kimsis, Ilva Pole, Egija Zole, Viktorija Leonova, and Renate Ranka. 2019. "Analysis of the bacterial communities in ancient human bones and burial soil samples: Tracing the impact of environmental bacteria." *Journal of Archaeological Science* 109: 104989.

Keegan, William F. 1996. "West Indian archaeology. 2. After Columbus." *Journal of Archaeological Research* 4 (4): 265–294.

Keegan, William F. 1992. *The People Who Discovered Columbus: The Prehistory of the Bahamas*. Gainesville: University of Florida Press.

Keegan, William F., and Corinne Hofman. 2017. *The Caribbean Before* Columbus: Oxford University Press.

Kerner, Gaspard, Etienne Patin, and Lluis Quintana-Murci. 2021. "New insights into human immunity from ancient genomics." *Current Opinion in Immunology* 72: 116–125.

Kim, Julie Chun. 2013. "The Caribs of St. Vincent and indigenous resistance during the age of revolutions." *Early American Studies* 11(1): 117–132.

Kimura, Birgitta K., Michelle J. LeFebvre, Susan D. deFrance, Hilary I. Knodel, Michelle S. Turner, Natalie S. Fitzsimmons, Scott M. Fitzpatrick, and Connie J. Mulligan. 2016. "Origin of pre-Columbian guinea pigs from Caribbean archeological sites revealed through genetic analysis." *Journal of Archaeological Science: Reports* 5: 442–452.

Kistler, Logan, Vanessa C. Bieker, Michael D. Martin, Mikkel Winther Pedersen, Jazmín Ramos Madrigal, and Nathan Wales. 2020. "Ancient plant genomics in archaeology, herbaria, and the environment." *Annual Review of Plant Biology* 71: 605–629.

Kistler, Logan, Roselyn Ware, Oliver Smith, Matthew Collins, and Robin G. Allaby. 2017. "A new model for ancient DNA decay based on paleogenomic meta-analysis." *Nucleic Acids Research* 45 (11): 6310–6320.

Kocher A, Papac L, Barquera R, Key FM, Spyrou MA, Hübler R, et al. 2021. "Ten millennia of hepatitis B virus evolution." *Science* 374: 182–188.

Korlević, Petra, Sahra Talamo, and Matthias Meyer. 2018. "A combined method for DNA analysis and radiocarbon dating from a single sample." *Scientific Reports* 8 (1): 1–10.

Krause, Johannes, Qiaomei Fu, Jeffrey M. Good, Bence Viola, Michael V. Shunkov, Anatoli P. Derevianko, and Svante Pääbo. 2010. "The complete mitochondrial DNA genome of an unknown hominin from southern Siberia." *Nature* 464 (7290): 894–897.

Kulstad González, Pauline. 2020. *Hispaniola-Hell or Home?: Decolonizing Grand Narratives about Intercultural Interactions at Concepción de la Vega (1494–1564)*. Leiden: Sidestone Press.

Laffoon, Jason E. 2013. "New perspectives from Isotope Analysis." In Keegan, W.F., Hofman, C.L., and Rodriguez Ramos, R. (Eds.), *The Oxford Handbook of Caribbean Archaeology*. Oxford: Oxford University Press, 418–435.

Lalueza-Fox, Carles, F. Luna Calderon, Francesc Calafell, Bernal Morera, and Jaume Bertranpetit. 2001. "MtDNA from extinct Tainos and the peopling of the Caribbean." *Annals of Human Genetics* 65 (2): 137–151.

Lalueza-Fox, C., M.T. Gilbert, A.J. Martinez-Fuentes, F. Calafell, and J. Bertranpetit. 2003. "Mitochondrial DNA from pre-Columbian Ciboneys from Cuba and the prehistoric colonization of the Caribbean." *American Journal of Biological Anthropology* 121 (2): 97–108. https://doi.org/10.1002/ajpa.10236.

Lawson, Daniel J., Lucy Van Dorp, and Daniel Falush. 2018. "A tutorial on how not to over-interpret STRUCTURE and ADMIXTURE bar plots." *Nature Communications* 9 (1): 1–11.

Lazaridis, Iosif, Nick Patterson, Alissa Mittnik, Gabriel Renaud, Swapan Mallick, Karola Kirsanow, Peter H. Sudmant, Joshua G. Schraiber, Sergi Castellano, and Mark Lipson. 2014. "Ancient human genomes suggest three ancestral populations for present-day Europeans." *Nature* 513 (7518): 409–413.

LeBlanc, Steven A., Lori S. Cobb Kreisman, Brian M. Kemp, Francis E. Smiley, Shawn W. Carlyle, Anna N. Dhody, and Thomas Benjamin. 2007. "Quids and aprons: Ancient DNA from artifacts from the American Southwest." *Journal of Field Archaeology* 32 (2): 161–175.

LeFebvre, Michelle J., Christina M. Giovas, and Jason E. Laffoon. 2018. "Advancing the Study of Amerindian Ecodynamics in the Caribbean: Current Perspectives". *Environmental Archaeology* 24(2): 107–114.

Lenik, Stephan. 2012. "Carib as a colonial category: Comparing ethnohistoric and archaeological evidence from Dominica, West Indies." *Ethnohistory* 59 (1): 79–107.

Leonardi, Michela, Pablo Librado, Clio Der Sarkissian, Mikkel Schubert, Ahmed H. Alfarhan, Saleh A. Alquraishi, Khaled A.S. Al-Rasheid, Cristina Gamba, Eske Willerslev, and Ludovic Orlando. 2017. "Evolutionary patterns and processes: Lessons from ancient DNA." *Systematic Biology* 66 (1): e1–e29.

Liaw, Yun-Fan, and Chia-Ming Chu. 2009. "Hepatitis B virus infection." *The Lancet* 373 (9663): 582–592.

Lindahl, Tomas. 1993. "Instability and decay of the primary structure of DNA." *Nature* 362 (6422): 709–715.

Lindahl, Tomas, and Annika Andersson. 1972. "Rate of chain breakage at apurinic sites in double-stranded deoxyribonucleic acid." *Biochemistry* 11 (19): 3618–3623.

Lipson, Mark, Pontus Skoglund, Matthew Spriggs, Frederique Valentin, Stuart Bedford, Richard Shing, Hallie Buckley, Iarawai Phillip, Graeme K. Ward, and Swapan Mallick. 2018. "Population turnover in Remote Oceania shortly after initial settlement." *Current Biology* 28 (7): 1157–1165. e7.

Llamas, Bastien, Guido Valverde, Lars Fehren-Schmitz, Laura S. Weyrich, Alan Cooper, and Wolfgang Haak. 2017. "From the field to the laboratory: Controlling DNA contamination in human ancient DNA research in the high-throughput sequencing era." *STAR: Science & Technology of Archaeological Research* 3 (1): 1–14.

Lleonart, Ricardo, Eileen Riego, Roberto Rodríguez Suárez, Rafael Travieso Ruiz, and José De la Fuente. 1999. "Analyses of DNA from ancient bones of a pre-columbian Cuban woman and a child." *Genetics and Molecular Biology* 22 (3): 285–289.

Lord, Edana, Catherine Collins, Susan deFrance, Michelle J. LeFebvre, and Elizabeth Matisoo-Smith. 2018. "Complete mitogenomes of ancient Caribbean Guinea pigs (Cavia porcellus)." *Journal of Archaeological Science: Reports* 17: 678–688.

Madrilejo, Nicole, Holden Lombard, and Jada Benn Torres. 2015. "Origins of marronage: Mitochondrial lineages of Jamaica's Accompong Town Maroons." *American Journal of Human Biology* 27 (3): 432–437.

Mann, Allison E., Susanna Sabin, Kirsten Ziesemer, Åshild J. Vågene, Hannes Schroeder, Andrew T. Ozga, Krithivasan Sankaranarayanan, Courtney A. Hofman, James A. Fellows Yates, and Domingo C. Salazar-García. 2018. "Differential preservation of endogenous human and microbial DNA in dental calculus and dentin." *Scientific Reports* 8 (1): 1–15.

Marcheco-Teruel, Beatriz, Esteban J. Parra, Evelyn Fuentes-Smith, Antonio Salas, Henriette N. Buttenschøn, Ditte Demontis, María Torres-Español, Lilia C. Marín-Padrón, Enrique J. Gómez-Cabezas, and Vanesa Álvarez-Iglesias. 2014. "Cuba: Exploring the history of admixture and the genetic basis of pigmentation using autosomal and uniparental markers." *PLoS Genetics* 10 (7): e1004488.

Mardulyn, Patrick. 2012. "Trees and/or Networks to Display Intraspecific DNA Sequence Variation?" *Molecular Ecology* 21 (14): 3385–3390.

Maricic, Tomislav, Mark Whitten, and Svante Pääbo. 2010. "Multiplexed DNA sequence capture of mitochondrial genomes using PCR products." *PloS One* 5 (11): e14004.

Martinez-Cruzado, Juan C. 2010. "The history of Amerindian mitochondrial DNA lineages in Puerto Rico." In Fitzpatrick, S.M., and Ross, A.H. (Eds.), *Island Shores, Distant Pasts: Archaeological and Biological Approaches to the Pre-Columbian Settlement of the Caribbean*. Gainsville-Florida, FL: University Press of Florida: 54–80.

Martínez-Cruzado, Juan C., Gladys Toro-Labrador, Jorge Viera-Vera, Michelle Y. Rivera-Vega, Jennifer Startek, Magda Latorre-Esteves, Alicia Román-Colón, Rebecca Rivera-Torres, Iris Y. Navarro-Millán, and Enid Gómez-Sánchez. 2005. "Reconstructing the population history of Puerto Rico by means of mtDNA phylogeographic analysis." *American Journal of Physical Anthropology* 128 (1): 131–155.

Mathieson, Iain, and Aylwyn Scally. 2020. "What is ancestry?" *PLoS Genetics* 16 (3): e1008624.

Mathieson, Iain, Songül Alpaslan-Roodenberg, Cosimo Posth, Anna Szécsényi-Nagy, Nadin Rohland, Swapan Mallick, Iñigo Olalde, Nasreen Broomandkhoshbacht, Francesca Candilio, and Olivia Cheronet. 2018. "The genomic history of southeastern Europe." *Nature* 555 (7695): 197–203.

Matisoo-Smith, Elizabeth. 2015. "Ancient DNA and the human settlement of the Pacific: A review." *Journal of Human Evolution* 79: 93–104.

McColl, Hugh, Fernando Racimo, Lasse Vinner, Fabrice Demeter, Takashi Gakuhari, J. Víctor Moreno-Mayar, George Van Driem, Uffe Gram Wilken, Andaine Seguin-Orlando, and Constanza De la Fuente Castro. 2018. "The prehistoric peopling of Southeast Asia." *Science* 361 (6397): 88–92.

McDonough, Molly M., Lillian D. Parker, Nancy Rotzel Mclnerney, Michael G. Campana, and Jesús E. Maldonado. 2018. "Performance of commonly requested destructive museum samples for mammalian genomic studies." *Journal of Mammalogy* 99 (4): 789–802.

Mendisco, Fanny, Christine Keyser, Veronica Seldes, Clara Rivolta, Pablo Mercolli, Pablo Cruz, Axel E. Nielsen, Eric Crubezy, and Bertrand Ludes. 2014. "Genetic diversity of a late prehispanic group of the Quebrada de Humahuaca, northwestern Argentina." *Annals of Human Genetics* 78 (5): 367–380.

Mendisco, Fanny, Marie-Hélène Pemonge, E. Leblay, T. Romon, Gérard Richard, Patrice Courtaud, and Marie-France Deguilloux. 2015. "Where are the Caribs? Ancient DNA from ceramic period human remains in the Lesser Antilles." *Philosophical Transactions of the Royal Society B: Biological Sciences* 370 (1660): 20130388.

Mendisco, Fanny, Marie-Hélène Pemonge, Thomas Romon, Gérard Lafleur, Gérard Richard, Patrice Courtaud, and Marie-France Deguilloux. 2019. "Tracing the genetic legacy in the French Caribbean islands: A study of mitochondrial and Y-chromosome lineages in the Guadeloupe archipelago." *American Journal of Physical Anthropology* 170 (4): 507–518.

Mendizabal, I., K. Sandoval, G. Berniell-Lee, F. Calafell, A. Salas, A. Martinez-Fuentes, and D. Comas. 2008. "Genetic origin, admixture, and asymmetry in maternal and paternal human lineages in Cuba." *BMC Evolutionary Biology* 8: 213. https://doi.org/10.1186/1471-2148-8-213.

Meyer, M., and M. Kircher. 2010. "Illumina sequencing library preparation for highly multiplexed target capture and sequencing." *Cold Spring Harbor Protocols* 2010 (6): pdb prot5448. https://doi.org/10.1101/pdb.prot5448.

Moilanen, Ulla, Tuija Kirkinen, Nelli-Johanna Saari, Adam B. Rohrlach, Johannes Krause, Päivi Onkamo, and Elina Salmela. 2021. "A woman with a sword?–Weapon Grave at Suontaka Vesitorninmäki, Finland." *European Journal of Archaeology*. 1–19.

Molak, Martyna, Eline D. Lorenzen, Beth Shapiro, and Simon Y.W. Ho. 2012. "Phylogenetic estimation of timescales using ancient DNA: The effects of temporal sampling scheme and uncertainty in sample ages." *Molecular Biology and Evolution* 30 (2): 253–262.

Monroy Kuhn, J.M., M. Jakobsson, and T. Gunther. 2018. "Estimating genetic kin relationships in prehistoric populations." *PLoS One* 13 (4): e0195491. https://doi:10.1371/journal.pone.0195491.

Moore, Richard B. 1973. "Carib "cannibalism": A study in anthropological stereotyping." *Caribbean Studies* 13 (3): 117–135.

Moreno-Estrada, Andrés, Simon Gravel, Fouad Zakharia, Jacob L. McCauley, Jake K. Byrnes, Christopher R. Gignoux, Patricia A. Ortiz-Tello, Ricardo J. Martínez, Dale J. Hedges, and Richard W. Morris. 2013. "Reconstructing the population genetic history of the Caribbean." *PLoS Genetics* 9 (11): e1003925.

Myers, Robert A. 1984. "Island carib cannibalism." *New West Indian Guide/Nieuwe West-Indische Gids* 58 (3–4): 147–184.

Nägele, Kathrin, Cosimo Posth, Miren Iraeta Orbegozo, Yadira Chinique De Armas, Silvia Teresita Hernández Godoy, Ulises M. González Herrera, Maria A. Nieves-Colón, Marcela Sandoval-Velasco, Dorothea Mylopotamitaki, Rita Radzeviciute. Jason Laffoon, William J. Pestle, Jazmin Ramos-Madrigal, Thiseas C. Lamnidis, William C. Schaffer, Robert S. Carr, Jane S. Day, Carlos Arredondo Antúnez, Armando Rangel Rivero, Antonio J. Martínez-Fuentes, Edwin Crespo-Torres, Ivan Roksandic, Anne C. Stone, Carlos Lalueza-Fox, Menno Hoogland, Mirjana Roksandic, Corinne L. Hofman, Johnannes Krause, and Hannes Schroeder.. 2020. "Genomic insights into the early peopling of the Caribbean." *Science* 369 (6502): 456–460.

Napolitano, Matthew F., Robert J. DiNapoli, Jessica H. Stone, Maureece J. Levin, Nicholas P. Jew, Brian G. Lane, John T. O'Connor, and Scott M. Fitzpatrick. 2019. "Reevaluating human colonization of the Caribbean using chronometric hygiene and Bayesian modeling." *Science Advances* 5 (12): eaar7806.

Narasimhan, Vagheesh M., Nick J. Patterson, Priya Moorjani, Iosif Lazaridis, Lipson Mark, Swapan Mallick, Nadin Rohland, Rebecca Bernardos, Alexander M. Kim, Nathan Nakatsuka, Iñigo Olalde, Mark Lipson, Alexander Kim, Luca Olivieri, Alfredo Coppa, Massimo Vidale, James Mallory, Vyacheslav Moiseyev, Egor Kitov, Janet Monge, Nicole Adamski,

Neel Alex, Nasreen Broomandkhoshbacht, Francesca Candilio, Kimberly Callan, Olivia Cheronet, Brendan Culleton, Matthew Ferry, Daniel Fernandes, Suzanne Freilich, Beatriz Gamarra, Daniel Gaudio, Mateja Hajdinjak, Eadaoin Harney, Thomas Harper, Denise Keating, Ann Lawson, Matthew Mah, Kirsten Mandl, Megan Michel, Mario Novak, Jonas Oppenheimer, Niraj Rai, Kendra Sirak, Viviane Slon, Kristin Stewardson, Fatma Zalzala, Zhao Zhang, Gaziz Akhatov, Anatoly Bagashev, Alessandra Bagnera, Bauryzhan Baitanayev, Julio Bendezu-Sarmiento, Arman Bissembaev, Gian Bonora, Temirlan Chargynov, Tatiana Chikisheva, Petr Dashkovskiy, Anatoly Derevianko, Miroslav Dobeš, Katerina Douka, Nadezhda Dubova, Meiram Duisengali, Dmitry Enshin, Andrey Epimakhov, Alexey Fribus, Dorian Fuller, Alexander Goryachev, Andrey Gromov, Sergey Grushin, Bryan Hanks, Margaret Judd, Erlan Kazizov, Aleksander Khokhlov, Aleksander Krygin, Elena Kupriyanova, Pavel Kuznetsov, Donata Luiselli, Farhod Maksudov, Aslan Mamedov, Talgat Mamirov, Christopher Meiklejohn, Deborah Merrett, Roberto Micheli, Oleg Mochalov, Samariddin Mustafokulov, Ayushi Nayak, Davide Pettener, Richard Potts, Dmitry Razhev, Marina Rykun, Stefania Sarno, Tatyana Savenkova, Kulyan Sikhymbaeva, Sergey Slepchenko, Oroz Soltobaev, Nadezhda Stepanova, Svetlana Svyatko, Kubatbek Tabaldiev, Maria Teschler-Nicola, Alexey Tishkin, Vitaly Tkachev, Sergey Vasilyev, Petr Velemínský, Dmitriy Voyakin, Antonina Yermolayeva, Muhammad Zahir, Valery Zubkov, Alisa Zubova, Vasant Shinde, Carles Lalueza-Fox, Matthias Meyer, David Anthony, Nicole Boivin, Kumarasamy Thangaraj, Douglas Kennett, Michael Frachetti, Ron Pinhasi, and David Reich. 2019. "The formation of human populations in South and Central Asia." *Science* 365 (6457): eaat7487. https://doi:10.1126/science.aat7487

Nelson, Elizabeth A., Jane E. Buikstra, Alexander Herbig, Tiffiny A. Tung, and Kirsten I. Bos. 2020. "Advances in the molecular detection of tuberculosis in pre-contact Andean South America." *International Journal of Paleopathology* 29: 128–140.

Newman, Simon P., Michael L. Deason, Yannis P. Pitsiladis, Antonio Salas, and Vincent A. Macaulay. 2013. "The West African ethnicity of the enslaved in Jamaica." *Slavery & Abolition* 34 (3): 376–400.

Nieves-Colón, M.A. 2022. "Genetic Insights on Caribbean Population History." *Evolutionary Anthropology* 31 (3): 118–137.

Nieves-Colón, María A., Kelly E. Blevins, Miguel Ángel Contreras-Sieck, and Miriam Bravo López. 2021. "Paleogenómica y bioarqueología en México." *Cuicuilco. Revista de Ciencias Antropológicas* 28 (81): 187–223.

Nieves-Colón, Maria A., William J. Pestle, Austin W. Reynolds, Bastien Llamas, Constanza de la Fuente, Kathleen Fowler, Katherine M. Skerry, Edwin Crespo-Torres, Carlos D. Bustamante, and Anne C. Stone. 2020. "Ancient DNA reconstructs the genetic legacies of precontact Puerto Rico communities." *Molecular Biology and Evolution* 37 (3): 611–626. https://doi.org/10.1093/molbev/msz267.

Obregon-Tito, Alexandra J., Raul Y. Tito, Jessica Metcalf, Krithivasan Sankaranarayanan, Jose C. Clemente, Luke K. Ursell, Zhenjiang Zech Xu, Will Van Treuren, Rob Knight, and Patrick M. Gaffney. 2015. "Subsistence strategies in traditional societies distinguish gut microbiomes." *Nature Communications* 6 (1): 1–9.

Olalde, Iñigo, Selina Brace, Morten E. Allentoft, Ian Armit, Kristian Kristiansen, Thomas Booth, Nadin Rohland, Swapan Mallick, Anna Szécsényi-Nagy, and Alissa Mittnik. 2018. "The Beaker phenomenon and the genomic transformation of northwest Europe." *Nature* 555 (7695): 190–196.

Oliveira, Sandra., Kathrin Nägele, Selina Carlhoff, Irina Pugach, Toetik Koesbardiati, Alexander Hübner, Matthias Meyer, Adhi Agus Oktaviana, Masami Takenaka, Chiaki Katagiri, Delta Bayu Murti, Rizky Sugianto Putri, Mahirta, Fiona Petchey, Thomas Higham,

Charles F. W. Higham, Sue O'Connor, Stuart Hawkins, Rebecca Kinaston, Peter Bellwood, Rintaro Ono, Adam Powell, Johannes Krause, Cosimo Posth, and Mark Stoneking. 2022. "Ancient genomes from the last three millennia support multiple human dispersals into Wallacea." *Nature Ecology & Evolution* 6: 1024–1034.

Orlando, Ludovic, Aurélien Ginolhac, Guojie Zhang, Duane Froese, Anders Albrechtsen, Mathias Stiller, Mikkel Schubert, Enrico Cappellini, Bent Petersen, and Ida Moltke. 2013. "Recalibrating Equus evolution using the genome sequence of an early Middle Pleistocene horse." *Nature* 499 (7456): 74–78.

Orlando, Ludovic, Robin Allaby, Pontus Skoglund, Clio Der Sarkissian, Philipp W. Stockhammer, María C. Ávila-Arcos, Qiaomei Fu, Johannes Krause, Eske Willerslev, and Anne C. Stone. 2021. "Ancient DNA analysis." *Nature Reviews Methods Primers* 1 (1): 1–26.

Oskam, Charlotte L., James Haile, Emma McLay, Paul Rigby, Morten E. Allentoft, Maia E. Olsen, Camilla Bengtsson, Gifford H. Miller, Jean-Luc Schwenninger, and Chris Jacomb. 2010. "Fossil avian eggshell preserves ancient DNA." *Proceedings of the Royal Society B: Biological Sciences* 277 (1690): 1991–2000.

Pääbo, Svante. 1989. "Ancient DNA: Extraction, characterization, molecular cloning, and enzymatic amplification." *Proceedings of the National Academy of Sciences* 86 (6): 1939–1943.

Pääbo, Svante, Hendrik Poinar, David Serre, Viviane Jaenicke-Després, Juliane Hebler, Nadin Rohland, Melanie Kuch, Johannes Krause, Linda Vigilant, and Michael Hofreiter. 2004. "Genetic analyses from ancient DNA." *Annual Review of Genetics* 38 (1): 645–679.

Pagán Jiménez, Jaime R., and Reniel Rodríguez Ramos. 2008. "Sobre arqueologías de liberación en una "colonia postcolonial" (Puerto Rico)." *Revista de Ciencias Sociales* 19: 8–41.

Palmer, Colin A. 2003. "The slave trade, African slavers and the demography of the Caribbean to 1750." In Knight, F.W. (Ed.), *General History of the Caribbean: The Slave Societies of the Caribbean*, Vol. 3. New York: Palgrave Macmillan, pp. 9–44.

Parker, Cody, Adam B. Rohrlach, Susanne Friederich, Sarah Nagel, Matthias Meyer, Johannes Krause, Kirsten I. Bos, and Wolfgang Haak. 2020. "A systematic investigation of human DNA preservation in medieval skeletons." *Scientific Reports* 10 (1): 1–16.

Patterson, Nick, Alkes L. Price, and David Reich. 2006. "Population structure and eigenanalysis." *PLoS Genetics* 2 (12): e190. https://doi.org/10.1371/journal.pgen.0020190.

Patterson, Nick, Priya Moorjani, Yontao Luo, Swapan Mallick, Nadin Rohland, Yiping Zhan, Teri Genschoreck, Teresa Webster, and David Reich. 2012. "Ancient admixture in human history." *Genetics* 192 (3): 1065–1093. https://doi.org/10.1534/genetics.112.145037.

Peltzer, Alexander., Günter Jäger, Alexander Herbig, Alexander Seitz, Christian Kniep, Johannes Krause, and Kay Nieselt. 2016. "EAGER: Efficient ancient genome reconstruction." *Genome Biology* 17: 60. https://doi.org/10.1186/s13059-016-0918-z.

Perri, Angela R., Tatiana R. Feuerborn, Laurent A.F. Frantz, Greger Larson, Ripan S. Malhi, David J. Meltzer, and Kelsey E. Witt. 2021. "Dog domestication and the dual dispersal of people and dogs into the Americas." *Proceedings of the National Academy of Sciences* 118 (6): e2010083118.

Pickrell, Joseph K., and David Reich. 2014. "Toward a new history and geography of human genes informed by ancient DNA." *Trends in Genetics* 30 (9): 377–389.

Pickrell, Joseph, and Jonathan Pritchard. 2012. "Inference of population splits and mixtures from genome-wide allele frequency data." *PLoS Genetics* 8 (11): e1002967.

Pinhasi, Ron, Daniel Fernandes, Kendra Sirak, Mario Novak, Sarah Connell, Songül Alpaslan-Roodenberg, Fokke Gerritsen, Vyacheslav Moiseyev, Andrey Gromov, and Pál Raczky.

2015. "Optimal ancient DNA yields from the inner ear part of the human petrous bone." *PloS One* 10 (6): e0129102.

Ponce de León, Marcia S., Toetik Koesbardiati, John David Weissmann, Marco Milella, Carlos S. Reyna-Blanco, Gen Suwa, Osamu Kondo, Anna-Sapfo Malaspinas, Tim D. White, and Christoph P.E. Zollikofer. 2018. "Human bony labyrinth is an indicator of population history and dispersal from Africa." *Proceedings of the National Academy of Sciences* 115 (16): 4128–4133.

Posth, Cosimo, Kathrin Nägele, Heidi Colleran, Frédérique Valentin, Stuart Bedford, Kaitip W. Kami, Richard Shing, Hallie Buckley, Rebecca Kinaston, and Mary Walworth. 2018. "Language continuity despite population replacement in Remote Oceania." *Nature Ecology & Evolution* 2 (4): 731–740.

Posth, Cosimo, Nathan Nakatsuka, Iosif Lazaridis, Pontus Skoglund, Swapan Mallick, Thiseas C. Lamnidis, Nadin Rohland, Katrin. Nägele, Nicole Adamski, Emilie Bertolini, Nasreen Broomandkhoshbacht, Alan Cooper, Brendan J. Culleton, Tiago Ferraz, Matthew Ferry, Anja Furtwangler, Wolfgang Haak, Kelly Harkins, Thomas K. Harper, Tábita Hünemeier, Ann Marie Lawson, Bastien Llamas, Megan Michel, Elizabeth Nelson, Jonas Oppenheimer, Nick Patterson, Stephan Schiffels, Jakob Sedig, Kristin Stewardson, Sahra Talamo, Chuan-Chao Wang, Jean-Jacques Hublin, Mark Hubbe, Katerina Harvati, Amalia Nuevo Delaunay, Judith Beier, Michael Francken, Peter Kaulicke, Hugo Reyes-Centeno, Kurt Rademaker, Willa R. Trask, Mark Robinson, Said M. Gutierrez, Keith M. Prufer, Domingo C. Salazar-Garcia, Eliane N. Chim, Lisiane Muller Plumm Gomes, Marcony L. Alves, Andersen Liryo, Mariana Inglez, Rodrigo E. Oliveira, Danilo V. Bernardo, Alberto Barioni, Veronica Wesolowski, Nahuel A. Scheifler, Mario A. Rivera, Claudia R. Plens, Pablo G. Messineo, Levy Figuti, Daniel Corach, Clara Scabuzzo, Sabine Eggers, Paulo DeBlasis, Markus Reindel, César Méndez, Gustavo Politis, Elsa Tomasto-Cagigao, Douglas J. Kennett, André Strauss, Lars Fehren-Schmitz, Johannes Krause, and David Reich. 2018. "Reconstructing the deep population history of Central and South America." *Cell* 175 (5): 1185–1197 e22. https://doi.org/10.1016/j.cell.2018.10.027.

Prendergast, Mary E., and Elizabeth Sawchuk. 2018. "Boots on the ground in Africa's ancient DNA 'revolution': Archaeological perspectives on ethics and best practices." *Antiquity* 92 (363): 803–815.

Pugach, Irina, Alexander Hubner, Hsiao-Chun Hung, Matthias Meyer, Mike T. Carson, and Mark Stoneking. 2021. "Ancient DNA from Guam and the peopling of the Pacific." *Proceedings of the National Academy of Sciences* 118 (1): e2022112118.

Rabinow, Sophie. 2020. "A novel commensal proxy for tracing indigenous interaction in the Ceramic Age Lesser Antilles, Caribbean: Ancient mitochondrial DNA of Agouti (Dasyprocta sp.)." MA thesis. Simon Fraser University.

Racimo, Fernando, Martin Sikora, Marc Vander Linden, Hannes Schroeder, and Carles Lalueza-Fox. 2020. "Beyond broad strokes: Sociocultural insights from the study of ancient genomes." *Nature Reviews Genetics* 21 (6): 355–366.

Raghavan, Maanasa, Matthias Steinrücken, Kelley Harris, Stephan Schiffels, Simon Rasmussen, Michael DeGiorgio, Anders Albrechtsen, Cristina Valdiosera, María C. Ávila-Arcos, Anna-Sapfo Malaspinas, Anders Eriksson, Ida Moltke, Mait Metspalu, Julian R Homburger, Jeff Wall, Omar E Cornejo, J Víctor Moreno-Mayar, Thorfinn S Korneliussen, Tracey Pierre, Morten Rasmussen, Paula F Campos, Peter de Barros Damgaard, Morten E Allentoft, John Lindo, Ene Metspalu, Ricardo Rodríguez-Varela, Josefina Mansilla, Celeste Henrickson, Andaine Seguin-Orlando, Helena Malmström, Thomas Stafford Jr,

Suyash S Shringarpure, Andrés Moreno-Estrada, Monika Karmin, Kristiina Tambets, Anders Bergström, Yali Xue, Vera Warmuth, Andrew D Friend, Joy Singarayer, Paul Valdes, Francois Balloux, Ilán Leboreiro, Jose Luis Vera, Hector Rangel-Villalobos, Davide Pettener, Donata Luiselli, Loren G Davis, Evelyne Heyer, Christoph P E Zollikofer, Marcia S Ponce de León, Colin I Smith, Vaughan Grimes, Kelly-Anne Pike, Michael Deal, Benjamin T Fuller, Bernardo Arriaza, Vivien Standen, Maria F Luz, Francois Ricaut, Niede Guidon, Ludmila Osipova, Mikhail I Voevoda, Olga L Posukh, Oleg Balanovsky, Maria Lavryashina, Yuri Bogunov, Elza Khusnutdinova, Marina Gubina, Elena Balanovska, Sardana Fedorova, Sergey Litvinov, Boris Malyarchuk, Miroslava Derenko, M J Mosher, David Archer, Jerome Cybulski, Barbara Petzelt, Joycelynn Mitchell, Rosita Worl, Paul J Norman, Peter Parham, Brian M Kemp, Toomas Kivisild, Chris Tyler-Smith, Manjinder S Sandhu, Michael Crawford, Richard Villems, David Glenn Smith, Michael R Waters, Ted Goebel, John R Johnson Ripan S Malhi, Mattias Jakobsson, David J Meltzer, Andrea Manica, Richard Durbin, Carlos D Bustamante, Yun S Song, Rasmus Nielsen, and Eske Willerslev 2015. "Genomic evidence for the Pleistocene and recent population history of Native Americans." *Science* 349 (6250): aab3884.

Ramos-Madrigal, Jazmín, Anne Kathrine Wiborg Runge, Laurent Bouby, Thierry Lacombe, José Alfredo Samaniego Castruita, Anne-Françoise Adam-Blondon, Isabel Figueiral, Charlotte Hallavant, José M. Martínez-Zapater, and Caroline Schaal. 2019. "Palaeogenomic insights into the origins of French grapevine diversity." *Nature Plants* 5 (6): 595–603.

Rasmussen, Morten, Yingrui Li, Stinus Lindgreen, Jakob Skou Pedersen, Anders Albrechtsen, Ida Moltke, Mait Metspalu, Ene Metspalu, Toomas Kivisild, and Ramneek Gupta. 2010. "Ancient human genome sequence of an extinct Palaeo-Eskimo." *Nature* 463 (7282): 757–762.

Reich, David, Richard E. Green, Martin Kircher, Johannes Krause, Nick Patterson, Eric Y. Durand, Bence Viola, Adrian W. Briggs, Udo Stenzel, and Philip L.F. Johnson. 2010. "Genetic history of an archaic hominin group from Denisova Cave in Siberia." *Nature* 468 (7327): 1053.

Reid, Basil A. 2009. *Myths and Realities of Caribbean History*. Tuscaloosa, AL: University of Alabama Press.

Rieux, Adrien, and François Balloux. 2016. "Inferences from tip-calibrated phylogenies: A review and a practical guide." *Molecular Ecology* 25 (9): 1911–1924.

Ringbauer, Harald, John Novembre, and Matthias Steinrücken. 2021. "Parental relatedness through time revealed by runs of homozygosity in ancient DNA." *Nature Communications* 12 (1): 1–11.

Rivera-Collazo, Isabel. 2011. "The ghost of Caliban: Island archaeology, insular archaeologists and the Caribbean." In Curet, L.A. and Hauser, M.W. (Eds.), *Islands at the Crossroads:Migration, Seafaring, and Interaction in the Caribbean*,Tuscaloosa, AL: University of Alabama Press, pp. 22–40.

Rivera-Pagán, Luis N. 2003. "Freedom and servitude: Indigenous slavery and the Spanish conquest of the Caribbean." In Sued-Badillo, J. (Ed.), *General History of the Caribbean*. New York: Palgrave Macmillan, pp. 316–362.

Rivollat, Maïté, Choongwon Jeong, Stephan Schiffels, İşil Küçükkalıpçı, Marie-Hélène Pemonge, Adam Benjamin Rohrlach, Kurt W. Alt, Didier Binder, Susanne Friederich, and Emmanuel Ghesquière. 2020. "Ancient genome-wide DNA from France highlights the complexity of interactions between Mesolithic hunter-gatherers and Neolithic farmers." *Science Advances* 6 (22): eaaz5344.

Rodríguez Ramos, Reniel. 2010. *Rethinking Puerto Rican Precolonial History.* Tuscaloosa, AL: University of Alabama Press.

Rodríguez Ramos, Reniel, Jaime R. Pagán-Jiménez, and Corinne L. Hofman. 2013. "The humanization of the insular Caribbean." In Keegan, W.F., Hofman, C., and Rodriguez Ramos, R. (Eds.), *The Oxford Handbook of Caribbean Archaeology*. Oxford: Oxford University Press, pp. 126–140.

Rohland, Nadin, Eadaoin Harney, Swapan Mallick, Susanne Nordenfelt, and David Reich. 2015. "Partial uracil-DNA-glycosylase treatment for screening of ancient DNA." *Philosophical Transactions of the Royal Society B: Biological Sciences* 370 (1660): 20130624. https://doi.org/10.1098/rstb.2013.0624.

Rohland, Nadin, and Michael Hofreiter. 2007. "Ancient DNA extraction from bones and teeth." *Nature Protocols* 2 (7): 1756–1762. https://doi.org/10.1038/nprot.2007.247.

Rosenberg, Noah A., and Mattias Jakobsson. 2008. "The relationship between homozygosity and the frequency of the most frequent allele." *Genetics* 179 (4): 2027–2036.

Ross, Ann H., William F. Keegan, Michael P. Pateman, and Colleen B. Young. 2020. "Faces divulge the origins of caribbean prehistoric inhabitants." *Scientific Reports* 10 (1): 1–9.

Rouse, Irving. 1960. "The classification of artifacts in archaeology." *American Antiquity* 25 (3): 313–323. https://doi.org/doi:10.2307/277514.

Rouse, Irving. 1992. *The Tainos: Rise and Decline of the People Who Greeted.* Columbus: Yale University Press.

Roux, Benoît. 2011. *De insulis Karaybicis relationes manuscriptœ: Adrien Le Breton, The Last Jesuit Missionary in the "Carib Island" of St. Vincent*. Leiden: Sidestone.

Sabin, Susanna, Hui-Yuan Yeh, Aleks Pluskowski, Christa Clamer, Piers D. Mitchell, and Kirsten I. Bos. 2020. "Estimating molecular preservation of the intestinal microbiome via metagenomic analyses of latrine sediments from two medieval cities." *Philosophical Transactions of the Royal Society B* 375 (1812): 20190576.

Sanger, Frederick, Steven Nicklen, and Alan R. Coulson. 1977. "DNA sequencing with chain-terminating inhibitors." *Proceedings of the National Academy of Sciences* 74 (12): 5463–5467.

Sankatsing Nava, Tibisay, and Corinne L. Hofman. 2018. "Engaging Caribbean island communities with indigenous heritage and archaeology research." *JCOM: Journal of Science Communication* 17 (4): 1–10.

Schablitsky, Julie M., Kelsey E. Witt, Jazmín Ramos Madrigal, Martin R. Ellegaard, Ripan S. Malhi, and Hannes Schroeder. 2019. "Ancient DNA analysis of a nineteenth century tobacco pipe from a Maryland slave quarter." *Journal of Archaeological Science* 105: 11–18.

Scheib, C.L., Hongjie Li, Tariq Desai, Vivian Link, Christopher Kendall, Genevieve Dewar, Peter William Griffith, Alexander Mörseburg, John R. Johnson, and Amiee Potter. 2018. "Ancient human parallel lineages within North America contributed to a coastal expansion." *Science* 360 (6392): 1024–1027.

Schroeder, Hannes., Martin Sikora, Shyam Gopalakrishnan, Lara M. Cassidy, Pierpaolo Maisano Delser, Marcela Sandoval Velasco, Joshua G. Schraiber, Simon Rasmussen, Julian R. Homburger, Maria C. Avila-Arcos, Morten. E. Allentoft, J. Victor. Moreno-Mayar, Gabriel. Renaud, Alberto. Gomez-Carballa, Jason. E. Laffoon, Rachel. J. A. Hopkins, Thomas. F. G. Higham, Robert. S. Carr, William. C. Schaffer, Jane. S. Day, Menno. Hoogland, Antonio. Salas, Carlos D. Bustamante, Rasmus. Nielsen, Daniel. G. Bradley, Corinne. L. Hofman, and E. Willerslev. 2018. "Origins and genetic legacies of the Caribbean Taino." *Proceedings of the National Academy of Sciences of the United States of America* 115 (10): 2341–2346. https://doi.org/10.1073/pnas.1716839115.

Schuenemann, Verena J., Alexander Peltzer, Beatrix Welte, W. Paul Van Pelt, Martyna Molak, Chuan-Chao Wang, Anja Furtwängler, Christian Urban, Ella Reiter, and Kay Nieselt. 2017. "Ancient Egyptian mummy genomes suggest an increase of Sub-Saharan African ancestry in post-Roman periods." *Nature Communications* 8 (1): 1–11.

Schurr, Theodore G., Jada Benn Torres, Miguel G. Vilar, Jill B. Gaieski, and Carlalynne Melendez. 2016. "An emerging history of indigenous Caribbean and circum-Caribbean populations: Insights from archaeological, ethnographic, genetic, and historical studies." *New Directions in Biocultural Anthropology* 1: 385–402.

Seguin-Orlando, Andaine, Cristina Gamba, Clio Der Sarkissian, Luca Ermini, Guillaume Louvel, Eugenia Boulygina, Alexey Sokolov, Artem Nedoluzhko, Eline D. Lorenzen, and Patricio Lopez. 2015. "Pros and cons of methylation-based enrichment methods for ancient DNA." *Scientific Reports* 5 (1): 1–15.

Sirak, Kendra A., Daniel M. Fernandes, Olivia Cheronet, Mario Novak, Beatriz Gamarra, Tímea Balassa, Zsolt Bernert, Andrea Cséki, János Dani, and József Zsolt Gallina. 2017. "A minimally-invasive method for sampling human petrous bones from the cranial base for ancient DNA analysis." *BioTechniques* 62 (6): 283–289.

Sirak, Kendra A., and Jakob W. Sedig. 2019. "Balancing analytical goals and anthropological stewardship in the midst of the paleogenomics revolution." *World Archaeology* 51 (4): 560–573.

Skoglund, Pontus, Helena Malmström, Maanasa Raghavan, Jan Storå, Per Hall, Eske Willerslev, M. Thomas P. Gilbert, Anders Götherström, and Mattias Jakobsson. 2012. "Origins and genetic legacy of Neolithic farmers and hunter-gatherers in Europe." *Science* 336 (6080): 466–469.

Skoglund, Pontus, Jan Storå, Anders Götherström, and Mattias Jakobsson. 2013. "Accurate sex identification of ancient human remains using DNA shotgun sequencing." *Journal of Archaeological Science* 40 (12): 4477–4482.

Slatkin, Montgomery, and Fernando Racimo. 2016. "Ancient DNA and human history." *Proceedings of the National Academy of Sciences* 113 (23): 6380–6387.

Slon, Viviane, Jamie L. Clark, David E. Friesem, Meir Orbach, Naomi Porat, Mathias Meyer, Andrew W. Kandel, and Ron Shimelmitz. 2022. "Extended longevity of DNA preservation in Levantine Paleolithic sediments." *Scientific Reports* 12 (1): 14528

Slon, Viviane, Charlotte Hopfe, Clemens L. Weiß, Fabrizio Mafessoni, Marco De La Rasilla, Carles Lalueza-Fox, Antonio Rosas, Marie Soressi, Monika V. Knul, and Rebecca Miller. 2017. "Neandertal and Denisovan DNA from Pleistocene sediments." *Science* 356 (6338): 605–608.

Spyrou, Maria A., Rezeda I. Tukhbatova, Chuan-Chao Wang, Aida Andrades Valtueña, Aditya K. Lankapalli, Vitaly V. Kondrashin, Victor A. Tsybin, Aleksandr Khokhlov, Denise Kühnert, and Alexander Herbig. 2018. "Analysis of 3800-year-old Yersinia pestis genomes suggests Bronze Age origin for bubonic plague." *Nature Communications* 9 (1): 1–10.

Stone, Anne C., and Mark Stoneking. 1993. "Ancient DNA from a pre-Columbian Amerindian population." *American Journal of Physical Anthropology* 92 (4): 463–471.

Stone, Anne C., George R. Milner, Svante Pääbo, and Mark Stoneking. 1996. "Sex determination of ancient human skeletons using DNA." *American Journal of Physical Anthropology: The Official Publication of the American Association of Physical Anthropologists* 99 (2): 231–238.

Stone, Erin Woodruff. 2013. "America's first slave revolt: Indians and African slaves in Española, 1500–1534." *Ethnohistory* 60 (2): 195–217.

Stoneking, Mark, and Johannes Krause. 2011. "Learning about human population history from ancient and modern genomes." *Nature Reviews Genetics* 12 (9): 603–614.

Sued-Badillo, Jalil. 2007. "Guadalupe:¿ Caribe o Taina? La isla de Guadalupe y su cuestionable identidad caribe en la epoca pre-Colombina: una revision etnohistorica y arqueologica preliminar." *Caribbean Studies* 35 (1) 37–85.

Sued-Badillo, Jalil. 2020a. *Caribe taíno: ensayos históricos sobre el siglo XVI*. San Juan: Editorial Luscinia C.E.

Sued-Badillo, Jalil. 2020b. *Caribe taíno: ensayos históricos sobre el siglo XVI.* San Juan: Editorial Luscinia C.E.

Sullivan, Alexis P., Stephanie Marciniak, Aaron O'Dea, Thomas A. Wake, and George H. Perry. 2021. "Modern, archaeological, and paleontological DNA analysis of a human-harvested marine gastropod (Strombus pugilis) from Caribbean Panama." *Molecular Ecology Resources* 21 (5): 1517–1528.

Tankersley, Kenneth Barnett, Nicholas P. Dunning, Lewis A. Owen, and Janine Sparks. 2018. "Geochronology and palaeoenvironmental framework for the oldest Achaeological Site (7800–7900 cal BP) in the West Indies, Banwari Trace, Trinidad." *Latin American Antiquity* 29 (4): 681–695.

Teasdale, Matthew D., Nienke L. van Doorn, Sarah Fiddyment, Christopher C. Webb, Terry O'Connor, Michael Hofreiter, Matthew J. Collins, and Daniel G. Bradley. 2015. "Paging through history: Parchment as a reservoir of ancient DNA for next generation sequencing." *Philosophical Transactions of the Royal Society B: Biological Sciences* 370 (1660): 20130379.

Twinn, Paul. 2006. "Land ownership and the construction of Carib identity in St. Vincent." In Forte, M.C. (Ed.), *Indigenous Resurgence in the Contemporary Caribbean: Amerindian Survival and Revival*, New York: Peter Lang, pp. 89–106.

Ulloa Hung, J., and S. de Ruiter. 2011. "Arqueología en la línea noroeste de la República Dominicana. Un esbozo del paisaje arqueológico y las interacciones." *El Caribe Arqueologico* 12: 60–57.

Vai, Stefania, Carlos Eduardo G. Amorim, Martina Lari, and David Caramelli. 2020. "Kinship determination in archeological contexts through DNA analysis." *Frontiers in Ecology and Evolution* 8: 83.

Valcárcel Rojas, Roberto, and J. Ulloa Hung. 2018. *De la desaparición a la permanencia: indígenas e indios en la reinvención del Caribe*. Santo Domingo: Instituto Tecnológico de Santo Domingo y Fundación García Arévalo.

Vallebueno-Estrada, Miguel, Isaac Rodríguez-Arévalo, Alejandra Rougon-Cardoso, Javier Martínez González, Angel García Cook, Rafael Montiel, and Jean-Philippe Vielle-Calzada. 2016. "The earliest maize from San Marcos Tehuacán is a partial domesticate with genomic evidence of inbreeding." *Proceedings of the National Academy of Sciences* 113 (49): 14151–14156.

van der Valk, Tom, Patrícia Pečnerová, David Díez-del-Molino, Anders Bergström, Jonas Oppenheimer, Stefanie Hartmann, Georgios Xenikoudakis, Jessica A. Thomas, Marianne Dehasque, Ekin Sağlıcan, Fatma Rabia Fidan, Ian Barnes, Shanlin Liu, Mehmet Somel, Peter D. Heintzman, Pavel Nikolskiy, Beth Shapiro, Pontus Skoglund, Michael Hofreiter, Adrian M. Lister, Anders Götherström, and Love Dalén. 2021. "Million-year-old DNA sheds light on the genomic history of mammoths." *Nature* 591: 265–269. https://doi.org/10.1038/s41586-021-03224-9

Via, Marc, Christopher R. Gignoux, Lindsey A. Roth, Laura Fejerman, Joshua Galanter, Shweta Choudhry, Gladys Toro-Labrador, Jorge Viera-Vera, Taras K. Oleksyk,

Kenneth Beckman, Elad Ziv, Neil Risch, Esteban Gonzalez Burchard, and Juan Carlos Martinez-Cruzado. 2011. "History shaped the geographic distribution of genomic admixture on the island of Puerto Rico." *PloS One* 6 (1): e16513.

Vilar, Miguel G., Carlalynne Melendez, Akiva B. Sanders, Akshay Walia, Jill B. Gaieski, Amanda C. Owings, Theodore G. Schurr, and Consortium Genographic. 2014. "Genetic diversity in Puerto Rico and its implications for the peopling of the Island and the West Indies." *American Journal Physical Anthropology* 155 (3): 352–368. https://doi.org/10.1002/ajpa.22569.

Warinner, Christina, Camilla Speller, and Matthew J. Collins. 2015. "A new era in palaeomicrobiology: Prospects for ancient dental calculus as a long-term record of the human oral microbiome." *Philosophical Transactions of the Royal Society B: Biological Sciences* 370 (1660): 20130376.

Warinner, Christina, Jessica Hendy, Camilla Speller, Enrico Cappellini, Roman Fischer, Christian Trachsel, Jette Arneborg, Niels Lynnerup, Oliver E. Craig, and Dallas M. Swallow. 2014. "Direct evidence of milk consumption from ancient human dental calculus." *Scientific Reports* 4 (1): 1–6.

West, Katrina, Catherine Collins, Olga Kardailsky, Jennifer Kahn, Terry L. Hunt, David V. Burley, and Elizabeth Matisoo-Smith. 2017. "The Pacific rat race to Easter Island: Tracking the prehistoric dispersal of Rattus exulans using ancient mitochondrial genomes." *Frontiers in Ecology and Evolution* 5: 52.

Wilson, Samuel M., Harry B. Iceland, and Thomas R. Hester. 1998. "Preceramic connections between Yucatan and the Caribbean." *Latin American Antiquity* 9 (4): 342–352.

Woods, Roseina, Ian Barnes, Selina Brace, and Samuel T. Turvey. 2021. "Ancient DNA suggests single colonization and within-archipelago diversification of Caribbean Caviomorph Rodents." *Molecular Biology and Evolution* 38 (1): 84–95.

Yaremko, Jason M. 2009. ""Obvious Indian"—missionaries, anthropologists, and the "Wild Indians" of Cuba: Representations of the Amerindian presence in Cuba." *Ethnohistory* 56 (3): 449–477.

Žegarac, Aleksandra, Laura Winkelbach, Jens Blöcher, Yoan Diekmann, Marija Krečković Gavrilović, Marko Porčić, Biljana Stojković, Lidija Milašinović, Mona Schreiber, and Daniel Wegmann. 2021. "Ancient genomes provide insights into family structure and the heredity of social status in the early Bronze Age of southeastern Europe." *Scientific Reports* 11 (1): 1–11.

Ziesemer, Kirsten A., Allison E. Mann, Krithivasan Sankaranarayanan, Hannes Schroeder, Andrew T. Ozga, Bernd W. Brandt, Egija Zaura, Andrea Waters-Rist, Menno Hoogland, and Domingo C. Salazar-Garcia. 2015. "Intrinsic challenges in ancient microbiome reconstruction using 16S rRNA gene amplification." *Scientific Reports* 5 (1): 1–20.

Ziesemer, Kirsten A., Jazmín Ramos-Madrigal, Allison E. Mann, Bernd W. Brandt, Krithivasan Sankaranarayanan, Andrew T. Ozga, Menno Hoogland, Courtney A. Hofman, Domingo C. Salazar-García, and Bruno Frohlich. 2019. "The efficacy of whole human genome capture on ancient dental calculus and dentin." *American Journal of Physical Anthropology* 168 (3): 496–509.

2

THE GENETIC LANDSCAPE OF THE GUADELOUPE ARCHIPELAGO

Fanny Mendisco, Patrice Courtaud, Thomas Romon, Marie-Hélène Pemonge and Marie-France Deguilloux

Introduction

Like the history of the Caribbean region, the history of the Guadeloupe archipelago is marked by numerous and various successive migration waves that were involved to different degrees in the constitution of present-day populations and cultures. The Guadeloupe archipelago, located in the Leeward Islands in the northern region of the Lesser Antilles, has a current population of more than 400,000 inhabitants and includes two main islands, Basse-Terre and Grande-Terre, as well as several smaller islands such as Marie-Galante, La Désirade and Les Saintes (Figure 2.1).

Christopher Columbus, during his second trip to the "New Word", was the first European to land on Guadeloupe in November 1493. This event marked the beginning of the drastic transformation of the genetic diversity, population structure, ancestry and admixture of Caribbean indigenous communities. As a consequence, the understanding of the origin of these ancient communities and their genetic relationship to present-day islanders remains largely unclear. Research on this topic has been greatly influenced by the interpretation of the texts from the early Colonial period. The narrative that emerges from these colonial writings offers a biased and simplistic view of Caribbean indigenous peoples, classifying them as "noble Indians" of the Greater Antilles (assimilated to the Taino) and "fierce Caribs" of the Lesser Antilles (Nieves-Colón 2022). Setting aside this questionable classification and referring to archaeological, linguistic and genetic evidence, it is now broadly accepted that the Caribbean Ceramic Age peoples originated in the Orinoco River delta region of South America (Benn Torres et al. 2019; Fortes-Lima et al. 2018; Granberry 2013; Rouse 1992; Moreno-Estrada et al. 2013; Nieves-Colón et al. 2020; Nägele et al. 2020; Schroeder et al. 2015, 2018). This Ceramic Age began 2,500–2,300 years ago, with the arrival of Arawak speakers who introduced

DOI: 10.4324/9780429444449-4

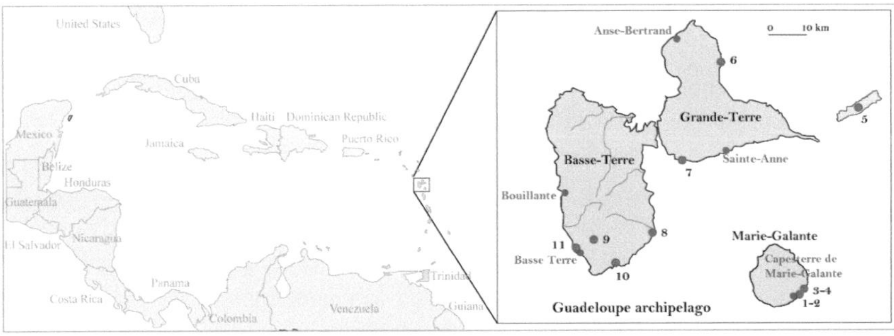

FIGURE 2.1 Location of the archaeological sites studied are numbered: (1) Grotte Cadet 2; (2) Grotte Blanchard; (3) Stade José Bade (Capesterre); (4) Grotte du Morne Rita; (5) Voute à Pin; (6) Anse St Marguerite; (7) Ilet Gosier; (8) Roseau; (9) Saint Claude; (10) Grande Anse (Trois-Rivières); (11) Gare-Maritime (Basse-Terre). Cities whose current population was sampled have names listed on the map.

large-scale agriculture and intensive pottery production in the Caribbean. In the Lesser Antilles and especially in Guadeloupe, the nature and degree of interactions between these newcomers and the previous indigenous Archaic peoples remains debated (Antczak and Hofman 2019). Our knowledge of the occupations of the Archaic Age in the Lesser Antilles is based essentially on indirect evidence (Bérard 2019), but recent paleogenomic publications bring very interesting elements of discussion for our understanding of the initial peopling of the Caribbean (Fernandes et al. 2021; Nägele et al. 2020).

The Caribbean indigenous communities experienced dramatic transformations soon after European settlement (Bérard 2011; Madrigal 2006; Rouse 1992). In 1635, with the establishment of the Compagnie des Isles d'Amérique, the first French colonists settled on the Guadeloupe archipelago. After a brief period of functional relationships between settlers and Amerindian groups, war was declared against indigenous communities and lasted until 1639. This episode resulted in the drastic reduction of the numbers of indigenous people from the archipelago, affected by conflicts, epidemics and alcohol (Madrigal 2006; Rouse 1992). In 1660, a peace treaty was finally established between the last surviving Amerindians and the French colonists, forcing indigenous peoples to leave the Guadeloupe archipelago and to join the Carib Territory of Dominica (Watts 1987). Even though the Amerindian indigenous population of Guadeloupe is supposed to be almost extinct today (Bérard 2011), there is evidence that indigenous populations maintained links across the islands of the southern Lesser Antilles (Honychurch 1997). These exchanges and the potential early interbreeding with Europeans and enslaved Africans may have left a genetic imprint of the indigenous groups in the present-day population of Guadeloupe.

The French colonists settled in Guadeloupe actively took part in the transatlantic slave trade to provide enslaved Africans, primarily as agricultural laborers, to the white plantation operators. The exact origin of a small fraction of the enslaved Africans could be documented, indicating that the most represented African ethnic group in Guadeloupe was the Ibos, native to Nigeria, followed by the Congos, coming from the south of Benin and part of Angola, the Aradas, coming from Togo and the Kingdom of Dahomey, the Caplaous, coming from the Ivory Coast, the Sossos, from current Guinea, Sierra Leone and Liberia and finally the Mocos, who are Bantu (Belenus 1998; Frisch and Frisch 1987). Until the successful and definitive abolition of slavery in May 1848, approximately 254 convoys disembarked roughly 70,410 enslaved people, forcibly transported to Guadeloupe (The Trans-Atlantic Slave Trade Database 2020). Following the abolition of slavery, many of the emancipated slaves left the plantations, leading to a crisis of labor. To compensate for the lack of workforce, immigration to Guadeloupe was encouraged and led to the arrival of indentured laborers originating from Congo (approximately 6,000 laborers) and especially from India (approximately 45,000) (Schnakenbourg 2005). At the end of the nineteenth century, Chinese, Lebanese and Syrians, who were mainly traders, immigrated to Guadeloupe and contributed to the mosaic of the present-day population of the island (Boutin 2006).

The inclusion of genetic data in interdisciplinary studies may facilitate the elucidation of past demographic and evolutionary events, such as human migrations and admixture processes (Barbujani and Colonna 2010; Jobling 2012; Underhill and Kivisild 2007). Several modern populations throughout the Caribbean have been genetically studied using mitochondrial and Y-chromosomal uniparental markers (Benn Torres et al. 2007, 2015, 2019; Cintado et al. 2009; Martínez-Cruzado et al. 2001, 2005; Deason et al. 2012; Madrilejo et al. 2015; Marcheco-Teruel et al. 2014; Martinez- Tajima et al. 2004; Mendizabal et al. 2008; Miljkovic-Gacic et al. 2005; Simms et al. 2010, 2012, 2013; Vilar et al. 2014; Wilson et al. 2012) and more recently using genome-wide data (Bryc et al. 2010; Gravel et al. 2013; Via et al. 2011). These studies reveal considerable variability in the degree, sources and modalities of admixture but also in patterns of sex-biased gene flow according to the islands and their colonial history (Moreno-Estrada et al. 2013). However, the early peopling of the Caribbean islands cannot be elucidated by these modern data. The indigenous component of these admixed populations is sometimes very small (Moreno-Estrada et al. 2013), and its origin is uncertain, as it may be the result of recent migrations. To overcome the limits induced by lineage extinction and/or recent admixture, one option can be ancient DNA (aDNA) studies. Dealing directly with ancient Caribbean indigenous groups, aDNA studies provide some of the most direct evidence of past evolutionary history but are proving difficult in this region due to preservation issues related to tropical environmental conditions. For a long time, aDNA analyses were extremely limited, both in terms of genetic markers analyzed (often limited to mitochondrial genome (mtDNA) data) (Lalueza-Fox et al. 2001, 2003; Mendisco et al. 2015) and individuals involved (Schroeder et al. 2015, 2018), making interpretations difficult and restricted. With the latest technical improvements for aDNA recovery (Pinhasi et al. 2015, 2019) and the advent of next-generation sequencing, we recently entered a new paleogenomic era. It is now easier to obtain genome-wide genotype data on difficult ancient samples, bringing exciting new perspectives for the study of the demographic history of ancient Caribbean communities (Fernandes et al. 2021; Nägele et al. 2020; Nieves-Colón et al. 2020).

The lasting collaboration we have established for over ten years between archaeologists, anthropologists and geneticists has allowed us to initiate analyses and reflect on the origin of the indigenous communities of the Lesser Antilles (Deguilloux et al. 2015). We continued and strengthened these collaborations in order to enlarge the sample of human remains analyzed and to combine aDNA analyses with the characterization of the genetic pool of the present-day Guadeloupe population. The acquisition and combination of these different types of data were used to discuss the diversity of indigenous pre-Contact communities from Guadeloupe, their response to colonization and their relationship to present-day islanders. Therefore, a part of our study focuses on the description of aDNA sequences obtained from human remains originating from the Guadeloupe archipelago and dating from the Ceramic Age. The initial objectives were (i) to discuss the origin of these ancient individuals, (ii) at the regional level, to potentially assess the extent of genetic divergence between groups marked by different material cultures and finally, (iii) at the local level, to provide clues concerning the relationship between individual identity and funerary practices. Complementary to this paleogenetic component, the other part of our study aimed to understand the dynamics of admixture in the Guadeloupean archipelago within the historical context of European conquest and colonization. Through the genetic analysis of the present-day Guadeloupe population, we aim to survey the geographic origin of current Guadeloupean ancestors and identify how the transatlantic slave trade would have impacted the population structure and admixture.

Material and Methods

Collection and Analysis of Ancient Samples

Bone or dental remains of 38 individuals were collected from 11 pre-Columbian archaeological sites from the Guadeloupe archipelago (Figure 2.1). Archaeological evidence and several direct radiocarbon dates (Mendisco et al. 2015) have associated all the ancient human remains with Ceramic (BC 500–600 AD) and late ceramic (600–1500 AD) contexts. Thanks to a strong collaboration with the archaeologists responsible for the excavations, the sampling conditions were optimized and carried out as much as possible *in situ*. The funerary contexts of the archaeological sites were extremely variable (see Mendisco et al. 2015 for more details), from scattered and transformed remains found within caves (Voute à Pin at La Désirade and Grotte Cadet 2, Grotte Blanchard and Morne Rita at Marie-Galante) (Courtaud 2011) to primary burials discovered in open-air sites (Van den Bel and Romon 2010).

Sample processing and analyses were performed in a laboratory dedicated to analyses of aDNA (UMR PACEA, Bordeaux University) following all established aDNA guidelines to minimize contamination (e.g., Cooper and Poinar 2000; Gilbert et al. 2005). After a treatment of bleach and ultraviolet radiation, each sample was reduced to powder and incubated overnight in a lysis solution (0.5 M EDTA, NaOH, 1–2 mg/mL proteinase K and 0.5% N-Lauryl Sarkosy). For some samples, collected and studied at the very beginning of this project, DNA was extracted using a phenol-chloroform protocol (Deguilloux et al. 2011), while for other samples, DNA extraction was realized using the NucleoSpin® Extract II kit (Macherey-Nagel, Düren, Germany) (Mendisco et al. 2011). The amplification of a 393-bp (base pair) fragment of the mitochondrial Hypervariable Region I (HVRI) (nps 16.008–16.401) was realized using four overlapping primer pairs (Deguilloux et al. 2011; Gabriel et al. 2001; Jehaes et al. 2001). In addition, a combination of 16 single nucleotide polymorphisms (SNPs) diagnostic of the principal founding mitochondrial and Y-chromosome Native American haplogroups were genotyped with the iPLEX® Gold technology (Sequenom Inc., San Diego, CA, USA), as described in a previous work (Mendisco et al. 2011).

Our data were compared with data from ancient groups from the Greater Antilles (Fernandes et al. 2021; Lalueza-Fox et al. 2001, 2003; Nägele et al. 2020; Nieves-Colón et al. 2020) and data from contemporary Caribbean, Mesoamerican and South American populations available in the literature (see the detail of the comparison populations used in Mendisco et al. 2015).

Sampling and Analysis of Modern Samples

Buccal swabs were collected and processed from 198 unrelated volunteers, randomly sampled in five different counties of the Guadeloupe archipelago (Figure 2.1). All ethical guidelines were followed, and written informed consent was obtained from all participants subsequent to project approval from the competent authorities (deliberation 2015-339 from the Commission Nationale de L'Informatique et des Libertés after positive recommendation 13.677bis from the Comité Consultatif sur le Traitement de L'information en matière de Recherche dans le domaine de la Santé).

Genomic DNA was isolated and extracted from each buccal swab using the DNeasy blood and tissue kit (Qiagen Ltd., Crawley, UK) following the manufacturer's instructions. Mitochondrial DNA hypervariable regions 1 and 2 (HVRI, HVRII) were PCR amplified (L15989/H16410 and L29/H408 primer pairs) and sequenced as described by Mendisco and colleagues (2019). The obtained sequences were manually aligned to the revised Cambridge Reference Sequence (Anderson et al. 1981; Andrews et al. 1999) and edited using the BioEdit software v7.2.5 (Hall 1999). Mitochondrial haplogroup assignment, based on HVRI and HVRII nucleotide positions, was confirmed or enhanced through the genotyping of 27 diagnostic SNPs in the coding region using the iPLEX® Gold technology (Sequenom Inc., San Diego, CA, USA) (Mendisco et al. 2011). This same genotyping technology was used to analyze 29 Y-chromosomal SNPs, diagnostic of the main worldwide haplogroups and sub-haplogroups, for the 94 male individuals of the sample.

Standard diversity and differentiation indices were calculated using Arlequin 3.5 (Excoffier et al. 2005). We established databases (based on mtDNA and Y-chromosome diversity) compiled from the literature for targeted reference populations from America, Africa, Europe and Asia to conduct inter-population and phylogenetic comparisons (see Mendisco et al. 2019 for the details of the comparison populations). Finally, an estimation of the admixture proportions was measured according to the mitochondrial and Y-chromosome haplogroup frequencies using the ADMIX2.0 software (Bertorelle and Excoffier 1998; Dupanloup and Bertorelle 2001).

Results and Discussion

The Caribbean Area: A Challenge for Ancient DNA Studies

It is widely acknowledged that working with aDNA is associated with various challenges such as contamination, postmortem damage, small amounts of data and few samples. Sampling and analysis protocols are constantly improving to alleviate these difficulties; however, some parts of the world, such as the Caribbean area, have long remained almost inaccessible to aDNA studies. Despite almost ideal sampling conditions, e.g., the samples were collected *in situ* and directly deposited into hermetically sealed sterile bags and immediately stored at -20°C, we obtained relatively low amplification efficiency for the ancient samples. From the 38 analyzed individuals, only 13 exploitable mitochondrial SNP profiles could be obtained, permitting the characterization of 13 maternal lineages for these late Ceramic Age individuals from Guadeloupe. Among these 13 individuals, 11 complete and authenticated mitochondrial haplotypes (HVRI) could be characterized. Interestingly, we observed heterogeneous DNA conservation among the ancient samples studied according to their specific geographic locations, highlighting a correlation between DNA preservation and micro-environmental conditions. We obtained more easily reproducible sequences from the samples collected in caves (Grotte Cadet 2, Grotte Blanchard and Voute à Pin archaeological sites), where environmental conditions are assumed to be more stable than in open-air sites (Elsner et al. 2014). In contrast, it was more difficult to obtain reproducible results from samples collected from open-air sites, as they demonstrated poor macroscopic conservation (due to the acidic volcanic sediments of the islands). The tropical nature of the Caribbean area, particularly the hot and humid conditions, hinders the preservation of human remains and DNA (Allentoft et al. 2012; Hofreiter et al. 2015; Kistler et al. 2017; Nieves-Colón et al. 2018). This would explain, in part, why until recently there were very few aDNA studies carried out in the Caribbean region

(Lalueza-Fox et al. 2001, 2003; Mendisco et al. 2015; Schroeder et al. 2015). As already mentioned, the situation has changed with the discovery of significantly better preservation of endogenous aDNA in the petrous bone (Pinhasi et al. 2015, 2019) together with the considerable advances in sequencing techniques, leading to a rapid expansion of the Caribbean paleogenomic studies (Fernandes et al. 2021; Nägele et al. 2020; Nieves-Colón et al. 2020).

Ancient DNA Perspectives in the Lesser Antilles on Cultural Data

DNA is one tool among many that can be used in interdisciplinary studies that deal with the history of human populations. It is important to integrate and interrogate all available data and not overinterpret the genetic data. Our study described aDNA sequences obtained from indigenous remains originating from the Guadeloupe archipelago and associated with late Ceramic Age contexts. In this Guadeloupean Late Ceramic group, we could detect three different Amerindian mitochondrial lineages (A2 = 5, C1 = 5, D1 = 3) (Figure 2.2), concordant with the diversity of extant Amerindian groups of the circum-Caribbean region. The statistical analyses conducted (genetic distances estimated based on the HVR1 data) showed that the closest affinities were found with inland indigenous populations [Native Colombians or Amazonians (Carvalho et al. 2008; Ribeiro-dos-Santos et al. 2007; Torres et al. 2006). We are aware of the small sample size as well as the lack of resolution of the mitochondrial lineages alone. However, these genetic relationships are consistent with the hypothesis of migrations along the waterways of Arawak speakers from the Amazonian region to the Antilles, which favored the emergence of the Caribbean Ceramic Age (Moreno-Estrada et al. 2013; Rouse 1986, 1992). A higher molecular resolution based on the analysis of genome-wide data would have been more appropriate and useful to refine phylogeographic inferences. The technological development of massively parallel sequencing in the last decades has facilitated the acquisition of ancient genomes, contributing to a better understanding of past population dynamics (Fernandes et al. 2021; Nägele et al. 2020). To make the most of these valuable remains, we attempted to obtain paleogenomic data using a capture/enrichment technique (through a collaboration with the Geogenetics laboratory at the University of Copenhagen). Unfortunately, these attempts were unsuccessful. Nevertheless, these mitochondrial data remain of great interest, being the first available from Ceramic Age individuals of the Lesser Antilles.

The mtDNA diversity observed for this ancient Ceramic group from Guadeloupe caught our attention since it is relatively higher than that observed for other communities from the Greater Antilles (Nieves-Colón et al. 2020). To discuss this divergent pattern, we focused more specifically on the archaeological and funerary contexts of each ancient individual.

One hypothesis we can propose is that several distinct communities inhabited Guadeloupe before the arrival of Europeans. We can group the archaeological sites

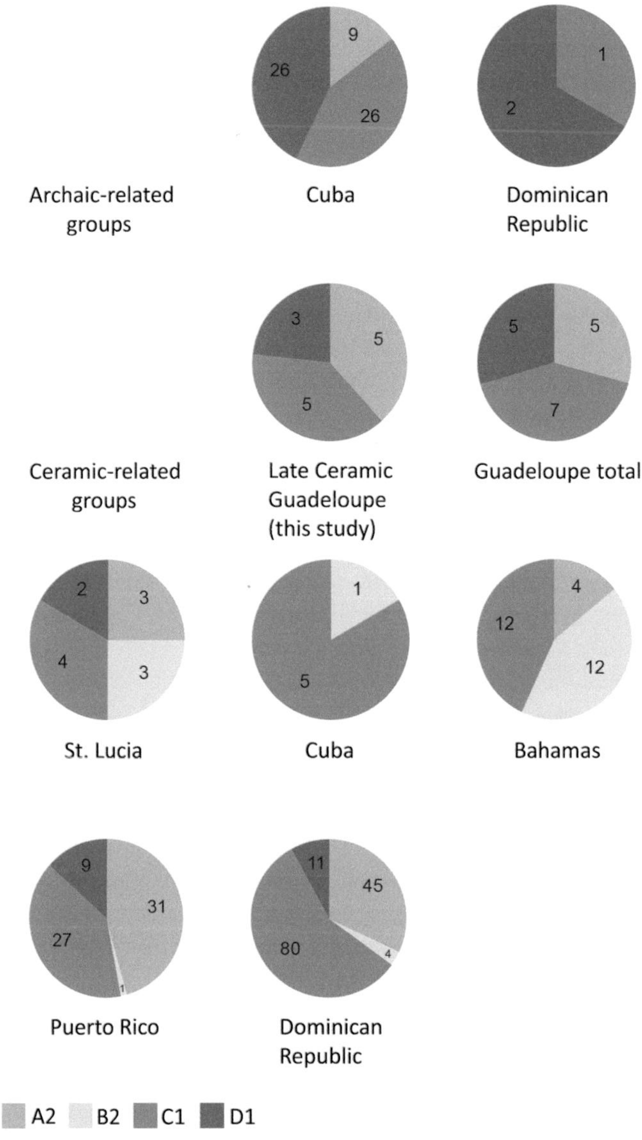

FIGURE 2.2 Proportions of mitochondrial haplogroups found in the ancient Guadeloupe samples.

studied into two categories, and it seems that each type of context is associated with a distinct mtDNA diversity. On one hand, individuals originating from mainland Guadeloupe (i.e., the archaeological sites of Grande Anse and Saint Claude) and collected from primary burials associated with habitations in open-air sites (Van den Bel and Romon 2010) belonged to the mitochondrial haplogroups A2

(5 individuals) and D1 (1 Individual) (Figure 2.2). On the other hand, human remains retrieved from the islands of La Désirade and Marie-Galante (i.e., Grotte Cadet, Grotte Blanchard and Voute à Pin) corresponded to scattered remains within caves and belonged to C1 (5 individuals) and D1 (2 individuals) mitochondrial haplogroups (Figure 2.2), echoing the genetic diversity observed for ancient communities of the Greater Antilles. Of course, this is a first glimpse of the genetic diversity of the ancient communities of Guadeloupe. Additional samples as well as a higher resolution are necessary in the future to test this hypothesis and to know if distinct communities, with potentially different demographic histories, occupied Guadeloupe during the Ceramic Age.

The other hypothesis that we can put forward to explain the observed mtDNA diversity in our sample is that the archaeological sites from the islands of La Désirade and Marie-Galante were used for particular functions or special rituals, as previously proposed for the site of Morne Cybèle on La Désirade (Hofman 1995; Hoogland and Hofman 1999). In addition to human skeletal remains being found scattered in caves, as in Grotte Cadet, bioarchaeologists could demonstrate the presence of numerous cutmarks, traces of disarticulation and defleshing, as well as intentional perforations (Courtaud 2011). These particular and uncommon treatments suggest the use of these sites for particular events, such as ceremonies or rituals.

The correlation between archaeological and maternal genetic diversity in the Caribbean indigenous communities is consistent with an increase in the number of ceramic styles in time and space, which suggested cultural fragmentation during the late Ceramic Age (Hofman et al. 2007). This cultural regionalism may have been correlated with decreased gene flow between these different communities, which may have resulted in the genetic differentiation supposed between the ancient communities of the Guadeloupe archipelago. We hope that more data, from numerous and distinct contexts, will be obtained in the near future to gain a more accurate and complex view of the genetic identity of the individuals who occupied the Guadeloupe archipelago during pre-Contact periods.

Colonization: A Major Impact on the Modern Gene Pool

The analysis of the current Guadeloupean population's genetic diversity presented a unique opportunity to (i) fill a gap, as the modern French populations of the Lesser Antilles have been studied very little, (ii) to understand the impact of the colonization on the evolutionary history of local lineages and (iii) to question the link between island history and gene pool composition. Present-day Guadeloupe stands as the perfect place to study the impact of French colonial power on admixture characteristics through the comparison of biological, ethnological and historical data.

As expected for populations that experienced numerous migration waves and admixture events, the present-day Guadeloupean population presents a high level

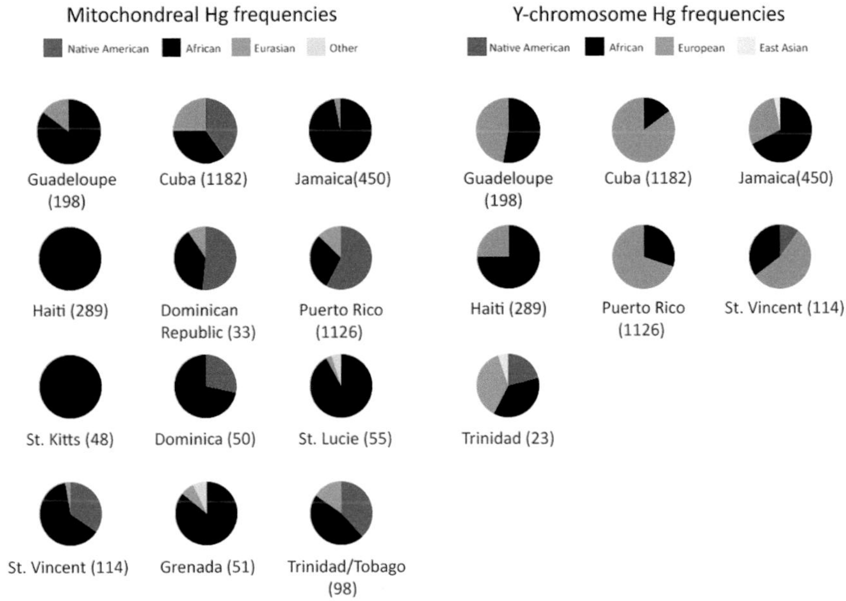

FIGURE 2.3 Summary of mitochondrial and Y-chromosome haplogroup frequencies described for different present-day Caribbean populations.

of diversity with a total of 169 distinct mtDNA haplotypes characterized for the 198 individuals typed (Mendisco et al. 2019). Based on the phylogeographic analysis of the maternal haplogroups, we were able to determine that the Guadeloupean gene pool results from a blend of 85% of African lineages, 12.5% of Eurasian lineages and 0.5% of Amerindian lineages (Figure 2.3). Considering the 94 male individuals, 53% present paternal lineages associated with an African origin and 44% to a West Eurasian origin.

The first striking conclusion from our data is that the estimated Native American component inferred in Guadeloupe is significantly low (only one maternal lineage and no paternal lineage detected). It is interesting to note that the only Amerindian maternal lineage was found in the Anse-Bertrand community, where a group declaring itself as descending from Amerindian ancestors has been recognized by local authorities since the end of the nineteenth century (Lasserre 1978). Even if our sample remains small, our genetic data suggest that the ancestral indigenous legacy was massively reduced by the arrival of Europeans and enslaved Africans and that potential admixture has not been able to leave a significant imprint on the present-day population of Guadeloupe. Historical data indicate that despite the good relationships initially maintained between the first European settlers and the indigenous people, a war was quickly declared against indigenous communities and lasted until 1639 (Madrigal 2006; Rouse 1992). These conflicts, along with epidemics, drastically reduced the number of indigenous people on the island and finally forced many of the survivors to join the Carib Territory of Dominica (Watts 1987). As can be seen in Figure 2.3, the Amerindian genetic contribution through the Caribbean region is far from homogeneous, with an almost complete absence of Amerindian lineages in some English-speaking islands, compared to substantial frequencies observed in the Spanish-speaking islands. This variation is correlated with both dominant colonial power and unique island histories, as across the Caribbean different colonial policies regarding social hierarchies and mate choice were decreed (Benn Torres et al. 2015; Bukhari et al. 2017). The identification of the Native American ancestral component in the Caribbean populations could be extremely challenging, especially when dealing with uniparental markers. The analysis of autosomal markers (such as ancestry-informative markers, for example) or genome-wide data could provide a more precise and contrasted picture concerning admixture rates (Schroeder et al. 2015; Shriver and Kittles 2004; Tian et al. 2006).

The second striking element emphasized is the large preponderance of African maternal and paternal lineages (representing 85% and 53% of lineages, respectively) in the present-day population of Guadeloupe, as regularly encountered in the Caribbean region (see, for example, Benn Torres et al. 2007; Martínez -Cruzado et al. 2005) (Figure 2.3). The putative origin of maternal/paternal African lineages in the Guadeloupe archipelago can be mainly traced to Western and West-central Africa. More precisely, our data indicate that the African ancestry of Guadeloupean individuals shows particular affinities with current populations from Benin, Nigeria, Sierra Leone, Mali and Senegal (Mendisco et al. 2019). This is in agreement with the proposed origin of enslaved Africans who disembarked in the Guadeloupe archipelago and can be traced specifically to the Guinean Gulf (Belenus 1998; Curtin 1991; Frisch and Frisch 1987). It is also worth noting the sharing of African lineages between Guadeloupe and other extant populations from the Caribbean region and also between Guadeloupe and African-American or Afro-Brazilian groups. This result reinforces the idea of a global shared genetic African legacy between all regions impacted by the transatlantic slave trade (Boutin 2006; Butel 2002), even if there is a rough differentiation of admixed New World groups linked to the specific point of embarkation, number of founders, or life history constraints (Gravel et al. 2013; Moreno-Estrada et al. 2013; Schroeder et al. 2015). When focusing on Eurasian maternal/paternal lineages found in this Guadeloupean sample, the highest affinities are observed from Western and South-Western European populations, particularly from Spain, Sicily and France. These data, once again, corroborate the known origin of colonizers established in the archipelago as historical documentation testifies to an overwhelming European emigration from France (Boutin 2006; Butel 2002). From these genetic data, we calculated an estimation of admixture proportions based on mitochondrial (0,879 ± 0,045 of Western African ancestry and 0,121 ± 0,045 of European ancestry) and Y-chromosomal data (0,605 ± 0,097 of Western African ancestry and 0,395 ± 0,097 of European ancestry). These admixture rates highlight a sex-biased pattern among European colonizers and enslaved people of African origin, in agreement with historical documentation and typical of admixed populations of the area (Benn Torres et al. 2007; Martínez -Cruzado et al. 2001, 2005; Mendizabal et al. 2008). Guadeloupe received a higher proportion of enslaved women and children than men and was often in need of new slave arrivals (Dorigny and Gainot 2006). This result reflects the predominant social dynamics of French slave societies, as mating between European men/masters and enslaved women was not uncommon (Frisch and Frisch 1987). From the beginning of the eighteenth century, the enslaved people born on Guadeloupe were more numerous than those who arrived directly from Africa (Belenus 1998), which facilitated the "creolization" of the population and the appearance of new social strata.

Finally, the maternal lineages of about ten individuals (5% of the sample studied) could be traced back to India (M-derived haplogroups) or East Asia (Mendisco et al. 2019), which echoes the historically well-known immigration of indentured laborers to Guadeloupe, following the emancipation of the enslaved population in 1848. Historical accounts indicate that India represented the origin of the first laborers immigrating to Guadeloupe during the nineteenth century (with a total of 45,000 immigrants; Schnakenbourg 2005). In sum, the genetic analyses conducted on the Guadeloupe archipelago demonstrate that all historically known major migratory events left a visible genetic imprint in the contemporary Caribbean population.

Conclusion

Our study provided an opportunity to study the peopling of the Guadeloupe archipelago through complementary ancient and modern genetic data. Our findings reconfirm the dramatic reshaping of the indigenous Amerindian gene pool and highlight the necessity of broadening aDNA analyses in the Caribbean region, both in time and space, to allow for a better understanding of early peopling dynamics. We were able to evoke how a combination of cultural (funerary) and biological contexts helps to discuss individual identity. Moreover, thanks to modern data, we were able to demonstrate that in Guadeloupe, the genetic legacy of each historical event was revealed to be highly variable. The Central-West African slaves and Western European colonizers primarily contributed to the genetic composition of the present-day Guadeloupe population through a model of sex-biased gene flow in which African women, African men and European men were the primary contributors. Importantly, whereas a noticeable genetic imprint of indentured laborers is visible, our results also indicate that indigenous populations made minimal genetic contributions to the present-day Guadeloupean population.

We believe that the analysis of groups/islands with different colonial histories must be advanced to gain a better understanding of the interplay between social and biological factors in shaping the genetic diversity of Caribbean communities. The evolution of human population genetic research toward the genomic/paleogenomic era gives us hope in obtaining unprecedented resolution in future studies, despite the major difficulties associated with the environment and climate of the Caribbean region (Fernandes et al. 2021; Nägele et al. 2020; Nieves-Colón et al. 2020; Schroeder et al. 2015). Today, one of the most limiting factors may be the scarcity of human remains found in the Caribbean region, making every piece of bone incredibly precious and valuable. The resolution of key issues concerning the multifaceted history of the peopling of the Caribbean region will only be possible with an interdisciplinary effort where scientists work closely together with the co-operation of local populations.

Acknowledgements

This research was made possible by funding from the Guadeloupe Region (Conseil regional de la Guadeloupe), the INRAP (Institut National de Recherches en Archéologie Préventives) and the CNRS (Centre National de la Recherche Scientifique) and a ministerial grant from the Research National Agency as a program of prospects investments ANR-10-LABX52 (CAP project; dir: M.F.D.; University of Bordeaux, LaScArBxANR; 2012–2014). Part of the presented experiments (SNPs analyses) were performed at the Genomic and Sequencing Facility of Bordeaux, grants from the Conseil Regional d'Aquitaine n°20030304002FA and 20040305003FA, https://www.nouvelle-aquitaine.fr/, and from the European Union, FEDER n°2003227 and from Investissements d'avenir, Convention attributive d'aide N°ANR-10-EQPX-16-01).

We thank everyone who participated directly in the excavations of the human remains submitted for the palaeogenetic analyses. We also acknowledge and thank each study participant for their crucial role in the study.

References

Allentoft ME, Collins M, Harker D, Haile J, Oskam CL, Hale ML, Campos PF, Samaniego JA, Gilbert MT, Willerslev E, Zhang G, Scofield RP, Holdaway RN, and Bunce M. 2012. The half-life of DNA in bone: Measuring decay kinetics in 158 dated fossils. *Proc. Biol. Sci.* 279(1748): 4724–4733.

Anderson S, Bankier AT, Barrell BG, de Bruijn MH, Coulson AR, Drouin J, Eperon IC, Nierlich DP, Roe BA, Sanger F, Schreier PH, Smith AJ, Staden R, and Young IG. 1981. Sequence and organization of the human mitochondrial genome. *Nature* 290(5806): 457–465.

Andrews RM, Kubacka I, Chinnery PF, Lightowlers RN, Turnbull DM, and Howell N. 1999. Reanalysis and revision of the Cambridge reference sequence for human mitochondrial DNA. *Nat. Genet.* 23(2): 147.

Antczak AT, and Hofman CL. 2019. Dearchaizing the Archaic. In Hofman CL, Antczak AT (Eds.), *Early Settlers of the Insular Caribbean* (pp. 29–42). Leiden: Sidestone Press.

Barbujani G, and Colonna V. 2010. Human genome diversity: Frequently asked questions. *Trends Genet.* 26: 285–295.

Belenus R. 1998. *L'Esclave en Guadeloupe et en Martinique du XVIIème au XIXème siècle*, Point-à-Pitre: Éditions Jasor.

Benn Torres J, Kittles RA, and Stone AC. 2007. Mitochondrial and Y chromosome diversity in the English-speaking Caribbean. *Ann. Hum. Genet.* 71: 782–790.

Benn Torres J, Martucci V, Aldrich MC, Vilar MG, MacKinney T, Tariq M, Gaieski JB, Bharath Hernandez R, Browne ZE, Stevenson M, Walters W, Schurr TG, and Genographic Consortium. 2019. Analysis of biogeographic ancestry reveals complex genetic histories for indigenous communities of St. Vincent and Trinidad. *Am. J. Phys. Anthropol.* 169(3): 482–497.

Benn Torres J, Vilar MG, Torres GA, Gaieski JB, Bharath Hernandez R, Browne ZE, Stevenson M, Walters W, Schurr TG, and Genographic Consortium. 2015. Genetic diversity in the Lesser Antilles and its implications for the settlement of the Caribbean basin. *PLoS One* 10(10): e0139192.

Bérard, B. 2011. A la recherche des Caraïbes archéologiques. In Grunberg B (Ed.), *Les Indiens de Petites Antilles. Des Premiers Peuplements aux Débuts de la Colonisation Européenne*, Cahiers d'Histoire de l'Amérique Coloniale, n°5 (pp. 27–36). Paris: L'Harmattan.

Bérard B. 2019. About Boxes and Labels: A periodization of the Amerindian occupation of the West Indies. *J. Caribb. Archaeol.* 19.

Bertorelle G, and Excoffier L. 1998. Inferring admixture proportions from molecular data. *Mol. Biol. Evol.* 15(10): 1298–1311.

Boutin R. 2006. *La population de la Guadeloupe: De l'émancipation à l'assimilation (1848–1946): aspects démographiques et sociaux.* Matoury, Guyane: Ibis Rouge Editions.

Bryc K, Velez C, Karafet T, Moreno-Estrada A, Reynolds A, Auton A, Hammer M, Bustamante CD, and Ostrer H. 2010. Colloquium paper: Genome-wide patterns of population structure and admixture among Hispanic/Latino populations. *Proc. Natl. Acad. Sci.* 107(2): 8954–8961.

Bukhari A, Luis JR, Alfonso-Sanchez MA, Garcia-Bertrand R, and Herrera RJ. 2017. Taino and African maternal heritage in the Greater Antilles. *Gene* 637: 33–40.

Butel P. 2002. *Histoire des Antilles françaises: XVIIe–XXe siècles*. Paris: Perrin.

Carvalho BM, Bortolini MC, Batista dos Santos SE, and Campos Ribeiro-dos-Santos AC. 2008. Mitochondrial DNA mapping of social-biological interactions in Brazilian Amazonian African-descendant populations. *Genet. Mol. Biol.* 31: 12–22.

Cintado A, Companioni O, Nazabal M, Camacho H, Ferrer A, De Cossio ME, Marrero A, Ale M, Villarreal A, Leal L, Casalvilla R, Benitez J, Novoa L, Diaz-Horta O, and Dueñas M. 2009. Admixture estimates for the population of Havana City. *Ann. Hum. Biol.* 36(3): 350–360.

Cooper A, and Poinar HN. 2000. Ancient DNA: Do it right or not at all. *Science* 289(5482): 1139.

Courtaud P. 2011. La cavité sépulcrale de Cadet 2 (Capesterre-de-Marie-Galante, Guadeloupe) et la question du cannibalisme. In Grunberg B (Ed.), *Les Indiens des Petites Antilles: Des premiers peuplements aux débuts de la colonisation européenne. Cahiers d'histoire de l'Amérique Latine*. 5: 49. Paris: L'Harmattan.

Curtin PD. 1991. *The Tropical Atlantic in the Age of the Slave Trade, essays on global and comparative History*, Washington: American Historical Association.

Deason ML, Salas A, Newman SP, Macaulay VA, St A Morrison EY, and Pitsiladis YP. 2012. Interdisciplinary approach to the demography of Jamaica. *BMC Evol. Biol.* 12: 24.

Deguilloux MF, Pemonge MH, Courtaud P, Romon T, Lafleur G, Richard G. 2015. Le peuplement amérindien céramique des Petites Antilles: apport de l'analyse génétique des vestiges humains et de groupes actuels. In Grunberg B (Ed.), *A la recherche du Caraïbe perdu: les populations amérindiennes des Petites Antilles de l'époque précolombienne à la période colonial* (pp.79–98). Paris: L'Harmattan.

Deguilloux MF, Soler L, Pemonge MH, Scarre C, Joussaume R, Laporte L. 2011. News from the west: Ancient DNA from a French megalithic burial chamber. *Am. J. Phys. Anthropol.* 144(1): 108–118.

Dorigny M, and Gainot B. 2006. *Atlas des esclavages. Traites, sociétés coloniales, abolitions, de l'Antiquité à nos jours*. Paris: Éditions Autrement, 80 p., ISBN 2–7467–0878-7.

Dupanloup I, and Bertorelle G. 2001. Inferring admixture proportions from molecular data: Extension to any number of parental populations. *Mol. Biol. Evol.* 18(4): 672–675.

Elsner J, Schibler J, Hofreiter M, and Schlumbaum A. 2014. Burial condition is the most important factor for mtDNA PCR amplification success in Palaeolithic equid remains from the Alpine foreland. *Archaeol. Anthropol. Sci.* 7(4): 505–515.

Excoffier L, Laval G, and Schneider S. 2005. Arlequin (version 3.0): An integrated software package for population genetics data analysis. *Evol. Bioinform Online* 1: 47–50.

Fernandes DM, Sirak KA, Ringbauer H, Sedig J, Rohland N, Cheronet O, Mah M, Mallick S, Olalde I, Culleton BJ, Adamski N, Bernardos R, Bravo G, Broomandkhoshbacht N, Callan K, Candilio F, Demetz L, Carlson KSD, Eccles L, Freilich S, George RJ, Lawson AM, Mandl K, Marzaioli F, McCool WC, Oppenheimer J, Özdogan KT, Schattke C, Schmidt R, Stewardson K, Terrasi F, Zalzala F, Antúnez CA, Canosa EV, Colten R, Cucina A, Genchi F, Kraan C, La Pastina F, Lucci M, Maggiolo MV, Marcheco-Teruel B, Maria CT, Martínez C, París I, Pateman M, Simms TM, Sivoli CG, Vilar M, Kennett DJ, Keegan WF, Coppa A, Lipson M, Pinhasi R, and Reich D. 2021. A genetic history of the pre-contact Caribbean. *Nature* 590(7844): 103–110.

Fortes-Lima C, Bybjerg-Grauholm JB, Marin-Padrón LC, Gomez-Cabezas EJ, Bækvad-Hansen M, Søholm Hansen C, Le P, Hougaard DM, Verdu P, Mors O, Parra EJ and Marcheco-Teruel B. 2018. Exploring Cuba's population structure and demographic history using genome-wide data. *Scientific Reports* 8:11422.

Frisch M, and Frisch N. 1987. Les esclaves de la Guadeloupe à la fin de l'Ancien Régime: du code noir aux codes numériques. *Histoire & Mesure* 2: 93–115.

Gabriel MN, Huffine EF, Ryan JH, Holland MM, and Parsons TJ. 2001. Improved MtDNA sequence analysis of forensic remains using a "mini-primer set" amplification strategy. *J. Forensic Sci.* 46(2): 247–53.

Gilbert MT, Bandelt HJ, Hofreiter M, and Barnes I. 2005. Assessing ancient DNA studies. *Trends Ecol. Evol.* 20(10): 541–544.

Granberry J. 2013. Indigenous languages of the Caribbean. In Keegan WF, Hofman CL, and Rodríguez Ramos R (Eds.), *The Oxford Handbook of Caribbean Archaeology*, (pp. 61–69). Oxford: Oxford University Press.

Gravel S, Zakharia F, Moreno-Estrada A, Byrnes JK, Muzzio M, Rodriguez-Flores JL, Kenny EE, Gignoux CR, Maples BK, Guiblet W, Dutil J, Via M, Sandoval K, Bedoya G; 1000 Genomes Project, Oleksyk TK, Ruiz-Linares A, Burchard EG, Martinez-Cruzado JC, and Bustamante CD. 2013. Reconstructing Native American migrations from whole-genome and whole-exome data. *PLoS Genet.* 9(12): e1004023.

Hall TA. 1999. BioEdit: A User-Friendly biological sequence alignment editor and analysis program for windows 95/98/NT. *Nucl. Acids Symp. Ser.* 41: 95–98.

Hofman CL. 1995. *Three late prehistoric sites in the periphery of Guadeloupe Grande Anse*. Terre de Bas, les Saintes, and Morne Cybèle 1 et 2, La Désirade. Proccedings of the XVIth congress for the Caribbean Archaeology, Basse-Terre, Guadeloupe: Conseil Régional de la Guadeloupe.

Hofman CL, Bright AJ, Boomert A, and Knippenberg S. 2007. Island Rhythms: The web of social relationships and interaction networks in the Lesser Antillean archipelago between 400 B.C. and A.D. 1492. *Latin Am. Antiq.* 18: 243–268.
Hofreiter M, Paijmans JL, Goodchild H, Speller CF, Barlow A, Fortes GG, Thomas JA, Ludwig A, and Collins MJ. 2015. The future of ancient DNA: Technical advances and conceptual shifts. *Bioessays* 37(3): 284–293.
Honychurch L. 1997. Crossroads in the Caribbean: A site of encounter and exchange on Dominica. *World Archaeol.* 28(3): 291–304.
Hoogland MLP, and Hofman CL. 1999. Expansion of the Taino cacicazgos towards the Lesser Antilles. *J. Soc. Am.* 85: 93–113.
Jehaes E, Toprak K, Vanderheyden N, Pfeiffer H, Cassiman JJ, Brinkmann B, and Decorte R. 2001. Pitfalls in the analysis of mitochondrial DNA from ancient specimens and the consequences for forensic DNA analysis: The historical case of the putative heart of Louis XVII. *Int. J. Legal Med.* 115(3): 135–141.
Jobling MA. 2012. The impact of recent events on human genetic diversity. *Philos. Trans. R. Soc. B Biol. Sci.* 367: 793–799.
Kistler L, Ware R, Smith O, Collins M, and Allaby RG. 2017. A new model for ancient DNA decay based on paleogenomic meta-analysis. *Nucl. Acids Res.* 45(11): 6310–6320.
Lalueza-Fox C, Calderón FL, Calafell F, Morera B, and Bertranpetit J. 2001. MtDNA from extinct Tainos and the peopling of the Caribbean. *Ann. Hum. Genet.* 65(Pt 2): 137–151.
Lalueza-Fox C, Gilbert M T, Martínez-Fuentes A J, Calafell F, and Bertranpetit J. 2003. Mitochondrial DNA from pre-Columbian Ciboneys from Cuba and the prehistoric colonization of the Caribbean. *Am. J. Phys. Anthropol.* 121(2): 97–108.
Lasserre G. 1978. La Guadeloupe, Etude géographique. Tome 1: La nature et les hommes. (Thèse). Ed. Kolodziej. Fort-de-France, Edition et Diffusion de la Culture Antillaise.
Madrigal L. 2006. *Human Biology of Afro-Caribbean Populations*. Cambridge: Cambridge University Press.
Madrilejo N, Lombard H, and Torres JB. 2015. Origins of marronage: Mitochondrial lineages of Jamaica's Accompong Town Maroons. *Am. J. Hum. Biol.* 27(3): 432–437.
Marcheco-Teruel B, Parra EJ, Fuentes-Smith E, Salas A, Buttenschøn HN, Demontis D, Torres-Español M, Marín-Padrón LC, Gómez-Cabezas EJ, Alvarez-Iglesias V, Mosquera-Miguel A, Martínez-Fuentes A, Carracedo A, Børglum AD, and Mors O. 2014. Cuba: Exploring the history of admixture and the genetic basis of pigmentation using autosomal and uniparental markers. *PLoS Genet.* 10(7): e1004488.
Martínez-Cruzado JC, Toro-Labrador G, Ho-Fung V, Estévez-Montero MA, Lobaina-Manzanet A, Padovani-Claudio DA, Sánchez-Cruz H, Ortiz-Bermúdez P, and Sánchez-Crespo A. 2001. Mitochondrial DNA analysis reveals substantial Native American ancestry in Puerto Rico. *Human Biol.* 73: 491–511.
Martínez-Cruzado JC, Toro-Labrador G, Viera-Vera J, Rivera-Vega MY, Startek J, Latorre-Esteves M, Román-Colón A, Rivera-Torres R, Navarro-Millán IY, Gómez-Sánchez E, Caro-González HY, Valencia-Rivera P. 2005. Reconstructing the population history of Puerto Rico by means of mtDNA phylogeographic analysis. *Am. J. Phys. Anthropol.* 128(1): 131–155.
Mendisco F, Keyser C, Hollard C, Seldes V, Nielsen AE, Crubézy E, Ludes B. 2011. Application of the iPLEX™ Gold SNP genotyping method for the analysis of Amerindian ancient DNA samples: Benefits for ancient population studies. *Electrophoresis* 32(3–4): 386–393.
Mendisco F, Pemonge MH, Leblay E, Romon T, Richard G, Courtaud P, Deguilloux MF. 2015. Where are the Caribs? Ancient DNA from ceramic period human remains in the Lesser Antilles. *Philos. Trans. R Soc. Lond. B Biol. Sci.* 370(1660): 20130388.

Mendisco F, Pemonge MH, Romon T, Lafleur G, Richard G, Courtaud P, Deguilloux MF. 2019. Tracing the genetic legacy in the French Caribbean islands: A study of mitochondrial and Y-chromocome lineages in the Guadeloupe archipelago. *Am. J. Phys. Anthropol.* 170(4): 507–518

Mendizabal I, Sandoval K, Berniell-Lee G, Calafell F, Salas A, Martínez-Fuentes A, Comas D. 2008. Genetic origin, admixture, and asymmetry in maternal and paternal human lineages in Cuba. *BMC Evol. Biol.* 8: 213.

Miljkovic-Gacic I, Ferrell RE, Patrick AL, Kammerer CM, Bunker CH. 2005. Estimates of african, European and native american ancestry in afro-Caribbean men on the island of tobago. *Hum. Hered.* 60: 129–133.

Moreno-Estrada A, Gravel S, Zakharia F, McCauley JL, Byrnes JK, Gignoux CR, Ortiz-Tello PA, Martínez RJ, Hedges DJ, Morris RW, Eng C, Sandoval K, Acevedo-Acevedo S, Norman PJ, Layrisse Z, Parham P, Martínez-Cruzado JC, Burchard EG, Cuccaro ML, Martin ER, Bustamante CD. 2013. Reconstructing the population genetic history of the Caribbean. *PLoS Genet.* 9(11): e1003925.

Nägele K, Posth C, Orbegozo MI, Chinique de Armas Y, Hernández Godoy ST, González Herrera UM, Nieves-Colón MA, Sandoval-Velasco M, Mylopotamitaki D, Radzeviciute R., Laffoon J, Pestle WJ, Ramos-Madrigal J, Lamnidis TC, Schaffer WC, Carr RS, Day JS, Arredondo Antúnez C, Rangel Rivero A, ... and Schroeder H. 2020. Genomic insights into the early peopling of the Caribbean. *Science* 369(6502): 456-460.

Nieves-Colón MA. 2022. Anthropological genetic insights on Caribbean population history. *Evol. Anthropol.* 31(3): 1–20.

Nieves-Colón MA, Ozga AT, Pestle WJ, Cucina A, Tiesler V, Stanton TW, Stone AC. 2018. Comparison of two ancient DNA extraction protocols for skeletal remains from tropical environments. *Am. J. Phys. Anthropol.* 166(4): 824–836.

Nieves-Colón MA, Pestle WJ, Reynolds AW, Llamas B, de la Fuente C, Fowler K, Skerry KM, Crespo-Torres E, Bustamante CD, Stone AC. 2020. Ancient DNA reconstructs the genetic legacies of precontact Puerto Rico communities. *Mol. Biol. Evol.* 37(3): 611–626.

Pinhasi R, Fernandes DM, Sirak K, and Cheronet O. 2019. Isolating the human cochlea to generate bone powder for ancient DNA analysis. *Nat. Protoc.* 14(4): 1194–1205.

Pinhasi R, Fernandes D, Sirak K, Novak M, Connell S, Alpaslan-Roodenberg S, Gerritsen F, Moiseyev V, Gromov A, Raczky P, Anders A, Pietrusewsky M, Rollefson G, Jovanovic M, Trinhhoang H, Bar-Oz G, Oxenham M, Matsumura H, and Hofreiter M. 2015. Optimal ancient DNA yields from the inner ear part of the Human petrous bone. *PLoS One* 10(6): e0129102.

Ribeiro-dos-Santos AK, Carvalho BM, Feio-dos-Santos AC, dos Santos SE. 2007. Nucleotide variability of HV-I in Afro-descendents populations of the Brazilian Amazon region. *Forensic Sci. Int.* 167(1): 77–80.

Rouse I. 1986. *Migrations in Prehistory; Inferring Population Movement from Cultural Remains*. New Haven, CT: Yale University Press.

Rouse I. 1992. *The Tainos: Rise and Decline of the People who Greeted Columbus.* New Haven, CT: Yale University Press.

Schnakenbourg C. 2005. L'immigration indienne en Guadeloupe (1848–1923) Histoire d'un flux migratoire, 1490 p. *Bull. Soc. Hist. Guadeloupe* 155: 63–82.

Schroeder H, Ávila-Arcos MC, Malaspinas AS, Poznik GD, Sandoval-Velasco M, Carpenter ML, Moreno-Mayar JV, Sikora M, Johnson PL, Allentoft ME, Samaniego JA, Haviser JB, Dee MW, Stafford TW Jr, Salas A, Orlando L, Willerslev E, Bustamante CD, and Gilbert MT. 2015. Genome-wide ancestry of 17th-century enslaved Africans from the Caribbean. *Proc. Natl. Acad. Sci. USA* 112: 3669–3673.

Schroeder H, Sikora M, Gopalakrishnan S, Cassidy LM, Maisano Delser P, Sandoval Velasco M, Schraiber JG, Rasmussen S, Homburger JR, Ávila-Arcos MC, Allentoft ME, Moreno-Mayar JV, Renaud G, Gómez-Carballa A, Laffoon JE, Hopkins RJA, Higham TFG, Carr RS, Schaffer WC, Day JS, Hoogland M, Salas A, Bustamante CD, Nielsen R, Bradley DG, Hofman CL, and Willerslev E. 2018. Origins and genetic legacies of the Caribbean Taino.*Proc. Natl. Acad. Sci. USA* 115(10): 2341–2346.

Shriver MD, and Kittles RA. 2004. Genetic ancestry and the search for personalized genetic histories. *Nat. Rev. Genet.* 5: 611–618.

Simms TM, Rodriguez CE, Rodriguez R, and Herrera RJ. 2010. The genetic structure of populations from Haiti and Jamaica reflect divergent demographic histories. *Am. J. Phys. Anthropol.* 142(1): 49–66.

Simms TM, Wright MR, Hernandez M, Perez OA, Ramirez EC, Martinez E, and Herrera RJ. 2012. Y-chromosomal diversity in Haiti and Jamaica: Contrasting levels of sex-biased gene flow. *Am. J. Phys. Anthropol.* 148(4): 618–631.

Simms TM, Wright MR, Martinez E, Regueiro M, McCartney Q, and Herrera RJ. 2013. Y-STR diversity and sex-biased gene flow among Caribbean populations. *Gene* 516(1): 82–92.

Tajima A, Hamaguchi K, Terao H, Oribe A, Perrotta VM, Baez CA, Arias JR, Yoshimatsu H, Sakata T, and Horai S. 2004. Genetic background of people in the Dominican Republic with or without obese type 2 diabetes revealed by mitochondrial DNA polymorphism. *J. Hum. Genet.* 49(9): 495–499.

The Trans-Atlantic Slave Trade Database. 2020. *SlaveVoyages.* https://www.slavevoyages.org (accessed February 18, 2022)

Tian C, Hinds DA, Shigeta R, Kittles R, Ballinger DG, Seldin MF. 2006. A genomewide single-nucleotide-polymorphism panel with high ancestry information for African American admixture mapping. *Am. J. Hum. Genet.* 79(4): 640–649.

Torres MM, Bravi CM, Bortolini MC, Duque C, Callegari-Jacques S, Ortiz D, Bedoya G, Groot de Restrepo H, and Ruiz-Linares A. 2006. A revertant of the major founder Native American haplogroup C common in populations from northern South America. *Am. J. Hum. Biol.* 18(1): 59–65.

Underhill PA, and Kivisild T. 2007. Use of Y chromosome and mitochondrial DNA population structure in tracing human migrations. *Annu. Rev. Genet.* 41: 539–564.

Van den Bel M, and Romon T. 2010. A troumassoid site at Trois-Rivières, Guadeloupe FWI. Funerary pratices and house patterns at La Pointe de Grande Anse. *J. Carib.Archaeol.* 10: 1–17.

Via M, Gignoux CR, Roth LA, Fejerman L, Galanter J, Choudhry S, Toro-Labrador G, Viera-Vera J, Oleksyk TK, Beckman K, Ziv E, Risch N, Burchard EG, and Martínez-Cruzado JC. 2011. History shaped the geographic distribution of genomic admixture on the island of Puerto Rico. *PLoS One* 6(1): e16513.

Vilar MG, Melendez C, Sanders AB, Walia A, Gaieski JB, Owings AC, Schurr TG, and Genographic Consortium. 2014. Genetic diversity in Puerto Rico and its implications for the peopling of the Island and the West Indies. *Am. J. Phys. Anthropol.* 155(3): 352–368.

Watts D. 1987. *The West Indies: Patterns of Development, Culture, and Environmental Change Since 1492*. Cambridge: Cambridge University Press. Cambridge Studies in Historical Geography: 609 p.

Wilson JL, Saint-Louis V, Auguste JO, Jackson BA. 2012. Forensic analysis of mtDNA haplotypes from two rural communities in Haiti reflects their population history. *J. Forensic Sci.* 57(6): 1457–1466.

3

WHEN HISTORY MEETS GENOMICS

Tracing the Complex Demographic Histories of African-Descendant Communities in the Caribbean

Jada Benn Torres and Hannes Schroeder

Introduction

The origins of African descendants in the Caribbean reflect a diverse history of population dispersions, displacement, migrations, and immigration. Though the origins of African descendants in the Caribbean are generally attributed primarily to the African continent, community histories and ethnological, archaeological, historical, and genomic evidence paint a more complex picture of the emergence of these contemporary communities that extends beyond Africa. Scholarly research on the topic of African-Caribbean origins generally centres around the logistics and repercussions of the trans-Atlantic slave trade and European colonisation. Undoubtedly, these historical phenomena were significant in uprooting the lives of millions of Africans and displacing innumerable Indigenous Caribbean peoples. However, as noted by mid-twentieth-century academics such as Eric Williams, Melville Herskovits, and Sidney Mintz, among others, the origins of contemporary African descendant communities in the Caribbean are results of serial admixture events of different African populations as well as admixture with peoples from Asia, the Americas, and Europe. The diverse origins of Afro-Caribbean peoples are reflected in everyday life, from food, language, religion, music, science, and other cultural practices (Mintz, 1989). The impacts of these varied heritages can also be observed in the genetic diversity within contemporary populations. In this chapter, we attempt to synthesise some of the more recent genetic and genomic efforts that examine the origins of African descendants with the purpose of illustrating the immense heterogeneity of African descendants throughout the Caribbean. We review both modern and ancient DNA studies, drawing on several cases throughout the Americas and how they have informed our understanding of the origins, trajectories, and demographic histories of African-descendant populations in the

DOI: 10.4324/9780429444449-5

Caribbean. Drawing on these case studies, we discuss the potential as well as the limitations of genomic approaches for offering new perspectives on the study of the trans-Atlantic slave trade and the complex population history of the Caribbean within the broader context of the Americas.

Africa and the African Diaspora

Before we can delve deeper into questions relating to the ethnogenesis of African descendants in the Caribbean, it is crucial to first consider the history and movements of people within Africa immediately prior to the trans-Atlantic slave trade. Some understanding of pre-colonial Africa is also essential for making sense of which nations and peoples were the most likely ancestors of those transported to the Caribbean. Hence, we start this chapter with a brief overview of West African history prior to the trans-Atlantic slave trade.

Before the trans-Atlantic slave trade, West and West-Central Africa was home to a variety of empires and other nation-states that each had developed complex social systems inclusive of extensive trade networks (Diop & Salemson, 2012). For example, some of western Africa's greatest empires date from the eighth to sixteenth centuries and include the empires of Ghana, Mali, and Songahi. Though sometimes characterised as having arisen due to foreign interventions, more recent historical and archaeological perspectives suggest that these empires were the result of local developments (Parker & Rathbone, 2007). Moreover, western African peoples also developed trade networks that extended throughout western Africa, south toward Central Africa, and north to the Atlantic. These transregional trade networks influenced the politics and economies of western Africa. Like any other highly organised political complex, western African nation-states were subject to local idiosyncrasies that detailed how social, economic, and political structures were formed and functioned. Consequently, some historians argue that among nation-states regionalism was present and appropriately characterises pre-colonial Africa. If one adopts a constructivist approach to regionalism, regions "must possess the element of 'identity', interests, learning, communication, shared knowledge, as well as how cooperation and communities emerge" (Izuagie, 2014, p. 24). Though regionalism may have been an integral part of pre-colonial West Africa, there was competition within these regions for material goods, labour, and other resources. Member communities competed for these resources in a variety of ways, including diplomacy, annexation, and even warfare (Smith, 1989). Furthermore, at times, regional competition for resources included competition for human labour. This demand for human labour expanded with expanded trade with the Portuguese and later other European merchants (Havik & Green, 2012). This burgeoning demand for labour resulted in the enslavement and exploitation of neighbouring peoples to meet European market demands for slaves in their colonies in the Americas (Eltis & Jennings, 1988). According to Falola and Fleming (2009), the trade in enslaved Africans deeply impacted not only those who were enslaved but also the African traders who engaged in procuring people to be enslaved. The Asante, Dahomey, and Oyó, and other coastal populations, for example, became embroiled in violence, sometimes civil war, as a result of the social and political instability created by the trans-Atlantic slave trade.

Though slavery existed in a variety of forms since antiquity, racialised slavery, characteristic throughout colonial-era Americas, was initiated with the European exploration of western Africa and fully institutionalised with the establishment of chattel slavery in the Americas. In particular, the Portuguese, with the aid of several African nations, developed and then extended the trade in enslaved peoples, initially between Africa and Europe, then later including the Caribbean and other regions in the Americas (Thornton, 1998). From the late fifteenth century through the early nineteenth century, trade in enslaved Africans was substantial. Roughly 36,000 people a year were enslaved during the early period of trade, culminating in the mid-1700s at 80,000 people a year. Trade in enslaved Africans emanated from ports within six main trading regions scattered along the western African coast: Senegambia, Sierra Leone, Gold Coast, Bight of Benin, Bight of Biafra, and West Central Africa (Thornton, 1998). According to Thornton (1998), in the eighteenth century, nearly 75% of enslaved peoples originated from ports in West Central Africa, the Bight of Benin or the Bight of Biafra. While ports in the other regions were also important embarkation regions for enslaved people, these three regions were the most substantial suppliers (Thornton, 1998). Though some African nations were complicit and facilitated the enslavement and trade of other African peoples (Nafafé, 2012), over time, these nations could not sustain equitability with their European counterparts. As articulated by Hall, "African power was undermined and factionalized by the slave trade and the warfare, social disorganization, and population loss that it involved" (Hall, 2005, p. 11). Thus, the impact of the trans-Atlantic slave trade was profound for Africans and African descendants on both sides of the Atlantic.

Genetic Perspectives on African Population History

Contemporary genetic studies of western African populations often consider the influence of both ancient and modern demographics on patterns of genetic variation within the region. In the most general terms, many of these studies address at least one of the following themes: (1) human origins and dispersals within Africa (e.g. Chan et al., 2019; Schlebusch et al., 2017); (2) present-day genetic diversity and admixture patterns (e.g. Busby et al., 2016; Petersen et al., 2013); and (3) local adaptations (e.g. Choudhury, Aron, Sengupta, Hazelhurst, & Ramsay, 2018; Martin et al., 2017; Patin et al., 2017). In addition to these themes, efforts are being made to integrate biological and cultural approaches to the study of African population history (Fortes-Lima, Mtetwa, & Schlebusch, 2021). The details of these studies highlight the power of combined approaches in shedding new light on the complex demographic history of the African continent and its central role in human evolution.

The question of where our species originated and subsequently spread around the globe has always been a central one in biological anthropology (Stringer, 1992). The now classic study by Rebecca Cann et al. (1987) was the first to use DNA (mitochondrial DNA) as opposed to classical markers to consider the origin of our species. Despite interpretive limitations of the study, the conclusion that the deepest phylogenetic root for human mitochondrial lineages was among African populations, inferring an African origin of our species, has held over time. Along with paleoanthropological and archaeological evidence, the current consensus dates the African origin of our species to 300 kya–250 kya (Henn, Steele, & Weaver, 2018; Schlebusch & Jakobsson, 2018; Schlebusch et al., 2017). Contemporary studies about human origins and early dispersals within Africa now seek to address the potential for introgression with other now-extinct hominin species (Durvasula & Sankararaman, 2020; Lorente-Galdos et al., 2019; Santander, Montinaro, & Capelli, 2019). Overall, these studies suggest a complex demographic history of our species and African populations characterised by variable levels of introgression, admixture, and periodic isolation. Additional data drawn from a wider range of both ancient and modern African populations will help to piece together a more comprehensive understanding of human evolutionary history within the African continent.

Given the African origin of our species, it follows that humans have a long history of migration within, outside, and back to the continent. Genomic perspectives offer nuanced and sometimes novel views on migration patterns as well as broad patterns of genetic variation within Africa (Fan et al., 2019). Genomic data can also complement or challenge linguistic, paleoanthropological, and archaeological lines of evidence regarding early human demography (Fortes-Lima et al., 2021). Over the last two decades, researchers have considered numerous African populations dispersed across the continent to make sense of migration patterns and diversity inclusive of population substructure (e.g. Bryc et al., 2010; Choudhury et al., 2020; Lachance et al., 2012; Tishkoff et al., 2009). Moreover, beginning with a 2015 publication by Gallego Llorente and colleagues, researchers have also turned to ancient DNA to consider long-term patterns of variation within and between African populations (Lipson et al., 2020; Llorente et al., 2015; Prendergast et al., 2019; Skoglund et al., 2017). These and related studies point to a complex population history within the continent in which the ancestors of Khoe-San populations form the oldest branches of humanity (divergence dates to around 270 kya), followed by the ancestors of Central African foragers, then the ancestors of eastern and western Africans (Hollfelder, Breton, Sjödin, & Jakobsson, 2021; Schlebusch & Jakobsson, 2018). Furthermore, these studies suggest variable demographic histories in which effective population sizes contract or expand and populations experience periods of relative isolation or admixture (Busby et al., 2016; Hollfelder et al., 2021). Researchers have also studied the signatures of migration (or expansions) and subsequent admixture of African populations, including the large-scale Bantu-speaker expansion. This expansion began about 5 kya and entailed the dispersion of peoples and associated farming and iron technologies from West Africa toward southern Africa. Genomic signatures of this expansion are marked by evidence of admixture between Bantu-speaking migrants and local hunter-gatherer populations (Choudhury et al., 2020; Choudhury, Sengupta, Ramsay, & Schlebusch, 2021; Tallman, Sungo, Saranga, & Beleza, 2022). Overall, studies about ancient human demography in Africa generally illustrate that both geography and language can but do not always act as barriers to gene flow within continental populations (Busby et al., 2016; Patin et al., 2017; Skoglund et al., 2017).

With regard to local adaptation, various studies have found genomic evidence of genetic adaptations to environmental stressors (Fan, Hansen, Lo, & Tishkoff, 2016; Pereira, Mutesa, Tindana, & Ramsay, 2021). Additionally, some of these adaptations are suggestive of having arisen from ancient introgression events (Xu et al., 2017). Many studies on genetic adaptation have found signals of positive selection for genes involved in pathways related to dietary and immune function. In particular, research around genetic adaptations to dietary factors related to dairy consumption as well as adaptations to malaria parasites highlight the importance of variable ecologies in shaping human genomes (Choudhury et al., 2018; Gomez, Hirbo, & Tishkoff, 2014; Patin et al., 2017). Other researchers have examined additional loci associated with different stressors like altitude, UV exposure, and height (Mulindwa et al., 2020; Rees, Castellano, & Andrés, 2020). These studies illustrate how convergent evolution can be a significant factor in shaping genetic diversity across populations. As additional genome-wide data from both modern and ancient populations becomes available, studies about genetic adaptation will continue to aid in increasing anthropological understandings of adaptive processes. Beyond anthropology, understanding these types of genetic adaptations also has significance in the biomedical sciences. In applied fields such as precision medicine and pharmacogenomics, genetic adaptations can be of use to understand disease processes as well as in the development of treatments (Gurdasani et al., 2015; Ibrahim & Rotimi, 2019; Rotimi et al., 2017). Within the continent, the development of these types of genomic applications to medicine and health is underway in large-scale consortiums such as H3Africa (Choudhury, Sengupta, Aron, & Ramsay, 2021; Mulder et al., 2018; Ramsay, Sankoh, & as members of the AWI-Gen study and the H3Africa Consortium, 2016) and the Consortium on Asthma among African-Ancestry Populations in the Americas (CAAPA) (Mathias et al., 2016). Collectively, these studies are contributing to a more profound understanding of the demographic history and changing genetic landscape of sub-Saharan Africa up until the colonial period and beyond, which is crucial for studying the genetic consequences of the trans-Atlantic slave trade and the connections between African-descendant populations in the Americas and ancestral source populations in Africa.

The Caribbean and the Trans-Atlantic Slave Trade

Similar to how genomic data have provided insights into early demographic histories of African populations, DNA studies have also contributed to the understanding of dispersions and movements of Indigenous Caribbean peoples and later Europeans and Africans. The origins and dispersions of Indigenous Caribbean peoples are relevant to this discussion of African demographic histories in the Caribbean because of the momentous impacts of European colonisation on both populations within the region. As discussed in Chapter 1 by Nägele and Nieves-Colón, an increasing number of ancient DNA studies are shedding light on the precolonial history of the Caribbean (Fernandes et al., 2021; Lalueza-Fox, Calderón, Calafell, Morera, & Bertranpetit, 2001; Lalueza-Fox, Gilbert, Martínez-Fuentes, Calafell, & Bertranpetit, 2003; Mendisco et al., 2014; Nägele et al., 2020; Nieves-Colón, 2019; Nieves-Colón et al., 2020; Schroeder et al., 2015, 2018; Shamoon-Pour, Fehren-Schmitz, Merriwether, & Reid, 2018). In conjunction with archaeological data, these studies provide increasingly detailed insights into the origins and demographic histories of Indigenous populations in the Caribbean, and future studies are likely to add even more detail to our understanding of interactions across the region throughout prehistory and the wider impact of European colonisation, which still remains to be explored using genetic data. Despite these open questions, the overall negative impact of European colonisation on Indigenous Caribbean peoples is not debated. The arrival of European colonists predicated rapid population declines and significant disruptions to existing political, social, and economic systems on the islands (Sued-Badillo, 2003). Though Indigenous populations suffered incredible losses, there were efforts to resist and persist in spite of European colonisation (Benn Torres, 2014). As is detailed below, the ongoing legacies of Indigenous Caribbean populations remain noteworthy, especially in particular regions of the Caribbean where Indigenous and African Caribbean communities continue to coexist.

The Indigenous peoples of the Caribbean were the first to be pressed into labouring to support European colonial endeavours. However, due to acts of genocide, abuse, and the introduction by colonists of diseases novel to the Americas, the Indigenous population was demographically diminished (Cameron, Kelton, & Swedlund, 2015; Newson, 1993; Valcárcel Rojas, Laffoon, Weston, Hoogland, & Hofman, 2020). Consequently, some colonists no longer saw Caribbean Indigenous populations as viable for large-scale exploitation and instead turned to Africa as a source of exploitative labour (Lampe, 2018). By the early sixteenth century, the trans-Atlantic slave trade was well established, and enslaved Africans were transported to the Caribbean by the thousands to aid in European efforts to colonise the region. Over the course of 400 years, an estimated 12.5 million Africans were uprooted and forcibly transported across the Atlantic to be pressed into labour building and producing wealth for European imperialist endeavours throughout the Americas.[1] In her extensive study of African ethnicities in the Americas, Gwendolyn Midlo Hall argues that despite the reality that enslaved peoples were often purposefully scattered across regions in the Americas to mitigate rebellion risk, in response to market demand as well as the logistics of maritime trade, "there were predominant countervailing patterns that tended to cluster new Africans from the same ethnicities and regions" within the Americas (Hall, 2005, p. 78–79). According to Hall, this clustering of ethnic groups proved to be influential in shaping the distribution of African and African descendants in the Americas. For example, Hall argues that early Spanish Caribbean colonies were populated with enslaved people from the Greater Senegambia region due to the relatively shortened voyage time between this region of Africa and the Caribbean. Peoples identified as Igbo in various historical records from the Bight of Biafra (the southeastern portion of Lower Guinea) were listed as having made up a significant proportion of the enslaved population in early nineteenth-century Trinidad, St. Lucia, St. Kitts, Berbice, and Anguilla (Hall, 2005). Related archival research by other scholars (Burnard, 2001; Higman, 1995; Stark, 2009) illustrates that the African origins of African descendants in the Caribbean were varied as a result of serial replacement and admixture events involving people from across western and central Africa. Such historical sources are critical for identifying potential source populations as well as contextualising genetic signatures in ancestry tests of African descendants in the Americas. However, due to the political, economic, and social factors surrounding the identification of African ethnicities of enslaved peoples, there are irregularities in these records that require careful interrogation and interpretation (Hall, 2010). Given these limitations inherent in archival sources, genetic perspectives offer some potential to more critically consider interpretations of the past.

Modern DNA Perspectives on the Population History of African Descendant Populations in the Caribbean

With the advent of genetic testing, researchers began to use DNA studies to address themes related to continental and local biogeographic origins as well as admixture among African descendants in the Caribbean and more broadly, in the Americas (Fortes-Lima, 2021; Fortes-Lima & Verdu, 2021). The earliest attempts to use DNA to unravel the biogeographic origins of African descendants in the Caribbean generally relied on uniparental ancestry tests. Uniparental tests trace ancestry along one parental lineage, where mitochondrial DNA follows the direct maternal lineage and the Y chromosome follows the paternal lineage. While these types of tests are somewhat informative about continental origins, sex-biased gene flow, and migration, they generally cannot be used to trace the biogeographic origins of individuals except in some cases as they lack resolution (Ely, Wilson, Jackson, & Jackson, 2006; Salas, Carracedo, Richards, & Macaulay, 2005; Salas et al., 2004). However, on the whole, uniparental ancestry tests have illustrated variable but significant amounts of maternal and paternal ancestry from Africa in modern Caribbean populations on a scale from around 50% to greater than 90%. They have also shown that European males and Indigenous American females have contributed to contemporary African descendant populations to much greater extents relative to European females and Indigenous American males (Benn Torres, Kittles, & Stone, 2007; Deason et al., 2012; Mendisco et al., 2019; Mendizabal et al., 2008;

Nieves Colón & Martínez Cruzado, 2012; Simms, Rodriguez, Rodriguez, & Herrera, 2010; Simms et al., 2012).

Beyond biogeographic continental ancestry, both uniparental and genome-wide markers have been used to explore the regional African origins of contemporary Caribbean peoples. This question about which African regions and, by extension, ethnicities, contemporary Caribbean peoples may trace their ancestry is a complex one and one that genetic perspectives do not necessarily resolve. As discussed elsewhere, there are several interpretive limitations and repercussions of applying genetic data to questions of social identities (Benn Torres, 2014; Bolnick et al., 2007). For example, one runs the risk of reinscribing biological notions of race and ethnicity when attempting to attach ethnic labels to genetic data. Additionally, the names used to identify ethnicities have changed over time and space, making it sometimes difficult to accurately link the present to a specific ancestral group. With these caveats in mind, researchers have attempted to use genetic data to fill gaps unaddressed by archival sources about who were the primary ancestors of contemporary African Caribbean peoples. In a 2013 study by Newman and colleagues, mtDNA was examined in 400 contemporary Jamaicans, revealing that an appreciable amount of African ancestry derived from populations found in Lower Senegambia (Gold Coast, the Bight of Benin, and the Bight of Biafra). However, they reported that the Gold Coast region, "was the largest single source of Jamaican slaves who arrived, remained and survived in Jamaica" (Newman, Deason, Pitsiladis, Salas, & Macaulay, 2013, p. 376).

Other researchers have turned to whole genome analyses, which can provide more detail on the genetic origins and admixture patterns of human populations. Beyond the use of genomic data, sophisticated statistical approaches such as ancestry-specific principle components analysis (ASPCA), patterns of linkage disequilibrium, and ancestry tract analysis have also been incorporated into studies to extrapolate richer details about the local ancestral origins, sex-biased gene flow, as well as the timing of admixture events (e.g. Fortes-Lima et al., 2017; Hellenthal et al., 2014; Keith et al., 2021; Moreno-Estrada et al., 2013). For example, Moreno-Estrada and colleagues, in their study about the ancestries of people from populations in Cuba, Puerto Rico, and Hispaniola, detected evidence of two major influxes from African peoples. The earlier pulse, circa mid-sixteenth century, was from coastal northwestern African peoples, and the latter, circa the late eighteenth century, was from people that are most similar to modern-day West-Central Africans. These influxes were congruent with the introduction and culmination of the trans-Atlantic slave trade (Moreno-Estrada et al., 2013). In a more recent study by Micheletti et al. (2020), genome-wide data from consumers of the commercial ancestry testing company 23&Me were examined to consider local African ancestry. They found that around 82% of consumers from French and British Caribbean islands have ancestry that derives from the Atlantic region of Africa: Nigeria; Senegambia (Gambia, Guinea, Guinea-Bissau, Senegal); Coastal West Africa (Sierra Leone, Ghana, Côte d'Ivoire, Liberia); and Congolese (Angola,

Democratic Republic of the Congo). For these consumers, Nigeria tended to be the most common source of African ancestry. They also noted that among consumers from the Spanish-speaking Caribbean, African ancestry also tended to come from the same four regions, though the majority of consumers from these islands had ancestries that traced to Senegambia (northwestern coastal Africa).

In a detailed follow-up to the Micheletti et al. study, Jackson (2021) provided the necessary historical context to Micheletti et al.'s genomic findings. In her paper, Jackson notes how historical factors related to economics impacted where enslaved Africans were sourced. In addition, Jackson notes differential treatments of enslaved populations brought about by the end of the trans-Atlantic trade also impacted patterns of genetic variation in the studied populations. Notably, she explains that the overrepresentation of peoples from Nigeria as ancestral to African North Americans is a matter of both biases in comparative databases as well as consequences of the intensification of intra-American slave trades. For example, within North America, enslaved Africans who happened to be from Nigerian populations became deeply embroiled in markets that commodified enslaved offspring to feed the market for enslaved peoples. This historical fact helps to explain the abundance of Nigerian ancestry observed in contemporary African North Americans. Similar factors could also have been at work in African Caribbean populations. Based on this historical contextualisation of the findings, it appears that a deeper understanding of the experiences of African descendants in the diaspora is critical for making sense of observed genetic variation. To put it in Jackson's own words, "Only when armed with the proper historical and cultural context data can we rely upon the genetics and genomics to faithfully reconstruct a people's population biology" (Jackson, 2021, p. 207).

In addition to learning more details about the biogeographical origins of African Caribbean and populations within the African diaspora, researchers have also attempted to quantify ancestral components within these groups by estimating admixture in order to learn more about the relative contributions of both African and non-African peoples to contemporary African Caribbean populations (Benn Torres et al., 2019; Keith et al., 2021; Mathias et al., 2016; Micheletti et al., 2020; Moreno-Estrada et al., 2013; Ongaro et al., 2019). Collectively, these studies show that African Caribbean populations are genetically admixed to varying degrees with peoples from across Africa, Eurasia, and the Americas. For example, in a review of several of the aforementioned papers, Fortes-Lima and Verdu (2021) report African ancestry to be around 89% in Jamaicans, 88% in Bajans, 70% in Garifuna from St. Vincent, and 38% in eastern Cubans. Similarly, Micheletti et al. (2020) reported African ancestry to be about 76% in Anglophone Caribbean populations, 71% in the Guianas, and 20% in the Hispanic Caribbean in their consumer sample population.

While admixture estimates vary across ancestral components, some broad patterns are consistent across the Caribbean regions. Namely, sex-biased admixture generally follows a pattern in which maternal lineages are enriched for primarily

African ancestry and paternal lineages are enriched for both African and European ancestry. This suggests that African females, males, and European males were the main contributors to contemporary African Caribbean populations. However, depending on the island, signatures of Indigenous gene flow from mainly females and, to a lesser extent, males are also present. In one island in particular, Trinidad, admixture from South Asia also figures prominently among contemporary populations (Benn Torres et al., 2019, 2015). This observation aligns with the history of Trinidad, in which indentured labourers from India migrated to the island in the post-emancipation period through the early twentieth century (Roopnarine, 2007). In the broadest terms, these patterns of sex-biased gene flow are insightful for understanding more about the biocultural impacts of colonisation. These observations are illustrative of how social structures regarding power dynamics, gender roles, mate choice, and the like shape patterns of genetic variation. In this case, genetic data serve the purpose of critically examining the intersections of social hierarchies and biology.

Genetic approaches have also been applied to specific subpopulations, such as maroons, who have unique demographic histories within the Caribbean. Briefly, the term "maroon" refers to people and their descendants who eschewed enslavement and established semi-autonomous communities. Wherever there was slavery in the Americas, there were maroon communities (de Groot, Christen, & Knight, 2003; Price, 1996). As expected, genetic studies with maroon communities tend to show similar origins as the adjacent greater populace. In some maroon communities, like the Noir Marron of French Guiana and Suriname, there appears to be very little evidence of non-African admixture (Brucato et al., 2010, 2009; Fortes-Lima et al., 2017). However, in some island communities, such as the Accompong Town Maroons of Jamaica and the Garifuna of St. Vincent, there also appears to be an ancestral influx from Indigenous American females and males (Benn Torres et al., 2015; Fuller & Benn Torres, 2018; Madrilejo, Lombard, & Torres, 2015). The finding of Indigenous paternal ancestry contradicts commonly held assertions among scholars that Indigenous men did not generally survive European colonisation in numbers enough to impact contemporary populations. Though generally rare within the English-speaking Caribbean and more common in Spanish-speaking islands, there is increasing evidence of Indigenous ancestry persisting within Caribbean communities. For example, Indigenous ancestry was detected among the First People's Community of Arima in Trinidad (Benn Torres et al., 2019), in eastern Cuba (Fortes-Lima et al., 2018; Marcheco-Teruel et al., 2014), and among communities adjacent to the Kalinago Reserve in Dominica (Benn Torres, Stone, & Kittles, 2013; Keith et al., 2021). Within Puerto Rico, several studies have shown appreciably high levels of Indigenous ancestry among contemporary populations (Martínez-Cruzado et al., 2005; Via et al., 2011; Vilar et al., 2014) as well as in preliminary studies with Afro-Puerto Ricans (Benn Torres & Torres Colón, 2020). The presence of Indigenous ancestry within living groups in the Caribbean, in addition to long-lasting community-based resurgence movements (Feliciano-Santos, 2021), suggests that notions of widespread extinction of Native Caribbean populations were overblown (see also Schroeder et al. 2018 for an ancient DNA perspective). Consequently, there may be more to learn about interactions of African and Indigenous Caribbean populations within the Caribbean. In sum, modern genetic studies of African Caribbean populations indicate that their African origins were variable and diverse. This is in agreement with historical sources; however, genetics can provide more nuanced understandings of how different African source populations contributed to contemporary populations in the Caribbean and beyond.

Ancient DNA Perspectives on the Population History of African Descendant Populations in the Caribbean

Ancient DNA (aDNA) has become another source of information about the demographic history of the Caribbean and the trans-Atlantic trade in enslaved Africans in recent years. Rather than having to rely on patterns of extant human genetic variation to infer past processes, the use of ancient DNA opens up a window into the past and enables us to test specific hypotheses directly. In this section, we draw on case studies to illustrate how ancient DNA research, along with archaeological and other types of evidence, has started to inform our understanding of the history of the trans-Atlantic slave trade and the movements of people across the Atlantic. Due to the relatively small number of published studies for the Caribbean, we will also draw on case studies from the other parts of the Americas that mirror processes in the Caribbean and serve to illustrate the potential and limitations of using ancient DNA to reconstruct historical processes relating to the trans-Atlantic slave trade.

One of the first large-scale projects to include ancient DNA as part of their research design was the New York African Burial Ground Project, which began in 1991 when construction workers discovered the skeleton of the first of more than 400 African and African-descendant men, women, and children who had been buried at the site during the seventeenth and eighteenth centuries (Blakey, 1998).[2] Following an extensive mitigation and public consultation process involving the local descendant community, the investigations that followed focused on four main topics, including origins, daily life, resistance to enslavement, and the development of African American identities (Mack & Blakey, 2004). The researchers on the project used a wide range of scientific methods to explore these topics, including isotopic and aDNA analysis. However, while groundbreaking in terms of their overall scientific approach and dedication to public engagement and community involvement, the project was limited by the DNA technologies available at the time and a lack of human genetic reference data from Africa (Ely et al., 2006; Jackson et al., 2009).

Thanks to advances in DNA sequencing technologies and laboratory methods, the field of ancient DNA has changed significantly since the African Burial Ground Project was completed. The single most important change has been the introduction of next-generation or high-throughput sequencing, which started to replace earlier forms of DNA sequencing technologies around 2005 (Metzker, 2010). Among other things, this technological innovation led to a flurry of activity trying to map human genetic variation in different regions across the globe, including Africa (Bryc et al., 2010; Tishkoff et al., 2009). Taking advantage of these new developments, Schroeder et al. (2015) used ancient DNA analyses to trace the genetic origins of three African individuals who had died on the Caribbean island of St. Martin sometime during the late seventeenth century. The three individuals – two men and a woman – had culturally modified teeth, and the strontium isotope signatures in their dental enamel suggested that they had been born in Africa, as opposed to St. Martin (Schroeder, Haviser, & Price, 2014). However, where in Africa they had come from was unclear. The ancient DNA analyses revealed that the three individuals had likely originated in different parts of Africa, as they had different genetic affinities with various modern reference populations from West and West Central Africa, including Bantu speakers from present-day Cameroon and non-Bantu speakers from present-day Ghana and Nigeria (Schroeder et al., 2015).

In a later study, Barquera et al. (2020) analysed the skeletal remains of three African individuals who had been buried in a mass grave near the Royal Hospital of San José de los Naturales in Mexico City, Mexico. Like the individuals from St. Martin, these three individuals also had culturally modified teeth, and closer inspection of the remains revealed a history of physical hardship and violence. For example, one of the individuals exhibited gunshot wounds, while another suffered from a poorly aligned complete fracture of the right fibula and tibia, which resulted in associated joint changes of the knee that must have severely affected their ability to move during their lifetime. In addition, several of the individuals displayed skeletal changes that result from intense physical activity or manual labour, including enthesopathies, as well as osteophytic lipping on several joint surfaces, indicating that the three individuals had been engaged in hard physical labour during life. Strontium isotope and ancient DNA analyses further suggested that the individuals had been born in West Africa, but for reasons that we will discuss below, it was not possible to narrow down their origins any further at this point (Barquera et al., 2020).

In 2023, three additional studies were published that provided new insights into the origins and lives of enslaved Africans and their descendants. The first study by Fleskes et al. (2023) focused on a group of burials from the Avery's Rest archaeological site (circa 1675–1725 CE) in Delaware (see Figure 3.1). The burials had been physically separated into two groups, and ancient DNA analyses revealed that they had been separated based on their genetic ancestry. The eight individuals in one group were of European descent, while the three individuals in the other had African genetic ancestry. European colonists imported thousands of Africans to the wider Chesapeake area, and it is estimated that around 500 enslaved Africans lived in Delaware by 1700. However, there is little to no information as to where in Africa they had come from. The DNA analyses revealed that, like the three African individuals from St. Martin, the African-descended individuals from the Avery's Rest site were related to different modern African reference populations,

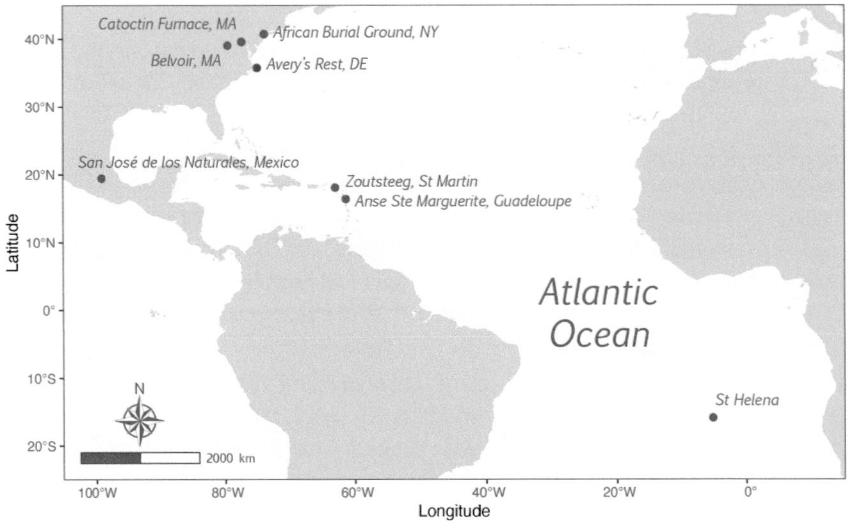

FIGURE 3.1 Map of the Atlantic world showing the locations of some of the historic sites discussed in this chapter (Barquera et al., 2020; Blakey, 1998; Fleskes et al., 2023; Guzmán-Solís et al., 2021; Harney, Micheletti et al., 2023; Lösch et al., 2015; Sandoval-Velasco et al., 2023; Schablitsky et al., 2019; Schroeder et al., 2015).

suggesting that they or their ancestors had originated in different parts of Africa. The genetic analyses also revealed a parent-offspring (father-son) relationship between two of the individuals with African ancestry. Together, the results shed new light on the origins and familial relationships of African-descended communities in late seventeenth- and early eighteenth-century North America.

The second study by Harney, Micheletti et al. (2023) focused on an enslaved African American community who lived and laboured at an iron-working site near Thurmont, Maryland (see Figure 3.1). The DNA analyses revealed that all but one of the individuals had African ancestry, with variable levels of European admixture. Drawing on genome-wide analyses focused on identical-by-descent (IBD) segments, they concluded that the individuals were most closely related to Wolof or Kongo speakers in West and West Central Africa, while their European ancestry could be traced to Great Britain and Ireland. They also found that the majority of the individuals belonged to five family groups and that some of them carried risk factors for sickle cell anaemia and glucose-6-phosphate dehydrogenase deficiency, genetic diseases that disproportionally affect African Americans today. Finally, using 23&Me's commercially derived database that contains genotype data of over 9 million customers, Harney, Micheletti et al. (2023) were able to identify nearly 42,000 close and distant relatives of the 27 people buried at the site. When looking at close relatives, Harney, Micheletti et al. (2023) observed the highest rates of sharing in Maryland, suggesting that at least some descendants of the individuals or their family members remained in or returned to the Maryland area after emancipation. Harney, Micheletti et al.'s (2023) study was the first to make use of a large-scale, commercially derived genetic database to link peoples of the past and their living descendants. Although this approach will undoubtedly become more commonplace in the future, it raises some important ethical questions that still have to be explored more fully regarding the ownership of genetic information and how, if at all, to disclose research results to study participants (Harney, Sirak et al., 2023).

The third study by Sandoval-Velasco et al. (2023) centred on the island of St. Helena in the South Atlantic. St. Helena is probably best known as the island to which French Emperor Napoleon Bonaparte was exiled in 1815 after his defeat at the Battle of Waterloo. What is less well known is that St. Helena played an important role in the history of the trans-Atlantic slave trade and its abolition. Strategically located in the middle of the South Atlantic, the island served as a way station and refuelling point for the Royal British Navy throughout the eighteenth and nineteenth centuries. Following the Slave Trade Act of 1807, St. Helena took on another role as a reception point for enslaved Africans who had been "liberated" from slave ships intercepted by the British Navy. Between 1840 and 1867, approximately 27,000 liberated Africans were brought to St. Helena. Written sources suggest that the majority of these individuals came from West Central Africa, with a smaller number possibly coming from Mozambique, but their precise origins are unknown. As part of a wider effort to commemorate and learn more about St. Helena's liberated Africans, Sandoval-Velasco et al. (2023) analysed genome-wide data from 20 individuals who had been among the thousands that had been brought to St. Helena. Sandoval-Velasco et al. (2023) compared their genomes with genotype data for over 3,000 present-day individuals from 90 populations across sub-Saharan Africa and concluded that the individuals most likely originated from different source populations within the general area between northern Angola and Gabon. They also found that the majority (17/20) of the individuals were male – consistent with a well-documented sex bias in the latter phase of the trans-Atlantic slave trade. All three studies illustrate the power of ancient DNA studies to expand our understanding of the biogeographic origins and genetic identities of enslaved Africans and their descendants and provide novel perspectives on the lives of people most impacted by the trans-Atlantic slave trade.

While human remains can be very informative, they are not the only source of ancient DNA. For example, Schablitsky et al. (2019) reported ancient DNA results recovered from two clay tobacco pipe stems (Figure 3.2) collected from a colonial-era tobacco plantation in Maryland, USA, where generations of enslaved Africans had lived and worked between 1736 and 1864. The pipe stems had been excavated in one of the slave quarters on the plantation. Ancient DNA analysis on one of them yielded DNA from a woman of African descent, which showed the characteristic damage patterns of ancient DNA, suggesting that it was indeed ancient or historical DNA as opposed to modern contamination (Schablitsky et al.,

FIGURE 3.2 A clay tobacco pipe stem, unearthed at the Belvoir plantation in Maryland, USA, was used for DNA extraction. The analysis revealed that the individual who smoked the pipe was female and shared close genetic ties with people from present-day Sierra Leone (Schablitsky et al., 2019).

2019). DNA has a tendency to bind to silica surfaces, which may explain why it was preserved in the clay pipes. Further analyses revealed that the woman was genetically closely related to present-day Mende from Sierra Leone, suggesting that she or her ancestors may have originated there and highlighting the link between Africa's "Rice Coast", the traditional rice-growing region between Senegambia and Ivory Coast, and the American South (Carney, 2002). Clay pipe stems are ubiquitous on colonial-era sites in the Caribbean (e.g. Brown, 2016; Fox, 2002; Handler & Norman, 2007; Orihuela & Viera, 2016), and this case study highlights the potential for similar studies in the region – without the need to sample and analyze human remains.

In cases where no human remains or artefacts are available, genetic information about historical individuals may also be recovered from the DNA of their living descendants. This was the approach by Jagadeesan et al. (2018), who used SNP chip data and whole genome sequences from living descendants of a historically known individual to reconstruct part of his genome in silico and to trace his maternal ancestry back to Africa. The individual was Hans Jonatan, who, according to historical sources, was born on the Caribbean island of St. Croix in 1784 to an enslaved African mother and a European father. In a tumultuous life, which is the subject of a biography by Gisli Palsson (2016), Hans Jonatan escaped to

Copenhagen, where he fought in the Napoleonic Wars, before emigrating to Iceland in 1802, where he settled down, married an Icelandic woman, and had two children. His children had children of their own, and today, almost 800 Icelanders can trace their ancestry to him. Like pieces in a jigsaw puzzle, Jagadeesan et al. (2018) managed to reconstruct 38% of Hans Jonatan's maternal genome and inferred that his mother (or her ancestors) probably originated from the region spanning present-day Benin, Nigeria, and northern Cameroon (Jagadeesan et al., 2018).

Another area where ancient DNA has started to make real contributions is the study of ancient pathogens (Barquera et al., 2020; Bravo-Lopez et al., 2020; Guzmán-Solís et al., 2021; Jaeger et al., 2013; Lösch et al., 2015). It is well known that European colonisation had devastating effects on the Indigenous populations of the Americas, and it has long been assumed that infectious diseases played a decisive role (Black, 1992; Crosby, 1972, 1976; Newson, 1993). However, most research in this area has focused on diseases introduced by Europeans and only recently started to factor in the trans-Atlantic slave trade as a vector for transmission (Pimenoff & Houldcroft, 2021). Hygienic conditions aboard slave ships were appalling, and although it is difficult to determine how many Africans died during the infamous Middle Passage, it is now estimated that over 1.8 million people lost their lives (Eltis & Richardson, 1997). Those who survived the transport to the colonies carried with them unique strains of disease-causing pathogens that introduced new diseases to the Americas, and it has been argued that it was these diseases – alongside those introduced by European colonists – that contributed to the drastic population declines of the Indigenous population, as they lacked the innate immunity to these novel diseases (Black, 1992; Crosby, 1972, 1976).

Early studies used PCR gene amplification to screen remains from 18th and 19th slave cemeteries in Guadeloupe and Brazil for the presence of *Mycobacterium tuberculosis* complex (MTBC) DNA (Jaeger et al., 2013; Lösch et al., 2015). In both cases, the researchers managed to obtain amplification products and concluded that some of the individuals who had been buried at the sites had been infected with tuberculosis. However, the detection of MTBC DNA in archaeological samples using PCR-based methods is complicated by the presence of mycobacteria other than *M. tuberculosis* (MOTT) in the burial environment, which can easily lead to false positive results (Campana et al., 2014; Müller et al., 2016). In fact, previous studies have shown that even weak homology with MOTT can cause false-positive PCR results for unrelated microbes (Borówka et al., 2019; Savelkoul et al., 2006). As a result, PCR-based detection methods have largely been superseded by high-throughput sequencing technologies that allow for the validation and fuller characterisation of ancient pathogens (Spyrou et al., 2019). For example, Barquera et al. (2020) used high-throughput sequencing to recover complete hepatitis B virus (HBV) and *Treponema pallidum* sub. *pertenue* genomes from the skeletal remains of three individuals of African descent from a colonial-era cemetery in Mexico City. Phylogenetic analysis of both pathogens revealed that they were closely related to strains circulating in current West African populations, highlighting the role of the trans-Atlantic slave trade in the introduction and spread of major human pathogens into the Americas. In a separate study, Guzmán-Solís et al. (2021) recovered three ancient human parvovirus B19 genomes and another ancient HBV genome from different African individuals from the same colonial-era cemetery in Mexico City. Interestingly, the team also found the same strain of parvovirus in the remains of a local individual with Native American ancestry, suggesting that the new virus may have spread to the local population and contributed to the disease outbreaks that affected millions of people. Both studies highlight the potential of ancient genomics to shed new light on the health status of enslaved Africans and their descendants and the spread of infectious pathogens with the trans-Atlantic slave trade.

Discussion and Concluding Remarks

Modern and ancient DNA studies have started to provide some fresh insights into the population history of the Caribbean and the demographic impact of the trans-Atlantic slave trade. However, the work is not without its challenges. In the past, genetic studies of African-descendant populations in the Americas were limited by inadequate reference data from modern and historical African populations – both in terms of geographic coverage as well as genomic coverage. Africa is the place with the greatest human genetic diversity on earth and, therefore, arguably the most interesting place to study from a genetic perspective. And yet, Africa has long been grossly underrepresented in studies of human genetic variation (Popejoy & Fullerton, 2016; Sirugo, Williams, & Tishkoff, 2019). There are over 2,000 ethno-linguistic groups in sub-Saharan Africa (Heine & Nurse, 2000), and only a handful of those have been included in genetic studies. This underrepresentation not only limits our understanding of human evolution and population history but also the successful application of precision medicine to individuals of African ancestry (Popejoy & Fullerton, 2016). However, as discussed above, the picture is slowly changing as more and more studies are starting to improve our understanding of African population history and present-day human genetic diversity in Africa, which will ultimately enable us to reconstruct the genetic links between Africa and African-descendant populations in the Americas more accurately (Abel & Schroeder, 2020; Tucci & Akey, 2019).

While it appears to be just a matter of time before the lack of reference data from Africa will be overcome, there are other fundamental issues that are much more difficult to resolve. For example, it has long been acknowledged that while genome-wide studies of human genetic variation in sub-Saharan Africa have revealed extensive genetic substructure (e.g. Bryc et al., 2010; Sengupta et al., 2021), it is unclear how long this structure has been in place, and it seems likely that the present-day diversity does not necessarily reflect historic patterns due to more recent population movements and displacements during the period of the trans-Atlantic slave trade and the colonial period (Jackson et al., 2009). As a result, it can happen that there is no close match between historic or modern genomes from the Americas and present-day African populations. For example, Barquera et al. (2020) found that none of the three African individuals from the mass grave at the site of the Royal Hospital of San José de los Naturales in Mexico City showed a specific genetic affinity with any of the modern African populations for which reference data are available. They concluded that this could be either due to the fact that the original source population was not represented in the reference dataset (i.e., it has not been genotyped or sequenced yet) or that it does not exist any longer as a distinct genetic grouping in Africa (Barquera et al., 2020).

It should also be recognised that while the genetic structure of human populations is very real, how we label these structures and clusters can be contentious (Abel & Schroeder, 2020; Benn Torres, 2019). The concept that is most often used to describe the pre-slavery identities of Africans forcibly displaced by the trans-Atlantic trade in the historical literature is "ethnicity" (e.g. Chambers, 2001; Hall, 2005). However, ethnic boundaries are often fluid and changeable, and it is clear that group membership is not solely, or even primarily, defined by shared ancestry or genetic kinship (Barth, 1998). Precolonial African societies generally comprised overlapping social networks that allowed for multiple group membership and flexible boundaries (Lentz, 1995). In fact, it has been argued that many contemporary ethnic groups are colonial "inventions" by European explorers and administrators who, in their attempts to map "native" peoples, depicted them as discrete, bounded entities rather than dynamic and interlocking networks (Southall, 1970; Vail, 1991). As MacEachern puts it,

> African ethnic units are not bounded, homogeneous monoliths either frozen in place since before A.D. 1492 or caromming around the continent like culture-bearing billiard balls. They are dynamic entities, manipulated in response to changes in the natural and social environment, and their composition can change very quickly indeed.
>
> *(MacEachern, 2000, p. 370)*

The fluid, permeable nature of ethnicity implies, however, that any attempt to recover historical African identities using genetic data is fraught with difficulties.

Another complicating factor when trying to disentangle the demographic histories of African-descendant populations in the Americas is that they are all – perhaps with the exception of some isolated maroon communities – admixed to a greater or lesser extent. While some scholars have argued that historical processes and migratory patterns led to a clustering of African ethnicities in the Americas that persisted over centuries (Hall, 2005), there is ample evidence for large movements of people within the Americas as a result of the Intra-American slave trade and later population movements. For example, Deyle (2005) has estimated that up to 1 million enslaved people were transported across the southeastern United States as a result of the domestic slave trade in the United States between 1790 and 1860. Later on, the "Great Migration" that occurred between 1916 and 1970 would bring an estimated 6 million African Americans from the American South to all parts of the Northeast and Midwest United States in one of the greatest demographic shifts in US history (Gregory, 2006). Similar population movements also occurred elsewhere in the Americas, such as Brazil (Bethell, 1970). Large migrations such as these likely obscured pre-existing local population structure in African-descendant communities as a result of admixture (Baharian et al., 2016; Micheletti et al., 2020).

The work on ancient DNA comes with its own set of challenges. The biggest problem is that ancient DNA does not preserve well in tropical settings as it degrades more quickly than in colder environments (Hofreiter et al., 2015; Nieves-Colón et al., 2018). Previous studies suggest that ancient DNA in the Caribbean degrades around 5–10 times faster than in more temperate settings like North America (Schroeder et al., 2015, 2018). In practice, this means that it can be very challenging to generate enough data for population genetic analysis and that many, especially older samples, might fail to yield any useful DNA at all. Having said that, there appears to be a large amount of variation in DNA preservation across the region, and, hopefully, future studies will be able to shed more light on this aspect in order to guide future studies and help researchers target specimens that are more likely to yield enough ancient DNA where destructive analysis is unavoidable. That aside, and given the poor preservation overall, it appears that targeted enrichment of ancient DNA libraries using hybridisation capture is a necessary step to reduce sequencing costs and to recover enough data for population genetic analysis (Fernandes et al., 2021; Nägele et al., 2020).

Despite the interpretive and practical challenges, genetic data have been leveraged by scholars and entrepreneurs alike to illuminate the African origins of African descendants (Abel, 2022; Benn Torres & Torres Colón, 2020; Nelson, 2016). The use of DNA to address questions surrounding African identities among African descendants carries the potential to attend to voids created and expanded by the inhumanity of chattel slavery. In the face of such history, there are many stakeholders that are motivated to resist the denigration of Africa, Africans, and African descendants (Hartman, 2008). Genetic ancestry, despite its limitations, therefore becomes one means of redress. The ability to use genetic material to reimagine the past and connect with broader communities, both past and present, is an alluring idea because it offers alternatives to existing narratives that were created to uphold racist justifications for slavery and discrimination (Fehler, 2011). The Direct to Consumer (DTC) ancestry testing market has responded to consumer demand for ancestry tests specific to African ancestry. There are a variety of commercial enterprises that provide genetic ancestry testing services to consumers who wish to investigate their genetic connections to presumed African ancestors (Nelson, 2008; Shriver & Kittles, 2004). More recently, researchers have already used data from DTC consumers to connect past and living people through shared genomic segments (Harney, Sirak et al., 2023). These efforts, however, are not without critics due to the aforementioned technological and interpretive limitations, as well as newly emerging ethical issues in contacting genetically related DTC consumers (Abel & Schroeder, 2020; Duster, 2011; Harney,

Sirak et al., 2023; Royal et al., 2010). Moreover, scholars have questioned the power of this technology to modify existing notions of race and identity among consumers, and Caribbean perspectives have been included in some of these studies (Abel, 2018; Benn Torres, 2020; Etkins, 2016; Nelson & Hwang, 2011; Peters, 2021). On the whole, researchers report that consumers' identities are not generally impacted by the results of ancestry tests. However, this finding requires an understanding of the historical, social, and economic context of the test taker as well as some understanding of the motivations for taking the ancestry test in the first place (Roth & Ivemark, 2018). Among Caribbean ancestry testing consumers specifically, testing seems to uphold existing ideas of hybridity rather than challenge it (Abel, 2018; Benn Torres, 2014). How these technologies will be used or set aside by various African-descendant communities as the DTC ancestry testing market grows and develops more sophisticated algorithms to infer finer-scale ancestry estimations remains to be seen.

As highlighted by the studies mentioned in this chapter, genomic data have introduced novel perspectives to the understanding of African and African-descendant experiences in the Caribbean and across the African diaspora more broadly. By utilising population genetic approaches, the intersections of biology and culture are evident and explanatory for making sense of the patterns of variation observed across human groups. As it pertains to the origins and demographic histories of African-descendant communities in the Caribbean, genomic perspectives have only just begun to illuminate the complexities of serial migrations, replacements, and admixture of African and African-descendant peoples within the region. As noted by Jackson (2021), these data cannot simply live in the laboratory. Instead, they must be contextualised with a comprehensive understanding of the histories and cultures of the people from which the data originate. By extension, this means that genomic researchers of both extant and past peoples should be cognizant of the implications and ethical obligations to studied communities. While a detailed discussion about the bioethics of genomic work is beyond the scope of this chapter, it is available elsewhere (Argüelles, Fuentes, & Yáñez, 2022; Bardill et al., 2018; Harney, Sirak et al., 2023; Marks, 2005; Matimba et al., 2022; Radin, 2018; Wagner et al., 2020). These and other related publications highlight the need for funding institutions, researchers, participants, and descendant communities to be in communication regarding the expectations, limitations, and entanglements of genomic research. This body of scholarship also addresses the importance of understanding the social, historical, economic, and political contexts of stakeholder communities. These issues are not unique to genetic anthropology but instead overlap with other branches of biological anthropology, inclusive of bioarchaeology and forensic anthropology. These fields are also in the midst of grappling with the legacies of scientific racism as well as the ethics and obligations associated with research on human remains (Association of Black Anthropologists, 2021; DiGangi & Bethard, 2021). These and related publications also highlight the increased call for protection of human remains, inclusive of ancient DNA, from minoritised communities akin to provisions present in the Native American Graves and Repatriation Act

(NAGPRA) (Dunnavant, Justinvil, & Colwell, 2021; Fox & Hawks, 2019; Squires & García-Mancuso, 2021). Other scholars have called for the integration of principles derived from public health, known as community-engaged research in genomic research (Hindorff et al., 2018; Vries et al., 2015). This approach places the needs of the study community at the centre of all stages of research. It aims to eliminate extractive and exploitative aspects amplified by social and power hierarchies between the researched, and the researcher (Mikesell, Bromley, & Khodyakov, 2013). Based on these types of ongoing controversies and discussions, it is clear that there must be continued dialogue around ethical contemporary genetic anthropological research and that these discussions must include perspectives representing the whole gamut of the discipline, its practitioners, and stakeholders.

Notes

1 All estimates relating to the trans-Atlantic slave trade are taken from the trans-Atlantic slave trade Database https://www.slavevoyages.org/assessment/estimates (accessed 8 March 2022).
2 Although the project is now also referred to as the Upper Manhattan African Burial Ground project, we have decided to use its original title for consistency. The final reports for the NYABG Project can be accessed at https://www.gsa.gov/africanburialground (accessed 8 March 2023).

References

Abel, S. (2018). Of African descent? Blackness and the concept of origins in cultural perspective. *Genealogy*, 2(1), 11. https://doi.org/10.3390/genealogy2010011

Abel, S. (2022). *Permanent markers: Race, Ancestry, and the body after the genome*. University of North Carolina Press. https://www.jstor.org/stable/10.5149/9781469665177_abel

Abel, S., & Schroeder, H. (2020). From country marks to DNA markers: The genomic turn in the reconstruction of African identities. *Current Anthropology*, 61 (Supplement 22), 1–12. https://doi.org/10.1086/709550

Argüelles, J. M., Fuentes, A., & Yáñez, B. (2022). Analyzing asymmetries and praxis in aDNA research: A bioanthropological critique. *American Anthropologist* 124, 130–140. https://doi.org/10.1111/aman.13692

Association of Black Anthropologists. (2021, April 28). *Collective statement concerning the possession and unethical use of remains.* https://aba.americananthro.org/collective-statement-concerning-the-possession-and-unethical-use-of-remains/

Baharian, S., Barakatt, M., Gignoux, C. R., Shringarpure, S., Errington, J., Blot, W. J., Bustamente, C.D., Kenny, E. E., Williams, S. M., Aldrich, M. C., & Gravel, S. (2016). The Great Migration and African-American genomic diversity. *PLoS Genetics*, 12(5), e1006059–27. https://doi.org/10.1371/journal.pgen.1006059

Bardill, J., Bader, A. C., Garrison, N. A., Bolnick, D. A., Raff, J. A., Walker, A., Malhi, R. S., & The Summer internship for Indigenous peoples in Genomics (SING) Consortium. (2018). Advancing the ethics of paleogenomics. *Science*, 360(6387), 384–385. https://doi.org/10.1126/science.aaq1131

Barquera, R., Lamnidis, T. C., Lankapalli, A. K., Kocher, A., Hernández-Zaragoza, D. I., Nelson, E. A., Zamora-Herrera, A.C., Ramallo, P., Bernal-Felipe, N., Immel, A., Bos,

K., Acuña-Alonzo, V., Barbieri, C., Roberts, P., Herbig, A., Kühnert, D., Márquez-Morfin, L., & Krause, J. (2020). Origin and health status of first-generation Africans from Early Colonial Mexico. *Current Biology* 30(11), 2078–2091.e11. https://doi.org/10.1016/j.cub.2020.04.002

Barth, F. (Ed.). (1998). *Ethnic groups and boundaries: The social organization of culture difference*. Waveland Press.

Benn Torres, J. (2014). Prospecting the past: Genetic perspectives on the extinction and survival of indigenous peoples of the Caribbean. *New Genetics and Society*, 33(1), 21–41. https://doi.org/10.1080/14636778.2013.873245

Benn Torres, J. (2019). Race, rare genetic variants, and the science of human difference in the post-genomic age. *Transforming Anthropology*, 27(1), 37–49. https://doi.org/10.1111/traa.12144

Benn Torres, J. (2020). Anthropological perspectives on genomic data, genetic ancestry, and race. *American Journal of Physical Anthropology*, 171 (Suppl 70), 74–86. https://doi.org/10.1002/ajpa.23979

Benn Torres, J., Kittles, R. A., & Stone, A. C. (2007). Mitochondrial and Y chromosome diversity in the English-speaking Caribbean. *Annals of Human Genetics*, 71(Pt 6), 782–790. https://doi.org/10.1111/j.1469-1809.2007.00380.x

Benn Torres, J., Martucci, V., Aldrich, M. C., Vilar, M. G., MacKinney, T., Tariq, M., Gaieski, J. B., Bharath Hernandez, R., Browne, Z. E., Stevenson, M., Walters, W., Schurr, T. G., & The Genographic Consortium. (2019). Analysis of biogeographic ancestry reveals complex genetic histories for indigenous communities of St. Vincent and Trinidad. *American Journal of Physical Anthropology*, 169(3), 482–497. https://doi.org/10.1002/ajpa.23859

Benn Torres, J., Stone, A. C., & Kittles, R. (2013). An anthropological genetic perspective on Creolization in the Anglophone Caribbean. *American Journal of Physical Anthropology*, 151(1), 135–143. https://doi.org/10.1002/ajpa.22261

Benn Torres, J., & Torres Colón, G. A. (2020). *Genetic ancestry: Our stories, our pasts*. Routledge. https://doi.org/10.4324/9780429398650

Benn Torres, J., Vilar, M. G., Torres, G. A., Gaieski, J. B., Bharath Hernandez, R., Browne, Z. E., Stevenson, M., Walters, W., Schurr, T. G., & The Genographic Consortium. (2015). Genetic diversity in the Lesser Antilles and its implications for the settlement of the Caribbean Basin. *PloS One*, 10(10), e0139192. https://doi.org/10.1371/journal.pone.0139192

Bethell, L. (1970). *The abolition of the Brazilian slave trade: Britain, Brazil and the slave trade question*. Cambridge University Press. https://doi.org/10.1017/CBO9780511759734

Black, F. L. (1992). Why did they die? *Science*, 258(5089), 1739–1740. https://doi.org/10.1126/science.1465610

Blakey, M. L. (1998). The New York African Burial Ground Project: An examination of enslaved lives, a construction of ancestral ties. *Transforming Anthropology*, 7(1), 53–58. https://doi.org/10.1525/tran.1998.7.1.53

Bolnick, D. A., Fullwiley, D., Duster, T., Cooper, R. S., Fujimura, J. H., Kahn, J., Kaufman, J. S., Marks, J., Morning, A., Nelson, A., Ossorio, P., Reardon, J., Reverby, S. M., & TallBear, K. (2007). The science and business of genetic ancestry testing. *Science*, 318(5849), 399–400. https://doi.org/10.1126/science.1150098

Borówka, P., Pułaski, Ł., Marciniak, B., Borowska-Strugińska, B., Dziadek, J., Żądzińska, E., Lorkiewicz, W., & Strapagiel, D. (2019). Screening methods for detection of ancient *Mycobacterium tuberculosis* complex fingerprints in next-generation sequencing data derived from skeletal samples. *GigaScience* 8(6), 1–14. https://doi.org/10.1093/gigascience/giz065

Bravo-Lopez, M., Villa-Islas, V., Rocha Arriaga, C., Villaseñor-Altamirano, A. B., Guzmán-Solís, A., Sandoval-Velasco, M., Wesp, J. K., Alcantara, K., López-Corral, A., Gómez-Valdéz, J., Mejía, E., Herrera, A., Meraz-Moreno, A., de la Luz Moreno-Cabrera, M., Moreno-Estrada, A., Nieves-Colón, M., Olvera, J., Pérez-Pérez, J, Højholt Iversen, K., … Ávila-Arcos, M. C. (2020). Paleogenomic insights into the red complex bacteria *Tannerella forsythia* in Pre-Hispanic and Colonial individuals from Mexico. *Philosophical Transactions of the Royal Society of London. Series B, Biological Sciences*, 375(1812), 20190580. https://doi.org/10.1098/rstb.2019.0580

Brown, B. (2016). Reexamining English clay pipes in captive African burials on the Island of Barbados. *Journal of African Diaspora Archaeology and Heritage*, 5(3), 245–262. https://doi.org/10.1080/21619441.2016.1245539

Brucato, N., Cassar, O., Tonasso, L., Tortevoye, P., Migot-Nabias, F., Plancoulaine, S., Guitard, E., Larrouy, G., Gessain, A., & Dugoujon, J.-M. (2010). The imprint of the slave trade in an African American population: Mitochondrial DNA, Y chromosome and HTLV-1 analysis in the Noir Marron of French Guiana. *BMC Evolutionary Biology*, 10, 314. https://doi.org/10.1186/1471-2148-10-314

Brucato, N., Tortevoye, P., Plancoulaine, S., Guitard, E., Sanchez-Mazas, A., Larrouy, G., Gessain, A., & Dugoujon, J.-M. (2009). The genetic diversity of three peculiar populations descending from the slave trade: Gm study of Noir Marron from French Guiana. *Comptes Rendus Biologies*, 332(10), 917–926. https://doi.org/10.1016/j.crvi.2009.07.005

Bryc, K., Auton, A., Nelson, M. R., Oksenberg, J. R., Hauser, S. L., Williams, S., Froment, A., Bodo, J.-M., Wambebe, C., Tishkoff, S. A., & Bustamante, C. D. (2010). Genome-wide patterns of population structure and admixture in West Africans and African Americans. *Proceedings of the National Academy of Sciences of the United States of America*, 107(2), 786–791. https://doi.org/10.1073/pnas.0909559107

Burnard, T. (2001). E pluribus plures: African ethnicities in seventeenth and eighteenth century Jamaica. *Jamaican Historical Review*, 21, 8–22.

Busby, G. B., Band, G., Si Le, Q., Jallow, M., Bougama, E., Mangano, V. D., Amenga-Etego, L. N., Enimil, A., Apinjoh, T., Ndila, C. M., Manjurano, A., Nyirongo, V., Doumba, O., Rockett, K. A., Kwiatkowski, D. P., Spencer, C. C. A., & Malaria Genomic Epidemiology Network. (2016). Admixture into and within sub-Saharan Africa. *eLife*, 5, e15266. https://doi.org/10.7554/eLife.15266

Cameron, C. M., Kelton, P., & Swedlund, A. C. (2015). *Beyond germs: Native depopulation in North America*. University of Arizona Press.

Campana, M. G., Robles García, N., Rühli, F. J., Tuross, N. (2014)., False positives complicate ancient pathogen indentifications using high-throughput shotgun sequencing. BMC Research Notes 7, 111. https://doi.org/10.1186/1756-0500-7-111

Cann, R. L., Stoneking, M., & Wilson, A. C. (1987). Mitochondrial DNA and human evolution. *Nature*, 325(6099), 31–36. https://doi.org/10.1038/325031a0

Carney, J. A. (2002). *Black rice: The African origins of rice cultivation in the Americas*. Harvard University Press.

Chambers, D. B. (2001). Ethnicity in the diaspora: The slave-trade and the creation of African "nations" in the Americas. *Slavery & Abolition*, 22(3), 25–39. https://doi.org/10.1080/714005214

Chan, E. K. F., Timmermann, A., Baldi, B. F., Moore, A. E., Lyons, R. J., Lee, S.-S., Kalsbeek, A. M. F., Petersen, D. C., Rautenbach, H., Förtsch, H. E. A., Riana Bornman, M. S., & Hayes, V. M. (2019). Human origins in a southern African palaeo-wetland and first migrations. *Nature*, 575(7781), 185–189. https://doi.org/10.1038/s41586-019-1714-1

Choudhury, A., Aron, S., Botigué, L. R., Sengupta, D., Botha, G., Bensellak, T., Wells, G., Kumuthini, J., Shriner, D., Fakim, Y. J., Ghoorah, A. W., Dareng, E., Odia, T., Falola, O., Adebiyi, E., Hazelhurst, S., Mazandu, G., Nyangiri, O. A., Mbiyavanga, M., … Hanchard, N. A. (2020). High-depth African genomes inform human migration and health. *Nature*, 586(7831), 741–748. https://doi.org/10.1038/s41586-020-2859-7

Choudhury, A., Aron, S., Sengupta, D., Hazelhurst, S., & Ramsay, M. (2018). African genetic diversity provides novel insights into evolutionary history and local adaptations. *Human Molecular Genetics*, 27(R2), R209–R218. https://doi.org/10.1093/hmg/ddy161

Choudhury, A., Sengupta, D., Aron, S., & Ramsay, M. (2021). The H3Africa consortium: Publication outputs of a Pan-African genomics collaboration (2013 to 2020). In C. Fortes-Lima, E. Mtetwa, & C. Schlebusch (Eds.), *Africa, the cradle of human diversity: Cultural and biological approaches to uncover African diversity* (pp. 257–304). Brill. https://doi.org/10.1163/9789004500228_011

Choudhury, A., Sengupta, D., Ramsay, M., & Schlebusch, C. (2021). Bantu-speaker migration and admixture in southern Africa. *Human Molecular Genetics*, 30(R1), R56–R63. https://doi.org/10.1093/hmg/ddaa274

Crosby, A. W. Jr. (1972). *The Columbian exchange: Biological and cultural consequences of 1492*. Greenwood Publishing.

Crosby, A. W. (1976). Virgin soil epidemics as a factor in the aboriginal depopulation in America. *The William and Mary Quarterly*, 33, 289–299. https://doi.org/10.2307/1922166

Deason, M. L., Salas, A., Newman, S. P., Macaulay, V. A., St A Morrison, E. Y., & Pitsiladis, Y. P. (2012). Interdisciplinary approach to the demography of Jamaica. *BMC Evolutionary Biology*, 12, 24. https://doi.org/10.1186/1471-2148-12-24

de Groot, S. W., Christen, C. A., & Knight, F. W. (2003). Maroon communities in the circum-Caribbean. In F. W. Knight (Ed.), *General history of the Caribbean* (pp. 169–193). Palgrave Macmillan. https://doi.org/10.1007/978-1-349-73770-3_6

Deyle, S. (2005). *Carry me back: The domestic slave trade in American life*. Oxford University Press. https://doi.org/10.1093/oso/9780195160406.001.0001

DiGangi, E. A., & Bethard, J. D. (2021). Uncloaking a lost cause: Decolonizing ancestry estimation in the United States. *American Journal of Physical Anthropology*, 175(2), 422–436. https://doi.org/10.1002/ajpa.24212

Diop, C. A., & Salemson, H. (2012). *Precolonial Black Africa*. Chicago Review Press.

Dunnavant, J., Justinvil, D., & Colwell, C. (2021). Craft an African American graves protection and repatriation act. *Nature*, 593(7859), 337–340. https://doi.org/10.1038/d41586-021-01320-4

Durvasula, A., & Sankararaman, S. (2020). Recovering signals of ghost archaic introgression in African populations. *Science Advances*, 6(7), eaax5097. https://doi.org/10.1126/sciadv.aax5097

Duster, T. (2011). Ancestry testing and DNA uses, limits, and caveat emptor. In S. Krimsky & K. Sloan (Eds.), *Race and the genetic revolution: Science, myth and culture* (pp. 99–115). Columbia University Press. https://doi.org/10.7312/krim15696-006

Eltis, D., & Jennings, L. C. (1988). Trade between Western Africa and the Atlantic world in the Pre-Colonial Era. *The American Historical Review*, 93(4), 936–959. https://doi.org/10.1086/ahr/93.4.936

Eltis, D., & Richardson, D. (Eds.) (1997). *Routes to slavery: Direction, ethnicity and mortality in the transatlantic slave trade*. Routledge.

Ely, B., Wilson, J. L., Jackson, F., & Jackson, B. A. (2006). African-American mitochondrial DNAs often match mtDNAs found in multiple African ethnic groups. *BMC Biology*, 4, 34. https://doi.org/10.1186/1741-7007-4-34

Etkins, O. (2016). Pick a box, any box: The influence of commercialized genetics on Caribbean biosociality. *Intersect: The Stanford Journal of Science, Technology, and Society,* 9(3), 1–11. https://ojs.stanford.edu/ojs/index.php/intersect/article/view/865

Falola, T., & Fleming, T. (2009). African civilizations: From the pre-colonial to the modern day. In R. Holton & W. R. Nasson (Eds.), *World civilization and history of human development,* vol 1 (pp. 123–140). Eolss Publishers.

Fan, S., Hansen, M. E. B., Lo, Y., & Tishkoff, S. A. (2016). Going global by adapting local: A review of recent human adaptation. *Science*, 354(6308), 54–59. https://doi.org/10.1126/science.aaf5098

Fan, S., Kelly, D. E., Beltrame, M. H., Hansen, M. E. B., Mallick, S., Ranciaro, A., Hirbo, J., Thompson, S., Beggs, W., Nyambo, T., Omar, S. A., Wolde Meskel, D., Belay, G., Froment, A., Patterson, N., Reich, D., & Tishkoff, S. A. (2019). African evolutionary history inferred from whole genome sequence data of 44 indigenous African populations. *Genome Biology*, 20(1), 82. https://doi.org/10.1186/s13059-019-1679-2

Fehler, B. O. (2011). (Re)constructing roots: Genetics and the "return" of African Americans to Ghana. *Mobilities*, 6(4), 585–600. https://doi.org/10.1080/17450101.2011.603949

Feliciano-Santos, S. (2021). *A contested Caribbean indigeneity: Language, social practice and identity within Puerto Rican Taíno activism*. Rutgers University Press. https://doi.org/10.36019/9781978808218

Fernandes, D. M., Sirak, K. A., Ringbauer, H., Sedig, J., Rohland, N., Cheronet, O., Mah, M., Mallick, S., Olalde, I., Culleton, B. J., Adamski, N., Bernardos, R., Bravo, G., Broomandkhoshbacht, N., Callan, K., Candilio, F., Demetz, L, Duffett Carlson, K. S., Eccles, L., … Reich, D. (2021). A genetic history of the pre-contact Caribbean. *Nature*, 590(7844), 103–110. https://doi.org/10.1038/s41586-020-03053-2

Fleskes, R. E., Owsley, D. W., Bruwelheide, K. S., Barca, K. G, Griffith, D. R., Cabana, G. S., & Schurr, T.G. (2023). Historical genomes elucidate European settlement and the African diaspora in Delaware. *Current Biology* 33(11), 2350–2358.E7. https://doi.org/10.1016/j.cub.2023.04.069

Fortes-Lima, C. A. (2021). Disentangling the impact of the transatlantic slave trade in African diaspora populations from a genomic perspective. In C. Fortes-Lima, E. Mtetwa, & C. Schlebusch (Eds.), *Africa, the cradle of human diversity: Cultural and biological approaches to uncover African diversity* (pp. 305–328). Brill. https://doi.org/10.1163/9789004500228_012

Fortes-Lima, C., Bybjerg-Grauholm, J., Marin-Padrón, L. C., Gomez-Cabezas, E. J., Bækvad-Hansen, M., Søholm Hansen, C., Le, P., Hougaard, D. M., Verdu, P., Mors, O., Parra, E. J., & Marcheco-Teruel, B. (2018). Exploring Cuba's population structure and demographic history using genome-wide data. *Scientific Reports*, 8(1), 11422. https://doi.org/10.1038/s41598-018-29851-3

Fortes-Lima, C., Gessain, A., Ruiz-Linares, A., Bortolini, M.-C., Migot-Nabias, F., Bellis, G., Moreno-Mayar, J. V., Restrepo, B. N., Rojas, W., Avendaño-Tamayo, E., Bedoya, G., Orlando, L, Salas, A., Helgason, A., Gilbert, M. T. P., Sikora, M., Schroeder, H., & Dugoujon, J.-M. (2017). Genome-wide ancestry and demographic history of African-descendant Maroon communities from French Guiana and Suriname. *American Journal of Human Genetics*, 101(5), 725–736. https://doi.org/10.1016/j.ajhg.2017.09.021

Fortes-Lima, C., Mtetwa, E., & Schlebusch, C. (Eds.). (2021). *Africa, the cradle of human diversity: Cultural and biological approaches to uncover African diversity*. Brill.

Fortes-Lima, C., & Verdu, P. (2021). Anthropological genetics perspectives on the transatlantic slave trade. *Human Molecular Genetics*, 30(R1), R79–R87. https://doi.org/10.1093/hmg/ddaa271

Fox, G. L. (2002). Interpreting socioeconomic changes in 17th-century England and Port Royal, Jamaica, through analysis of the Port Royal kaolin clay pipes. *International Journal of Historical Archaeology*, 6(1), 61–78. https://doi.org/10.1023/A:1014833426432

Fox, K., & Hawks, J. (2019). Use ancient remains more wisely. *Nature*, 572(7772), 581–583. https://doi.org/10.1038/d41586-019-02516-5

Fuller, H., & Benn Torres, J. (2018). Investigating the "Taíno" ancestry of the Jamaican Maroons: A new genetic (DNA), historical, and multidisciplinary analysis and case study of the Accompong Town Maroons. *Canadian Journal of Latin American and Caribbean Studies/Revue Canadienne des Études Latino-Américaines et Caraïbes*, 43(1), 47–78. https://doi.org/10.1080/08263663.2018.1426227

Gomez, F., Hirbo, J., & Tishkoff, S. A. (2014). Genetic variation and adaptation in Africa: Implications for human evolution and disease. *Cold Spring Harbor Perspectives in Biology*, 6(7), a008524. https://doi.org/10.1101/cshperspect.a008524

Gregory, J.N. (2006). *The southern diaspora: How the great migrations of Black and White southerners transformed America*. University of North Carolina Press.

Gurdasani, D., Carstensen, T., Tekola-Ayele, F., Pagani, L., Tachmazidou, I., Hatzikotoulas, K., Karthikeyan, S., Iles, L., Pollard, M. O., Choudhury, A., Ritchie, G. R. S., Xue, Y., Asimit, J., Nsubuga, R. N., Young, E. H., Pomilla, C., Kivinen, K., Rockett, K., Kamali, A., … Sandhu, M. S. (2015). The African Genome Variation Project shapes medical genetics in Africa. *Nature*, 517(7534), 327–332. https://doi.org/10.1038/nature13997

Guzmán-Solís, A. A., Villa-Islas, V., Bravo-López, M. J., Sandoval-Velasco, M., Wesp, J. K., Gómez-Valdés, J. A., de la Luz Moreno-Cabrera, M., Meraz, A., Solís-Pichardo, G., Schaaf, P., TenOever, B. R., Blanco-Melo, D., & Ávila Arcos, M. C. (2021). Ancient viral genomes reveal introduction of human pathogenic viruses into Mexico during the transatlantic slave trade. *eLife*, 10, e68612. https://doi.org/10.7554/eLife.68612

Hall, G. M. (2005). *Slavery and African ethnicities in the Americas: Restoring the links*. University of North Carolina Press.

Hall, G. M. (2010). Africa and Africans in the African diaspora: The uses of relational databases. *The American Historical Review*, 115(1), 136–150. https://doi.org/10.1086/ahr.115.1.136

Handler, J., & Norman, N. (2007). From West Africa to Barbados: A rare pipe from a plantation slave cemetery. *African Diaspora Archaeology Newsletter*, 10(3), 1–12.

Harney, E., Micheletti, S., Bruwelheide, K.S., Freyman, W.A., Bryc, K., Akbari, A., Jewitt, E., Comer, E., Gates Jr., H. L., Heywood, L., Thornton, J., Curry, R., Ancona Esselmann, S., Barca, K. G., Sedig, J., Sirak, K., Olalde, I., Adamski, N., Bernardos, R., ... 23andMe Research Team. (2023a). The genetic legacy of African Americans from Catoctin Furnace. *Science*, 381 (6657), eade4995. https://doi.org/10.1126/science.ade4995

Harney, E., Sirak, K., Sedig, J., Micheletti, S., Curry, R., Ancona Esselmann, S., & Reich, D. (2023). Ethical considerations when co-analyzing ancient DNA and data from private genetic databases. *American Journal of Human Genetics*, 110(9), 1447–1453. https://doi.org/10.1016/j.ajhg.2023.06.011

Hartman, S. (2008). *Lose your mother: A journey along the Atlantic slave route*. Farrar, Staus and Giroux.

Havik, P. J., & Green, T. (2012). Introduction: Brokerage and the role of Western Africa in the Atlantic world. In T. Green (Ed.), *Brokers of change: Atlantic commerce and culture in Pre-Colonial Western Africa, proceedings of the British Academy* (pp. 16–26). Oxford University Press. https://doi.org/10.5871/bacad/9780197265208.003.0001

Heine, B., & Nurse, D. (2000). *African languages: An introduction*. Cambridge University Press.

Hellenthal, G., Busby, G. B. J., Band, G., Wilson, J. F., Capelli, C., Falush, D., & Myers, S. (2014). A genetic atlas of human admixture history. *Science*, 343(6172), 747–751. https://doi.org/10.1126/science.1243518

Henn, B. M., Steele, T. E., & Weaver, T. D. (2018). Clarifying distinct models of modern human origins in Africa. *Current Opinion in Genetics & Development*, 53, 148–156. https://doi.org/10.1016/j.gde.2018.10.003

Higman, B. W. (1995). *Slave populations of the British Caribbean, 1807–1834*. University of the West Indies Press.

Hindorff, L. A., Bonham, V. L., Brody, L. C., Ginoza, M. E. C., Hutter, C. M., Manolio, T. A., & Green, E. D. (2018). Prioritizing diversity in human genomics research. *Nature Reviews Genetics*, 19(3), 175–185. https://doi.org/10.1038/nrg.2017.89

Hofreiter, M., Paijmans, J. L. A., Goodchild, H., Speller, C. F., Barlow, A., Fortes, G. G., Thomas, J. A., Ludwig, A., & Collins, M. J. (2015). The future of ancient DNA: Technical advances and conceptual shifts. *BioEssays*, 37(3), 284–293. https://doi.org/10.1002/bies.201400160

Hollfelder, N., Breton, G., Sjödin, P., & Jakobsson, M. (2021). The deep population history in Africa. *Human Molecular Genetics*, 30(R1), R2–R10. https://doi.org/10.1093/hmg/ddab005

Ibrahim, M. E., & Rotimi, C. N. (Eds.). (2019). *The genetics of African populations in health and disease*. Cambridge University Press. https://doi.org/10.1017/9781139680295

Izuagie, L. (2014). Pre-colonial regionalism in West Africa. *OFO: Journal of Transatlantic Studies*, 4(1), 21–42.

Jackson, F. L. C. (2021). So many Nigerians: Why is Nigeria overrepresented as the ancestral genetic homeland of legacy African North Americans? *American Journal of Human Genetics*, 108(1), 202–208. https://doi.org/10.1016/j.ajhg.2020.10.010

Jackson, F. L. C., Mayes, A., Mack, M. E., Froment, A., Keita, S. O. Y., Kittles, R. A., George, M., Shujaa, K., Blakey, M. L., & Rankin-Hill, L. M. (2009). Origins of the New York African Burial Ground Population: Biological evidence of geographical and macroethnic affiliations using craniometrics, dental morphology, and preliminary genetic analyses. In M. L. Blakey & L. M. Rankin-Hill (Eds.), *Skeletal biology of the New York African Burial Ground* (pp. 69–94). Howard University Press.

Jaeger, L. H., de Souza, S. M. F. M., Dias, O. F., Iñiguez, A. M., (2013). Mycobacterium tuberculosis complex in remains of 18th-19th century slaves, Brazil. *Emerging Infectious Diseases* 19(5), 837-839. https://doi.org/10.3201/eid1905.120193

Jagadeesan, A., Gunnarsdóttir, E. D., Ebenesersdóttir, S. S., Guðmundsdóttir, V. B., Thordardottir, E. L., Einarsdóttir, M. S., Jónsson, H., Dugoujon, J.-M., Fortes-Lima, C., Migot-Nabias, F., Massougbodji, A., Bellis, G., Pereira, L., Másson, G., Kong, A., Stefánsson, K., & Helgason, A. (2018). Reconstructing an African haploid genome from the 18th century. *Nature Genetics*, 50(2), 199–205. https://doi.org/10.1038/s41588-017-0031-6

Keith, M. H., Flinn, M. V., Durbin, H. J., Rowan, T. N., Blomquist, G. E., Taylor, K. H., Taylor, J.F., & Decker, J. E. (2021). Genetic ancestry, admixture, and population structure in rural Dominica. *PLoS One*, 16(11), e0258735. https://doi.org/10.1371/journal.pone.0258735

Lachance, J., Vernot, B., Elbers, C. C., Ferwerda, B., Froment, A., Bodo, J.-M., Lema, G., Fu, W., Nyambo, T. B., Rebbeck, T. R., Zhang, K., Akey, J. M., & Tishkoff, S. A. (2012). Evolutionary history and adaptation from high-coverage whole-genome sequences of diverse African hunter-gatherers. *Cell*, 150(3), 457–469. https://doi.org/10.1016/j.cell.2012.07.009

Lalueza-Fox, C., Luna Calderón, F., Calafell, F., Morera, B., & Bertranpetit, J. (2001). MtDNA from extinct Tainos and the peopling of the Caribbean. *Annals of Human Genetics*, 65(2), 137–151. https://doi.org/10.1046/j.1469-1809.2001.6520137.x

Lalueza-Fox, C., Gilbert, M. T. P., Martínez-Fuentes, A. J., Calafell, F., & Bertranpetit, J. (2003). Mitochondrial DNA from pre-Columbian Ciboneys from Cuba and the prehistoric colonization of the Caribbean. *American Journal of Physical Anthropology*, 121(2), 97–108. https://doi.org/10.1002/ajpa.10236

Lampe, A. (2018). Las Casas and African slavery in the Caribbean: A third conversion. In D. T. Orique O. P., & R. Roldán-Figueroa (Eds.), *Bartolomé de las Casas, O.P.: History, philosophy, and theology in the age of European expansion* (pp. 421–436). Brill. https://doi.org/10.1163/9789004387669_018

Lentz, C. (1995). "Tribalism" and ethnicity in Africa: A review of four decades of anglophone research. *Cahiers Des Sciences Humaines*, 31(2), 303–328.

Lipson, M., Ribot, I., Mallick, S., Rohland, N., Olalde, I., Adamski, N., Broomandkhoshbacht, N., Lawson, A. M., López, S., Oppenheimer, J., Stewardson, K., Asombang, R. N., Bocherens, H., Bradman, N., Culleton, B. J., Cornelissen, E., Crevecoeur, I., de Maret, P., Fomine, F. L. M., … Reich, D. (2020). Ancient West African foragers in the context of African population history. *Nature*, 577(7792), 665–670. https://doi.org/10.1038/s41586-020-1929-1

Llorente, M. G., Jones, E. R., Eriksson, A., Siska, V., Arthur, K. W., Arthur, J. W., Curtis, M. C., Stock, J. T., Coltorti, M., Pieruccini, P., Stretton, S., Brock, F., Higham, T., Park, Y., Hofreiter, M., Bradley, D. G., Bhak, J., Pinhasi, R., & Manica, A. (2015). Ancient Ethiopian genome reveals extensive Eurasian admixture throughout the African continent. *Science*, 350(6262), 820–822. https://doi.org/10.1126/science.aad2879

Lorente-Galdos, B., Lao, O., Serra-Vidal, G., Santpere, G., Kuderna, L. F. K., Arauna, L. R., Fadhlaoui-Zid, K., Pimenoff, V. N., Soodyall, H., Zalloua, P., Marques-Bonet, T., & Comas, D. (2019). Whole-genome sequence analysis of a Pan African set of samples reveals archaic gene flow from an extinct basal population of modern humans into sub-Saharan populations. *Genome Biology*, 20(1), 77. https://doi.org/10.1186/s13059-019-1684-5

Lösch, S., Kim, M.-R., Dutour, O., Courtaud, P., Maixner, F., Romon, T., Sola, C., & Zink, A. (2015) *Tuberculosis* 95(S1), S65–S68. https://doi.org/10.1016/j.tube.2015.02.006

MacEachern, S. (2000). Genes, tribes, and African history. *Current Anthropology*, 41(3), 357–384. https://doi.org/10.1086/300144

Mack, M. E., & Blakey, M. L. (2004). The New York African Burial Ground Project: Past biases, current dilemmas, and future research opportunities. *Historical Archaeology*, 38(1), 10–17. https://doi.org/10.1007/BF03376629

Madrilejo, N., Lombard, H., & Torres, J. B. (2015). Origins of marronage: Mitochondrial lineages of Jamaica's Accompong Town Maroons. *American Journal of Human Biology*, 27(3), 432–437. https://doi.org/10.1002/ajhb.22656

Marcheco-Teruel, B., Parra, E. J., Fuentes-Smith, E., Salas, A., Buttenschøn, H. N., Demontis, D., Torres-Español, M., Marín-Padrón, L. C., Gómez-Cabezas, E. J., Alvarez-Iglesias, V., Mosquera-Miguel, A., Martínez-Fuentes, A., Carracedo, A., Børglum, A. D., & Mors, O. (2014). Cuba: Exploring the history of admixture and the genetic basis of pigmentation using autosomal and uniparental markers. *PLoS Genetics*, 10(7), e1004488. https://doi.org/10.1371/journal.pgen.1004488

Marks, J. (2005). Your body, my property: The problem of colonial genetics in a postcolonial world. In L. Meskell & P. Pels (Eds.), *Embedding ethics* (pp. 29–45). Routledge. https://doi.org/10.4324/9781003085249

Martin, A. R., Lin, M., Granka, J. M., Myrick, J. W., Liu, X., Sockell, A., Atkinson, E. G., Werely, C. J., Möller, M., Sandhu, M. S., Kingsley, D. M., Hoal, E. G., Liu, X., Daly, M. J., Feldman, M. W., Gignoux, C., R., Bustamante, C. D., & Henn, B. M. (2017). An unexpectedly complex architecture for skin pigmentation in Africans. *Cell*, 171(6), 1340–1353.e14. https://doi.org/10.1016/j.cell.2017.11.015

Martínez-Cruzado, J. C., Toro-Labrador, G., Viera-Vera, J., Rivera-Vega, M. Y., Startek, J., Latorre-Esteves, M., Román-Colón, A., Rivera-Torres, R., Navarro-Millán, I. Y., Gómez-Sánchez, E., Caro-González, H. Y., & Valencia-Rivera, P. (2005). Reconstructing the population history of Puerto Rico by means of mtDNA phylogeographic analysis. *American Journal of Physical Anthropology*, 128(1), 131–155. https://doi.org/10.1002/ajpa.20108

Mathias, R. A., Taub, M. A., Gignoux, C. R., Fu, W., Musharoff, S., O'Connor, T. D., Vergara, C., Torgerson, D. G., Pino-Yanes, M., Shringarpure, S. S., Huang, L., Rafaels, N., Boorgula, M. P., Johnston, H. R., Ortega, V. E., Levin, A. M., Song, W., Torres, R., Padhukasahasram, B. … & Barnes, K. C. (2016). A continuum of admixture in the Western Hemisphere revealed by the African diaspora genome. *Nature Communications*, 7, 12522. https://doi.org/10.1038/ncomms12522

Matimba, A., Ali, S., Littler, K., Madden, E., Marshall, P., McCurdy, S., Nembaware, V., Rodriquez, L., Seeley, J., Tindana, P., Yakubu, A., de Vries, J., & H3Africa Ethics and Community Engagement Working Group. (2022). Guideline for feedback of individual genetic research findings for genomics research in Africa. *BMJ Global Health*, 7(1), e007184. https://doi.org/10.1136/bmjgh-2021-007184

Mendisco, F., Pemonge, M. H., Leblay, E., Romon, T., Richard, G., Courtaud, P., & Deguilloux, M. F. (2014). Where are the Caribs? Ancient DNA from ceramic period human remains in the Lesser Antilles. *Philosophical Transactions of the Royal Society of London. Series B, Biological Sciences*, 370(1660), 20130388–20130388. https://doi.org/10.1098/rstb.2013.0388

Mendisco, F., Pemonge, M.-H., Romon, T., Lafleur, G., Richard, G., Courtaud, P., & Deguilloux, M.-F. (2019). Tracing the genetic legacy in the French Caribbean islands: A study of mitochondrial and Y-chromosome lineages in the Guadeloupe archipelago. *American Journal of Physical Anthropology*, 170(4), 507–518. https://doi.org/10.1002/ajpa.23931

Mendizabal, I., Sandoval, K., Berniell-Lee, G., Calafell, F., Salas, A., Martínez-Fuentes, A., & Comas, D. (2008). Genetic origin, admixture, and asymmetry in maternal and paternal human lineages in Cuba. *BMC Evolutionary Biology*, 8, 213. https://doi.org/10.1186/1471-2148-8-213

Metzker, M. L. (2010). Sequencing technologies - the next generation. *Nature Reviews Genetics*, 11(1), 31–46. https://doi.org/10.1038/nrg2626

Micheletti, S. J., Bryc, K., Ancona Esselmann, S. G., Freyman, W. A., Moreno, M. E., Poznik, G. D., Shastri, A. J., 23andMe Research Team, Beleza, S., & Mountain, J. L. (2020). Genetic consequences of the transatlantic slave trade in the Americas. *American Journal of Human Genetics*, 107(2), 265–277. https://doi.org/10.1016/j.ajhg.2020.06.012

Mikesell, L., Bromley, E., & Khodyakov, D. (2013). Ethical community-engaged research: A literature review. *American Journal of Public Health*, 103(12), e7–e14. https://doi.org/10.2105/ajph.2013.301605

Mintz, S. W. (1989). *Caribbean transformations*. Columbia University Press.

Moreno-Estrada, A., Gravel, S., Zakharia, F., McCauley, J. L., Byrnes, J. K., Gignoux, C. R., Ortiz-Tello, P. A., Martínez, R. J., Hedges, D. J., Morris, R. W., Eng., C., Sandoval, K., Acevedo-Acevedo, S., Normal, P. J., Layrisse, Z., Parham, P., Martínez-Cruzado, J.

C., González Burchard, E., Cuccaro, M. L., … Bustamante, C. D. (2013). Reconstructing the population genetic history of the Caribbean. *PLoS Genetics*, 9(11), e1003925. https://doi.org/10.1371/journal.pgen.1003925

Mulder, N., Abimiku, A., Adebamowo, S. N., de Vries, J., Matimba, A., Olowoyo, P., Ramsay, M., Skelton, M. & Stein, D. J. (2018). H3Africa: Current perspectives. *Pharmacogenomics and Personalized Medicine*, 11, 59. https://doi.org/10.2147/pgpm.s141546

Mulindwa, J., Noyes, H., Ilboudo, H., Pagani, L., Nyangiri, O., Kimuda, M. P., Ahouty, B., Asina, O. F., Ofon, E., Kamoto, K., Kabore, J. W., Koffi, M., Ngoyi, D. M., Simo, G., Chisi, J., Sidibe, I., Enyaru, J., Simuunza, M., Alibu, P., … TrypanoGEN Research Group of the H3Africa Consortium. (2020). High levels of genetic diversity within Nilo-Saharan populations: Implications for human adaptation. *American Journal of Human Genetics*, 107(3), 473–486. https://doi.org/10.1016/j.ajhg.2020.07.007

Müller, R., Roberts, C. A., & Brown, T. A. (2016). Complications in the study of ancient tuberculosis: Presence of environmental bacteria in human archaeological remains. *Journal of Archaeological Science* 68, 5–11. http://doi.org/10.1016/j.jas.2016.03.002

Nafafé, J. L. (2012). Challenges of the African voice: Autonomy, commerce, and resistance in precolonial Western Africa. In T. Green (Ed.), *Brokers of change: Atlantic commerce and cultures in Pre-Colonial Western Africa* (pp. 71–88). Oxford University Press. https://doi.org/10.5871/bacad/9780197265208.003.0004

Nägele, K., Posth, C., Iraeta Orbegozo, M., Chinique de Armas, Y., Hernández Godoy, S. T., González Herrera, U. M., Nieves-Colón, M. A., Sandoval-Velasco, M., Mylopotamitaki, D., Radzeviciute, R., Laffoon, J., Pestle, W. J., Ramos-Madrigal, J., Lamnidis, T. C., Schaffer, W. C., Carr, R. S., Day, J. S., Arredondo Antúnez, C., Rangel Rivero, A., … Schroeder, H. (2020). Genomic insights into the early peopling of the Caribbean. *Science*, 369(6502), 456–460. https://doi.org/10.1126/science.aba8697

Nelson, A. (2008). Bio science: Genetic genealogy testing and the pursuit of African ancestry. *Social Studies of Science*, 38(5), 759–783. https://doi.org/10.1177/0306312708091929

Nelson, A. (2016). *The social life of DNA: Race, reparations, and reconciliation after the genome*. Beacon Press.

Nelson, A., & Hwang, J. W. (2011). Roots and revelation: Genetic ancestry testing and the YouTube generation. In L. Nakamura, & P. Chow-White (Eds.), *Race after the Internet* (pp. 271–290). Routledge. https://doi.org/10.4324/9780203875063

Newman, S. P., Deason, M. L., Pitsiladis, Y. P., Salas, A., & Macaulay, V. A. (2013). The West African ethnicity of the enslaved in Jamaica. *Slavery & Abolition*, 34(3), 376–400. https://doi.org/10.1080/0144039X.2012.734054

Newson, L. A. (1993). The demographic collapse of native peoples of the Americas, 1492–1650. In W. Bray (Ed.), *The meeting of two worlds: Europe and the Americas 1492–1650. Proceedings of the British Academy*, 81, 247–288. Oxford University Press.

Nieves-Colón, M. A. (2019). Ancient DNA and Caribbean archaeology. *The SAA Archaeological Record*, 19(2), 26–31.

Nieves Colón, M. A., & Martínez Cruzado, J. C. (2012). La contribución genética del África Sub Sahariana a la población de la República Dominicana según el estudio de su ADN mitocondrial. *Cuadernos de Investigación Histórica* 7, 28–41.

Nieves-Colón, M. A., Ozga, A. T., Pestle, W. J., Cucina, A., Tiesler, V., Stanton, T. W., & Stone, A. C. (2018). Comparison of two ancient DNA extraction protocols for skeletal remains from tropical environments. *American Journal of Physical Anthropology*, 166(4), 824–836. https://doi.org/10.1002/ajpa.23472

Nieves-Colón, M. A., Pestle, W. J., Reynolds, A. W., Llamas, B., de la Fuente, C., Fowler, K., Skerry, K. M., Crespo-Torres, E., Bustamente, C. D., & Stone, A. C. (2020). Ancient DNA reconstructs the genetic legacies of precontact Puerto Rico communities. *Molecular Biology and Evolution*, 37(3), 611–626. https://doi.org/10.1093/molbev/msz267

Ongaro, L., Scliar, M. O., Flores, R., Raveane, A., Marnetto, D., Sarno, S., Gnecchi-Ruscone, G. A., Alarcon-Riquelme, M. E., Patin, E., Wangkumhang, P., Hellenthal, G., Gonzalez-Santos, M., King, R. J., Kouvatsi, A., Balanovsky, O., Balanovska, E., Atramentova, L., Turdikulova, S., Mastana, S., … Montinaro, F. (2019). The genomic impact of European colonization of the Americas. *Current Biology*, 29(23), 3974–3986.e4. https://doi.org/10.2139/ssrn.3417946

Orihuela, J., & Viera, R. A. (2016). Clay tobacco pipes from a colonial refuse deposit in Fort San Severino, Matanzas Province, Cuba. *International Journal of Historical Archaeology*, 20(2), 378–404. https://doi.org/10.1007/s10761-016-0337-6

Palsson, G. (2016). *The man who stole himself: The slave odyssey of Hans Jonathan*. University of Chicago Press.

Parker, J., & Rathbone, R. (2007). *African history: A very short introduction*. Oxford University Press. https://doi.org/10.1093/actrade/9780192802484.001.0001

Patin, E., Lopez, M., Grollemund, R., Verdu, P., Harmant, C., Quach, H., Laval, G., Perry, G. H., Barreiro, L. B., Froment, A., Heyer, E., Massougbodji, A., Fortes-Lima, C., Migot-Nabias, F., Belis, G., Dugoujon, J.-M., Perira, J. B., Fernandes, V., Pereira, L., … Quintana-Murci, L. (2017). Dispersals and genetic adaptation of Bantu-speaking populations in Africa and North America. *Science*, 356(6337), 543–546. https://doi.org/10.1126/science.aal1988

Pereira, L., Mutesa, L., Tindana, P., & Ramsay, M. (2021). African genetic diversity and adaptation inform a precision medicine agenda. *Nature Reviews: Genetics*, 22(5), 284–306. https://doi.org/10.1038/s41576-020-00306-8

Peters, C. (2021). "Add up all my Black": Understanding race and genetic ancestry through critical interpersonal and family communication. *Review of Communication*, 21(3), 223–230. https://doi.org/10.1080/15358593.2021.1961851

Petersen, D. C., Libiger, O., Tindall, E. A., Hardie, R.-A., Hannick, L. I., Glashoff, R. H., Mukerji, M., Indian Genome Variation Consortium, Fernandez, P., Haacke, W., Schork, N. J., … Hayes, V. M. (2013). Complex patterns of genomic admixture within southern Africa. *PLoS Genetics*, 9(3), e1003309. https://doi.org/10.1371/journal.pgen.1003309

Pimenoff, V. N., & Houldcroft, C. J. (2021). How infectious diseases arrived in the colonial Americas. *eLife*, 10, e72791. https://doi.org/10.7554/elife.72791

Popejoy, A. B., & Fullerton, S. M. (2016). Genomics is failing on diversity. *Nature*, 538(7624), 161–164. https://doi.org/10.1038/538161a

Prendergast, M. E., Lipson, M., Sawchuk, E. A., Olalde, I., Ogola, C. A., Rohland, N., Sirak, K. A., Adamski, N., Bernardos, R., Broomandkhoshbacht, N., Callan, K., Culleton, B. J., Eccles, L., Harper, T., K., Lawson, A. M., Mah, M., Oppenheimer, J., Stewardson, Kl, Zalzala, F., … Reich, D. (2019). Ancient DNA reveals a multistep spread of the first herders into sub-Saharan Africa. *Science*, 365(6448), eaaw6275. https://doi.org/10.1126/science.aaw6275

Price, R. (Ed.). (1996). *Maroon societies: Rebel slave communities in the Americas* (3rd ed.). JHU Press.

Radin, J. (2018). Ethics in human biology: A historical perspective on present challenges. *Annual Review of Anthropology*, 47(1), 263–278. https://doi.org/10.1146/annurev-anthro-102317-045922

Ramsay, M., Sankoh, O.; as members of the AWI-Gen study and the H3Africa Consortium. (2016). African partnerships through the H3Africa Consortium bring a genomic dimension to longitudinal population studies on the continent. *International Journal of Epidemiology*, 45(2), 305–308. https://doi.org/10.1093/ije/dyv187

Rees, J. S., Castellano, S., & Andrés, A. M. (2020). The genomics of human local adaptation. *Trends in Genetics*, 36(6), 415–428. https://doi.org/10.1016/j.tig.2020.03.006

Roopnarine, L. (2007). *Indo-Caribbean indenture: Resistance and accommodation, 1838–1920*. University of the West Indies Press.

Roth, W. D., & Ivemark, B. (2018). Genetic options: The impact of genetic ancestry testing on consumers' racial and ethnic identities. *The American Journal of Sociology*, 124(1), 150–184. https://doi.org/10.1086/697487

Rotimi, C. N., Bentley, A. R., Doumatey, A. P., Chen, G., Shriner, D., & Adeyemo, A. (2017). The genomic landscape of African populations in health and disease. *Human Molecular Genetics*, 26(R2), R225–R236. https://doi.org/10.1093/hmg/ddx253

Royal, C. D., Novembre, J., Fullerton, S. M., Goldstein, D. B., Long, J. C., Bamshad, M. J., & Clark, A. G. (2010). Inferring genetic ancestry: Opportunities, challenges, and implications. *American Journal of Human Genetics*, 86(5), 661–673. https://doi.org/10.1016/j.ajhg.2010.03.011

Salas, A., Carracedo, A., Richards, M., & Macaulay, V. (2005). Charting the ancestry of African Americans. *American Journal of Human Genetics*, 77(4), 676–680. https://doi.org/10.1086/491675

Salas, A., Richards, M., Lareu, M. V., Scozzari, R., Coppa, A., Torroni, A., Macaulay, V., & Carracedo, Á. (2004). The African diaspora: Mitochondrial DNA and the Atlantic slave trade. *American Journal of Human Genetics*, 74(3), 454–465. https://doi.org/10.1086/382194

Sandoval-Velasco, M., Jagadeesan, A., Ramos-Madrigal, J., Ávila-Arcos, M. C., Fortes-Lima, C. A., Watson, J., Johannesdóttir, E., Cruz-Dávalos, D. I., Gopalakrishnan, S., Moreno-Mayar, J. V., Niemann, J., Renaud, G., Robson Brown, K. A., Bennett, H., Pearson, A., Helgason, A., Gilbert, M. T. P., & Schroeder, H. (2023). The ancestry and geographical origins of St Helena's liberated Africans. *American Journal of Human Genetics*, 110(9), 1590–1599. https://doi.org/10.1016/j.ajhg.2023.08.001

Santander, C., Montinaro, F., & Capelli, C. (2019). Searching for archaic contribution in Africa. *Annals of Human Biology*, 46(2), 129–139. https://doi.org/10.1080/03014460.2019.1624823

Savelkoul, P. H. M., Catsburg, A., Mulder, S., Oostendorp, L., Schirm, J., Wilke, H., van der Zanden, A. G. M., & Noordhoek, G. T. (2006). Detection of *Mycobacterium tuberculosis* complex with Real Time PCR: Comparison of different primer-probe sets based on the IS6110 element. *Journal of Microbiological Methods* 66(1), 177–180. http://doi.org/10.1016/j.mimet.2005.12.003

Schablitsky, J. M., Witt, K. E., Madrigal, J. R., Ellegaard, M. R., Malhi, R. S., & Schroeder, H. (2019). Ancient DNA analysis of a nineteenth century tobacco pipe from a Maryland slave quarter. *Journal of Archaeological Science*, 105, 11–18. https://doi.org/10.1016/j.jas.2019.02.006

Schlebusch, C. M., & Jakobsson, M. (2018). Tales of human migration, admixture, and selection in Africa. *Annual Review of Genomics and Human Genetics*, 19, 405–428. https://doi.org/10.1146/annurev-genom-083117-021759

Schlebusch, C. M., Malmström, H., Günther, T., Sjödin, P., Coutinho, A., Edlund, H., Munters, A. R., Vicente, M., Steyn, M., Soodyall, H., Lombard, M., & Jakobsson, M. (2017). Southern African ancient genomes estimate modern human divergence to 350,000 to 260,000 years ago. *Science*, 358(6363), 652–655. https://doi.org/10.1126/science.aao6266

Schroeder, H., Ávila-Arcos, M. C., Malaspinas, A.-S., Poznik, G. D., Sandoval-Velasco, M., Carpenter, M. L., Moreno-Mayar, J. V., Sikora, M., Johnson, P. L. F., Allentoft, M. E., Samaniego, J. A., Haviser, J. B., Dee, M. W., Stafford Jr., T. W., Salas, A., Orlando, L., Willerslev, E., Bustamante, C. C., & Gilbert, M. T. P. (2015). Genome-wide ancestry of 17th-century enslaved Africans from the Caribbean. *Proceedings of the National Academy of Sciences of the United States of America*, 112(12), 3669–3673. https://doi.org/10.1073/pnas.1421784112

Schroeder, H., Haviser, J. B., & Price, T. D. (2014). The Zoutsteeg three: Three new cases of African types of dental modification from Saint Martin, Dutch Caribbean. *International Journal of Osteoarchaeology*, 24(6), 688–696. https://doi.org/10.1002/oa.2253

Schroeder, H., Sikora, M., Gopalakrishnan, S., Cassidy, L. M., Maisano Delser, P., Sandoval Velasco, M., Schraiber, J. G., Rasmussen, S., Homburger, J. R., Ávila-Arcos, M., Allentoft, M. E., Morenso-Mayar, J. V., Renaud, G., Gómez-Carballa, A., Laffoon, J. E., Jopkins, R. J. A., Higham, T. F. G., Carr, R. S., Schaffer, W. C., … Willerslev, E. (2018). Origins and genetic legacies of the Caribbean Taino. *Proceedings of the National Academy of Sciences of the United States of America*, 115(10), 2341–2346. https://doi.org/10.1073/pnas.1716839115

Sengupta, D., Choudhury, A., Fortes-Lima, C., Aron, S., Whitelaw, G., Bostoen, K., Gunnink, H., Chousou-Polydouri, N., Delius, P., Tollman, S., Gómez-Olivé, F. X., Norris, S., Mashinya, F., Alberts, M., AWI-Gen Study, H3Africa Consortium, Hazelhurst, S., Schlebusch, C., M, & Ramsay, M. (2021). Genetic substructure and complex demographic history of South African Bantu speakers. *Nature Communications*, 12(1), 2080. https://doi.org/10.1038/s41467-021-22207-y

Shamoon-Pour, M., Fehren-Schmitz, L., Merriwether, D. A., & Reid, B. (2018). Mitochondrial DNA analysis of pre-contact human remains from the Red House Site. In B. A. Reid (Ed.), *An archaeological study of the Red House, Port of Spain, Trinidad and Tobago* (pp. 40–53). University of the West Indies Press.

Shriver, M. D., & Kittles, R. A. (2004). Genetic ancestry and the search for personalized genetic histories. *Nature Reviews: Genetics*, 5(8), 611–618. https://doi.org/10.1038/nrg1405

Simms, T. M., Rodriguez, C. E., Rodriguez, R., & Herrera, R. J. (2010). The genetic structure of populations from Haiti and Jamaica reflect divergent demographic histories. *American Journal of Physical Anthropology*, 142(1), 49–66. https://doi.org/10.1002/ajpa.21194

Simms, T. M., Wright, M. R., Hernandez, M., Perez, O. A., Ramirez, E. C., Martinez, E., & Herrera, R. J. (2012). Y-chromosomal diversity in Haiti and Jamaica: Contrasting levels of sex-biased gene flow. *American Journal of Physical Anthropology*, 148(4), 618–631. https://doi.org/10.1002/ajpa.22090

Sirugo, G., Williams, S. M., & Tishkoff, S. A. (2019). The missing diversity in human genetic studies. *Cell*, 177(1), 26–31. https://doi.org/10.1016/j.cell.2019.02.048

Skoglund, P., Thompson, J. C., Prendergast, M. E., Mittnik, A., Sirak, K., Hajdinjak, M., Salie, T., Rohland, N., Mallick, S., Peltzer, A., Heinze, A., Olalde, I., Ferry, M., Harney, E., Michel, M., Stewardson, K., Cerezo-Román, J. I., Chiumia, C., Crowther, A., … Reich, D. (2017). Reconstructing prehistoric African population structure. *Cell*, 171(1), 59–71.e21. https://doi.org/10.1016/j.cell.2017.08.049

Smith, R. S. (1989). *Warfare & diplomacy in Pre-colonial West Africa* (2nd ed.). Univeristy of Wisconsin Press.

Southall, A. W. (1970). The illusion of tribe. *Journal of Asian and African Studies*, 5(1–2), 28–50. https://doi.org/10.1163/15685217-90007037

Spyrou, M. A., Bos, K. I., Herbig, A., & Krause, J. (2019). Ancient pathogen genomics as an emerging tool for infectious disease research. *Nature Reviews Genetics* 20(6), 323–340. http://doi.org/10.1038/s41576-019-0119-1

Squires, K., & García-Mancuso, R. (2021). Desafíos éticos asociados al estudio y tratamiento de restos humanos en las ciencias antropológicas en el siglo XXI. *Revista Argentina de Antropología Biológica*, 23(2), 1–22. https://doi.org/10.24215/18536387e034

Stark, D. M. (2009). A new look at the African slave trade in Puerto Rico through the use of parish registers: 1660–1815. *Slavery & Abolition*, 30(4), 491–520. https://doi.org/10.1080/01440390903245083

Stringer, C. B. (1992). Replacement, continuity and the origin of *Homo sapiens*. In G. Brauer, F. H. Smith (Eds.), *Continuity or replacement* (pp. 9–24). CRC Press. https://doi.org/10.1201/9781003078968

Sued-Badillo, J. (2003). The Indigenous societies at the time of conquest. In J. Sued-Badillo (Ed.), *General history of the Caribbean: Volume I: Autochthonous societies* (pp. 259–291). Palgrave Macmillan US. https://doi.org/10.1007/978-1-349-73764-2_8

Tallman, S., Sungo, M. das D., Saranga, S., & Beleza, S. (2022). Whole-genome sequencing of Bantu-speakers from Angola and Mozambique reveals complex dispersal patterns and interactions throughout sub-Saharan Africa. *bioRxiv*, 2022.02.07.478793. https://doi.org/10.1101/2022.02.07.478793

Thornton, J. (1998). *Africa and Africans in the making of the Atlantic world, 1400–1800* (2nd ed.). Cambridge University Press. https://doi.org/10.1017/CBO9780511800276

Tishkoff, S. A., Reed, F. A., Friedlaender, F. R., Ehret, C., Ranciaro, A., Froment, A., Hirbo, J. B., Awomoyi, A. A., Bodo, J.-M., Doumbo, O., Ibrahim, M., Juma, A. T., Kotze, M. J., Lema, G., Moore, J. H., Mortensen, H., Nyambo, T. B., Omar, S. A., Powell, K., … Williams, S. M. (2009). The genetic structure and history of Africans and African Americans. *Science*, 324(5930), 1035–1044. https://doi.org/10.1126/science.1172257

Tucci, S. & Akey, J. M. (2019). The long walk to African genomics. *Genome Biology* 20, 130. https://doi.org/10.1186/s13059-019-1740-1

Vail, L. (Ed.). (1991). *The creation of tribalism in southern Africa*. University of California Press.

Valcárcel Rojas, R., Laffoon, J. E., Weston, D. A., Hoogland, M. L. P., & Hofman, C. L. (2020). Slavery of indigenous people in the Caribbean: An archaeological perspective. *International Journal of Historical Archaeology*, 24(3), 517–545. https://doi.org/10.1007/s10761-019-00522-x

Via, M., Gignoux, C. R., Roth, L. A., Fejerman, L., Galanter, J., Choudhry, S., Toro-Labrador, G., Viera-Vera, J., Oleksyk, T. K., Beckman, K., Ziv, E., Risch, N., González Burchard, E., & Martínez-Cruzado, J. C. (2011). History shaped the geographic distribution of genomic admixture on the island of Puerto Rico. *PloS One*, 6(1), e16513. https://doi.org/10.1371/journal.pone.0016513

Vilar, M. G., Melendez, C., Sanders, A. B., Walia, A., Gaieski, J. B., Owings, A. C., Schurr, T. G., & The Genographic Consortium. (2014). Genetic diversity in Puerto Rico and its implications for the peopling of the island and the West Indies. *American Journal of Physical Anthropology*, 155(3), 352–368. https://doi.org/10.1002/ajpa.22569

Vries, J. de, Abayomi, A., Littler, K., Madden, E., McCurdy, S., Oukem-Boyer, O. O. M., Seeley, J., Staunton, C., Tangwa, G., Tindana, P., Troyer, J., & The H3Africa Working

Group on Ethics. (2015). Addressing ethical issues in H3Africa research – The views of research ethics committee members. *The HUGO Journal*, 9(1), 1. https://doi.org/10.1186/s11568-015-0006-6

Wagner, J. K., Colwell, C., Claw, K. G., Stone, A. C., Bolnick, D. A., Hawks, J., Brothers, K. B., & Garrison, N. A. (2020). Fostering responsible research on ancient DNA. *American Journal of Human Genetics*, 107(2), 183–195. https://doi.org/10.1016%2Fj.ajhg.2020.06.017

Xu, D., Pavlidis, P., Taskent, R. O., Alachiotis, N., Flanagan, C., DeGiorgio, M., Blekman, R., Ruhl, S., & Gokcumen, O. (2017). Archaic hominin introgression in Africa contributes to functional salivary MUC7 genetic variation. *Molecular Biology and Evolution*, 34(10), 2704–2715. https://doi.org/10.1093/molbev/msx206

Deathways, Demography and Health

4

A SYNTHESIS OF BIOARCHAEOLOGICAL RESEARCH ON CARRIACOU, SOUTHERN GRENADINES

Jessica H. Stone, Scott M. Fitzpatrick, Scott Burnett and John Krigbaum

Introduction

Bioarchaeological research in the Caribbean has shed light on the health, lifeways, and mobility of precontact people in the region, spanning periods from initial human settlement through European arrival. However, in the southern Caribbean, comparatively little bioarchaeological work has been undertaken, especially in the Lesser Antilles. The island of Carriacou in the Grenadines has yielded skeletal remains from more than two dozen burials recovered from Grand Bay, a large Amerindian village site dating to between ca. AD 400–1300. Since work at Grand Bay began in 2003, bioarchaeological research has focused on the health, nutrition, and mortuary practices undertaken at the site (Burnett et al. in prep; Fitzpatrick et al. 2009; Osborne 2013; Reeves 2006; Stone 2011). However, an up-to-date synthesis that includes burials excavated during the 2011 and 2014 field seasons is warranted. Here, we present an updated synthesis on the bioarchaeology of Carriacou that includes results from excavations at the Grand Bay site and integrates individuals excavated from other sites across the island. Additionally, we present new results of light stable isotopes for a broader picture of Late Ceramic Age paleodiet on Carriacou and provide a preliminary exploration of mobility using radiogenic strontium isotope ratios from tooth enamel at Grand Bay.

Background

Carriacou is the largest (13 square miles; 33 km^2) and southernmost island in the Grenadines chain that stretches between Grenada and St. Vincent (Figure 4.1). While archaeologists had visited and worked at several sites on Carriacou beginning in the 1960s with Ripley and Adelaide Bullen (1972), much of the work was

DOI: 10.4324/9780429444449-7

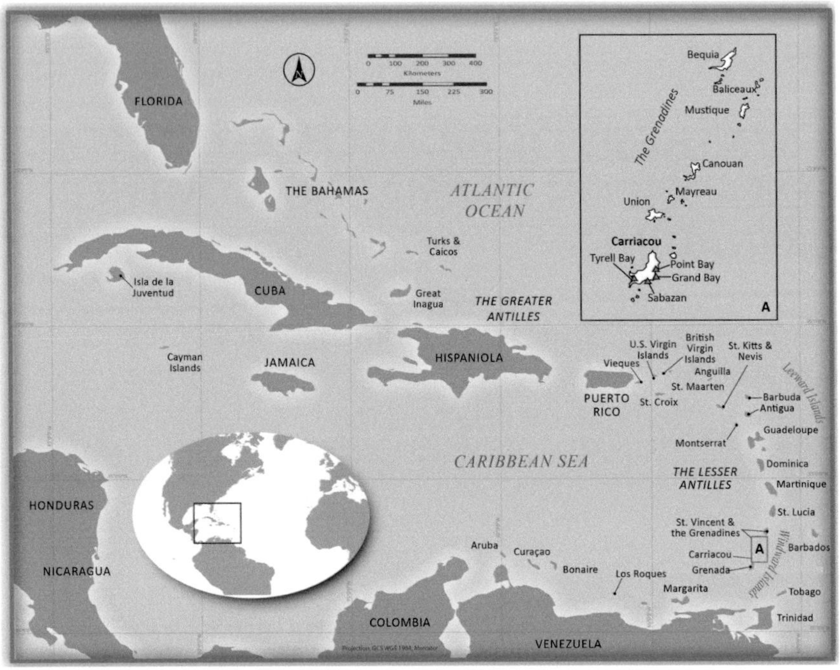

FIGURE 4.1 Map of the Caribbean with an inset of Carriacou (A) and site locations discussed in the text.

cursory, with no real attempt to systematically record the island's precontact sites or investigate any of them in detail. As such, the Carriacou Archaeological Project (CAP) was initiated in 2003 with one of the authors (SMF) and two colleagues, Quetta Kaye (University College London) and Michiel Kappers (InTerris - Registries/QLC Inc.). In this first year, the CAP surveyed and recorded or relocated a dozen sites with evidence of pre-contact occupation, including two major village sites—Grand Bay and Sabazan. The following year, in 2004, the project began large-scale excavation at the Grand Bay site and continued for five more field seasons (2005, 2007, 2008, 2011, 2014). During this time, Christina Giovas also began working at the Sabazan site as part of her Ph.D. research at the University of Washington, while a number of other sites, such as Point Bay and Harvey Vale, were investigated or had existing collections that were analyzed in more detail (see descriptions below).

The Grand Bay site is located on the east (windward) side of the island and stretches for more than 120 m along the coast. Excavation of 5×5 m units across the site in both midden and habitation areas revealed dense cultural deposits that in some areas extended more than a meter in depth. Numerous features were also recorded, including postholes, refuse pits, and human remains in a variety of contexts. Radiocarbon dates on charcoal, shell, and human bone from stratified deposits demonstrate that the site was occupied continuously from ca. AD 400–1300 during the terminal Saladoid period up through the Troumassan period (Fitzpatrick et al. 2009).

The Sabazan site is located a few km south of Grand Bay on the southern coast of Carriacou and also has remnants of a large Amerindian village. Given that both sites have succumbed to significant erosion over the years—and were much more expansive even a decade ago—a tremendous amount of cultural material has been lost to storms, tidal action, and sand mining (Fitzpatrick et al. 2006). Nonetheless, excavation of a series of 1×1 m units by Giovas (2013, 2018) recovered a rich collection of archaeological materials similar to that of Grand Bay, which was anchored by more than two dozen radiocarbon dates showing that peoples lived here between ca. AD 400–1400, more or less contemporaneous with Grand Bay (Giovas 2013).

The Harvey Vale area at Tyrell Bay is located on the southwestern side of Carriacou. Here, archaeological materials such as pottery have been found scattered over the years in different locations, but no substantial remnants of occupation have been recorded. Given that this area has long been used by sailors for mooring their boats because it provides the best protection of any bay on the island, and due to the density of residential and other structures, it is quite likely that much of the evidence for pre-contact settlement has been obscured or destroyed. The skeletal remains discussed below derive from fragments of three individuals that were recovered during the construction of a cistern near a private residence and submitted to the Carriacou Historical Society Museum.

The Point Bay site is located about 2 km north of Grand Bay and was first recorded in 2011 when a local farmer notified the CAP of human remains exposed on an eroded surface. Recovery of that burial, along with several others, took place in 2011 and 2014, but no other excavation was done. Based on examination of the area and the eroding coastal profile, deposits were thin (>20 cm) with faunal refuse (mostly shell) and pottery that likely dates back no earlier than the Late Ceramic Age based on observed ceramic types.

Bioarchaeology on Carriacou

Altogether, over two dozen individuals have been recovered from prehistoric contexts on Carriacou, with the majority having been found at Grand Bay. Additional sites yielding human remains include Sabazan (n = 2), Tyrell Bay (Harvey Vale) (n = 3), and Point Bay (n = 4). The majority of burials have been identified due to partial exposure resulting from coastal erosion or were salvaged during construction projects on the island; as a result, many are fragmented, incomplete, or could not be recovered for analysis (Fitzpatrick et al. 2009; Giovas et al. 2018). However, sufficient material has been recovered for bioarchaeological analyses of diet and health, mortuary practices, and mobility that reflect a comprehensive picture of early lifeways on Carriacou. Here, we present a synthesis of previous and current osteological research on burials recovered from across Carriacou, with a focus on Grand Bay.

Demography

The demography of the skeletal assemblage from Carriacou demonstrates a bias toward the recovery of adults and adolescents, which represents the overwhelming majority of the assemblage. Among individuals for whom sex could be estimated, males and females are represented relatively equally. What is most noticeable, however, is a paucity of infant and younger juvenile remains with the exception of two individuals: one classified as fetal or perinatal and a second child. Given the highly variable preservation and level of fragmentation often observed on Carriacou, especially at Grand Bay, the lack of juvenile remains may be due to preservation or recovery bias. Alternatively, there is a possibility of differential mortuary treatment for juveniles, which is supported by the presence of a similar demographic pattern observed at other sites in the Lesser Antilles (see discussion in Burnett et al. in prep).

Mortuary Practices on Carriacou

Mortuary patterns among adults and adolescents, however, are well documented on Carriacou and exhibit a variety of burial positions and contexts. A review of mortuary practices and burial taphonomy at Grand Bay has revealed an apparent pattern of sex-based burial positions that is similar to what has been observed elsewhere in the Lesser Antilles, whereby females are more likely to be interred laying on their backs while males tend to be interred in an upright and seated or semi-seated position (Burnett et al. in prep; Hoogland and Hofman 2013).

Burials have been recovered from a variety of contexts. A large portion of the remains recovered at Grand Bay, as well as a burial recovered from Point Bay and a burial feature identified at Sabazan, were exposed as surface remains due to high rates of coastal erosion, particularly at Grand Bay, where ~1 m of shoreline was lost each year, primarily due to sand mining by locals (Fitzpatrick et al. 2006). The recent cessation of this activity around 2014 seems to be resulting in some stabilization of the shoreline. Additional burials have been excavated from midden contexts, but the majority identified were associated with discrete subsurface features. While many appear to have been used exclusively for mortuary activity, others seem to have had a separate primary function (e.g., postholes for structure supports) and were later adapted for burial use (Figure 4.2). The distribution of burial features across the site of Grand Bay does not suggest a spatially distinct space for interments but instead may have been associated with specific habitation structures (Osborne 2013).

Burial positions are variable, but Burnett et al. (in prep) note that the majority of individuals encountered appear to have flexed legs at varying degrees, which has been observed at Grand Bay, Sabazan, and Point Bay (Giovas 2018; Osborne 2013). Otherwise, positions varied by sex, as noted above, while burial orientation was found to be variable. Interments represent mostly primary burials, but there is

FIGURE 4.2 Illustration reconstructing the use of postholes for human burials, a common mortuary tradition observed at Grand Bay. Illustration by John Swogger.

evidence for the possible reopening of burial features to add secondary burials later in time (discussed below), which parallels larger regional patterns.

Grave goods have been associated with burials from Grand Bay and Point Bay, though few exhibit clear association with a mortuary context. Two burial features, and possibly a third from Grand Bay, are associated with sea turtle (*Cheloniidae* sp.) remains. Other possible grave goods include spindle whorls, shell adzes, conch (*Lobatus* sp.) shells, and stones. Additionally, there is a high likelihood that perishable items may have been included and other items may have been disturbed or lost due to coastal or surface erosion and subsequent exposure. The taphonomy of multiple burial features at Grand Bay supports this possibility, as the tightly flexed nature of multiple individuals suggests that they may have been interred wrapped in textiles, possibly hammocks, as described in ethnohistoric accounts (Burnett et al. in prep).

A number of burials exhibit complex and possibly unique mortuary contexts, the majority of which were identified at Grand Bay. These are briefly described here but are presented in more detail by Burnett et al. (in prep) and Osborne (2013). Burial feature F0164 contains the primary interment of a male individual who was likely placed flexed and sitting upright, with infilling of the burial feature occurring later in time. Stratigraphic changes superior to the primary burial suggest the feature was reopened later in time, and additional skeletal material in the form of a complete adult cranium exhibiting artificial cranial deformation (parallel fronto-occipital) and disarticulated elements belonging to a fetal or perinatal individual were added to the interment. Similarly, feature F0177 contained the possible primary interment of an adult individual who could have been placed flexed and sitting upright in an open space but was later displaced. Elements of a lower limb belonging to a second individual were found in association and may represent another example of later secondary burial. Remains of a sea turtle, a spindle whorl, and charcoal and ash deposits, possibly representative of post-burial fires described in the ethnohistoric record, are also associated with the feature (Burnett et al. in prep; Kaye et al. 2012; Osborne 2013). Feature F0180 is a unique example of a clearly associated and intentional inclusion of grave goods in the form of four ceramic vessels placed surrounding the flexed burial. Lastly, at Point Bay, burial feature F0001 contained the remains of a single young adult female interred in a flexed position with her head facing the ground. The remains were surrounded by large stones, including one possibly placed on the lap and a second held in the arms, suggesting that these stones may have been used with the intention of holding the burial in place and implying that F0001 may contain the remains of a deviant individual. Alternatively, this individual may have been interred flexed and sitting upright, a burial position commonly encountered on Carriacou, with the stones falling inward post-deposition. Radiocarbon dates also place the individual toward the terminal occupation of nearby Grand Bay (AD 1400–1450). Therefore, this unique burial context could also represent a burial tradition dating to this later time period (Osborne 2013).

Skeletal Trauma and Pathology

Evidence for skeletal trauma and pathology on Carriacou is extremely limited. A total of five individuals from Grand Bay exhibit trauma in the form of fractures affecting the femur, fibula, humerus, ulna, radius, and clavicle (Fitzpatrick et al. 2009; Kraan n.d.; Reeves 2006). The distribution of these fractures suggests they are likely the result of accidental injury rather than interpersonal trauma (Fitzpatrick et al. 2009). Similarly, there appears to be evidence for a variety of pathological conditions, but all appear limited to discrete cases affecting one or a few individuals. For example, active periostitis was identified on the internal surface of the ribs associated with two adolescent individuals from Grand Bay, including on the sternum of one of the affected individuals. The location of the periosteal lesions suggests an inflammatory response to infection in both individuals and is perhaps indicative of a respiratory illness (Fitzpatrick et al. 2009). Additionally, an adolescent excavated from Grand Bay in 2014 exhibits localized porosity on the superior portion of the occipital in combination with evidence for craniosynostosis in the form of a nearly obliterated sagittal suture. The remaining cranial sutures are open, and the overall shape of the cranial vault, although somewhat fragmentary, suggests vertical fronto-occipital cranial deformation, which is a known practice across the Caribbean and has been observed previously at Grand Bay and elsewhere in the Lesser Antilles (Duijvenbode 2017). Therefore, the cranial pathologies observed in this individual could be effects of artificial cranial modification rather than disease or developmental processes. At Tyrell Bay, an individual displayed evidence of lytic lesions on the endocranial surface of the frontal and parietals, as well as localized areas on the limbs of the arms and legs (Reeves 2006).

The identification of degenerative changes is also noticeably limited and has only been identified from a few individuals from Grand Bay, fragments representing a single individual from Sabazan, and an individual from Point Bay. One individual excavated from Grand Bay in 2014 with more widely distributed evidence for degenerative changes possesses some minor lipping on the margins of the thoracic vertebral bodies and trochlea of the talus, but generally, degenerative changes appear mild and irregular in the form of lipping on one or a few elements. This low frequency can likely be attributed to the high prevalence of younger adult individuals present in the assemblage and may also be due to preservation biases where articular surfaces are often broken or eroded.

Analysis of Dental Remains

Because many of the burials on Carriacou are known primarily from dental remains, a comprehensive study of dental pathology, including linear enamel hypoplasia (LEH), caries rates, attrition, and patterns of antemortem dental chipping on remains from Grand Bay, Point Bay, and Tyrell Bay, was undertaken to examine the health and nutrition of Late Ceramic Age inhabitants (Burnett 2009; Heller and Burnett 2011; Stone 2011). LEH, an indicator of physiological stress occurring during enamel formation, was found to occur at a relatively high rate (affecting 10 of 25 individuals examined, or 40% of the dental assemblage), with LEH events most frequently occurring between the ages of two and four years. High frequencies of LEH are observed at other sites in the region, including Maisabel on Puerto Rico (28.13%) and Tutu on St. Thomas (21.7%). A region-wide study found an overall frequency of 8.52%, notably lower than what is observed on Carriacou (Larsen et al. 2002; Mickleburgh 2013). Regionally, a pattern of higher frequencies of LEH during the Late Ceramic Age can be observed, suggesting that regional conditions contributing to physiological childhood stress may have occurred during this time period. Given that the majority of burials on Carriacou date to the Late Ceramic Age, the high frequency of LEH may be another example of this pattern, though we must acknowledge that the comparison of macroscopic LEH analyses across different studies is notoriously challenging (Mickleburgh 2013).

The prevalence of LEH events affecting individuals between the ages of two and four could be due to nutritional shifts or compromised immune systems often observed as a result of the weaning process, which is a commonly cited reason for enamel disruption at this age (King et al. 2005; Moggi-Cecchi et al. 1994). Maize is a common food used for weaning, but no direct evidence for maize cultivation or consumption has yet been identified on Carriacou. Evidence suggests other softer carbohydrates, such as manioc or sweet potato, also common cultivars in the region, could have served as weaning foods, and the dietary shift and associated loss of nutrients during the weaning process may explain physiological stress associated with LEH formation.

Reliance on these starchy carbohydrates is also echoed in the analysis of caries frequencies. Carious lesions were identified at a noticeably high rate (30.1% of Grand Bay permanent teeth and 23.5% of Tyrell Bay permanent teeth) on Carriacou, which is even more unexpected given the young adult age of many of the individuals from Grand Bay in particular. Regionally, caries rates are highly variable, but when compared to a region-wide study, Grand Bay remains among the sites with the highest caries frequencies (Mickleburgh 2016). The combination of these high caries rates combined with rates of dental attrition, which are surprisingly low (mean Smith wear scores of 4.7 and 5.1 for the maxillary first incisor and molar, respectively) compared to other coastal sites suggests that carbohydrates or other processed foods with sticky consistencies were dietary staples (Stone et al., in prep). Ceramic griddles, evidence for the cooking of manioc and other plant foods, such as maize, have been recovered from Grand Bay (Fitzpatrick et al. 2014).

Antemortem dental chipping provides evidence for non-masticatory use of the dentition. Instances of chipping are high among individuals from Carriacou (47% of teeth scored); however, the majority (83.2%) were scored as minor. Higher frequencies and severity of dental chipping were observed on the maxillary and anterior dentitions among males, but overall chipping frequency was similar between males and females (Stone et al. in prep). Given the lack of evidence for dietary differentiation between males and females based on isotopic evidence (see below, Krigbaum et al. 2013), these differences can more likely be attributed to sex differences in non-dietary behaviors.

Isotope Analyses on Carriacou

Light Isotope Analysis

The use of stable isotope analysis to reconstruct early diets is an expanding area of study in the Caribbean but has received limited application in the southern Lesser Antilles. Paleodietary reconstruction is based on the premise that 'you are what you eat', meaning that consumer tissues may incorporate dietary elements that reflect the long-term average isotopic composition of foods consumed (Kohn 1999). Traditionally, this includes measuring the stable isotope ratios of bone collagen carbon ($^{13}C/^{12}C$) and nitrogen ($^{15}N/^{14}N$) with results reported using delta notation ($\delta^{13}C$ and $\delta^{15}N$, respectively) in parts per thousand, or per mil (‰). Samples are measured using isotope ratio mass spectrometry (IRMS) relative to an international standard of known value (Vienna Pee Dee Belemnite, or VPDB, for carbon and atmospheric nitrogen, or AIR, for nitrogen) (Katzenberg 2008; Pate 1994; Schwarcz and Schoeninger 1991). $\delta^{13}C$ can be used to differentiate between marine and terrestrial proteins that contribute to the average diet, as well as the consumption of plants that follow the C_3 or C_4 photosynthetic pathway. $\delta^{15}N$ values are a proxy for trophic level and tend to reflect stepwise enrichment between marine and terrestrial food resources (DeNiro and Epstein 1978; DeNiro and Epstein 1981; Minagawa and Wada 1984; Schoeninger et al. 1983). Different tissues analyzed reflect different dietary components. Bone collagen, for example, is primarily composed of amino acids and principally reflects dietary protein, while the mineral component of bone, bone apatite ('structural carbonate'), reflects all components of the diet, or 'total' diet (Ambrose and Norr 1993; Froehle et al. 2010; Krueger and Sullivan 1984; Tieszen and Fagre 1993). Carbon spacing between these two bone fractions ($\Delta^{13}C_{ap\text{-}co}$) provides a proxy to infer different patterns of carbohydrate input (C_3 vs. C_4) (Ambrose and Norr 1993; Kellner and Schoeninger 2007; Lee-Thorp et al. 1989).

To date, one paleodietary study has been published for Carriacou (Krigbaum et al. 2013), where results from bone collagen ($\delta^{13}C$ and $\delta^{15}N$) and bone apatite ($\delta^{13}C$) analysis were evaluated for 14 individuals recovered from Grand Bay. Krigbaum and colleagues demonstrated that the diet was focused on marine foods, with broadly similar results to those observed in the Bahamas (e.g., Keegan and DeNiro 1988; Stokes 1998), and suggest that proximity to productive reef environments contributed substantially to the marine protein component. Results also parallel the Grand Bay zooarchaeological record, which indicates a focus on marine protein (LeFebvre 2007; Fitzpatrick et al. 2009). However, questions remained regarding the possibility of dietary distinctions between burials exhibiting unique mortuary behaviors (described above) and in comparison to other pre-contact sites on Carriacou. To further explore this, an additional four individuals from the Grand Bay (n = 2) and the Point Bay (n = 2) sites were analyzed to explore potential dietary variability across sites on Carriacou. Bone samples were prepared and analyzed at the University of Florida's Bone Chemistry Lab using methods described elsewhere (Krigbaum et al. 2013) for bone collagen ($\delta^{13}C_{co}$ and $\delta^{15}N$) and bone apatite ($\delta^{13}C_{ap}$).

All four samples produced acceptable %N yields (~12.5 to 14.0) and atomic C:N (3.2 to 3.3), which are strong indicators for the integrity of derived bone collagen (DeNiro 1985; van Klinken 1999). Bone collagen $\delta^{13}C_{co}$ values for Grand Bay averaged -14.3 ± 0.03‰ and $\delta^{15}N$ values averaged 11.8 ± 0.23‰, while $\delta^{13}Cap$ values averaged -9.0 ± 2.69‰. For Point Bay, $\delta^{13}C_{co}$ values averaged -12.6 ± 0.06‰ and $\delta^{15}N$ values averaged 11.4 ± 0.41‰, while bone apatite $\delta^{13}C_{ap}$ values averaged -9.3 ± 0.6‰. When the two new Grand Bay samples are added to previously published values not including the single 'outlier' individual, mean bone collagen values shift slightly: $\delta^{13}C_{co}$ values average -12.8 ± 0.83‰ (compared to a previously published mean not including the single 'outlier' individual of -12.6‰), $\delta^{15}N$ values average 11.2‰ (compared to a previously published mean of 11.1‰), and $\delta^{13}C_{ap}$ values continue to average -8.7‰ (Krigbaum et al. 2013). When the new Point Bay values are compared to the pooled Grand Bay means, again, the values are nearly identical. Although they are different sites, Point Bay is located quite close to Grand Bay (~2 km north) and would have had similar access to nearby reefs and inshore taxa. As such, similar isotopic ratios between the two sites were not unexpected.

Taken together, these results continue to support previous research suggesting that pre-contact inhabitants on Carriacou took advantage of nearby productive reefs and coastal taxa, subsisting on marine protein. Consistent isotopic values between individuals of varying ages and sexes continue to support a lack of dietary differentiation between demographic groups. Interestingly, one of the two new Grand Bay samples presented here belongs to F0177, described earlier as an individual interred with a possible later secondary burial, sea turtle remains, a spindle whorl, and ash deposits. Although limited to one individual, these results also suggest that burials with more elaborate mortuary treatments have broadly similar diets and do not appear to have access to differential dietary resources.

Strontium Isotope Analysis

Given the mortuary variability observed at Grand Bay, questions regarding the identity of individuals with unique burial contexts, including potential migration to the site, have also emerged. To address this issue, we sampled six individuals (seven teeth) for strontium isotope ratios. Similar to light stable isotopes, heavy 'radiogenic' isotopes like strontium ($^{87}Sr/^{86}Sr$) are in part influenced by diet but principally vary as a result of underlying geological material that is absorbed into the food web through soil and groundwater, concentrated in plants, and eventually incorporated into human tissue (Bentley 2006; Ericson 1985; Faure and Powell 1972). Strontium is also present in the atmosphere, precipitation, and seawater. The combination of these signatures thus creates a local or 'biologically available' strontium signature that should be representative of a particular geographical area that forms the basis for mobility studies using strontium ratios derived from tooth enamel (Price et al. 2002). Because strontium substitutes for calcium in hydroxyapatite, samples of bone and tooth enamel will directly reflect the biologically available isotopic signature. However, over the course of an individual's lifetime, bone remodels in response to growth and development, disease, and activity (White et al. 2011). Therefore, the isotopic signature of bone will reflect the diet and location where an individual was last living. Because tooth enamel does not remodel, the isotopic signature of enamel will reflect an individual's place of residence at the time a sampled tooth was forming (Hillson 2005). Enamel is also more likely to preserve archaeologically and has been shown to be less susceptible to diagenesis, making it a preferable tissue to infer prehistoric mobility (Sillen and Kavanagh 1982).

Therefore, in order to track mobility over an individual's life course and evaluate potential migratory histories of individuals with unique mortuary contexts, we have sampled tooth enamel from first and third molars, when available, in order to examine signatures from infancy and late childhood, respectively (Hillson 2005). If the first or third molars were not available, a second molar was sampled instead, and a single individual exhibiting a unique mortuary context (F0177) was sampled for both the first and second molars. Sample methods are outlined in Giovas et al. (2016) and include mechanically removing a 30–40 mg chunk of enamel using a dental drill outfitted with a carbide tapered drill bit to produce a clean dental chunk. The enamel chunk was dissolved, and samples were processed for strontium separation using ion chromatography in a class 500 clean lab. Ratios were measured using a Nu-Plasma multicollector-inductively coupled plasma mass spectrometer (MC-ICP-MS).

Geologically, the inner arc of the Lesser Antilles, which includes Carriacou, is primarily composed of Cenozoic andesites and other volcanic rocks in combination with Late Tertiary and Quaternary limestone (Giovas et al. 2016). Strontium isotope mapping of the Caribbean has demonstrated overlap in the region as a result of observed geologic complexity, and Carriacou is no exception (Laffoon et al. 2012). The area surrounding Grand Bay, Point Bay, and Sabazan is primarily comprised of the Grand Bay and Carriacou formations, which are composed of Miocene limestones and volcanic conglomerates, sandstones, and siltstones above the Kendeace Formation, which is composed of basalt and andesite conglomerates with sedimentary rock dispersed throughout (Donovan et al. 2003; Giovas et al. 2016). Tyrell Bay lies on Quaternary sedimentary rocks interspersed with Neogenic magmatic rock along the southern margin (Donovan et al. 2003). Giovas et al. (2016) developed a strontium baseline for Carriacou using environmental and geologic samples from multiple locations around the island, including Grand Bay and Sabazan, to investigate the zoogeography of introduced mammalian taxa. Results demonstrated that agouti and possum were established as living populations on Carriacou despite a lack of agouti found on the island today. Archaeological and modern biological samples used to establish the local bioavailable strontium range for the island demonstrate the geologic variability complexity of estimating bioavailable strontium on Carriacou; however, the mean $^{87}Sr/^{86}Sr$ ratios of known local samples from Carriacou range from 0.70703 to 0.70892, with a mean ratio of 0.70807 (Giovas et al. 2016). Samples from Grand Bay and Sabazan in particular reflect the carbonate input of the Carriacou and Grand Bay formations described above.

Enamel ratios for individuals from Grand Bay range from 0.70728 to 0.70811, with a mean value of 0.70764, which falls within the range of the known local samples from Carriacou (Figure 4.3). All individual ratios fall within this known local range as well and are most similar to soil leachates and residues and a plant sample all collected from Grand Bay (Giovas et al. 2016). Values from volcanic rock are lower (less radiogenic) and lend further support to a carbonate contribution to the bioavailable strontium observed in the Grand Bay human results, likely due to the presence of carbonates present in the Carriacou and Grand Bay formations and similar to what Giovas et al. (2016) observed in fauna from the Grand Bay site.

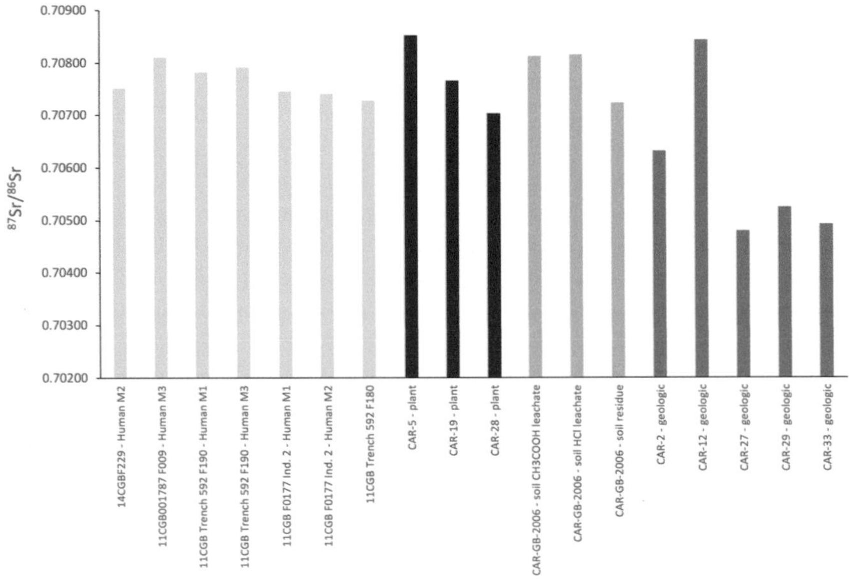

FIGURE 4.3 $^{87}Sr/^{86}Sr$ results of human tooth enamel compared to environmental samples from Giovas et al. (2016).

Together, these results are suggestive of the individuals sampled being local to Carriacou and more specifically to Grand Bay. Interestingly, the individuals sampled reflect a variety of mortuary treatments, including two samples from an individual associated with a number of grave goods (F0177), and demonstrate that mortuary variation reflects aspects of identity outside geographic origin or childhood residence. These results, in combination with dietary isotope ratios, suggest that the mortuary practices highlighted in this chapter are not reflective of migration to Carriacou nor do they reflect differential access to dietary resources. Despite the presence of a suite of translocated mammals (Giovas et al. 2012), non-local ceramics (Lawrence et al. 2016; Pavia et al. 2013), and exotic ritual objects (Fitzpatrick et al. 2009) that together suggest Carriacou was part of a widespread regional exchange system with other islands in the region and the South American mainland (Fitzpatrick et al. 2009), the inhabitants at Grand Bay appear to be locals, and more specifically, local to the Grand Bay site. However, given the preliminary nature of this analysis, the addition of enamel samples from a larger sample of individuals at Grand Bay may demonstrate a wider range of variation in strontium values.

Conclusions

Research on Carriacou since 2003 has provided substantial evidence for pre-contact lifeways in the southern Lesser Antilles, a relatively understudied area in comparison to the rest of the Caribbean. The Grand Bay site in particular, which contains a large Amerindian village site that was occupied for centuries, has shown that pre-contact life on Carriacou relied on productive marine resources from nearby reef environments (LeFebvre 2007; Giovas 2013; Krigbaum et al. 2013) and were engaged in an intensive and extensive regional exchange system with other islands and probably the South American mainland (Fitzpatrick et al. 2009; Giovas et al. 2012; Pavia et al. 2013).

Bioarchaeological research focused on more than two dozen burials from multiple sites on the island is generating complementary osteological data that informs the health, nutrition, and mortuary practices of pre-contact people living on Carriacou. Osteological and isotopic analyses are also lending support to archaeological and zooarchaeological evidence that suggests a strong reliance on marine resources for subsistence, while dental pathology suggests a significant starchy carbohydrate terrestrial component that contributed to a high rate of dental caries. Infectious diseases and traumatic injuries do not appear widespread, as paleopathological data show a variety of conditions that only affected a small number of people.

Mortuary activity on Carriacou shows a wide variety of contexts, positions, and associated grave goods, while demographic results suggest the possibility of differential mortuary treatment for juvenile individuals. Strontium isotope results demonstrate that regardless of mortuary treatment, individuals at Grand Bay exhibit strontium ratios consistent with local residence since childhood. Light isotope results likewise demonstrate evidence that individuals had broadly similar access to subsistence resources. Therefore, the variation in mortuary treatment reflects some other aspect of identity, including sex, rather than place of origin or some aspect of social status that would dictate access to certain types of dietary resources, as evidenced by the variation in burial position. Moving forward, additional research involving ancient DNA and expanded strontium isotope analyses may contribute to our understanding of inter-island mobility and interaction as well as biological relationships among Ceramic Age inhabitants on Carriacou and other islands in the region.

Acknowledgements

We thank the Carriacou Historical Society Museum for their help and assistance over the years with conducting this and other lines of research. Quetta Kaye and Michiel Kappers were instrumental in developing and implementing various research and public programs for the CAP. In the Department of Geological Sciences, University of Florida, we thank Jason Curtis for conducting the IRMS and George Kamenov for facilitating the strontium ratio analysis on the ICP-MS.

References

Ambrose, S.H., and Norr, L. 1993. Experimental evidence for the relationship of the carbon isotope ratios of whole diet and dietary protein to those of bone collagen and carbonate. In *Prehistoric Human Bone—Archaeology at the Molecular Level*, edited by J.B. Lambert, and G. Grupe, pp. 1–37. Berlin: Springer.

Bentley, A.R. 2006. Strontium isotopes from the earth to the archaeological skeleton: A review. *Journal of Archaeological Method and Theory* 13(3): 135–187.

Bullen, R.P., and Bullen, A.K. 1972. *Archaeological Investigations on St. Vincent and the Grenadines West Indies*. *American Studies Report Number* 8. Orlando, FL: The William L. Bryant Foundation.

Burnett, S.E. 2009. *Dental pathology as an indicator of diet on the island of Carriacou (West Indies)*. Poster presented at the 36th Annual Meeting of the Paleopathology Association, March 31–April 1, 2009, Chicago, IL.

Burnett, S.E., et al. in prep. Mortuary practices. In *Pre-Columbian Archaeology on Carriacou*, edited by S.M. Fitzpatrick, Q. Kaye, and M. Kappers. Tuscaloosa: University of Alabama Press.

DeNiro, M.J. 1985. Postmortem preservation and alteration of *in vivo* bone collagen isotope ratios in relation to palaeodietary reconstruction. *Nature* 317: 806–809.

DeNiro, M.J., and S. Epstein. 1978. Influence of diet on the distribution of carbon isotopes in animals. *Geochimica et Cosmochimica Acta* 42(5): 495–506.

DeNiro, M.J., and S. Epstein. 1981. Influence of diet on the distribution of nitrogen isotopes in animals. *Geochimica et Cosmochimica Acta* 45(3): 341–351.

Donovan, S.K., Pickerill, R.K., Portell, R.W., Jackson, T.A., and Harper, D.A.T. 2003. The Miocene paleobathymetry and paleoenvironments of Carriacou, the Grenadines, Lesser Antilles. *Lethaia* 36: 255–272.

Duijvenbode, A. van. 2017. *Facing Society: A Study of Identity through Head Shaping Practices among the Indigenous Peoples of the Caribbean in the Ceramic Age and Colonial Period*. Ph.D. diss., Leiden University.

Ericson, J.E. 1985. Strontium isotope characterization in the study of prehistoric human ecology. *Journal of Human Evolution* 14(5): 503–514.

Faure, G., and Powell, J.L. 1972. *Strontium Isotope Geology*. New York: John Wiley.

Fitzpatrick, S.M., Kappers, M., and Kaye, Q. 2006. Coastal erosion and site destruction on Carriacou, West Indies. *Journal of Field Archaeology* 31(3): 251–262.

Fitzpatrick, S.M., Kappers, M., Kaye, Q., Giovas, C.M., LeFebvre, M.J., Harris, M.H., Burnett, S.E., Pavia, J.A., Marsaglia, K., and Feathers, J. 2009. Precolumbian settlements on Carriacou, West Indies. *Journal of Field Archaeology* 34: 247–266.

Fitzpatrick, S.M., Kaye, Q., Feathers, J., Pavia, J.A., and Marsaglia, K.M. 2009. Evidence for inter-island transport of heirlooms: Luminescence dating and petrographic analysis of ceramic inhaling bowls from Carriacou, West Indies. *Journal of Archaeological Science* 36(3): 596–606.

Fitzpatrick S., Kaye, Q., Kappers, M., and Giovas, C. 2014. A decade of archaeological research on Carriacou, Grenadine Islands, West Indies. *Caribbean Journal of Science* 48(2–3): 151–161.

Froehle, A.W., Kellner, C.M., and Schoeninger, M.J.. 2010. FOCUS: Effect of diet and protein source on carbon stable isotope ratios in collagen: Follow up to Warinner and Tuross (2009). *Journal of Archaeological Science* 37: 2662–2670.

Giovas, C.M. 2013. *Foraging Variability in the Prehistoric Caribbean: Multiple Foraging Optima, Resource Use, and Anthropogenic Impacts on Carriacou, Grenada*. Ph.D. diss., University of Washington.

Giovas, C.M. 2018. Pre-Columbian Amerindian lifeways at the Sabazan Site, Carriacou, West Indies. *The Journal of Island and Coastal Archaeology* 13(2): 161–190.

Giovas, C.M., Kamenov, G.D., Fitzpatrick, S.M., and Krigbaum, J. 2016. Sr and Pb isotopic investigation of mammal introductions: Pre-Columbian zoogeographic records from the Lesser Antilles, West Indies. *Journal of Archaeological Science* 69: 39–53.

Giovas, C.M., LeFebvre, M.J., and Fitzpatrick, S.M. 2012. New records for prehistoric introduction of Neotropical mammals to the West Indies: Evidence from Carriacou, Lesser Antilles. *Journal of Biogeography* 39(3):476–487.

Heller, A.G., and Burnett, S.E. 2011. *Antemortem dental chipping in the prehistoric inhabitants of Carriacou (West Indies).* Poster presented at the 28th Annual Meeting of the Paleopathology Association, April 13, 2011, Minneapolis, MN.

Hillson, S. 2005. *Teeth.* Cambridge: Cambridge University Press.

Hoogland, M.L.P., and Hofman, C.L. 2013. From corpse taphonomy to mortuary behavior in the Caribbean: A case study from the Lesser Antilles. In *The Oxford Handbook of Caribbean Archaeology*, edited by W.F. Keegan, C.L. Hofman, and R. Rodriguez Ramos, pp. 452–469. Oxford: Oxford University Press.

Katzenberg, M.A. 2008. Stable isotope analysis: A tool for studying past diet, demography, and life history. In *Biological Anthropology of the Human Skeleton*, edited by M.A. Katzenberg, and S.R. Saunders, pp. 413–441. Second ed. Hoboken, NJ: John Wiley & Sons.

Kaye, Q., Fitzpatrick, S., Harris, M., and Kappers, M. 2012. Bowls and burials–An update from Grand Bay, Carriacou, West Indies: May–June 2011. *Papers from the Institute of Archaeology* 21: 91–100.

Keegan, W.F., and DeNiro, M.J. 1988. Stable carbon- and nitrogen-isotope ratios of bone collagen used to study coral-reef and terrestrial components of prehistoric Bahamian diet. *American Antiquity* 53(2): 320–336.

Kellner, C.M., and Schoeninger, M.J. 2007. A simple carbon isotope model for reconstructing prehistoric human diet. *American Journal of Physical Anthropology* 133(4): 1112–1127.

King T., Humphrey, L., and Hillson, S. 2005. Linear enamel hypoplasias as indicators of systemic physiological stress: Evidence from two known age-at-death and sex populations from postmedieval London. *American Journal of Physical Anthropology* 128: 547–559.

Kohn, M.J. 1999. You are what you eat. *Science* 283(5400): 335–336.

Krigbaum, J., Fitzpatrick, S.M., and Bankaitis, J. 2013. Human paleodiet at Grand Bay, Carriacou, Lesser Antilles. *The Journal of Island and Coastal Archaeology* 8(2): 210– 227.

Krueger, H.W., and C.H. Sullivan. 1984. Models for carbon isotope fractionation between diet and bone. In *Stable Isotopes in Nutrition*, edited by Judith R. Turnlund, and Phyllis E. Johnson, pp. 205–220. Washington, DC: American Chemical Society Symposium Series 258, American Chemical Society.

Laffoon, J.E., Davies, G.R., Hoogland, M.L., and Hofman, C.L. 2012. Spatial variation of biologically available strontium isotopes ($^{87}Sr/^{86}Sr$) in an archipelagic setting: A case study from the Caribbean. *Journal of Archaeological Science* 39: 2371–2384.

Larsen, C., Teaford, M., and Sandford, M. 2002. The Tutu teeth: Assessing prehistoric health and lifeway from St. Thomas. In *The Tutu Archaeological Village Site: A Multidisciplinary Case Study in Human Adaptation*, edited by Elizabeth Righter, pp. 230–249. London: Routledge.

Lawrence, J., Marsaglia, K., and Fitzpatrick, S.M. 2016. Petrographic analysis of pre-Columbian pottery from four islands in the lesser antilles and implications for inter-island transport and interactions. *Journal of Archaeological Science: Reports* 9: 663–680.

Lee-Thorp, J.A., Sealy, J.C., and van der Merwe, N.J. 1989. Stable carbon isotope ratio differences between bone collagen and bone apatite, and their relationship to diet. *Journal of Archaeological Science* 16(6): 585–589.

LeFebvre, M. 2007. Zooarchaeological analysis of prehistoric vertebrate exploitation at the Grand Bay Site, Carriacou, West Indies. *Coral Reefs* 26: 931–944.

Mickleburgh, H. 2013. *Reading the Dental Record: A Dental Anthropological Approach to Foodways, Health and Disease, and Crafting in the Pre-Columbian Caribbean*. Ph.D. diss., Leiden University.

Mickleburgh, H. 2016. Dental wear and pathology in the precolonial Caribbean: Evidence for dietary change in the Ceramic Age. *International Journal of Osteoarchaeology* 26(2): 290–302.

Minagawa, M., and Wada, E. 1984. Stepwise enrichment of^{15}N along food chains: Further evidence and the relation between (^{15}N and animal age. *Geochimica et Cosmochimica Acta* 48(5): 1135–1140.

Moggi-Cecchi, J., Pacciani, E., and Pinto-Cisternas, J. 1994. Enamel hypoplasia and age at weaning in 19th-Century Florence, Italy. *American Journal of Physical Anthropology* 93: 299–306.

Osborne, J. 2013. *Deciphering Patterns of Burial Distribution: A Mortuary Analysis of a Precontact Sample Population from the Island of Carriacou*. M.A. Thesis, North Carolina State University.

Pate, F.D. 1994. Bone chemistry and paleodiet. *Journal of Archaeological Method and Theory* 1(2): 161–209.

Pavia, J.A., Marsaglia, K.M., and Fitzpatrick, S.M. 2013. Petrography and provenance of sand temper within ceramic sherds from Carriacou, Southern Grenadines, West Indies. *Geoarchaeology* 28(5): 450–477.

Price, T.D., Burton, J.H., and Bentley, R.A. 2002. The characterization of biologically available strontium isotope ratios for the study of prehistoric human migration. *Archaeometry* 44(1): 117–135.

Reeves, R.R. 2006. *Bioarchaeology at Grand Bay and Tyrell Bay, Two Contemporary Amerindian Sites on Carriacou Island, Grenada, West Indies*. MA Thesis, University of Montana.

Schoeninger, M.J., DeNiro, M.J., and Tauber, H. 1983. Stable nitrogen isotope ratios of bone collagen reflect marine and terrestrial components of prehistoric human diet. *Science* 220(4604): 1381–1383.

Schwarcz, H.P., and Schoeninger, M.J. 1991. Stable isotope analyses in human nutritional ecology. *Yearbook of Physical Anthropology* 34: 283–321.

Sillen, A., and Kavanagh, M. 1982. Strontium and paleodietary research: A review. *Yearbook of Physical Anthropology* 25(S3): 67–90.

Stokes, A.V. 1998. *A Biogeographic Survey of Prehistoric Human Diet in the West Indies Using Stable Isotopes*. Ph.D. diss., University of Florida.

Stone, J.H. 2011. *Paleodiet and Nutrition at Grand Bay, Carriacou, West Indies*. MA Thesis, North Carolina State University.

Stone, J.H., Heller, A.G., Burnett, S.E., and Fitzpatrick, S.M. in prep. *An Analysis of Dental Pathology in a Pre-Columbian Population from Carriacou*. Southern Grenadines, West Indies.

Tieszen, L.L., and Fagre, T. 1993. Effect of diet quality and composition on the isotopic composition of respiratory CO_2, bone collagen, bioapatite and soft tissues. In *Prehistoric Human Bone—Archaeology at the Molecular Level*, edited by J.B. Lambert, and G. Grupe, pp. 121–155. Berlin: Springer.

van Klinken, G.J. 1999. Bone collagen quality indicators for palaeodietary and radiocarbon measurements. *Journal of Archaeological Science* 26: 687–695.

White, T.D., Black, M.T., and Folkens, P.A. 2011. *Human Osteology*. Third ed. Burlington, MA: Academic Press.

5

PRE-COLUMBIAN FUNERARY BIOARCHAEOLOGY OF PUERTO RICO AND THE LA GALLERA SITE, CEIBA, PR

Luz Veronica Muñoz Guevara

Introduction

Funerary customs are inherent to the cultural behaviors of human beings, as the material and symbolic link for interpreting belief systems and worldviews tied to the death of past populations. They can offer information about social organization and the place that the individual occupies within his community. The study of funerary customs in anthropology and archaeology has been approached in a multidisciplinary way (Childs 1997; Duday 2011; O'Shea 1984; McAnany 1995; Morin 1994; Thomas 1989, 1993) and can intertwine multivariate information in a way that relates both to the individual and society through the study of funerary patterns, biological characteristics, health and disease, subsistence, physical activity, and migratory patterns (Buikstra and Beck 2006; Crespo et al., 2013; Gonzalez and Marquez 2009; Larsen 1997; Pestle 2010; Nieves-Colón et al. 2020). To provide some background, we will first carry out a general review of the most important studies relating to the pre-Columbian funerary bioarchaeology of Puerto Rico, which, in their vast majority, were carried out by the forensic anthropologist and bioarchaeologist Edwin Crespo Torres. Likewise, we will present for the first time the funerary bioarchaeological findings of the pre-Columbian site of La Gallera, located in the town of Ceiba in eastern Puerto Rico.

Bioarchaeological Research in Puerto Rico

Bioarchaeological studies in Puerto Rico prior to the 1980s were practically non-existent. The findings of pre-Columbian human burials were described in a general way without acknowledging all the information that could be obtained through their osteobiological study and their funerary contexts. The main interest revolved

DOI: 10.4324/9780429444449-8

around analyses of ceramics and other materials to establish chrono-cultural sequences (Alegría et al., 1955; Fewkes 2009 [1907]; Rouse 1992). Later, both in Puerto Rico and in the rest of the Antillean Caribbean, archaeologists and anthropologists began to carry out work incorporating studies of human bone remains and their cultural context (Budinoff 1991; Crespo 2000; Curet and Oliver 1998; Goodwin 1979; Roe et al. 1990; Siegel 1992). This interest paralleled the development of the bioarchaeological discipline, both in the United States (Buikstra 1977; Buikstra and Beck 2006) and in other regions of Latin America (Gonzalez and Marquez 2009).

In the case of Puerto Rico in the 1990s, when work began to be carried out incorporating the study of human remains, researchers tried to answer specific research questions with the aim of learning more about the ancient populations and reconstructing their stories. Among these researchers was the forensic anthropologist and bioarchaeologist Edwin Crespo Torres, who stands out for being the first Puerto Rican specialist in this discipline. He incorporated a different perspective into Puerto Rican archaeology, applying bioarchaeology and the biocultural perspective (Buikstra 2006; Crespo 2000; Hernández et al. 2008; Martin 1998). He conducted multiple osteobiological analyses of human burials and their funerary contexts from pre-Columbian and colonial sites that contributed to a better overview of the ancient history of Puerto Rico (Crespo 1988, 1989, 1991, 1997, 1998, 2000; Crespo and Muñoz 2009; Crespo and Muñoz in Pantel 2011; Crespo 1988), setting a standard for understanding the funerary practices of the ancient Indigenous populations that inhabited the *Boriken*[1] *archipelago in* Puerto Rico.

Early Period: Pre-Arawak to Saladoid

According to chrono-cultural horizons and radiocarbon dating, it is estimated that different human groups began arriving in the Puerto Rican archipelago more than 6,000 years ago (Ayes 1988, Rodriguez 1999). These first groups have been defined as pre-Arawaks (Rodriguez Ramos 2010), previously identified as "Archaic" (Alegría et al. 1955; Rouse 1992: 62). We now know that the ancient Indigenous societies were more complex than previously thought and had extensive knowledge of navigation techniques, horticultural skills, and methods of domesticating plants for agricultural and medicinal purposes. In addition, they maintained social interactions with populations of the Circum-Caribbean region, exchanging raw materials, objects, technologies, and ideas (Pagan-Jimenez 2013; Rodríguez Ramos 2010, 2013: 159).

Among the archaeological sites where human burials associated with early period populations have been detected is the site of Cueva María de la Cruz in the town of Loíza, excavated in 1948 and 1954 (Alegría et al. 1955). The find consisted of several secondary skeletal remains and a primary burial, placed in an extended supine position (Rouse and Alegría 1990). This was thought to be one of the sites that defined pre-Arawak or Archaic period burials; however, subsequently, other sites were detected where more reliable dates could be obtained, such as Angostura, where human burials were found (Ayes 1989, cal. 5960 ± 250 BP [Beta-29778] in Rodriguez-Ramos et al. 2013: 127). For both sites, no osteobiological data are available for the excavated burials.

One site that did provide more information on pre-Arawak funerary contexts was Maruca in Ponce (2890 – 395 BP (uncalibrated) [in Rodriguez López 2004: 12]), where eleven adult burials were discovered, most in an extended dorsal decubitus position (face up) with arms placed at the sides, with the exception of one, which was in a flexed position. The funerary and osteobiological analysis was carried out by Crespo (in Rodriguez López 2004). Another outstanding find from the pre-Arawak period was at Puerto Ferro on the island municipality of Vieques. There was a primary flexed burial of an adult male between 35 and 40 years old. The osteopathological analysis indicated that this individual had severe osteoarthritis, dental wear, and other health conditions that were associated with the subsistence mode and physical activity of a fisher-gatherer and semi-nomad (Crespo 1998: 336). The project archaeologists named him "Puerto Ferro Man" (Chanlatte and Narganes 1991). One of the characteristics of early period burial contexts is the scarcity of associated grave goods.

From approximately 500 BC onwards, human groups associated with Saladoid and Huecoid traditions began to arrive on the island. In Saladoid contexts, we have different sites where burials have been detected, including the sites of Hacienda Grande in the town of Loíza (Rouse and Alegría 1990), Monserrate in Luquillo (Roe et al. 1990), Maisabel in Vega Baja (Siegel 1992), and Punta Candelero, located in the town of Humacao (Rodriguez-López 1991). At this last site, 85 burials were detected[2] with radiocarbon dating, placing them in the chronological range of 600–1010 cal. AD (Pestle 2010: 241) and the ceramic style corresponding to Cuevas (Rodriguez-López 1991, 2004). According to the bioarchaeological and osteobiological analyses, 78% were adults and 12% were subadult individuals (Crespo 2000: 156–157). Most of the burials were primary inhumations, while the most frequent anatomical position was dorsal flexed. One indirect[3] burial of an infant was detected, placed inside an urn or clay pot. In the case of the adults, seven individuals had a complete vessel placed on top, while in other cases they had a large vessel fragment as part of the funerary offering. In addition, this study included an extensive osteopathological analysis of health-disease conditions, diet, physical activity, and dental pathology (Crespo 1994, 2000).

One of the topics that Crespo researched was intentional cranial modification in pre-Columbian populations in Puerto Rico. Based on comparative studies, he identified the fronto-occipital oblique tabular type, according to the Dembo and Imbelloni (1938: 375) and Romano (1974) classifications. Eminent Antillean physical anthropologists, such as Rivero de la Calle (1960) in Cuba, also worked on the subject. Regarding this cultural trait, Crespo (2000: 225, 2002, 2005) proposed a hypothesis on which human group introduced this cultural trait to Puerto Rico. Crespo suggested that the influence came from the Huecoids[4] (Crespo 2000: 225, 2005:

59), based on representations of the heads that appear on the claws of ornithomorphic pendants representing the Andean condor (*Vultur gryphus*) (Chanlatte and Narganes 2005: 23). Some consider the heads in the vultures' talons to be symbolic of "trophy heads". However, Crespo's approach emphasized the importance of the practice of cranial modification as a hallmark of the ethnic identity of these groups, for which there was no gender distinction. He mentions that its influence was progressive and gradual as a result of biological and cultural hybridization so this practice became widespread among Caribbean-Antillean populations (Crespo 2000: 228). Condor figures were detected at the Sorcé site, La Hueca neighborhood in Vieques, and at the Punta Candelero site in Humacao. Interestingly, the practice of intentional cranial deformation is not observed in early pre-Arawak populations nor in early Saladoids. Part of what is conspicuous in this discussion is that no human burials associated with Huecoids have been recorded in Puerto Rico. According to Crespo (2000), this could be a consequence of different, non-inhumation-based funerary practices or even inadequate archaeological field methods. One example of comparative human osteological evidence mentioned by Crespo (2000, 2005: 61) comes from Guadeloupe in the Lesser Antilles, where a burial associated with a Huecoid context was detected, showing intentional cranial deformation (Durant and Roget 1991). Undoubtedly, this topic will continue to be a subject for debate among archaeologists.

On the other hand, some interpretations of Saladoid burial customs assume that plazas or centralized areas were used as human burial "cemeteries" (Curet and Oliver 1998; Siegel 1992, 1996). This has been established on the basis of a few analyzed sites, one of these being Punta Candelero. Later reassessment of the spatial distribution of the so-called "cemetery" at Punta Candelero indicated that most of the burials were near mounds or promontories, suggesting that it could be likely that the burials had been deposited in domestic areas (Crespo 2000: 143). A similar case was observed at the site of La Gallera, which we will discuss later.

Mid-late Period (Ostionoid)

We consider the Ostionan and the Elenan Ostionoid, or pre-Taíno, as the Middle-Late period. Here we include all the ceramic stylistic variants, as we consider that cultural behavior is more complex than ceramic tradition. This period is the one for which the best bioarchaeological information is available for Puerto Rico. Several of Crespo's works were definitive for the analysis of the funerary practices, bioarchaeology, and paleopathology of populations from this period (Crespo 1988, 1994, 1995, 2000, 2000a, 2001, 2008, 2010, 2010a, 2011, 2013; Crespo and Muñoz 2009). Undoubtedly, one of Crespo's most influential works comes from the pre-Columbian site of Paso del Indio, located in the town of Vega Baja in northern Puerto Rico. Radiocarbon dates place its pre-Taino component between 800 and 1200 cal AD (Pestle 2010: 247). At this site, a total of 141 burials (67 adults and 74 subadults) were recovered and analyzed. This population exhibited highly complex funerary practices, diversity in burial systems, unprecedented funerary offerings, and important osteopathological cases.

Among the most outstanding finds were three primary burials of adult males who "held" a secondary skull of another adult individual, and three other primary adult burials that also had a secondary skull but with no appearance of holding it. Cases were also found where the individual had secondary postcranial bone elements from another individual, which according to taphonomic analysis had been deposited intentionally. According to Crespo (2000: 153–156), this symbolic expression is associated with the "cult of the ancestor", highlighting the importance of retaining body parts that would later be placed or reburied as companions of their deceased descendants.

Another of the findings from Paso del Indio of great importance to bioarchaeology in Puerto Rico are two burials of women who died during childbirth. There was no precedent for this type of finding at any another archaeological site in Puerto Rico. These were primary burials placed in a dorsal flexed position. During the excavation process, it was possible to detect the small skeletons of the neonates located in the pelvic regions of both women (Burials P7U2-1a, 1b and P7U4–5d, 5e) (Crespo 2000: 352–353, 2002: 91). According to the taphonomic analysis, one of them (Ent. P7U4–5d) had her skull intentionally separated from her body, an event that likely occurred during the decomposition process (Crespo 2000). This action corresponded to the symbolic custom of retaining body elements for the cult of the ancestor, which was indicative that these individuals held an important rank within society, reflecting the complex organization and sophisticated cultural practices that were part of an incipient cacical organization system (Crespo 2000: 157, 2002, 2011).

Crespo (2008, 2019: 174) carried out a critical analysis of information from chronicles and ethnohistoric and historical sources from the Antillean Caribbean on the subject of paleopathology, proposing important analogies regarding the therapeutic treatments used by the Indigenous people to alleviate certain ailments (e.g. repair fractured bones, ease the discomfort of syphilitic buboes, among others). The first case of syphilitic treponema (*Treponema pallidum* subsp. *pallidum)* reported in a Puerto Rican pre-Columbian population comes from the site of Paso del Indio and corresponds to an adult female between 20 and25 years with characteristic tertiary stage cranial lesions (Crespo 2005a: 386).

Other bioarchaeological work carried out by Crespo (2008; Pantel 2011) was that of the Los Indios site, located in the municipality of Santa Isabel in the southern region of Puerto Rico. At this site, 130 burials were detected, of which 121 were adults, 6 were subadults, and 3 were undetermined due to their poor state of preservation. A total of 91.5% were primary burials (119). The most frequent body placement was dorsal flexed with the skull facing west (Crespo 2011 in Pantel 2011). Due to the spatial distribution of the burials and the location of postholes, the site was defined as a large Indigenous village. Radiocarbon dates place this site at approximately AD 783 to AD 1431 (uncalibrated) (Pantel 2011), a late chronology compared to other sites.

Another important Ostionoid period site is Punta Mameyes (Do-42), located on the north-central coast in the town of Dorado. This site is located 9.31 kilometers east of Paso del Indio, which suggests the possibility that there may have been some type of interaction between the inhabitants given their geographical proximity and contemporaneity. In Punta Mameyes, 40 human burials were recovered (Muñoz 2014). Among these, burial #9 stands out, which corresponds to an adult female between 30 and 45 years of age, placed in a dorsal flexed position. The dates place it between 990 and 1150 cal AD (2 sigma, Beta-324227). This individual exhibits intentional cranial modification; a snail (*Strombus gigas)* was placed under the left arm, and Monserrate-style ceramic fragments were placed on and under the body. The placement of fragmented pottery is part of the funerary customs of these groups, and we see it frequently in female burials, as well as those of males and infants. This could correspond to the symbolic conception of "killing or breaking the vessel", possibly as a sign of mourning for the loss of a family member (Muñoz 2014, 2016).

Funerary Bioarchaeology of the La Gallera Site, Ceiba PR

Following theoretical and methodological research guidelines established by Crespo, we conducted a bioarchaeological study of the human remains recovered from the pre-Columbian site of La Gallera. This archaeological site is located in the town of Ceiba, located on the east coast of the island of Puerto Rico, where 78 human burials were detected during the archaeological mitigation project *Stage IV. Boatyard improvements Marina Puerto del Rey. Bo. Machos, Ceiba* directed by archaeologist Hernán Ortiz Montañez.

Geographic Location

The Gallera deposit is located at UTM coordinates (20Q) 0221047 E /2013438 N, between the Demajagua Valley and the Luquillo Mountains, bordering the towns of Ceiba and Fajardo. The Demajagua River is located to the north and flows into the bay of the same name toward the Vieques Sound, a portion of the Atlantic Ocean. The natural geographical location of the site can be considered strategic, as the Demajagua River was available to supply fresh water and provide the necessary ecosystems for human subsistence. Toward the bay, less than a kilometer to the east, there is a highly diverse area of mangroves and reefs, where food resources such as crustaceans, molluscs, and fish were available. In addition, raw materials such as shells and other materials were available for the manufacture of tools and symbolic objects. Similarly, from the Sierra de Luquillo and the Demajagua River, various stones could be collected to create tools and lithic artifacts. The Gallera site has direct access to the coast, which allowed for convenient and effective maritime mobility to transport materials and exchange goods with other populations.

Materials and Methods

The site had an approximate area of 7,203.47 m^2. Human burials were found distributed throughout, with a slight concentration to the north. The burials were found at depths between 50 centimeters and 120 centimeters above surface level (m.a.s.l.). The stratum where they were primarily detected had a compact clay texture with *very fine to fine* granulometry (≥1–2mm). The area was bioturbated by flora and microfauna, which impacted the integrity of the skeletal remains; in general terms, preservation is fair to poor. Despite this, most of the finds were excavated and documented in great detail. The location and spatial distribution of the burials and features were georeferenced with the UTM system and documented with drawings and photographs (Muñoz 2020). For the descriptions of the burials, we used methods outlined by Duday (2011), O'Shea (1984), and Romano (1974), while Ubelaker (1997) and Tibbett and Carter (2008) methods were used for recording taphonomic alteration.

The artifacts recovered from the burial contexts consisted of ceramic vessels, burin fragments, shell ornaments, animal bone, lithic objects such as petaloid axes, tubular stone beads, coral tools, and food remains belonging to manatees, fish, crab, and molluscs. The ceramic component corresponded to the Saladoid, Cuevas, Monserrate, and Ostionoid traditions, with a slight predominance of the Saladoid with bell-shaped vessels and red paint.

Samples from three adult burials were submitted for AMS dating: individuals #3, #27, and #42. Selection criteria were based on spatial location; they were distant enough from each other to potentially definc parameters for the temporal occupation of the site and the complexity of their funerary contexts (Muñoz 2020).

Burial #3 was a primary burial of an adult male, whose body was positioned ventral extended (prone), with a skull-foot axis orientation of 92^{0}E. The dating placed it between 660 and 730 cal AD (IntCal13; Beta-501367). In terms of grave goods, a petaloid ax was found next to the body, along with Monserrate and Cuevas-type ceramic fragments around the body, as well as two stingray (*Mobula birostris*) stingers or points, 8 and 10 cm long. One was found under the left os coxa and the other over the left humerus. No cut marks were visible on the bones. The presence of these manta ray stingers is significant, as they were among the tools possibly used to make arrowheads, spears, or harpoons for hunting and fishing activities and may even have been used as weapons during interpersonal conflicts. The stingray stinger has two serrated edges, so it can be highly damaging or lethal. Interestingly, in this project, three other male burials were identified who had associated manta ray stingers (#12, #67, and #72). We cannot rule out the possibility of warlike activity or that these individuals may have been the warriors of the group, as well as hunter-fishermen (Figures 5.1 and 5.2).

Burial #27 was an adult female dating between 764–897 cal AD (IntCal13; Beta-501368) in a dorsal flexed position on the right side. The skull-foot axis orientation was 20^0 N and the facial skull 260^0 W. It was found in association with

FIGURE 5.1 Burial # 3, adult male with stingray tip from under os coxa.

FIGURE 5.2 Burial #12, adult male with manta ray tip on the left humerus.

the indirect secondary burial of a neonate (burial #31), placed inside two probably Cuevas style and Saladoid vessels, the inner one with straight walls where the skeleton was placed and the outer one a hemispherical vessel. The vessels that contained the bones of the newborn covered the skull and part of the neck of burial #27 (Figures 5.3 and 5.4). Likewise, around the body of the adult, fragments of vessels and plates and mollusk shells (i.e., *Cittarium pica*) were found. Additionally, 25 centimeters away from the skull, there was a set of large tortoise shell plates.

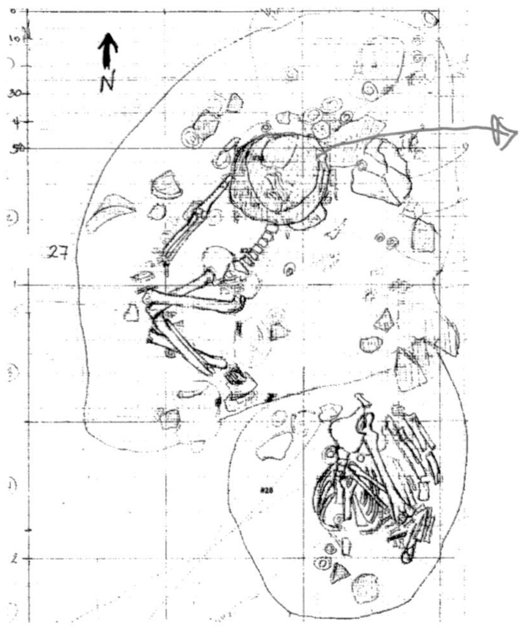

FIGURE 5.3 Burial plan drawing with burials #27 adult female, #31 neonate (upper), and #28 adult male (lower).

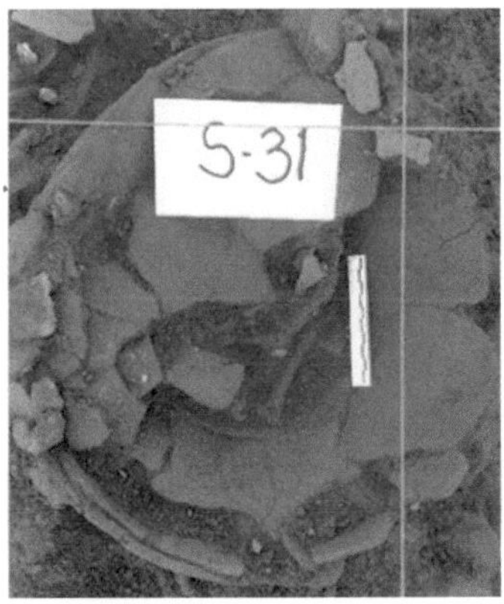

FIGURE 5.4 Burial #31 of an infant inside a vessel.

Burial #42 was an adult male placed in a strongly flexed dorsal position on the right side and dated between 892–1014 cal AD (intCal13; Beta-501369). This individual had a cylindrical lithic bead between the ribs and left humerus; it was part of a necklace that was probably worn when the burial was deposited and was subsequently displaced. Also close to the skull was a peccary tusk (4.4 cm long) that did not have a perforation as a pendant but could have been an amulet.

Among the collective or grouped burials, we have burials #71, #72, #76, and #77. We do not have radiocarbon dates for these at present. Of these, #72 stood out for its associated materials and good preservation. It was the primary burial of an adult male in a dorsal flexed position on his right side. The skull-foot axis orientation was 155⁰ S and the skull-facial axis was 5⁰ N. This individual had three elements that stood out among the associated grave goods: a globular vessel placed inverted, completely covering the skull; a sea turtle shell (27cm x 20cm), which covered part of the thorax up to the knees; and a manta ray stinger, which was found between the 8th and 9th thoracic vertebrae. With the tip of the stinger angled toward the interior of the body, it is reasonable to assume its placement was associated with a wound. The funerary context of the burial is very elaborate, which indicates that this individual was possibly a prominent member of this community (Figure 5.5).

Results

La Gallera has been known since the 1980s via archaeologist Ricardo Alegría, who was director of the Institute of Puerto Rican Culture (ICP). Later, the archaeologist

FIGURE 5.5 Primary burial #72 of an adult male, skull covered by a globular vessel and a tortoise shell on the body.

Miguel Rodriguez (1986, 1990) included this site in the inventory of archaeological sites of the Archeology Council of the Terrestrial Heritage of Puerto Rico, attached to the ICP. Rodríguez mentions that during the survey he observed numerous mounds and a large concentration of ceramics and Indigenous lithics on the site's surface. Subsequently, several site assessments were carried out near the site, originating through CRM contract archaeology[5] (Daubón 2002; Melendez 1991; Moline 1991; Rodriguez 1997). None report the discovery of human burials, with the exception of the *Mitigation Project for the Puerto del Rey Marina Electrical Substation at the La Gallera Site* carried out by archaeologist Luis Antonio Curet, who reports findings of "poorly preserved isolated human tooth and bone fragments" (1990: 23).

In 2015, during renovations of the Puerto del Rey Marina facilities, archaeological monitoring was conducted, and the presence of human remains was reported. Subsequently, between 2017 and 2019, archaeological mitigation was carried out under the Stage IV Boatyard Improvements project at Marina Puerto del Rey, and 78 human burials from La Gallera were recovered (Muñoz 2020; Ortiz et al. 2020).

The site of La Gallera presents us with a probable period of occupation between 660–1014 cal AD based on the three radiocarbon dates. Of the 78 burials, a total of 60 (77%) were adults and 18 (23%) were subadults (ranging from neonatal to approximately 12 years). With regards to burial type, primary (anatomically articulated) burials predominated with 64 individuals (82%), 8 (10.2%) were secondary burials, and mixed cases comprised six individuals (7.7%). Mixed cases refer to those with articulated bone segments and those without. Inhumations prevailed with a total of 73 (93.5%), and with respect to indirect burials, we detected two (2.5%) involving juveniles under six years of age, with both cases placed inside a vessel. We also detected three (3.8%) indirect burials of adults.

The majority (52 or 67%) are individual burials; however, several grouped burials were detected, classified as collective burials, with a total of 26 individuals (33%) distributed in eight groups (Table 5.1). This does not mean that they are simultaneous burials, but that they share the same funerary space. This practice could be due to several factors – to economize on available space or because there was a filial bond between them. We have very specific cases in which there is a clear intention to place one body on top of another, probably in a symbolic way, such as burials #4 A-B and burials #27 and #31.

With regards to the orientation of the body based on magnetic north, two variables are considered: the cranio-foot axis and the cranio-facial axis. Comparing both orientations, we observe that the most frequent orientation is toward the south. With respect to facial orientation, the most common position is facing southwest, and for the cranio-foot axis, it is orientation to the north (Tables 5.2 and 5.3).

For body positioning, the most common is right side dorsal flexed, followed by left side dorsal flexed. The lateral flexed positions on both the right and left sides are observed in adults but not in subadults. We have 13 (21%) individuals in the "frog" type dorsal flexed position (femur-coxal hyperflexion). This position is the

TABLE 5.1 Grouped burials

# Burial	*Subadult*	*Adult*	*Class*	*Type*	*Number of individuals*	*Variety*
4A, 4B		2	Prim	Direct	2	DFI, LFI
27, 31,28	1	2	Prim, secn	Direct, indirect	3	DFD, LFD
34A, 34B		2	Mixed, prim	Direct	2	SWIM
49, 50, 51, 52, 53	1	4	Prim	Direct	5	DFI, VFD, DFI, LD, LFI
58, 57, 59		3	Prim	Direct	3	DFD
60, 61, 63, 64, 65	1	5	Mixed	Mixed, direct	5	N/A, DFD, FROM
71, 72, 76, 77		4	Prim	Direct	4	DFI, DFD
73, 75	1	1	Secn	Mixed	2	N/A

Legend:
prim = primary
Secn= secondary
Variety position of the body in decubitus:
D*- dorsal (partially impacted)
DE- dorsal extended
DF- dorsal flexed
DFD- dorsal flexed right side
DFI- dorsal flexed left side
LFD- lateral flexed right side
LFI- lateral flexed left side
VFD- ventral flexed right side
VFI- ventral flexed left side
N/A- not applicable

TABLE 5.2 Distribution by craniofacial orientation of the burials

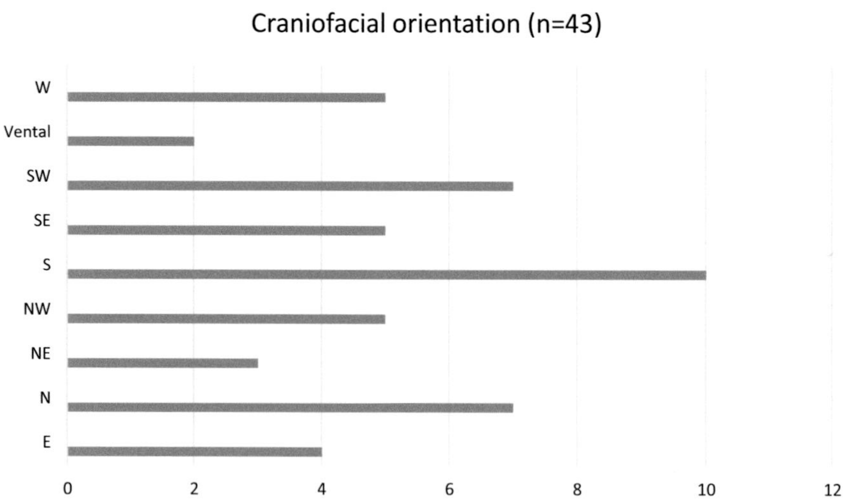

TABLE 5.3 Orientation by skull-feet axis

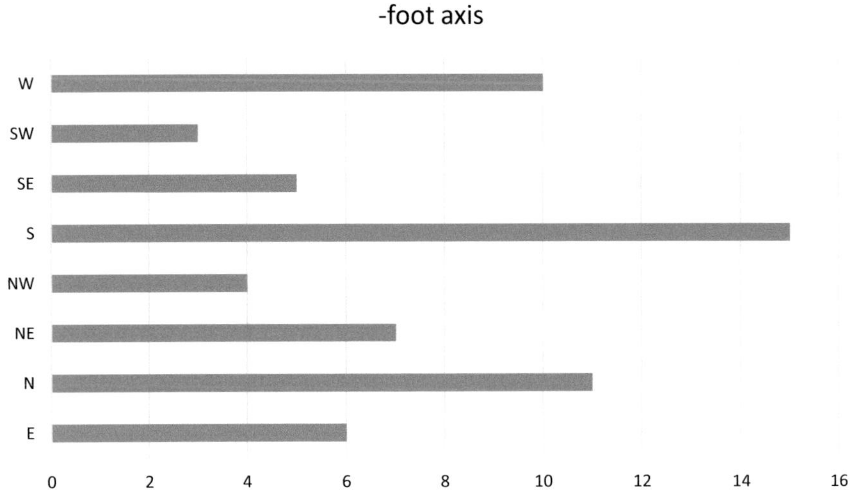

TABLE 5.4 Distribution by anatomical position of the burials

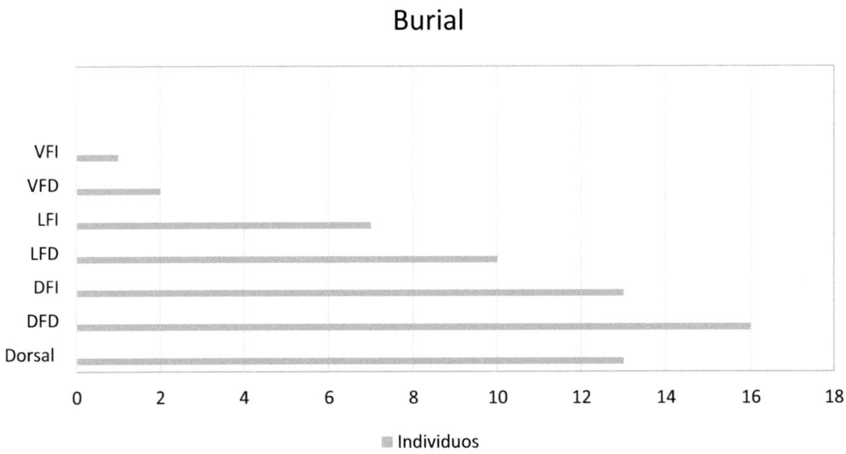

most frequently observed for subadult burials, with a total of eight. The least frequent adult position was ventral (prone), with three (4.8%) (Table 5.4).

Associated Materials

Different types of cultural objects associated with the burial contexts were detected at this site, undoubtedly part of a funerary offering. For example, we found several burials had vessels placed on the head and body, some complete or comprised of

TABLE 5.5 Distribution of materials associated with burials

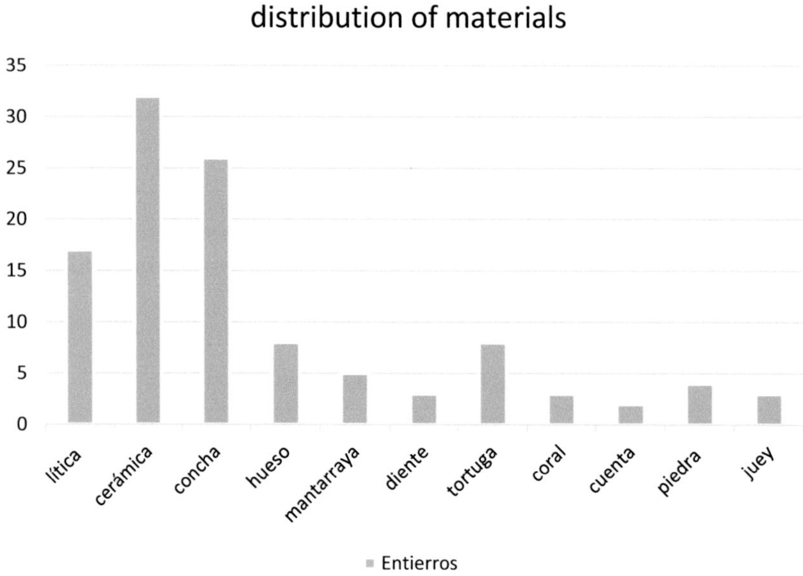

large fragments. We also found tortoise shells of different sizes, stone necklace beads, mollusk shells, bones, teeth, and parts of different animal species (e.g., peccary tusks, barracuda, and manta ray stingers). In other cases, we found within the pit and peripheral area of the skeleton, animal bone segments, and ceramic and stone fragments, which we considered to be part of the funerary context (Table 5.5). The diversity of objects found alerted us to carefully examine the burial finds, as on many occasions these objects were classified as part of the midden, which undermines the interpretation of the funerary customs of these pre-Columbian groups, both in Puerto Rico and in the Antilles.

Despite the fact that in some cases we found highly fragmented objects altered via bioturbation, climate, and other events affecting preservation and integrity, we tried to determine the intentionality with which these objects were deposited in a primary manner in an attempt to bring us closer to understanding the symbolism that accompanied the deceased. We must consider that there were third parties who prepared the body and that there was an intrinsic cultural meaning to the beliefs surrounding death.

On the other hand, we have cases where we did not detect artifactual materials – this does not mean that the burial lacked an offering or funerary object, but rather that it was not preserved, or at least not in an obvious way. These cases required other types of specialized analysis. The methodology we used for this work was the macroscopic evidence of funerary offerings. A total of 56 burials (adults and subadults) had associated material, and a total of 22 burials had no observable associated material.

Discussion and Conclusion

The La Gallera site was undoubtedly an important Indigenous settlement, both for its population density (considering that only a segment of the site was studied) and for its cultural activities, as attested by the burials and objects recovered. The evidence suggests that highly complex and diverse mortuary practices were carried out at this site, placing it on a par with another site of great importance in the eastern region of Puerto Rico, Punta Candelero in Humacao (Rodriguez López 1991; Crespo 2000). Due to the contemporaneity and spatial proximity of these sites, it is very likely that both villages had some kind of interrelationship and human, social, and material exchange.

In this investigation, we were able to identify that burials were distributed virtually throughout the entire assessed area, with a strong possibility that there were more burials in non-evaluated areas. Although we detected a concentration to the north of the site, this was largely due to the archaeological work occurring there, specifically during stage IV of the mitigation phase.

Likewise, we saw that the majority of the population detected were adults (60 individuals), but there was also the presence of 18 subadults, indicating that they did not have a specific or separate place to bury adults and children but treated them equally. Most of the burials were individual burials placed in primary inhumations, while others were grouped together collectively.

For body orientations, with respect to the cardinal points, we saw a slight preference for positioning both the face and the cranio-foot axis toward the south. Regarding body placement, they preferred dorsal flexed positions, with a slight preference for the right side. Interestingly, we saw cases of children and adults in a dorsal flexed position, bending their legs upwards (hyperflexion of the hip and legs), that is, in the form of a "frog", with a higher prevalence occurring in children.

Something that is undoubtedly worth mentioning is the cases of the vessels found on the bodies of individuals. Here we saw a preference for placing them covering the head, although not all were the same, particularly burial #27, which had two vessels containing the body of a neonate. Subsequent DNA analysis could be used to confirm a filial relationship.

Another aspect to highlight is the burials with stingray stingers. Although there are other archaeological cases reported (Siegel 1992), the site of La Gallera stands out for the number of individuals (n = 4) that presented stingray tips, which could well have been part of the paraphernalia needed to make certain tools such as arrowheads, spearheads, and harpoons. In other words, the presence of these manta ray stingers tells us about certain activities. The stingray tip is a unique weapon and undoubtedly had great symbolic meaning, possibly related to power.

Another of the findings that stand out are the turtle shells or carapaces. Although fragments of turtle bone were found in several of the burials, we have cases where the carapace is complete, such as burial #72, where an adult was partially covered by one, and burial #18, where a subadult had a tortoiseshell carapace placed in the abdominal area. Additionally, there is the case of Burial #27, associated with multiple large tortoise shell fragments. There is no doubt that turtle meat was an important food resource for these populations; even the Spanish chronicles refer to it (Fernandez 1959: 62–63). Turtle is widely represented artifactually and has great relevance to Indigenous cosmogony. Although the available narratives refer to much later populations, specifically the so-called Taínos (Robiou 2003: 141,217), there is no doubt that, for the earliest populations, this reptile had an important symbolic meaning, as a companion in the face of death.

With this work, we wanted to present some of the most important bioarchaeological funerary contexts in Puerto Rico, the vast majority of which were investigated by the distinguished researcher Edwin Crespo Torres, as well as present the bioarchaeological findings of the pre-Columbian site of La Gallera. It is pertinent to emphasize the importance of treating all funerary contexts ethically and with respect, giving them the importance and care that each one requires from the simplest to the most complex, recognizing their complexity and symbolism. Much remains to be investigated and reconstructed from the site of La Gallera, part of the valuable history of this pre-Columbian community, which forms part of the bioarchaeological heritage of the Puerto Rico archipelago.

Notes

1. Indigenous name for the island of Puerto Rico.
2. 78 individuals were analyzed, the remaining seven could not be analyzed osteologically due to their poor state of preservation.
3. Bodies that are placed in a burial receptacle or container, including cave burials (Romano 1974: 83).
4. Some archaeologists argue that it was a parallel migration with the Saladoid (Chanlatte 2013; Chanlatte and Narganes 2005; Rodríguez López 1991, 2004).
5. Cultural Resource Management.

References

Alegría, R., Nicholson, H.B., and Willey, G.R. 1955. "The Archaic tradition in Puerto Rico". *American Antiquity* 21 (2), 113–121

Ayes, Suárez Carlos M. 1988. *Phase 2 Type Archaeological Assessment.* Barceloneta, Puerto Rico: Angostura, Florida Afuera. Unpublished manuscript available in the Archives of the Council for the Protection of the Terrestrial Archaeological Heritage of Puerto Rico. PCI.

Budinoff, L.C. 1991. "An Osteological Analysis of Human Burials Recovered from and Early Ceramic Site Located on the North Coast of Puerto Rico". En *Proceedings of the Twelfth International Congress for Caribbean Archaeology,* edited by L.S. Robinson, pp. 117–133.. Martinique: International Association for Caribbean Archaeology.

Buikstra, Jaen E. 1977. "Biocultural Dimensions of Archaeological Study: A Regional Perspective." In *Biocultural Adaptation in Prehistoric America*, edited by R. L. Blakely, pp. 67–84. Athens, GA: University of Georgia Press.

Buikstra, Jane E. 2006. "People and Projects: Early Landmarks in American Bioarchaeology." In *Bioarchaeology. The Contextual Analysis of Human Remains*, edited by J. Beck, Jane Buikstra, pp. 1–6. Boston: Academic Press.

Buikstra, Jane E. and Beck Lane (eds.). 2006. *Bioarchaeology: The Contextual Analysis of Human Remains.* Boston: Elsevier Inc.

Chanlatte, Baik L. and Storde, Yvonne Narganes. 1991. El hombre de Puerto Ferro, Vieques – Puerto Rico. In *Proceedings of the Fourteenth Congress of the International Association for Caribbean Archaeology*, Barbados, edited by A. Cummings and P. King, pp. 599–611. Martinique: IACA. https://ufdc.ufl.edu/aa00061961/00516.

Chanlatte, Baik L and Storde Yvonne Narganes. 2005. *Hueca culture. Sorcé farm, La Hueca neighborhood, Vieques.* San Juan, Puerto Rico: Museum of History, Anthropology and Art University of Puerto Rico, Río Piedras Campus.

Crespo, Torres Edwin F. 1988. "Anthropophysical Study of Human Burials from the Archaeological Site, located in the Carlos Bonet Stadium in the Town of Comerío, Puerto Rico." *Requested by Jaime Velez for the Final Archaeological Report Phase 1A, 1B, and II.* Unpublished manuscript available in the Archives of the Council for the Protection of the Terrestrial Archaeological Heritage of Puerto Rico. PCI.

Crespo, Torres Edwin F. 1989. *Analysis of Sex, Age and Mortuary Practices of Burials Recovered from Punta Candelero: A Site on the Southeast Coast of Puerto Rico*. Manuscript on file, Tempe, AZ: Department of Anthropology, Arizona State University.

Crespo, Torres Edwin F. 1991. "Informe preliminar sobre los entierros humanos en el yacimiento de Punta Candelero, Puerto Rico." En *Proceedings of the Thirteenth International Congress for Caribbean Archaeology,* edited by Ayubi y J.B. Haviser, pp. 840–853. Curaçao: Reports of Archaeological-Anthropological Institute of the Netherlands Antilles.

Crespo, Torres Edwin F. 1994. "Dental Analysis of Human Burials Recoverend from Punta Candelero: A Prehistoric Site on the Southeast Coast of Puerto Rico." Master's thesis, Tempe, AZ: Department of Anthropology, Arizona State University.

Crespo, Torres Edwin F. 1997. "Osteological Analysis of Human Remains from Maruca: A Preceramic Site South of the Island of Puerto Rico." In *Final Archaeological Report Maruca Site in Ponce submitted by Miguel Rodríguez*. Unpublished report available in the Archives of the Council for the Protection of the Terrestrial Archaeological Heritage of Puerto Rico. PCI.

Crespo, Torres Edwin F. 2000. *Biocultural Comparative Study between Two Prehistoric Populations on the Island of Puerto Rico: Punta Candelero and Paso del Indio.* Doctoral thesis, Autonomous University of Mexico.

Crespo, Torres Edwin F. 2000a. Mortuary analysis of 18 human remains from the archaeological site of Luján, Vieques, Puerto Rico. Requested by Virginia Rivera and submitted to the Council for the Protection of the Terrestrial Archaeological Heritage of Puerto Rico. Unpublished manuscript available in the Archeology Council Archive, ICP.

Crespo, Torres Edwin F. 2001. "Final Report on Burials Discovered During Archaeological Excavations in the Esperanza Community in the Municipality of Vieques, Puerto Rico." Report for Marlene Ramos Vélez, Project Stage 1 Phase III, improvements to the health system in the Esperanza community. Unpublished manuscript available in the Archeology Council Archive, ICP.

Crespo, Torres Edwin F. 2002. "New Interpretations about the Belief about Death and Funerary Practices of the Indians of Boriquén." *Journal Institute of Puerto Rican Culture (ICP)* 3 (5): 83–94.

Crespo, Torres Edwin F. 2005. "The Huecoid Culture and its Connection to the Introduction of the Practice of Intentional Head Deformation in the Antilles." In *La Hueca Culture. Finca Sorcé, Barrio La Hueca, Vieques,* edited by Y. Nargane Storde and L. Chanlatte Baik, pp. 57–65. San Juan, Puerto Rico: Museum of History, Anthropology and Art. University of Puerto Rico, Río Piedras Campus.

Crespo, Torres Edwin F. 2005a. "Evidence of Pre-Columbian Tremonematosis from Paso del Indio: An Archaeological Site on the Island of Puerto Rico." In *The Myth of Syphilis: The Natural History of Treponematosis in North America*, edited by M.L. and D. Collins Cook Powell, pp. 386–4001. Gainesville: University Press of Florida.

Crespo, Torres Edwin F. 2008. "Illness in the Indians of Borinquén." *Journal of the Institute of Puerto Rican Culture* 16: 3–15.

Crespo, Torres Edwin F. 2010. "Bioarquelogia en resto humanos precolombinos en el Caribe insular". In *Encuentro con el Caribe Francés: Un encuentro entre literatos, historiadores, arqueólogos y antropólogos de Haití, Guadalupe, Martinica y Puerto Rico,* edited by L. Del Olmo Frese, pp. 130–139. San Juan, PR: Instituto de Cultura Puertorriqueña. https://www.icp.pr.gov/wp-content/uploads/2018/06/Encuentro-con-el-Caribe-Frances-768x768.pdf

Crespo, Torres Edwin F. 2010a. "Ancient Bones Tell Stories: Osteobiography of Human Remains from Tibes." In *Tibes People, Power and Ritual at the Center of Cosmos*, edited by A. Curet and L.M. Stringer, pp. 191–208. Tuscaloosa: University of Alabama Press.

Crespo, Torres Edwin F. 2011. "Bioarchaeological Evidence of Ancestor Worship from Paso del Indio: An Archaeological Site in the Island of Puerto Rico." In *University Contribution: Anthology of Interdisciplinary Essays in the Social Sciences*, edited by J. Rodríguez-Gómez, pp. 187–211. Río Piedras: University of Puerto Rico.

Crespo, Torres Edwin F. 2019. "The History of Paleopathology in the Caribbean Archipelago." In *Biological Anthropology of Latin America. Historical Development and Recent Advances*, edited by D. Ubelaker and S. Colantonio, pp. 173–188. Washington, DC: Smithsonian, Scholarly Press.

Crespo, Torres E. and Luz V. Muñoz 2009. "Voces desde la tumba: Estudio bioarqueológico de dos atrio cemeteries de finales del siglo 18 en la isla de Puerto Rico". In *6to Encuentro de Investigadores de Arqueología y Etnohistoria*, edited by L. Del Olmo Frese, pp. 20–29. San Juan, Puerto Rico: Instituto de Cultura Puertorriqueña.

Curet, Antonio and José Oliver. 1998. "Mortuary Practices, Social Development and Ideology in Precolumbian Puerto Rico." *Latin American Antiquity* 9 (3): 217–239 https://doi.org/10.2307/971729

Curet, Luis A. 1990. *Electric Substation Project. Marina Puerto del Rey*. Ceiba, PR: Mitigation Phase. Unpublished manuscript available in the Archeology Council Archive, ICP.

Daubón, Antonio. 2002. *Cultural Resources Study Phase 1B Village at the Hill Project.* Ceiba PR: La Gallera Field. Unpublished manuscript available in the Archeology Council Archive, ICP.

Dembo, Adolfo and Imbelloni, José. 1938. *Deformaciones intencionales del cuerpo humano de carácter étnico.* Buenos Aires: Jose Anesi.

Duday, Henri. 2011. *The Archeology of the Dead.* Oxford and Oakville: Oxbow Book.

Durant, Jean-François and Roget, Henri Petitjean. 1991. "A props d'un collier funeraire a Morel Guadaloupe: Les Huecoid sont-ils un mythe?" In *Proceedings of the Twelfth Congress of the Caribbean Archeology Association*, edited by L.S. Robinson, pp. 53–72. Martinique: IACA.

Fernandez, from Oviedo Gonzalo. 1959. *General and Natural History of the Indies. Volume 2,* edited by Juan Perez de Tutela Bueso. Vol. 2. 5 vols. Madrid, Spain: Library of Spanish Authors.

Fewkes, Jesse W. 2009. *The Aborigines of Puerto Rico and Neighboring Islands.* Originally published in 1907. Tuscaloosa: The University of Alabama Press.

Gonzalez, L. Ernesto and Lourdes Marquez. 2009. *Paradigms and Challenges of Mexican Bioarchaeology.* Mexico: INAH.

Goodwin, R. Christopher. 1978. "The History and Development of Osteology in the Caribbean Area." *Revista/Review Interamericana* 8 (3): 463–494.

Larsen, C.S. 1997. *Bioarchaeology: Interpreting behavior from the human skeleton.* Cambridge: Cambridge University Press.

Martin, Debra L. 1998. "Owning the Sins of the Past: Historical Trends, Misserd Opportunities, and New Directions in the Study of Human Remains." In *Building a New Biocultural Synthesis. Political-Economic Perspectives on Human Biology*, edited by T.L. Leatherman and A.H. Goodman, pp. 171–190. Ann Arbor: University of Michigan Press.

McAnany, P.A. 1995. *Living with the Ancestors: Kinship and Kingship in Ancient Maya Society.* Austin: University of Texas Press.

Melendez, M. Marisol. 1991. "Archaeological Evaluation Phase II. La Gallera Site, Maranatha Baptist Church. Bo. Males, Ceiba". Unpublished manuscript available in the Archeology Council Archive, ICP.

Moline, Daniel. 1991. "Evaluation of Cultural Resources in Lots A, B and D of Puerto del Rey Marina & Resort, Ceiba, PR". Unpublished manuscript available in the Archeology Council Archive, ICP.

Muñoz, Guevara Luz V. 2014. *Bioarchaeological Study of the Punta Mameyes Pre-Hispanic Site (DO-42) Dorado, Puerto Rico* Master's thesis, Center for Advanced Studies of Puerto Rico and the Caribbean.

Muñoz, Guevara Luz V. 2016. "Bioarchaeology of the Pre-Hispanic Site of Punta Mameyes DO-42." In *10th Meeting of Archeology and Ethnohistory Researchers*, edited by L. Del Olmo Frese, pp. 40–53. San Juan: Institute of Puerto Rican Culture.https://www.icp.pr.gov/wp-content/uploads/2018/06/Arqueologia-Interactivo.pdf

Muñoz, Guevara Luz V. 2020. "Field Bioarchaeological Final Report. La Gallera Site, Ceiba PR." As Part of the Project: Archaeological Evaluation Phase III, Construction Stage IV. Boatyard Improvements, Marina Puerto del Rey, Ceiba, Puerto Rico. Unpublished manuscript available in the Archeology Council Archive, ICP.

Nieves-Colón, María, William Pestle, Austin W Reynolds, Bastien Llamas, Costanza de la Fuente, Kathleen Skerry, Edwin Crespo-Torres, Carlos Bustamante, and Anne C. Stone. 2020. "Ancient DNA Reconstructs the Genetic Legacies of PRE-Contact Puerto Rico Communities." *Molecular Biology and Evolution* 37 (3): 611–626. https://doi.org/10.1093/molbev/msz267.

Ortiz, Hernan, Ivor Hernandez, and Peter Camacho. 2020. Field Completion Report Archaeological Evaluation Phase III, Site: La Gallera carr. PR-3 Km. 51.6, Los Machos neighborhood, Ceiba, Puerto Rico. Project: Boatyard Improvements Marina Puerto Del Rey. OGPE Case #2016–107562-REA-170939. Unpublished report available in the Archeology Council Archive, ICP

O'Shea, John M. 1984. *Mortuary Variability an Archaeological Investigation.* Orlando: Academic Press, INC.

Pagan-Jimenez, Jaime. 2013. "Human-Plant Dynamics in the Precolonial Antilles: A Synthetic Update." In *The Oxford Handbook of Caribbean Archaeology*, edited by William Keegan and R. Rodríguez Ramos, C. Hofman, pp. 391–406. Oxford: Oxford University Press.

Pantel, G. 2011. Informe Final. *Mitigación Arqueológica FaseIII del Conector Sur de Santa Isabel [AC-016104] Santa Isabel, Puerto Rico. San Juan. Barrett Hale & Alamo autoridad de Carreteras y Transportación*. Sometido al: Consejo para la Protección del Patrimonio Terrestre de Puerto Rico (unpublished manuscript).

Pestle, William J. 2010. *Diet and Society in Prehistoric Puerto Rico an Isotopic Approach.* Ph.D. dissertation, Chicago: University of Illinois.

Rivero de la Calle, M. 1960. "Cranial Deformation in the Aborigines of Cuba. Comparative Study." *Actes du Vie Congrés Internatinal des Sciences Anthropologiques et Ethnologique* 1: 251–260.

Robiou, Sebastian. 2003. *Tainos and Caribes. Antillean Aboriginal Cultures.* San Juan, Puerto Rico: Semicolon.

Rodriguez, Herminio. 1997. *Phase IA-IB Puerto de Rey Marina & Resort.* Unpublished manuscript available in the Archeology Council Archive, ICP.

Rodriguez, Lopez Miguel. 1986. *Archaeological Evaluation, Phase IB Marina Puerto del Rey*. Fajardo, PR: Demajagua Bay. Unpublished manuscript available in the Archeology Council Archive, ICP.

Rodriguez, Lopez Miguel. 1990. "Archaeological Inventory of the East Coast of Puerto Rico. Final Report Presented to the Institute of Puerto Rican Culture." Unpublished manuscript available in the ICP Archeology Council Archive.

Rodriguez, Lopez Miguel. 1991. "Archaeology of Punta Candelero." In *Proceedings of the Thirteenth International Congress of Caribbean Archaeology,* edited by E.N. Ayubi and J.B. Haviser, pp. 9: 605–627. Curacao: Netherlands Antilles.

Rodriguez, Lopez Miguel. 1999. "Excavations at Maruca, a Preceramic Site in Southern Puerto Rico." In *Proceedings of the Seventeenth Congress of the International Association for Caribbean Archaeology*, edited by J.H. Winter, pp. 166–180. Rockville Center, New York: Molly Collage.

Rodriguez, Lopez Miguel. 2004. *Excavations at the Archaic Site of* Maruca *Ponce*. Puerto Rico. Unpublished manuscript available in the Archeology Council Archive, ICP.

Rodríguez Ramos, R. 2010. "What is the Caribbean? An archaeological perspective". Journal of Caribbean Archaeology. Special Publication no 3, edited by C. Hofman and A. Bright, pp. 19–51.

Rodriguez-Ramos, Reniel. 2013. "Isthmo-Antillean Engagements." In *The Oxford Handbook of Caribbean Archaeology*, edited by William Keegan, Corinne Hofman, and Reniel Rodríguez Ramos, pp. 155–170. Oxford: Oxford University Press.

Roe, Peter, G. Pantel, and M. Hamilton. 1990. "Monserrate Reestudied: The 1978 Centro Field Season at Luquillo Beach Excavation Overview, Lithic and Physical Anthropological Remains." In *Proceedings of the XI International Congress of Caribbean Archaeology,* San Juan, Puerto Rico, edited by Agamemnon G. Pantel Tekakis, Iraida Vargas Arenas and Mario Sanoja Obediente. pp. 338–369. Martinique: IACA.

Romano, Arturo. 1974. "Sistema de enterramientos." In *Antropología Física Época Prehispánica. Panorama Histórico y Cultural III,* edited by J Comas, pp. 83–112. México: SEP/INAH.

Rouse, Irving. 1992. *The Tainos: Rise and Decline og the People who Greeted Columbus.* Yale, NH: Yale University Press.

Rouse, Irving and R. Alegría. 1990. *Excavations at María de la Cruz Cave and Hacienda Grande Village Site, Loíza Puerto Rico.* Yale University Publications in Anthropology, No. 80. New Haven: Yale University Press.

Siegel, Peter. 1992. *Ideology, Power and Social Complexity in Prehistoric Puerto Rico.* Ph.D. dissertation, Satate University of New York at Binghamton.

Siegel, Peter. 1996. "Ideology and Culture Change in Prehistoric Puerto Rico: A View from the Community." *Journal of Field Archaeology* 23: 313–333.

Tibbett, Mark and David Carter. 2008. *Soil Analysis in Forensic Taphonomy: Chemical and Biological Effects of Buried Human Remains*. Boca Ratón, FL: CRC Press.

Ubelaker, Douglas. 1997. "Taphonomic Applications in Forensic Anthropology." In *Forensic Taphonomy the Postmortem Fate of Human Remains*, edited by William D. Haglund and Marcella H. Sorg, pp. 77–87. Boca Ratón, FL: CRS Press.

6

PALEODEMOGRAPHY AND HEALTH AT JUAN DOLIO, DOMINICAN REPUBLIC

Darlene Weston

Introduction

The coastal site of Juan Dolio, in southeast Dominican Republic, is located approximately 70 km east of the capital, Santo Domingo (see Figure 6.1). The site was excavated and surveyed variously in the 1920s (Franco Bido), 1940s (Boyrie Moya and Herrera Fritot), 1950s (Boyrie Moya, Chanlatte and Cruxent), and 1970s (Morbán Laucer and Manuel Garcia, Veloz Maggiolo, Luna Calderón, Rimoli, and Ortega) (Boyrie Moya 1960; Boyrie Moya and Cruxent 1955; Chanlatte Baik 1954; Ortega 2002; Veloz Maggiolo 1972) and is known as one of the principal sites of the "Boca Chica" or Chicoid ceramic style in the southern Dominican Republic (Fernandes et al. 2021). Looting of the site in the 1950s and 1960s resulted in artefacts being sold to private collectors (Ortega 2002). Dating of the first phase of the site's occupation was based on the presence of Saladoid series ceramics; a subsequent Ostionoid phase was radiocarbon dated to 625, 820, and 825 CE; and a final prehistoric (Chicoid) phase was dated to 1130–1300 CE (Ortega 2002). More recent radiocarbon dates based on human bone samples from eight separate burial contexts have two dates encompassing 400–700 CE and six dates encompassing 1200–1700 CE, which span the Ostionoid, Chicoid, and Colonial periods (Fernandes et al. 2021).

In addition to Indigenous material culture, a profusion of Spanish and other European potsherds was found throughout the site, both on the surface and in lower levels (García Arévalo 1978; García Arévalo 1991), including hundreds of sherds typical of majolica, olive jars, and other glazed wares (Keehnen 2019). Several unique transcultural finds include two locally manufactured Spanish chamber pots (*bacines*), with Indigenous decoration and zoomorphic elements dating to circa 1520 CE (Veloz Maggiolo 1993), a typical Indigenous pot with an incised cross

DOI: 10.4324/9780429444449-9

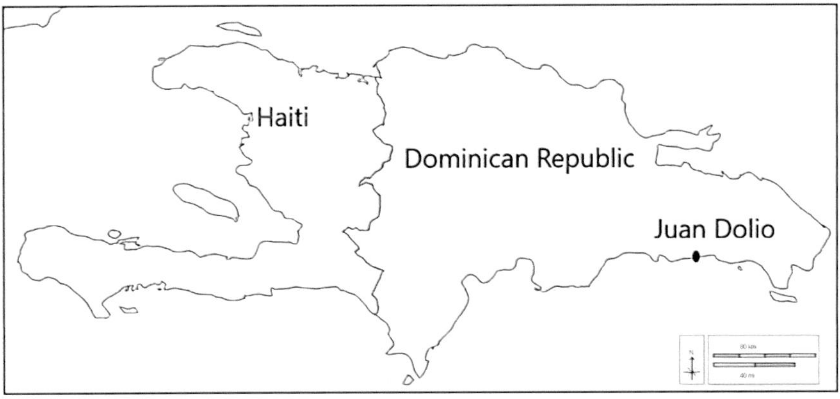

FIGURE 6.1 Map of Hispaniola illustrating location of site Juan Dolio within the Dominican Republic.

(Morbán Laucer and García Arévalo 1971), and two metal threepointers or *cemís*, one cast with a mould and one hammered (García Arévalo 1978; Keehnen 2019).

The site, once situated in the territory of the Maguana chiefdom (Drusini et al. 1987) and comprising an Indigenous settlement and associated cemetery, yielded 102 burial contexts, identified by the Dominican physical anthropologist, Fernando Luna Calderón in 1974 (Drusini et al. 1987). The burials, mainly single primary crouched interments, contained Indigenous grave goods (ceramics), and one particular burial of a male, 46+ years and exhibiting an artificially modified skull, in addition to serveral large Indigenous vessels placed next to his head, also contained a Spanish bowl of a ceramic type known as Columbia Plain (Keehnen 2019). A group of 10 burials, located closer to the beach, were buried in a straight line, in typical European Christian fashion – extended supine (Veloz Maggiolo 1980). These Indigenous individuals, as evidenced by their modified skulls, had associated grave goods, including Spanish jars and vessels, and one juvenile was interred with a bronze buckle, likely part of a belt (Keehnen 2019). In addition to the primary single interments, several double primary interments were present. Fragmentary skeletal elements were also found in association with several of the primary burials, likely a consequence of excavation practices that disturbed nearby graves.

The human skeletal remains from Juan Dolio have been the subject of several studies, including preliminary (Drusini et al. 1987; Veloz Maggiolo 1972) and more detailed reports of the paleodemography (Coppa et al. 1995), an investigation of their dental characteristics (Coppa et al. 1995), a study on the prevalence of treponemal disease in the population (Rothschild et al. 2000), stable isotope studies of population diet (Stokes 1998), an analysis of the artificial cranial modification (van Duijvenbode 2017), and most recently an investigation of their genetic origins (Fernandes et al. 2021).

The author undertook an investigation of the Juan Dolio human skeletons in the summer of 2015, summer of 2016, and autumn of 2016 at the Museo del Hombre Dominicano (Santo Domingo, Dominican Republic). It should be noted that the dentitions were not available for analysis, so all inferences were based on analyses of only the bones. The purpose of the study was to determine the paleodemography and health status of the population within the wider context of investigating the effects of European colonization on the paleodemography and health of Pan-Caribbean Indigenous peoples, as part of the wide-ranging Nexus 1492 project (https://www.universiteitleiden.nl/nexus1492). As the dating of the Juan Dolio site indicates it spans both the pre-contact and post-contact periods, only two individuals have precision ^{14}C dates (Fernandes et al. 2021), and much of the contextual information for individual burials is not currently available, only broad assessment of European impact on the site's Indigenous population is possible. Despite these issues, a great deal can still be gleaned from the skeletal remains of Juan Dolio's burial population. The human skeleton is a repository for an individual's life history – their daily activities, illnesses, nutritional status, as well as their physical characteristics and age at death. When pooled together, these life histories reveal a population history (Weston and Varcárcel Rojas 2016). By reading the population history of the Juan Dolio people, as written in their bones and teeth, a greater understanding of their lives in a rapidly changing world can be achieved.

Minimum Number of Individuals

Human remains from 96 primary burial contexts were examined, from which 104 individuals were identified. Eight of these burial contexts were multiple burials (see Table 6.1), each containing two individuals, most frequently an adult male and female. In addition, 39 burial contexts contained associated fragmentary human remains, likely inclusions from adjacent burials (Table 6.2). These associated human remains fragments were recorded but did not contribute to the palaeodemography of the site. None of the fragments exhibited pathological lesions.

TABLE 6.1 Juan Dolio multiple burial contexts

Burial Context	*Age & Sex*	*MNI*
1a & 1b	?M, 18+; infant, <1 yr	2
25a & 25b	F, 36–45;?M, 18+	2
37a & 37b	I, 36–45; M, 18+	2
40a & 40b	M, 18+; F, 18–25	2
42a & 42b	F, 18+; F, 18+	2
54a & 54b	?F, 18+; J, 4–6	2
66a & 66b	F, 18+; M, 18+	2
95a & 95b	M, 26–35; F, 26–35	2

M = male, ?M = possible male, F = female, ?F = possible female, I = Indeterminate, J = juvenile.

TABLE 6.2 Juan Dolio burial contexts with associated human remains

Burial context	*Bone/Tooth Elements*	*Burial Context*	*Bone/Tooth Elements*
3	L. temporal fragment	61	Juvenile – 2 x vertebral body fragments, MC fragment
8	Ethmoid fragment	62	14 tooth
9a	Occipital fragment	64	MT fragment
9b	R. patella, fibula fragment, 34 tooth	65	Frontal fragment, long bone fragments, C2, R. mastoid, L. clavicle fragment; Juvenile – MC4
10	MC shaft fragment	70	Dist. femur fragment, R. 1st rib, femur head fragment, lunate, cuneiform III,
13	Dist. manual phalanx, lunate; Juvenile – L. femur fragment, L. ischium fragment	74	MC shaft; Juvenile – femur shaft
16	Humerus fragment; Foetus – occipital fragment	75	Juvenile – femur shaft
20	MT4 fragment	76	Juvenile – dist. fibula shaft fragment, 2 x ilia fragments
22a	C1, C2, 2 x T1 vertebrae fragments, MC2, MC3, MC4, MC5, prox. manual phalanx, 2 x scapula fragments, occipital fragment, 1st rib, temporal fragment, 2 x MT2, 2 x MT4; Juvenile – MC4, MC5, carpal, dist. manual phalanx, 3 x prox. manual phalanx, 3 x int. manual phalanx, MC shaft, MT1, prox. pedal phalanx	79	Radius shaft fragment, fibula shaft fragment, 3 MC shaft fragments
25b	C2 fragment, 2 x L. scapula fragment	81	Radius/ulna shaft fragment
26	L. radius shaft fragment, L. ulna shaft fragment	82	MT5
28	L. zygomatic, R. distal radius shaft	83	C1 fragment, C2 fragment
34	MT shaft	89	MC shaft
38-11	R. radius shaft	90	3 x prox. manual phalanx fragments, manual int. phalanx, manual dist. phalanx, 4 x carpals, MT fragment, pedal prox. phalanx, 2 x sesamoids, L. patella
40a	R. temporal fragment, occipital fragment	92	Ulna shaft fragment

(*Continued*)

TABLE 6.2 (Continued)

Burial context	*Bone/Tooth Elements*	*Burial Context*	*Bone/Tooth Elements*
41	Juvenile – L. tibia with prox. epiphysis	93	Frontal fragment
43	L. femur shaft	100	2 x MC
47	2 x humerus fragments	D	Juvenile – sternebra, rib fragment
55a	Juvenile – MT5, MC shaft, manual int. phalanx, rib fragment	F	Juvenile – dist. femur fragment, thoracic vertebra fragment
58	MT5; Juvenile – manual phalanx		

Unless indicated, all bones, teeth or bone fragments are adult. Teeth are designated using FDI nomenclature. L = left, R = right, MC = metacarpal, MT = metatarsal, prox = proximal, int = intermediate, dist = distal, C = cervical, T = thoracic.

Preservation and Completeness

Skeletal preservation and completeness influence the ability to record data related to demography, stature, and health, and as such they are essential to record. For the Juan Dolio skeletons, McKinley's (2004) criteria for skeletal preservation were employed, grading skeletons as excellent, good, fair, poor, and very poor based on the integrity of the cortical bone surfaces. Skeletal completeness was recorded according to the percentage of the skeleton present and available for analysis: 0–25%, 25–50%, 50–75%, and 75–100%.

The Juan Dolio skeletons as a whole primarily exhibited fair preservation (42.7%), and almost 41% of the population (40.8%) exhibited good to excellent preservation, meaning there was relatively little erosion on the cortical surfaces of the bones. However, although the cortical bone surfaces were reasonably well preserved, joint surfaces, bone features, and elements largely composed of trabecular bone were frequently destroyed. In terms of completeness, the Juan Dolio skeletons were distributed nearly equally among the 50–75% complete (33.9%), 25–50% complete (24.3%), and 0–25% complete (31.1%) categories. Echoing the state of preservation, those skeletal elements comprised mainly of trabecular bone (e.g., vertebrae, ribs, pelvis, joint surfaces, etc.) tended to be missing or fragmented. Drusini et al. (1987) commented that coconut tree roots were responsible for much of the poor bone preservation on the site, frequently invading the medullary cavities of long bones. Table 6.3 outlines the overall preservation and completeness of the Juan Dolio burials.

Demography – Age and Sex Estimation

Adult skeletons were assigned age estimations based on morphological changes seen on the auricular surfaces of the os coxae (Lovejoy et al. 1985), the pubic

TABLE 6.3 Preservation and completeness of Juan Dolio Burials

Preservation	*Excellent*	*Good*	*Fair*	*Poor*	*Very Poor*	*Total*
Number of individuals	6	37	44	16	1	104
Percentage	5.8%	35.6%	42.3%	15.4%	1.0%	100%
Completeness	*0–25%*	*25–50%*	*50–75%*	*75–100*	*Total*	
Number of individuals	33	25	35	11	104	
Percentage	31.7%	24.0%	33.6%	10.6%	100%	

TABLE 6.4 Age distribution of Juan Dolio skeletons

Age	*Total*	*% of Total Population*
<0 (foetus)	0	0.0
<1 year	1	1.0
1–4 years	6	5.8
5–9 years	6	5.8
10–14 years	10	9.6
15–17 years	4	3.8
Juvenile < 18 years	0	0.0
18–25 years (Young Adult)	11	10.6
26–35 years (Young Middle Adult)	12	11.5
36–45 years (Old Middle Adult)	10	9.6
46+ years (Mature Adult)	8	7.7
Adult ≥ 18 years	36	34.6
Age indeterminate	0	0.0
Total	104	100.0

symphyses (Katz and Suchey 1986; Todd 1921a, b), and the sternal rib ends (Işcan and Loth 1986a, b), as well as cranial suture closure (Meindl and Lovejoy 1985). For juveniles aged less than 18 years, age was estimated based on long bone length (Sundick 1978; Ubelaker 1989) and degree of epiphyseal fusion (Scheuer and Black 2000). Individual skeletons were assigned to standard age groups rather than given precise ages. The age groups used in this analysis are presented in Tables 6.4–6.6 together with the age distributions of the individuals from the Juan Dolio burials.

The Juan Dolio burial population totalled 104 individuals: 77 adults and 27 juveniles. Unfortunately, a significant percentage of adults (46.7%) could not be assigned to a specific age group and were simply categorized as adults (18+ years). Excluding this group, the skeletal population was relatively evenly distributed among the various age categories, with no category predominating. Typical of Caribbean skeletal populations, infants were rare, with only one being present.

TABLE 6.5 Age distribution of adult Juan Dolio skeletons

Age	*Total*	*% of Adult Population*
18–25 years (Young Adult)	11	14.3
26–35 years (Young Middle Adult)	12	15.6
36–45 years (Old Middle Adult)	10	13.0
46+ years (Mature Adult)	8	10.4
Adult ≥ 18 years	36	46.7
Total	77	100.0

TABLE 6.6 Age distribution of juvenile Juan Dolio skeletons

Age	*Total*	*% of Juvenile Population*
<0 (foetus)	0	0.0
<1 year	1	3.7
1–4 years	6	22.2
5–9 years	6	22.2
10–14 years	10	37.0
15–17 years	4	14.8
Juvenile < 18 years	0	0.0
Total	27	100.0

The biological sex of the Juan Dolio population was determined using morphological and metric methods. Morphological features of the skull (Ascádi and Nemeskéri 1970; Buikstra and Ubelaker 1994) and pelvis (Buikstra and Ubelaker 1994; Phenice 1969) were recorded and analysed, as were measurements of such bones as the clavicle (Jit and Singh 1966), femur (Pearson and Bell 1917/1919; Stewart 1979), humerus (Stewart 1979), and scapula (Iordanidis 1961). As sex determination of the juvenile skeleton is challenging (Scheuer and Black 2000), it was not attempted.

Table 6.7 illustrates the biological sex distribution of the adult individuals from Juan Dolio. The majority (41.5%) of individuals were categorized as female. When the "possible" female and female categories are combined, this total rises to 55.8%, with combined male and "possible male" individuals totalling 39%. In contrast to the age estimates, where a large proportion of the adults in the population could not be assigned to more specific age categories, relatively few individuals could not be assigned a biological sex. This was due to the disparity in preservation status of pelvic and cranial bones. Pubic symphyses and auricular surfaces, frequently used for age determination, were not well preserved, while frontal, temporal, and occipital bones, even if they were fragmentary, could frequently be observed displaying features diagnostic of biological sex.

When the age and sex data for the entire skeletal sample are combined, the most commonly represented group by far is that of adults (18+ years) of indeterminate sex (34.6%), followed by adult females (18.2%), and adult males (14.4%).

TABLE 6.7 Biological sex distribution of Juan Dolio skeletons

Sex	*Total*	*Percentage*
Male	23	29.9
Male?	7	9.1
Female	32	41.5
Female?	11	14.3
Indeterminate	4	5.2
Total	**77**	**100**
Combined M+M?	30	39.0
Combined F+F?	43	55.8

TABLE 6.8 Age and sex distribution of Juan Dolio skeletal population

Age	*Male (M +?M)*	*%*	*Female (F +?F)*	*%*	*Indeterminate*	*%*	*Total*	*%*
>0 (foetus)	0	0.0	0	0.0	0	0.0	0	0.0
juvenile <18 yrs	0	0.0	0	0.0	0	0.0	0	0.0
<1 year	0	0.0	0	0.0	1	1.0	1	1.0
1–4 years	0	0.0	0	0.0	6	5.8	6	5.8
5–9 years	0	0.0	0	0.0	6	5.8	6	5.8
10–14 years	0	0.0	0	0.0	10	9.6	10	9.6
15–17 years	0	0.0	0	0.0	4	3.8	4	3.8
18–25 years	2	1.9	9	8.7	0	0.0	11	10.6
26–35 years	6	5.8	5	4.8	1	1.0	12	11.5
36–45 years	4	3.8	5	4.8	1	1.0	10	9.6
46+ years	3	2.9	5	4.8	0	0.0	8	7.7
Adult ≥ 18 years	15	14.4	19	18.2	2	1.9	36	34.6
Indeterminate	0	0.0	0	0	0	0.0	0	0.0
Total	30	28.8	43	41.3	30	28.8	104	100.0

The combined age and sex data for the Juan Dolio skeletal population is illustrated in Table 6.8.

As part of their study of the genetic history of the pre-contact Caribbean, Fernandes et al. (2021) genetically determined the biological sex of 8 individuals from Juan Dolio [6 (M), 11 (M), 19 (M), 32 (F), 38 (F), D (M), and E (F)]. The author's anthroposcopically determined biological sexes matched the genetically determined sexes in all cases.

Stature Estimation

The living stature of an adult individual can be estimated using regression equations based on long bone lengths (Trotter 1970); however, caution must be applied when using these regression equations, as it is necessary to employ population-specific formulae due to different body proportions related to varied geographic, ancestral, and temporal origins. Unfortunately, there are no regression equations specific to archaeological populations from the Caribbean. To compensate, many researchers use equations devised by Genovés (1967), which were derived from Mexican Amerindian and mixed Mexican (European and Amerindian ancestry) cadaver stature data. Genovés' original equations were revised by Del Angel and Cisneros (2004), and it was these revised equations that were used to estimate the stature for the Juan Dolio population.

The relatively high degree of fragmentation and the poor preservation of long bone joint surfaces in the Juan Dolio skeletal population prevented many stature estimations from being performed, and as a result, stature could only be estimated for 19 individuals. Statures with corresponding biological sex are presented in Table 6.9. The average male stature was 160.86 cm, and the average female stature was 150.16 cm. There appear to be two outliers in this group: a male measuring 172.71 cm and a possible female measuring 161.5 cm. Both of these individuals are considerably taller than the population averages.

Artificial Cranial Modification

Artificial cranial modification was observed in 28 Juan Dolio crania. All exhibited the fronto-occipital parallel style of cranial modification, which is characterized by

TABLE 6.9 Estimated statures (in cm) for the Juan Dolio skeletal population

Burial No.	*Sex*	*Stature (in cm)*
8	M	172.71
19	M	160.65
22a	M	162.53
25a	F	145.88
33	M	159.58
34	F	149.27
39	F	150.77
45	F	154.91
46	M	154.66
47	M	162.27
61	F	149.72
68	F	151.68
69	M	155.73
75	F	144.22
81	M	158.78
86	F	146.58
89	F	148.33
A	F	148.93
F	?F	161.69

TABLE 6.10 Juan Dolio crania with artificial cranial modification

Burial No.	*Sex*	*Burial No.*	*Sex*
3	?M	44	?M
9a	?F	47	M
10	F	50	J
12	J	55	M
19	M	57	F
21	M	62	J
28	F	63	?M
35b	J	69	M
37a	I	70	J
38	F	88	F
40a	M	92	F
41	J	94	M
42a	F	D	?M
43	F	F	F

M = male, ?M = possible male, F = female, ?F = possible female, I = indeterminate, J = juvenile.

flattened frontal and occipital bones and seemingly bulging parietal bones. Artificial cranial modification occurs through the alteration of cranial growth via the use of external appliances, which limit the growth of the cranial vault to particular dimensions and regions of the skull (Cheverud and Midkiff 1992). The shape of the skull can only be altered during infancy when the sutures are still open, growth is rapid, and the bones of the skull are still pliable (Henschen 1966). Bandages, pads, stones, boards, manual moulding, cradle boards, or various combinations of these may be used to attain the desired effect (Dingwall 1931). Cranial modification is a cultural practice whose goal is usually to achieve a societally defined form (Weston 1994). It is also part of the early socialization process and may represent an important rite of passage (van Duijvenbode 2017).

Table 6.10 lists the skulls that were artificially modified in the Juan Dolio skeletal population. Twenty-two of the modified crania belonged to adult individuals with 10 biological males, 12 biological females, and one individual of indeterminate biological sex. Six juvenile cranials were also affected. Calculating the frequency of artificial cranial modification in the skeletal sample is difficult due to the absence or fragmentary nature of many of the crania.

Skeletal Pathology

Of the 104 individuals present in the Juan Dolio population, 75 (72.1%) showed signs of skeletal pathology. Pathological conditions identified include non-specific inflammation/infection, specific infection, degenerative joint disease (DJD), enthesopathy (activity-related pathology), congenital anomalies, trauma, and neoplasm.

Non-specific Inflammation/Infection

Non-specific inflammatory/infectious lesions may materialize on the skeleton as periosteal/endosteal new bone formation, osteitis, or osteomyelitis, depending on which part of the bone is involved: the subperiosteal/subendosteal surface, the bone cortex, or the medullary cavity, respectively. Due to the generalized nature of these lesions, it is very difficult to determine their aetiologies (Weston 2008, 2012a); they may have infectious, metabolic, circulatory, traumatic, or neoplastic origins. The distribution of lesions throughout the skeleton and their specific anatomical locations on individual bones may help to refine possible differential diagnoses.

There were seven individuals who demonstrated inflammatory/infectious lesions: five biological female/possible females, one male, and one juvenile. One female individual (burial 28, 18–25 years) had diffuse, healed, and active grey-coloured striated periosteal reactions on the antero-medial mid-shaft femora, tibiae, and on the medial mid-shaft aspect of the right fibula. The distribution of the lesions suggests a systemic condition, possibly an infection, but as no features diagnostic of specific infections are indicated (i.e., rugose periosteal lesions with nodes and superficial cavitations suggestive of treponemal disease [Hackett 1976]) and no other aspects of the skeleton were affected, a more specific lesion aetiology cannot be proposed. Two individuals had bilaterally affected tibiae. Burial 76 (M, 18+ years) had healed, striated periosteal reactions located antero-laterally mid-shaft, and Burial F (?F, 46+ years) had slight, striated periosteal lesions located antero-medially midshaft. The remainder of the affected individuals were unilaterally affected. Burial 55b (?F, 18+ years) had healing, porous, striated periosteal reactions on the proximal third of the left tibia and fibula shafts; Burial 45 (F, 18–25) had a healing, striated periosteal reaction on the antero-medial aspect of the distal third of the right ulna shaft; and Burials 62 (J, 12–16 years) and C (?F, 18+ years) had right tibiae with striated periosteal reactions on the posterior and antero-medial mid-shaft areas, respectively. Due to the non-specificity of these unilateral lesions, it is not possible to give a more specific aetiology.

Specific Infection

One individual (Burial 3, ?M, 18+ years) had skeletal lesions that were suggestive of treponemal disease. This individual had bilateral lesions on his legs. The left femur had diffuse, rugose, well-remodelled periosteal lesions on the posterior aspect of the mid- and distal third of the shaft. The right femur had rugose, very porous, well-remodelled periosteal lesions on the postero-lateral-medial aspects of the proximal metaphysis. The posterior aspect of the left tibia shaft appeared thickened and expanded, exhibiting very rugose, heavily remodelled periosteal lesions that extended laterally and medially (see Figure 6.2). The bilateral, rugose nature of the periosteal lesions, and particularly the thickened and expanded shaft of the left tibia, are indicative of treponemal disease but not diagnostic, as per Hackett's (1976) criteria. Cranial lesions of the facial and frontal bones exhibiting caries

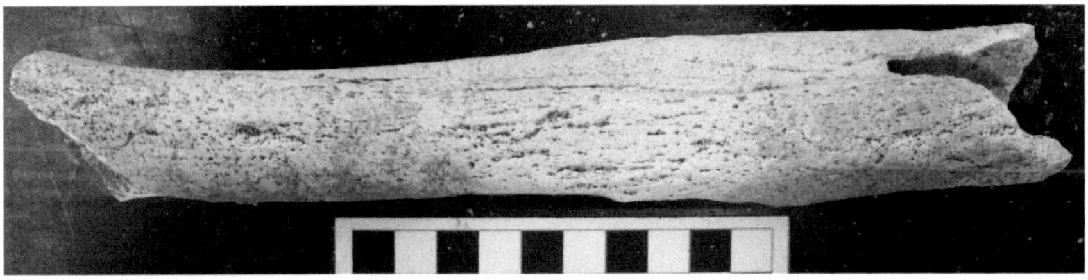

FIGURE 6.2 Left tibia of Juan Dolio burial 3 (?M, 18+ years), exhibiting thickened, expanded shaft and heavily remodelled, rugose periosteal new bone.

sicca and naso-palatine destruction would have helped with a diagnosis, but unfortunately, the anterior aspect of this individual's cranium was too fragmented to be of use. As such, only a cautious suggestion that this individual may have suffered from treponemal disease can be made.

Joint Disease

Eighteen individuals exhibited skeletal evidence for degenerative joint disease (DJD) – 15 in the spine, two in the ribs, one in the clavicle, one in the manual phalanges, and one in the pedal phalanges. It was possible for individuals to exhibit different types of DJD on various joint surfaces.

Seven individuals had very mild examples of degenerative joint disease (DJD). One male and five females have osteophytosis on the vertebral bodies of various thoracic and lumbar vertebrae, as well as the superior body of S1. Among these individuals, for those in their mid-20s and 30s, this is likely an initial sign of spinal degeneration, and for those 46+, it could potentially be an age-related phenomenon (Rogers and Waldron 1995). Additionally, one male had osteophytosis on a proximal pedal phalanx.

Intervertebral disc disease can be diagnosed when there is pitting on the inferior or superior surfaces of the vertebral bodies and marginal osteophytes, all of which occur due to degeneration of the intervertebral disc (Waldron 2009). Five individuals, three males and two females, demonstrated intervertebral disc disease via osteophytosis and porosity on the bodies of various cervical, thoracic, and lumbar vertebrae, with cervical vertebrae being particularly affected.

One individual (Burial 38–7, F, 18+ years; see Figure 6.3) demonstrated Schmorl's nodes on the superior and inferior bodies of the L2 and L3 vertebrae, on the inferior body of the L1 vertebra, and on the superior bodies of the L4 and L5 vertebrae. Schmorl's nodes on the inferior bodies tended to appear shallow (<1 mm deep), while those on the superior bodies tended to appear deep (>2 mm deep). Schmorl's nodes are a result of degenerative and/or traumatic alteration of the intervertebral disc, resulting in herniation of the disc cartilage through the vertebral endplate into the subchondral spaces of the vertebral body. They tend to occur in individuals over the age of 40 and are rare in juveniles (Ortner 2003).

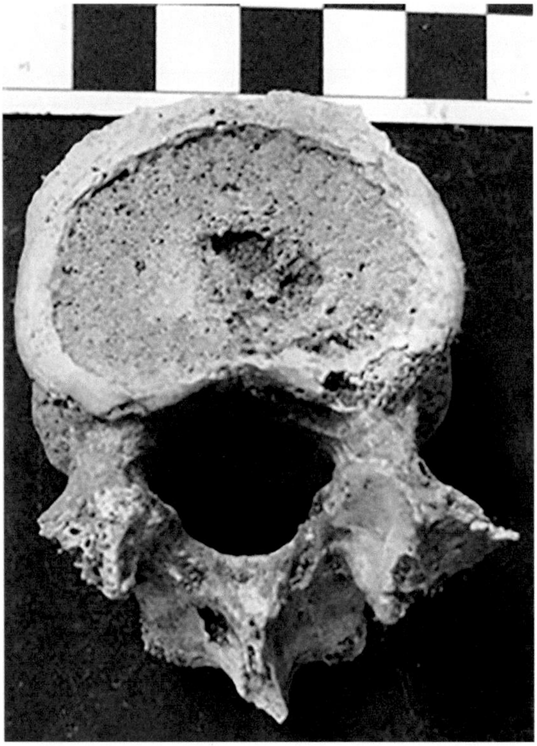

FIGURE 6.3 L3 vertebra of burial 38–7 (F, 18+ years) with Schmorl's node on superior vertebral body.

Six individuals exhibited osteoarthritis in the spine. In order to diagnose osteoarthritis, a number of features must be visible on a synovial joint surface: eburnation alone, or at least two of (1) osteophytosis; (2) porosity; (3) new bone on the joint surface; or (4) joint contour change (Waldron 2009). Those with spinal osteoarthritis primarily had affected inferior and superior articular facets, demonstrating eburnation, porosity, and joint contour change. One example involved Burial 27 (F, 46+ years), who had osteophytosis, joint contour change, and eburnation on the left inferior articular facet of the C1 vertebra and corresponding eburnation, joint contour change, and osteophytosis of the left superior articular facet of the C2 vertebra (see Figure 6.4). This individual also had a congenital defect on this vertebra (see text below). Another example involving Burial 33 (M, 36–45) has what may be a case of axial spondyloarthritis (Sieper and Poddubnyy 2017) or underdifferentiated spondyloarthritis (Waldron 2009). This individual has large osteophytes (>5mm) on the left side of the S1 body, osteophytes on the left costal facet and associated left rib head of T8, ossification of ligamentum flavum on T4, T6, and T7, and fusion of the C3–C7 vertebrae via the inferior and superior articular processes

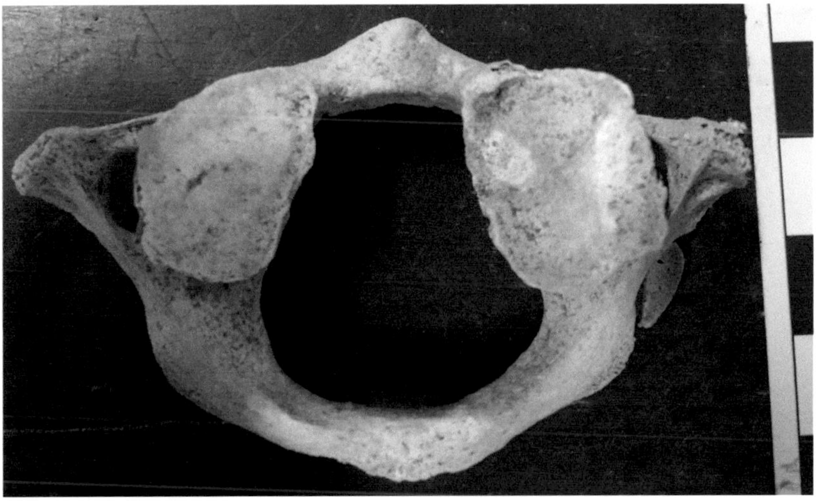

FIGURE 6.4 C1 vertebra of burial 27 (F, 46+) with osteophytosis, joint contour change and eburnation on the left inferior articular facet.

and bodies, with a significant amount of reactive bone formation on the anterior bodies of these vertebrae (see Figure 6.5). No reactive bone was present on the neural arches or in the vertebral foramen. In addition, the right superior articular facet of the C3 exhibited porosity and osteophytosis.

One individual (Burial 25b, M, 18+ years) exhibited an unusual case of erosive osteoarthritis in his right manual phalanges (see Figure 6.6). The right proximal manual phalanges from digits III, IV, and V demonstrated osteophytosis, joint contour change, and eburnation of their distal joint surfaces, with lesion severity reducing from the third digit to the fifth. The corresponding right intermediate manual phalanges exhibited similar lesions of their proximal joint surfaces. A peri-articular erosive lesion was present on the posterior aspect of the distal articular surface of the third proximal manual phalanx. The well-remodelled lesion had destroyed the central aspect of the posterior joint surface, exposing the trabecular bone below. The alteration of the posterior aspects of the affected interphalangeal joints is suggestive of a somewhat clawed positioning of the fingers. Erosive osteoarthritis differs from non-erosive osteoarthritis in that it involves inflammation of the synovial joint capsules and is usually limited to the distal and proximal interphalangeal joints (Peter et al. 1966).

Lastly, two individuals, one biological male (Burial 33, 36–45 years) and one possible biological female (Burial F, 46+ years) had ossification of the ligamentum flavum. Both had evidence for the ossified ligament adhering to mid-thoracic vertebrae. While the aetiology of this condition is not entirely certain, genetic, metabolic, hormonal, as well as degenerative and local biomechanical factors are thought to be involved (Ricciardi et al. 2021).

FIGURE 6.5 Fused C3–C7 vertebrae of burial 33 (M, 36–45), with a possible case of undifferentiated spondylarthritis.

Activity-Related Pathology

Activity-related pathologies, commonly termed enthesopathies, can manifest as either proliferative lesions (enthesophytes) or lytic lesions (cortical defects) (Knüsel 2000) and appear where ligaments and tendons attach to bones or joint capsules. Enthesopathies can be interpreted as a sign of increased muscle or ligament activity but cannot be correlated to specific actions or individual occupations. Only movements can be inferred. Table 6.11 outlines the prevalence of enthesopathies in the Juan Dolio population.

It can be seen that for the biological males, enthesopathies are most frequently observed in the right humerus (60.0%), right clavicle (57.7%), left humerus (56.0%), left clavicle (38.5%), right tibia (22.7%), and right ulna (20.0%). Among the biological females, enthesopathies are most frequently observed in the right clavicle (44.4%), left clavicle (33.3%), right humerus (33.3%), left humerus (27.7%), left ulna (16.2%), and left radius (12.1%). When the data from the adult

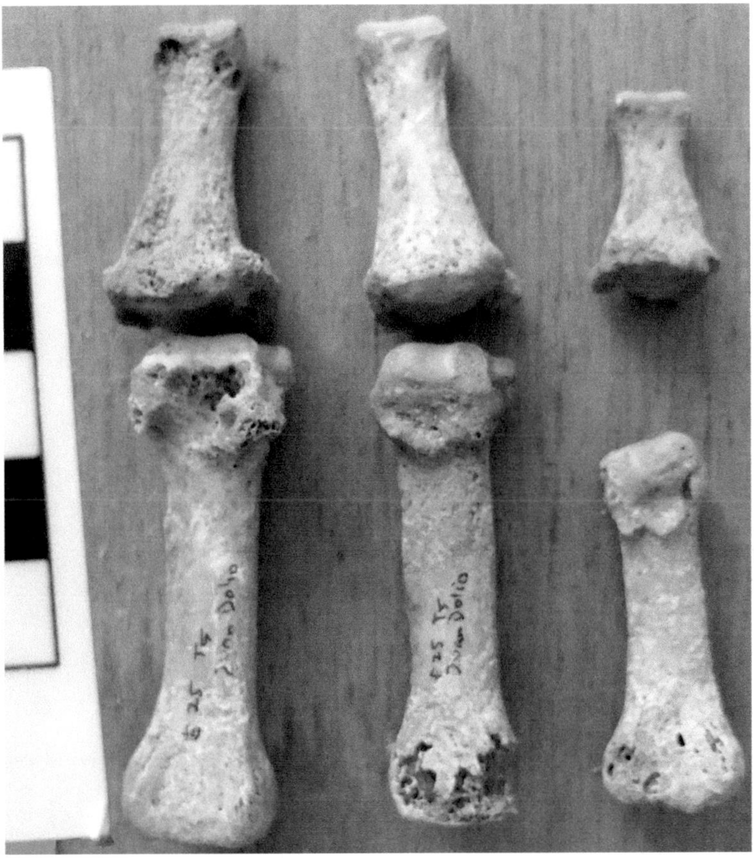

FIGURE 6.6 Right phalanges of burial 25b (M, 18+), with erosive osteoarthritis.

individuals are combined, the most commonly observed enthesopathies occurred in the right clavicle (50.0%), left humerus (40.0%), left clavicle (38.1%), right humerus (29.8%), left ulna (16.1%), and right ulna (14.3%). Among juveniles, the most commonly observed enthesopathies were in the left clavicle (11.1%), right clavicle (6.3%), and right tibia (4.2%). Finally, when evaluating the entire Juan Dolio burial population, the most commonly observed enthesopathies occurred in the right clavicle (41.3%), left clavicle (32.1%) (see Figure 6.7), left humerus (29.2%), right humerus (23.4%), left ulna (12%), and right ulna (10.5%). Based on the distribution of lesions, it appears that a number of individuals from Juan Dolio were engaged in activities that strained the neck, shoulders, elbows, and knees.

Looking at the muscles and ligaments involved in the enthesopathies, for the clavicles, muscle and ligament involvement included the deltoid, pectoralis major, posterior sternocleidomastoid, sternohyoid, subclavius, and trapezius muscles and the conoid, coracoclavicular, and costoclavicular ligaments. The anterior fibres of

TABLE 6.11 Prevalence of enthesopathies in the Juan Dolio burial population by skeletal element

	Male (M + M?)			*Female (F + F?)*			*Indeterminate*			*Total Adults*	*Juvenile*			*Total Population*
	n	*Total*	%	*n*	*Total*	%	*n*	*Total*	%	%	*n*	*Total*	%	%
R clavicle	15	26	57.7	16	36	44.4	1	2	50.0	50.0	1	16	6.3	41.3
L clavicle	10	26	38.5	11	33	33.3	3	4	75.0	38.1	2	18	11.1	32.1
R scapula	0	21	0	0	28	0	0	1	0	0	0	14	0	0
L scapula	0	22	0	0	28	0	0	3	0	0	0	16	0	0
R humerus	12	20	60.0	12	36	33.3	1	3	33.3	29.8	0	23	0	23.4
L humerus	14	25	56.0	10	36	27.7	2	4	50.0	40.0	0	24	0	29.2
R radius	0	23	0	1	35	2.9	0	0	0	1.7	0	21	0	1.3
L radius	0	27	0	4	33	12.1	0	2	0	6.5	0	20	0	4.9
R ulna	5	25	20.0	4	36	11.1	0	2	0	14.3	0	23	0	10.5
L ulna	4	23	17.4	6	37	16.2	0	2	0	16.1	0	21	0	12.0
R hand	1	25	4.0	2	34	5.9	0	2	0	4.9	0	16	0	3.9
L hand	0	25	0	2	40	5.0	0	2	0	3.0	0	20	0	2.3
R pelvis	0	19	0	0	27	0	0	2	0	0	0	17	0	0
L pelvis	0	18	0	0	27	0	0	3	0	0	0	18	0	0
R femur	0	17	0	0	37	0	0	3	0	0	0	21	0	0
L femur	2	23	8.7	0	37	0	0	4	0	3.1	0	24	0	2.3
R tibia	5	22	22.7	0	34	0	0	3	0	8.5	1	24	4.2	7.2
L tibia	3	24	12.5	2	39	5.1	0	3	0	7.6	0	18	0	6.0
R fibula	0	22	0	0	32	0	0	3	0	0	0	23	0	0
L fibula	0	21	0	0	34	0	0	3	0	0	0	18	0	0
R patella	1	13	7.7	0	14	0	0	1	0	3.6	0	5	0	0
L patella	0	8	0	0	17	0	0	0	0	0	0	7	0	0
R foot	1	21	4.8	0	28	0	0	2	0	2.0	0	16	0	1.5
L foot	0	20	0	0	32	0	0	3	0	0	0	15	0	0

FIGURE 6.7 Left and right clavicles of burial 33 (M, 35–45 years), with lytic enthesopathies.

the deltoid muscle work in combination with the pectoralis major muscle to produce flexion of the arm during running and walking and are also activated during medial rotation of the humerus (Soames et al. 2012). The sternocleidomastoid muscle acts during lateral flexion of the neck and elevation and rotation of the head. It also elevates the sternum and clavicle (Kennedy et al. 2017). The sternohyoid muscle depresses and stabilizes the hyoid bone and larynx, and as such is involved in the production of speech, swallowing (Hacking et al. 2016), and head and neck movements. The subclavius muscle stabilizes the clavicle at the sternoclavicular joint during movements of the shoulder and arm and depresses the sternal end of the clavicle and elevates the first rib during respiration. It also protects the brachial plexus and subclavian vessels in the event of a clavicle fracture (Standring 2021). The costoclavicular ligament preserves the stability of the clavicle and restricts the elevation of the shoulder girdle (Cave 1961). The conoid ligament, working in conjunction with the trapezoid ligament, forms a branch of the coracoclavicular ligament, which maintains the acromioclavicular joint, supports the scapula, and transmits and intervenes in the motion of the clavicle and scapula (Nobuhara 2003). Enthesopathies of these clavicular muscles and ligaments suggest regular straining of the neck and shoulders.

Enthesopathies of the humeri primarily involved hypertrophy of the deltoid muscle insertion points. The deltoid muscle abducts, adducts, extends, and flexes the shoulder joint (McMinn et al. 1993). In a few instances, enthesopathies involving pectoralis major and latissimus dorsi were also seen. Both muscles assist in the adduction and medial rotation of the humerus, while latissimus dorsi also extends the humerus (Standring 2021). Enthesopathies at these muscle insertion points again suggest regularly straining of the shoulders.

Enthesopathies of the ulnae involved the brachialis, extensor carpi ulnaris, flexor digitorum superficialis, pronator quadratus, pronator teres, and supinator muscles, as well as the oblique cord ligament. The brachialis muscle is the main muscle involved in flexion of the elbow joint. In addition, it helps to maintain the tension on the elbow joint capsule, which prevents damage in instances of hyperextension. Extensor carpi ulnaris works pull the hand medially via ulnar abduction.

Flexor digitorum superficialis is involved in flexion of the hand joints and is a weak flexor of the elbow joint. More specifically, it flexes the metacarpophalangeal and proximal interphalangeal joints of the second to fifth manual digits (Standring 2021). Pronator teres pronates the forearm, turning the hand to face posteriorly, and with the assistance of pronator quadratus, can turn the hand to face inferiorly when the elbow is flexed at a right angle. It is also a weak flexor of the elbow and assists other muscles in flexion of the elbow during periods of strong resistance (Standring 2021). The oblique cord ties the radius and ulna together, and under severe compressive loads, it may prevent anterior shearing of the proximal radius (Patel 2005). Enthesopathies of the radii involved the biceps brachii and abductor pollicis longus muscles. At the elbow joint, the biceps brachii muscle is involved in flexion and supination of the forearm, while abductor pollicis longus abducts and extends the thumb and helps to abduct the hand (Standring 2021). The presence of enthesopathies at all of these muscle and ligament sites suggests that some individuals were placing a great deal of stress on their elbow and hand joints.

With regards to the lower limbs, it appears that enthesopathy was primarily seen on the tibia, affecting the gracilis, sartorius, semitendinosus, and soleus muscles. At the knee joint, gracilis is involved in flexion and rotation of the leg and is particularly important during the initial swing phase of walking, while also helping to balance the trunk (Standring 2021). Sartorius is involved in flexion of the knee joint and medial rotation of the tibia and may play an important role in climbing (Pirie 2022). Semitendinosus at the knee is also involved in flexion and medial rotation of the leg, also acting to stabilize the joint (Vasković 2021). Lastly, the soleus muscle is involved in plantarflexion of the foot, playing an essential role in walking, running, and jumping, and helps to maintain standing posture (McMinn et al. 1993). The presence of enthesopathies at these muscle sites suggests that some individuals, particularly males, were placing significant stress on their knee joints.

Congenital Anomalies

All the congenital anomalies in the Juan Dolio population appear to be related to the spine. Five individuals were affected, with involvement of cervical or lumbar vertebrae. A biological female, 18–25 years old (Burial 10), exhibited non-fusion of the neural arch of the L5 vertebra at the spinous process. This vertebra also displayed healed trauma (see text below). A juvenile aged four to six years (Burial 13) had fusion of the neural arches of the C1 and C2 vertebrae on the left side (the right side was missing). A biological female, 46+ years (Burial 27), had non-fusion of the neural arch of the C2 vertebra at the spinous process (in addition to osteoarthritis in the cervical vertebrae previously mentioned above) (see Figure 6.8). A biological male, aged 26–35 years (Burial 95A), had partial sacralization of the L5 vertebra on the left side. Lastly, a juvenile aged 3–5 years (Burial 104) had non-fusion of the neural arch at the anterior junction of the C5 vertebra.

FIGURE 6.8 C2 vertebra of burial 27 (F, 46+ years), with non-fusion of the neural arch.

Congenital anomalies in the spine are not unexpected due to the complex nature of the development of the vertebral column (Barnes 2008). One of the most widespread anomalies involves border shifting among different types of vertebrae, where vertebrae develop transitional characteristics of the vertebra above (cranial shift) or below (caudal shift). Shifts at the lumbosacral border are a frequent occurrence, with sacralization of the L5 vertebra occurring bilaterally or unilaterally, as occurred with Burial 95A, and the sacrum appearing to have six segments instead of five (Barnes 2008).

The failure of the formation of transverse fissures between two precursor vertebral segments may result in what is termed a "block vertebra": two vertebral segments fused together as one (e.g., Burial 13). The fusion can be partial or complete and depends in part on the scale of failure in the formation of the fissure. A single-block vertebra is considered to be more anomalous than pathological. They typically occur among the cervical vertebrae and may have a genetic component, commonly occurring in family members. For block vertebrae to be pathological in nature, they must occur in multiples throughout the spine and appear with other vertebral and neurological anomalies, as in the case of Klippel–Feil syndrome (Barnes 2008). Unfortunately, it was not possible to assess the nature of Burial 13's condition, as only a few other fragments of the vertebral column survived.

When one or both halves of a developing neural arch suffer from hypoplasia and fail to fuse, a gap or cleft can be created. The two parts of the arch typically remained joined by fibrous tissue. Clefts are unusual in the thoracic vertebrae but are particularly common in the L5 and sacrum (Barnes 2008). Among the Juan Dolio population, lumbar and cervical vertebrae were affected.

Trauma

There were six cases of trauma in the Juan Dolio population, three affecting the spine, two affecting the forearm, and one affecting a rib. Burial 8 (M, 26–35), had a case of spondylolysis in the fifth lumbar vertebra (see Figure 6.9). Spondylolysis is a stress fracture that occurs in the vertebral neural arch, separating it from the vertebral body (Bennike 2008). The fracture usually occurs between the superior and inferior articular processes, and although any vertebra can be affected, spondylolysis of the fifth lumbar vertebra is most common, being present in 65% of all cases

FIGURE 6.9 L5 vertebra of burial 8 (M, 26–35 years), with spondylolysis.

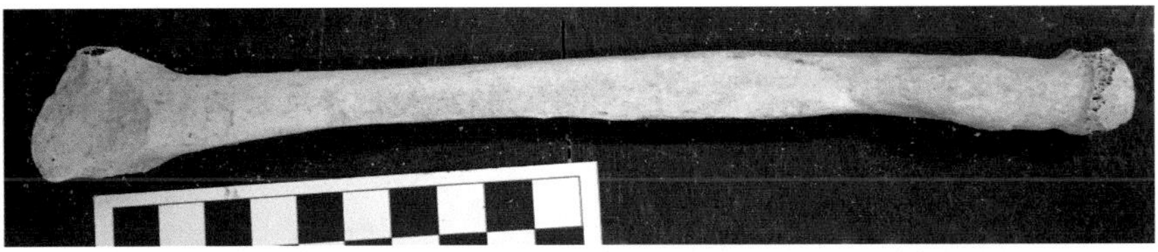

FIGURE 6.10 Colles' fracture of distal right radius of burial A (F, 35–45 years).

(Resnick et al. 1995). Some authors (i.e., Barnes 2008) consider spondylolysis to be a congenital condition occurring due to a genetic predisposition for weakness in the neural arches of some individuals, but ultimately its aetiology is traumatic in origin – a result of intense repeated and/or sustained stress on the neural arch that results in fracture. It occurs more frequently in males, athletes, and labourers involved in heavy lifting (Merbs 1989). As this individual exhibited a well-healed bi-lateral fracture of the neural arch of the L5, had osteophytosis on the inferior body of the affected vertebra, signs of osteophytosis and degeneration throughout the spine, and enthesopathies affecting his right clavicle, left ulna, and left femur, it is safe to assume that he was likely engaged in strenuous physical activities that resulted in these pathologies. The two other cases of spinal trauma included a fracture of the neural arch of a T3 vertebra (Burial 11, M, 18+) with associated osteoarthritis of the right inferior articular process and a fracture of a neural arch of a L5 vertebra (Burial 10, F, 18–25) with significant callus formation. This vertebra was also subject to a congenital defect (see text above).

The two well-healed forearm fractures were identified among the burials of two biological female individuals. Burial 89 (46+ years) exhibited a well-healed spiral fracture of the distal right radius shaft, located approximately 5 cm from the distal end. The individual's right ulna was fragmented at the distal end, so it was not possible to conclusively determine if there was an associated ulnar fracture, as is often the case when a radius is fractured. However, there was some widening in the medial-lateral plane of a distal ulnar shaft fragment, which could be evidence of a healed ulna fracture. The ulna shaft fragment would have been adjacent to the location of the radius fracture. Burial A (36–45 years) demonstrated a well-healed Colles' fracture of the distal right radius, with no accompanying ulna fracture (see Figure 6.10). Colles' fractures are a result of dynamic and static forces working in opposition and are typically the result of an accidental fall on an outstretched arm (Grauer and Roberts 1996). In modern clinical practice, these types of fractures are most commonly seen in elderly women, who sustain falls associated with ageing and have osteoporosis as an associated risk factor (Summers and Fowles 2022).

Neoplastic Disease

There were three individuals with neoplastic disease in the Juan Dolio population. A possible biological female (Burial 9A, 18+ years) and a biological female (Burial

FIGURE 6.11 Neoplastic lesion in rib fragment of burial 33 (M, 36–45 years).

39, 26–35 years) both had osteoid osteomas on their crania. Burial 9A's osteoma, measuring 6.7 mm x 9.2 mm, was on the left side of the occipital, just below the nuchal crest. Burial 39's osteoma, measuring 1.5 cm x 1.85 cm, was located on the right parietal adjacent to the sagittal suture, 1 cm from bregma. An osteoma is a benign, asymptomatic, compact bone tumour that continues to grow very slowly and increase in size throughout an individual's life. Most commonly affecting cranial bones, osteomas are found in individuals of all ages and sexes but most frequently affect individuals aged between 30 and 50 years. Females are affected three times more often than males (de Andrea and Hogendoorn 2009).

One biological male individual (Burial 33, 36–45 years) had a smooth-walled lytic lesion positioned mid-shaft on the inferior-ventral surface of an unsided rib fragment. There were no signs of fracture (see Figure 6.11). Neoplastic lesions are uncommon in the chest wall, with primary rib tumours comprising 5–7% of all primary bone neoplasms (Sabanathan et al. 1984). Most rib lesions are malignant, caused by metastases or invasion from adjacent malignancies like breast or lung cancer (Sakellaridis et al. 2014). Lesions of this nature are difficult to diagnose with certainty, but possible differential diagnoses include benign aneurysmal bone cysts (Sabanathan et al. 1984), fibrous dysplasia, or chondroma, which are all benign, or chondrosarcoma or osteosarcoma, which are malignant (Sakellaridis et al. 2014).

Discussion

With regards to the preservation and completeness of the human skeletal remains from Juan Dolio, the vast majority (77.9%) were in a state of fair to good preservation, though about a third of the individuals were composed of less than 25% of the skeletal elements (see Table 6.3). The relatively fragmentary nature of the skeletons had a negative effect on the ability to reconstruct the demography of the Juan Dolio population, particularly making ageing of the adult skeletons challenging and resulting in an inability to assign an age to over one-third of the burial population. There were 104 individuals examined from 96 burial contexts at Juan Dolio. This contrasts with the 108 reported by Coppa et al. (1995) and the 102 reported by Drusini et al. (1987).

Among the 77 adults and 27 juveniles, the most prevalent demographic group at Juan Dolio was that of adult female (18.26%), followed by adult male (14.4%) and juveniles 10–14 years (9.6%) (see Table 6.8). Excluding the general "adult" category, there was a relatively even distribution of individuals across the age categories, which suggests bias, as with a standard attritional burial assemblage one would expect a high number of infants (though this seems to be rarely the case in Caribbean burial assemblages due to differential mortuary practices for infants [Weston 2011a, b]), a dearth of adolescents, and then an increasing number of adults, peaking with those in old age (Paine 2000). More confidence can be had in the assessment of the biological sex ratio of the burial population, as only 5.2% of the adult skeletons could not be assigned biological sex and the author's anthroposcopic assessment of biological sex matched all the skeletons sexed via DNA analysis (Fernandes et al. 2021).

In terms of biological males and females, we see a definite bias, with biological females outnumbering males 1.4 to 1. As temporally the site spanned the period of interaction with Spanish colonizers, this sex bias could be influenced by the forced labour practices of the Spanish Crown, where males were required to leave their communities to engage in work assignments in other locations, a practice particularly specific to Indigenous populations under encomienda rule (Valcarcel Rojas et al. 2020). It is likely that many men would have died outside of their home villages, proportionally increasing the number of women buried in the village cemetery. Mira Caballos (1997) refers to historical data that indicate encomienda villages have a significantly higher mortality rate for women. The large number of crania exhibiting artificial modification indicates that the population was still carrying out traditional social practices at this time, as evidenced by the discovery of individuals with artificially modified skulls and European goods in burial contexts (Keehnen 2019).

It was possible to provide stature estimates for only 19 Juan Dolio individuals, eight biological males and 11 biological females, due to a lack of long bone preservation (see Table 6.9). The average height for biological females was 150.16 cm, and the average height for biological males was 160.86 cm. This compares favourably to average statures recorded for other Caribbean skeletal populations, including El Flaco, Dominican Republic (F = 149.95 cm; M = 159.25 cm) (Weston 2017), El Chorro de Maita, Cuba (F = 147 cm; M = 159 cm) (Weston 2012b), and Manzanilla, Trinidad (F = 149.7 cm; M = 162 cm) (Weston 2013). As these sites span both pre-contact and colonial periods in several places across the Caribbean, it suggests that stature may have been highly influenced by genetics as opposed to environmental conditions.

With 72.1% of the population demonstrating pathological skeletal lesions, Juan Dolio ranks as the most pathological Caribbean archaeological skeletal population analysed by the author to date. This can be contrasted with El Flaco (21.1%) (Weston 2017), El Chorro de Maita (32%) (Weston 2012b), Anse Lavoutte (33%) (Weston 2011b), Maisabel (40%) (Weston and Schats 2010), Anse à la Gourde (52%) (Weston 2011a), Kelbey's Ridge, Saba (54%) (Weston 2010), and Manzanilla (56%) (Weston 2013).

There were seven cases of non-specific inflammation/infection, one case of specific infection (treponemal disease), 18 cases of DJD, three examples of neoplasm, six examples of trauma, and five congenital spinal anomalies. Enthesopathies were seen primarily affecting the neck, shoulders, elbows, and knees. These results contrast significantly with Drusini et al.'s (1987) results, as this study did not encounter any examples of congenital hip dysplasia, cribra orbitalia, osteoporosis, or spina bifida occulta and recorded more instances of trauma.

These results also contrast with Rothschild et al.'s (2000) tally of three individuals affected by treponemal disease, which those authors propose to be of syphilitic origins. That treponemal disease was present on at least one individual from Juan Dolio, this author agrees, but a diagnosis of syphilis is not supported at this stage due to a lack of Hackett's (1976) diagnostic features visible on the affected bones, particularly a lack of pathognomonic carries sicca. It stands to reason that the treponemal disease most likely to have caused the lesions in Juan Dolio would have been yaws, as this disease was considered endemic in the Caribbean and South America. In 1956, 3,827 cases of yaws were reported in the Dominican Republic (St John 1985). As cited by Vargas Cuellar (1949), Don Gonzalo Fernandez de Oviedo y Valdez wrote in *Historia general de las Indias* (1535) of a yaws-like illness in the local Indigenous population, describing it thus:

> *The Indian naturally has the entire or major portion of the skin ulcerated with the skin lifted up in a hardened scar. They appear ugly but more commonly are robust with considerable strength until they become shrivelled and this wrinkling is the disease that is finished when the itching or disease has changed the whole skin of the person. It is confirmed that this disease is contagious and is taken in many ways such as a healthy person using the clothes of the sick, as in eating and drinking in his company, or using plates and cups which the sick have used for eating and drinking... The truth is that this is a true plague of this land and as common to the Indians as other diseases are common in other people.*

The presence of treponemal disease in the Juan Dolio population was certainly noteworthy, but by far the most striking aspect of the paleopathological assessment of the Juan Dolio skeletons was the inordinate amount of pathology seen in the spine as indicative of hard physical labour – osteoarthritis, degenerative disc disease, and several instances of trauma. Could these pathologies be records of forced labour written into bone? Spanish gold mines were some of the dominant sources of labour for enslaved Indigenous peoples in Hispaniola (Hofman et al.

2018; Valcarcel Rojas et al. 2020), and some of Juan Dolio's inhabitants may have been participants. The spinal pathology, together with the enthesopathies, paints a picture of a population engaged in extremely physically demanding work – even juveniles were involved, as evidenced by enthesopathies seen in the clavicles of at least two. Because the Juan Dolio skeletal population spans both prehistoric and post-colonial periods and the radiocarbon dating of many of the skeletons cannot be precise, unfortunately it will not be possible to determine when the affected individuals acquired their labour-based pathologies.

Conclusions

The Juan Dolio skeletal population, consisting of approximately 104 individuals, spanning dates from 400 to 1700 CE, was interred in primary burial contexts with Indigenous ceramics. European and transcultural artefacts were also discovered on the site in burial contexts with individuals with artificially modified skulls. While the fragmentary nature of the population precludes a detailed interpretation of the site's demography, it was discovered that biological females outnumbered biological males, a hallmark of communities subject to Spanish colonial rule where males were forced to leave their villages and toil as miners or hard labourers. Juan Dolio exhibited an extremely high number (72.1%) of individuals with pathological skeletal lesions, with a notable number of spinal pathologies and skeletal enthesopathies indicative of hard physical labour involving exertion of the muscles and ligaments of the neck, shoulders, elbows, and knees. Treponemal disease was also visible in the skeletal population, most probably a consequence of yaws, which was likely endemic.

Acknowledgements

This research was funded by the European Research Council under the European Union's Seventh Framework Programme (FP7/2007–2013)/ERC-NEXUS1492 grant agreement 319209. Thanks to Christian Martinez-Villaneuva, Director of the Museo del Hombre Dominicano in Santo Domingo, Dominican Republic, for permission to study the Juan Dolio skeletal collection and to Glenys Tavares for facilitating access to the collection.

References

Ascádi G, Nemeskéri J. 1970. *History of Human Life Span and Mortality*. Budapest: Akadémiai Kiadó.

Barnes E. 2008. Congenital anomalies. In: Pinhasi R, Mays S, eds.: *Advances in Human Palaeopathology*. Chichester: John Wiley & Sons, pp. 329–362.

Bennike P. 2008. Trauma. In: Pinhasi R, Mays S, eds.: *Advances in Human Palaeopathology*. Chichester: John Wiley & Sons, pp. 309–328.

Buikstra JE, and Ubelaker, DH. 1994. *Standards for Data Collection from Human Skeletal Remains*. Fayetteville: Arkansas Archaeological Survey.

Cave AJE. 1961. The nature and anatomy of the costoclavicular ligament. *Journal of Anatomy* 95: 170–179.

Chanlatte Baik LA. 1954. Una Exploracion Arqueologica a Juandolio. *Alma Mater, Periodico de Los estudiantes de la Universidad de Santo Domingo,* Nos. 1 and 2.

Cheverud JM, and Midkiff JE. 1992. Effects of fronto-occipital cranial reshaping on mandibular form. *American Journal of Physical Anthropology* 87: 167–171.

Coppa A, Cucina A, Chiarelli B, Luna Calderon F, and Mancinelli D. 1995. Dental anthropology and paleodemography of the Precolumbian populations of Hispaniola from the Third Millennium B.C. to the Spanish Conquest. *Human Evolution* 10: 153–167.

de Andrea CE, and Hogendoorn PCW. 2009. Bone: Osteoma. *Atlas of Genetics and Cytogenetics in Oncology and Haematology*. https://AtlasGeneticsOncology.org/Tumors/OsteomaID5427.html.

de Boyrie Moya E. 1960. *Cinco años de arqueología Dominicana*. Anales de la Universidad de Santo Domingo. Santo Domingo, DR: Universidad de Santo Domingo.

de Boyrie Moya, E, and Cruxent JM. 1955. Muestras arqueológicas de Juan Dolio, Rep. Dominicana. *Boletín del Museo de Ciencias Naturales* 1.

del Angel A, Cisneros HB. 2004. Technical note: Modification of regression equations used to estimate stature in Mesoamerican skeletal remains. *American Journal of Physical Anthropology* 125: 264–265.

Dingwall EJ. 1931. *Artificial Cranial Deformation: A Contribution to the Study of Ethnic Mutilations.* London: John Bale, Sons & Danielsson.

Drusini A, Businaro F, and Calderon FL. 1987. Skeletal biology of the Taino: A preliminary report. *International Journal of Anthropology* 2: 247–253.

Fernandes DM, Sirak KA, and Ringbauer H. et al. 2021. A genetic history of the pre-contact Caribbean. *Nature* 590: 103–110. https://doi.org/10.1038/s41586-020-03053-2

García Arévalo MA. 1978. La Arqueología Indo-Hispana En Santo Domingo. In: E Wagner and A Zucchi (eds.): *Unidad Y Variedad: Ensayos Antropológicos En Homenaje a José M. Cruxent*. San Antonio de Los Altos Miranda, Venezuela: Instituto Venezolano de Investigaciones Científicas, pp. 77–127.

García Arévalo MA. 1991. Influencias Hispánicas En La Alfarería Taína. In: EN Ayubi and JB Haviser (eds): *Proceedings of the 13th Congress of the International Association for Caribbean Archaeology*. Martinique: IACA, pp. 363–383.

Genovés S. 1967. Proportionality of the long bones and their relation to stature among Mesoamericans. *American Journal of Physical Anthropology* 26: 67–78.

Grauer A, and Roberts C. 1996. Paleoepidemiology, healing, and possible treatment of trauma in the medieval cemetery population of St. Helen-on-the Walls, York, England. *American Journal of Physical Anthropology* 100: 531–544.

Hackett CJ. 1976. *Diagnostic Criteria of Syphilis, Yaws and Treponarid (Treponematoses) and of Some Other Diseases in Dry Bones (for Use in Osteo-Archaeology).* Heidelberg: Springer-Verlag.

Hacking C, Weerakkody Y, and Ariyasinghe C. 2016. *Sternohyoid Muscle*. https://radiopaedia.org/articles/sternohyoid-muscle.

Henshen F. 1966. *The Human Skull: A Cultural History.* New York: Frederick A. Praeger.

Hofman C, Hung J, Malatesta E, Jean J, Sonnemann T, and Hoogland M. 2018. Indigenous Caribbean perspectives: Archaeologies and legacies of the first colonised region in the New World. *Antiquity* 92: 200–216. https://doi.org/10.15184/aqy.2017.247

Iordanidis, P. 1961. Détermination du sexe par les os du squelette (atlas, axis, clavicule, omoplate, sternum). *Annales de Médecine Légale* 41: 280–291.

Işcan M, and Loth S. 1986a. Determination of age from the sternal rib in white males: A test of the phase method. *Journal of Forensic Sciences* 31: 122–132.

Işcan M, and Loth S. 1986b. Determination of age from the sternal rib in white females: A test of the phase method. *Journal of Forensic Sciences* 31: 990–999.

Jit I, and Singh S. 1966. The sexing of adult clavicles. *Indian Journal of Medical Research* 54: 551–571.

Katz D, and Suchey JM. 1986. Age determination of the male os pubis. *American Journal of Physical Anthropology* 69: 427–435.

Keehnen FWM. 2019. Treating 'trifles': The Indigenous adoption of European material goods in early Colonial Hispaniola (1492–1550). In: CL Hofman and FWM Keehnen (eds): *Material Encounters and Indigenous Transformations in the Early Colonial Americas*. Leiden: Brill.

Kennedy E, Albert M, and Nicholson H. 2017. The fascicular anatomy and peak force capabilities of the sternocleidomastoid muscle. *Surgical and Radiologic Anatomy* 39: 629–645. https://doi.org/10.1007/s00276-016-1768-9.

Knüsel C. 2000. Bone adaptation and its relationship to physical activity in the past. In: Cox M, Mays S (eds.): *Human Osteology in Archaeology and Forensic Science*. London: Greenwich Medical Media, pp. 381–402.

Lovejoy CO, Meindl RS, Pryzbeck TR, and Mensforth RP. 1985. Chronological metamorphosis of the auricular surface of the ilium: A new method for the determination of age at death. *American Journal of Physical Anthropology* 68: 15–28.

McKinley JI. 2004. Compiling a skeletal inventory: Disarticulated and co-mingled remains. In: Brickley M, McKinley JI (eds.): *Guidelines to the Standards for Recording Human Remains*. Institute of Field Archaeologists Paper No. 7. Reading/Southampton: Institute of Field Archaeologists/BABAO, pp. 14–17.

McMinn RMH, Hutchings RT, Pegington J, and Abrahams P. 1993. *A Colour Atlas of Human Anatomy,* 3rd ed. London: Wolfe.

Meindl RS, and Lovejoy CO. 1985. Ectocranial suture closure: A revised method for the determination of skeletal age at death based on the lateral-anterior sutures. *American Journal of Physical Anthropology* 68: 57–66.

Merbs CF. 1989. Spondylolysis: Its nature and anthropological significance. *International Journal of Anthropology* 4: 163–169.

Mira Caballos E. 1997. *El Indio Antillano: Repartimiento, Encomienda y Esclavitud (1492–1542).* Munoz Moya Editor. Bogata: Sevilla.

Morbán Laucer F, and García Arévalo MA. 1971. Describen Hallazgos Objetos Indígenas. *Listín Diario, January* 21.

Nobuhara, K. 2003. *The Shoulder: Its Function and Clinical Aspects*. Singapore: World Scientific Publishing.

Ortega EJ. (2002) *Artefactos en Concha. Arqueologia en Coral Costa Caribe, Juan Dolio, R.D.* Santo Domingo, DR: Publicaciones de la Fundacion Ortega Alvarez, Inc., Vol. IX.

Ortner, DJ. 2003. *Identification of Pathological Conditions in Human Skeletal Remains*, 2nd Ed. London: Academic Press.

Paine, RR. 2000. If a population crashes in prehistory, and there is no paleodemographer there to hear it, does it make a sound? *American Journal of Physical Anthropology* 112: 181–190.

Patel BA. 2005. Form and function of the oblique cord (chorda oblique) in anthropoid primates. *Primates* 46: 47–57.

Pearson K, and Bell J. 1917/1919. A study of the long bones of the English skeleton. I. the femur. *Drapers' Company Research Memoirs. Biometric Series X*. London: Department of Applied Mathematics, University College, University of London.

Peter JB, Pearson CM, and Marmor L. 1966. Erosive osteoarthritis of the hands. *Arthritis & Rheumatism* 9: 365–388. https://doi.org/10.1002/art.1780090302

Phenice T. 1969. A newly developed visual method of sexing in the os pubis. *American Journal of Physical Anthropology* 30: 297–301.
Pirie, E. 2022. *Sartorius Muscle*. https://www.kenhub.com/en/library/anatomy/the-sartorius-muscle.
Resnick, Goergen TG, and Niwayama, G. 1995. Physical injury: Concepts and terminology. In: Resnick D (ed.): *Diagnosis of Bone and Joint Disorders*, 3rd ed. Philadelphia: WB Saunders, pp. 2561–2692.
Ricciardi GA, Garfinkel IG, Carrioli GG, and Ricciardi DO. 2021. Thoracic ligamentum flavum ossification: A rare cause of spinal cord injury without tomographic evidence of trauma in a Caucasian patient. Case report and literature review. *Spinal Cord Series and Cases* 7: 57. https://doi.org/10.1038/s41394-021-00422-5
Rogers J, and Waldron T. 1995. *A Field Guide to Joint Diseases in Archaeology*. Chichester: John Wiley and Sons.
Rothschild BM, Luna Calderon F, Coppa A, and Rothschild C. 2000. First European exposure to syphilis: The Dominican Republic at the time of Columbian contact. *Clinical Infectious Diseases* 31: 936–941.
Sabanathan S, Chen K, Robertson CS, and Salama FD. 1984. Aneurysmal bone cyst of the rib. *Thorax* 39: 125–130.
Sakellaridis T, Gaitanakis S, and Piyis A. 2014. Rib tumors: A 15-year experience. *General Thoracic and Cardiovascular Surgery* 62: 424–440.
Scheuer L, and Black S. 2000. *Developmental Juvenile Osteology*. London: Academic Press.
Sieper J, and Poddubnyy D. 2017. Axial spondyloarthritis. *The Lancet* 390: 73–84. https://doi.org/10.1016/S0140-6736(16)31591-4.
Soames R, Palastanga N, and Richardson P. 2012. *Anatomy and Human Movement*, 6th ed. Amsterdam: Elsevier.
Standring S. 2021. *Gray's Anatomy*, 42nd ed. Amsterdam: Elsevier.
Stewart TD. 1979. *Essentials of Forensic Anthropology Especially as Developed in the United States*. Springfield: Charles C Thomas.
St John RK. 1985. Yaws in the Americas. *Reviews of Infectious Diseases* 7: S266–S272.
Stokes AV. 1998. *A biogeographic survey of prehistoric human diet in the West Indies using stable isotopes.* PhD diss., University of Florida, Gainesville, FL.
Summers K, and Fowles SM. 2022. Colles' fracture. In: StatPearls [Internet]. StatPearls Publishing. https://www.ncbi.nlm.nih.gov/books/NBK553071
Sundick RI. 1978. Human skeletal growth and age determination. *Homo* 29: 228–249.
Todd TW. 1921a. Age changes in the pubic bone. I: The male white pubis. *American Journal of Physical Anthropology* 3: 285–334.
Todd TW. 1921b. Age changes in the pubic bone. III: The pubis of the white female. IV: The pubis of the female white-negro hybrid. *American Journal of Physical Anthropology* 4: 1–70.
Trotter M. 1970. Estimation of stature from intact long limb bones. In: Stewart TD (ed.): *Personal Identification in Mass Disasters*. Washington, DC: Smithsonian Institution, pp. 71–83.
Ubelaker DH. 1989. *Human Skeletal Remains: Excavation, Analysis, Interpretation*, 2nd ed. Washington, DC: Taraxacum.
Valcarcel Rojas R, Laffoon JE, Weston DA, Hooland MLP, and Hofman CL. 2020. Slavery of Indigenous people in the Caribbean: An archaeological perspective. *International Journal of Historical Archaeology* 24: 517–545. https://doi.org/10.1007/s10761-019-00522-x
van Duijvenbode A. 2017. *Facing Society: A Study of Identity through Head Shaping Practices among the Indigenous Peoples of the Caribbean in the Ceramic Age and Colonial Period.* Unpublished PhD thesis. The Netherlands: University of Leiden.

Vargas Cuellar P. 1949. Yaws in the Department of Cauca Valley. *Bulletin of the Pan American Sanitary Bureau* 20: 897–913.
Vasković J. 2021. *Semintendinosus muscle*. https://www.kenhub.com/en/library/anatomy/semitendinosus-muscle.
Veloz Maggiolo M. 1972. *Arqueología Prehistórica de Santo Domingo*. Singapore: McGraw-Hill Far Eastern Publishers.
Veloz Maggiolo M. 1980. *Vida y cultura en la prehistoria de Santo Domingo*. San Pedro de Macorís, DR: Universidad Central del Este.
Veloz Maggiolo M. 1993. *La isla de Santo Domingo antes de Colón*. Santo Domingo, DR: Banco Central de la Republica Dominicana.
Waldron T. 2009. *Palaeopathology*. Cambridge: Cambridge University Press.
Weston DA. 1994. *Prehistoric Artificial Cranial Deformation at the Pender Canal Site, British Columbia, Canada.* Unpublished MSc thesis. University of Bradford, UK.
Weston DA. 2008. Investigating the specificity of periosteal reactions in pathology museum specimens. *American Journal of Physical Anthropology* 137: 48–59.
Weston DA. 2010. *Human Skeletal Report: Kelbey's Ridge 2 and Spring Bay 1c, Saba.* Archive report prepared for NWO Project Communicating communities. Unravelling networks of human mobility and exchange of goods and ideas from a pre-colonial, pan-Caribbean perspective. Manuscript on file, Leiden: Leiden University.
Weston DA. 2011a. *Human Skeletal Report: Anse à la Gourde, Guadeloupe*. Archive report prepared for NWO Project Communicating communities. Unravelling networks of human mobility and exchange of goods and ideas from a pre-colonial, pan-Caribbean perspective. Manuscript on file, Leiden: Leiden University.
Weston DA. 2011b. *Human Skeletal Report: Anse Lavoutte, St Lucia.* Archive report prepared for NWO Project Communicating communities. Unravelling networks of human mobility and exchange of goods and ideas from a pre-colonial, pan-Caribbean perspective. Manuscript on file, Leiden: Leiden University.
Weston DA. 2012a. Non-specific infection in paleopathology: Interpreting periosteal reactions. In: Grauer AL (ed): *Companion to Paleopathology*. New York: Wiley-Blackwell, pp. 492–512.
Weston DA. 2012b. *Human Skeletal Report: El Chorro de Maita,* Cuba. Archive report prepared for NWO Project Communicating communities. Unravelling networks of human mobility and exchange of goods and ideas from a pre-colonial, pan-Caribbean perspective. Manuscript on file, Leiden: Leiden University.
Weston DA. 2013. *Human Skeletal Report: Manzanilla, Trinidad.* Archive report prepared for NWO Project: Communicating communities. Unravelling networks of human mobility and exchange of goods and ideas from a pre-colonial, pan-Caribbean perspective. Manuscript on file, Leiden: Leiden University.
Weston DA. 2017. *Human Skeletal Report: El Flaco, Dominican Republic*. Archive report prepared for NEXUS 1492 ERC-Synergy Project: New World Encounters in a Globalising World. Manuscript on file. Leiden: Leiden University.
Weston DA, and Schats R. 2010. *Human Skeletal Report: Maisabel, Puerto Rico*. Archive report prepared for NWO Project: Communicating communities. Unravelling networks of human mobility and exchange of goods and ideas from a pre-colonial, pan-Caribbean perspective. Manuscript on file, Leiden: Leiden University.
Weston, DA, and Valcárcel Rojas R. 2016. Communities in contact: Health and paleodemography at El Chorro de Maíta, Cuba. In: Roksandic, I. (ed.): *Cuban Archaeology in the Caribbean*. Gainesville, FL: University Press of Florida, pp. 83–105.

7

PALEODEMOGRAPHY, HEALTH, AND PHYSICAL ACTIVITY AT PLAYA DEL MANGO SITE, GRANMA, CUBA

Naila García Méndez, Dany Morales Valdés, Ulises M. González Herrera, and Yadira Chinique de Armas

Introduction

Recent paleodietary and genetic studies on early groups in Cuba have provided new directions for understanding the biological and cultural diversity within these populations (Chinique de Armas et al., 2016, 2022, 2025; Nägele et al., 2020). Regarding the diversity of dietary practices and lifestyles amongst early indigenous populations in Cuba, it has been observed that at least three different subsistence patterns coexisted in the archipelago in early precolonial times (Chinique de Armas et al., 2016, 2022, 2025). Significant differences in diet were observed between individuals from the archaeological site of Canímar Abajo (Matanzas), where most individuals had a greater dependence on marine resources, and those from Playa del Mango (Granma), where the diet was more oriented towards terrestrial resources. Additionally, the practice of horticulture was confirmed for both groups, although the plants identified and their frequency of consumption likely varied (Chinique de Armas et al., 2015, 2022). More recently, a diversity of mobility patterns, as evidenced by their isotope signals, was identified among early groups (Chinique de Armas et al., 2025).

The genomic studies conducted by Nägele et al. (2020), including samples from seven "Archaic Age" archaeological sites from Cuba (Cueva del Perico I, Canímar Abajo, Manuelito, Cueva Calero, Guayabo Blanco, and Playa del Mango), revealed that at least two possible ancestries coexisted on the island at around 250 BC to 550 BC (2σ). Since the physical characteristics of a population are a result of the combined action of genetics and environment, it could be assumed that groups with different ancestries, exploiting different ecosystems over generations, with different subsistence patterns and mobility, would have different phenotypical traits and traces of behavior reflected in their skeletal remains. While osteological

DOI: 10.4324/9780429444449-10

differences have been observed between archaeological populations in western Cuba (Bolufé, 2015), groups from the so-called Banwaroid tradition of the Antilles have not been examined in detail. In this paper, we conduct the osteological analysis of the individuals from Playa del Mango in eastern Cuba. This site represents the most comprehensive source of information about the individuals from the Banwaroid tradition who inhabited the lower basin of the Cauto River in eastern Cuba. This will provide the foundations for further exploration of phenotypic diversity found on the island and the origins and different lifeways of early Antillean groups.

The Archaeological Site of Playa del Mango

Playa del Mango is located near the lagoon of El Mango in the municipality of Río Cauto, Granma province. The site is 3.5 km from the eastern end of the Las Playas Lagoon System and 14 km from the coastal shoreline of the Gulf of Guacanayabo (Figure 7.1). The site covers an approximate area of 60,000 m^2 and consists of three mounds. The mounds 1 and 2 (M1 and M2), which are the largest, are funerary areas where human remains have been exhumed. The chronology of the burials on the periphery of M2 suggests that the cemetery was used continuously from at least 116 BC to 241 AD (2σ) (Chinique de Armas et al., 2020). According to the study, these dates coincide with and encompass the use of the domestic area of the mound, dated between 55 BC and 435 AD (2σ). The analysis of an isolated tooth from M1

FIGURE 7.1 Location of the Playa del Mango site in a riverine and estuarine environment.

suggested that its use as a funerary area began after the funerary area of M2: cal. 125 CE–435 CE (2σ) (Chinique de Armas et al., 2020). The material culture of the site resembles the typical artifacts of the Banwaroid tradition of the Antilles (Cruxent and Rouse, 1961; Tabío and Rey, 1979).

Archaeological work at the site began in 1941, when Bernardo Utset Macía, a physician, amateur archaeologist, and then-delegate of the Junta Nacional de Arqueología y Etnología, discovered a burial space containing between 30 and 40 skeletons on the periphery of M1. Despite poor preservation conditions, the remains were found intact, arranged in a triple row at a depth of 17 cm (Utset, 1941). In 1980, José M. Guarch and members of the Eastern Central Department of Archaeology excavated the site, and in 1986, a team from the Archaeology Department of the Academy of Sciences of Cuba, under the direction of Jorge Febles, conducted another archaeological excavation. In both cases, isolated human remains were found.

Systematic archaeological excavations were resumed at the site in 2015 under the direction of Dr. Ulises Miguel González Herrera (Cuban Institute of Anthropology) and Dr. Yadira Chinique de Armas (University of Winnipeg, Canada). Between 2015 and 2018, four archaeological campaigns were conducted at the site, with participation from the Casa de la Nacionalidad de Bayamo, the Municipal Museum of Río Cauto and other Cuban institutions. As a result of the surveys completed in 2015, a skull was recovered in M1 (González Herrera et al., 2015). During the excavations of 2016 and 2018, 20 anatomically articulated skeletal remains in an extended supine position were exhumed at the periphery of M2 (Chinique de Armas et al., 2020). Additionally, local residents recovered human remains, without stratigraphic control, from M1, and the archaeological team retrieved a total of 44 individual teeth from the same area. All the burials considered exhibited poor preservation conditions.

Materials and Methods

The skeletal sample is composed of osteological material recovered from M2 during the excavations of 2016 and 2018. The sample initially consisted of 20 primary burials identified in situ, as well as several isolated teeth and bones (Table 7.1). This sample, along with the remains from M1 analyzed by Morales et al. (2023), constitutes the most representative information on indigenous human remains from the lower Cauto River watershed recovered to date.

The demographic profile of this population was studied using the Minimum Number of Individuals (MNI) as a quantifiable measure. In the laboratory, individuals were verified based on the repeatability of pieces of the same laterality, age, size, robusticity, and other anatomical-morphological characteristics (Lambacher et al., 2016; White and Folkens, 2005). Stratigraphic data recorded during excavations and reports from previous excavations were also considered. Estimations of probable age at death for individuals from M2 were made following the considerations of Roksandic and Armstrong (2011), who propose eight age ranges based on

TABLE 7.1 Sample from Mound 2 of Playa del Mango (Granma, Cuba)

Sample ID	*Block/Grid*	*Level (m)*	*Chronology*[a]
PM2_E-1	B4/C-5; C-14; C-8	0.10–0.25	AD 48–235
PM2_E-2	B4/C-5; C-8	0.10–0.26	AD 15–225
PM2_E-3	B4/C-11; C-10	0.10–0.27	BC 116–AD 75
PM2_E-4	B4/C-11; C-4	0.10–0.28	BC 52–AD 58
PM2_E-5	B4	0.2	n/a
PM2_E-6	B4	0.15	n/a
PM2_E-7	B6/C-I; C-VI	0.10–0.25	AD 57–177
PM2_E-8	B6/C-H; C-K	0.10–0.20	n/a
PM2_E-9	B6/C-H; C-K	0.10–0.20	BC 114–AD 25
PM2_E-10	B6/C-J; C-III	0.10–0.25	AD 79–216
PM2_E-11	B6	0.10–0.25	n/a
PM2_E-12	B6/C-III; C-G	0.10–0.25	AD 15–91
PM2_E-13	B6; C-F; C-VI	0.10–0.25	AD 127–241
PM2_E-14	B6/C-III	0.10–0.20	AD 47–140
PM2_E-15	B6/C-J	0.10–0.20	BC 3–AD 81
PM2_E-16	B6/C-11	0.25	n/a
PM2_E-17	B5/C-L	0.2	n/a
PM2_E-18	B6	0.10–0.20	n/a
PM2_E-19	B6/C-III	0.10–0.20	AD 23–135
PM2_E-20	B4/C-4		n/a

[a] Chronology was taken from Chinique de Armas et al. (2020).

dental eruption, fusion of epiphyses in long bones, obliteration of cranial sutures, the spheno-occipital synchondrosis, fusion of vertebrae, and sacral segments, changes in the pubic symphysis, and degenerative changes, among other criteria.

Sex estimation in adult individuals was conducted following the recommendations of Buikstra and Ubelaker (1994), which are based on morphological indicators of the pelvis, skull, and mandible. The fragmentation of skeletal elements, particularly in skulls, made it impossible to analyze sexual characteristics in some cases. Therefore, the information gathered in the lab was supplemented with previous observations of each skeleton *in situ*, prior to the exhumation of the remains. Sex identification at the site and in the laboratory was corroborated in cases where genetic sex determinations were available (Chinique de Armas et al., 2022; Nägele et al., 2020).

The anatomical elements of the skeletons from M2 were measured according to their state of preservation. Measurements were taken for height calculation following the descriptions of Buikstra and Ubelaker (1994). Height estimates were made using regression equations proposed by Del Ángel and Cisneros (2004), measuring the maximum lengths of long bones. The recording of paleo-pathological lesions was done macroscopically, following procedures described by Buikstra and Ubelaker (1994) and White et al. (2012). Indicators of oral health, traumatic injuries (fractures), and infectious, metabolic, and degenerative diseases were identified.

TABLE 7.2 Entheses included in the study

Entheses	*Site of insertion*
M. triceps brachii	Scapula
Costoclavicular lig.	Clavicle
Conoid lig.	clavicle
Trapezoid lig.	clavicle
M. deltoideus	clavicle
M. pectoralis major	humerus
M. lat. dorsii/teres major	humerus
M. deltoideus	humerus
M. brachioradialis	humerus
M. triceps brachii	ulna
M. brachialis	ulna
M. supinator	ulna
M. biceps brachii	radius
M. pronator teres	radius
Interosseous membrane	radius
M. gluteus maximus	femur
M. iliopsoas	femur
M. vastus medialis	femur
Quadriceps tendon	tibia
Quadriceps tendon	patella
M. soleus	tibia
Achilles tendon	calcaneus

A total of 22 entheses were selected from the postcranial appendicular skeleton to relate their prevalence to possible patterns of physical activity and mobility in the population (Table 7.2). Individuals who were not adults or were senile, as well as those with pathological lesions, were excluded from the study. Therefore, the sample size for this analysis was 14 individuals.

Although there have been widely accepted methodological proposals for studying entheses (Henderson et al., 2013, 2016, 2017), there is no consensus regarding the most appropriate methodology. It has been suggested that analyzing entheses in isolation results in a loss of functional information since joint movements are coordinated by multiple muscles (Karakostis et al., 2019; Karakostis and Harvati, 2021). For this reason, other methodologies focus on reconstructing muscular synergies through three-dimensional and multivariate analyses among entheses (Karakostis and Havarti, 2019, 2021). In the case of Playa del Mango populations, which showed poor bone preservation, we followed Mariotti et al.'s (2004, 2007) visual descriptive methodology. We classified the entheses into three levels of development: weak, notable, and strong.

Results

Minimum Number of Individuals

A minimum number of 25 individuals from M2 were identified during the laboratory analysis, including four new non-adults and one adult. Although other isolated bones found in the context were analyzed, they could not be considered as additional individuals; these remains may correspond to some of the 20 identified skeletons due to their position in the record, size, side, and the absence of the same pieces in some skeletal remains.

Of the 25 individuals counted in M2, it was estimated that 18 were in adulthood at the time of death, representing 72%. Of them, 16 were identified as full adults (64%) and two as young adults (8%). One individual was identified as an adolescent (4%), three as early childhood (12%), and three as infants (12%) (Table 7.3).

TABLE 7.3 Parameters regarding the sex and age of the skeletons from Mound 2 of Playa del Mango, Granma, Cuba

Sample ID	*Sex*[a]	*Age*
PM2_E-1	Male	Full adulthood
PM2_E-2	Female (XX)	Full adulthood
PM2_E-3	Male (XY)	Full adulthood
PM2_E-4	Possible Female	Full adulthood
PM2_E-5	Male	Full adulthood
PM2_E-6	Female (XX)	Early childhood
PM2_E-7	Male	Full adulthood
PM2_E-8	Indeterminate	Adolescence
PM2_E-9	Female (XX)	Full adulthood
PM2_E-10	Male (XY)	Full adulthood
PM2_E-11	Male	Full adulthood
PM2_E-12	Possible Male	Full adulthood
PM2_E-13	Male	Full adulthood
PM2_E-14	Male (XY)	Full adulthood
PM2_E-15	Female (XX)	Young adulthood
PM2_E-16	Indeterminate	Full adulthood
PM2_E-17	Indeterminate	Infancy
PM2_E-18	Indeterminate	Young adulthood
PM2_E-19	Indeterminate	Full adulthood
PM2_E-20	Posible Female	Full adulthood
PM2_E-21	Indeterminate	Infancy
PM2_E-22	Indeterminate	Infancy
PM2_E-23	Indeterminate	Early childhood
PM2_E-24	Indeterminado	Early childhood
PM2_E-25	Indeterminate	Full adulthood

[a] Molecular sex is included within parenthesis when available.

None of the skeletons corresponded to mature or senile stages. Infant individuals were identified based on the absence (or partial development/eruption) of deciduous dentition. Individuals in early childhood were characterized by an intermediate stage between the eruption of deciduous teeth and the eruption of the first permanent tooth (either the first molar or central incisor). Adolescents exhibited long bone epiphyses that were still unfused. In the Young Adulthood category, individuals had fused long bone epiphyses, but epiphyseal lines in the arms were still visible, and the medial epiphyses of the clavicles had not yet fused. In the full adulthood category, individuals showed closure of secondary ossification centers in the long bones, though the fusion lines between the centrum and the annular rings of the vertebrae, especially in the thoracic and upper lumbar regions, were not yet fully obliterated. Additionally, they displayed no degenerative changes (Roksandic & Armstrong, 2011).

Sex Estimation

Of the 18 adult skeletons examined from M2, sex could not be identified in three individuals (PM2-E16, PM2-E18, and PM2-E25 [Table 7.3]) due to the lack of pelvis, skull, or other diagnostic areas. Sex was corroborated with genetic determination in seven individuals (PM2_E-2, PM2_E-3, PM2_E-6, PM2_E-9, PM2_E-10, PM2_E-14, PM2_E-15), achieving 100% agreement (Table 7.3). Overall, three skeletons were estimated and confirmed genetically as female and eight as male (three confirmed with genetic results). Four individuals were classified as possibly female (2) and possibly male (2), mainly based on mandibular characteristics, due to the absence of pelvis and skull.

Stature Estimation

Height was estimated for six adults: one female (PM2_E-2) and five males (PM2_E-1, PM2_E-3, PM2_E-7, PM2_E-11, PM2_E-13) (Table 7.4). The average height for the population was 159.82 cm. The average height for males was 162.68 cm, while the value for the only female, for whom it was possible to perform

TABLE 7.4 Estimated height in individuals from Playa del Mango, Granma, Cuba

Sample ID	*Sex*	*Height (cm)* *Del Ángel and Cisneros (2004)*
PM2_E-1	Male	158.46
PM2_E-2	Female	145.06
PM2_E-3	Male	164.85
PM2_E-7	Male	160.7
PM2_E-11	Male	162.5
PM2_E-13	Male	167.4

the calculation, was 145.06 cm. The difference between the average height of males and the female height is 17.62 cm.

Skeletal Pathologies and Trauma

Regarding pathologies, the analysis indicates that most individuals from this site did not exhibit diseases that left visible marks on the postcranial skeleton. The skeleton PM2_E-12 exhibits an active periosteal reaction manifested as porosity accompanied by thickening that wraps around the entire diaphysis of the right ulna and femur, the left tibia, and a fragment of the diaphysis of the fibula (Weston, 2008, 2012). The pathological signs observed in PM2_E-12, such as multiple periosteal reactions in most of its bones, superficial cavitation, bone remodeling, and cortical thickening, may be associated with infections caused by treponematosis (Hackett, 1976; Ortner and Putschar, 1985; Filippini et al., 2019). The treponematosis had previously been proposed at the sites Cueva del Perico I (Travieso et al., 1999) and Canímar Abajo (Vento, 2004; Campillo, 2009) in Cuba in the periods before the arrival of the Spanish colonizers.

On the medial surface of the right humerus of individual PM2_E-7, osteomyelitis was present, characterized by visible bone destruction, cavitation indicating a response to infection, and formation of new bone tissue around the affected areas (Ortner and Putschar, 1985). A healed spiral fracture was found in the proximal region of the diaphysis of the right fourth metacarpal of PM2_E-1 and in the body (middle of the diaphysis) of the right third metacarpal of PM2_E-13. Spiral fractures are usually caused by torsional forces on the bone and can lead to rotational misalignment (Suárez and Forero, 2012). A healed spiral fracture occurs when the bone, after breaking in a spiral line due to torsion, has fully healed. Healing occurs when the fracture is stable enough, completely mineralized, similar to its original state, and the process generally involves the formation of bone callus (Wollstein et al., 2020), as seen in the right fourth metacarpal of PM2_E-1 and in the right third metacarpal of PM2_E-13. Healing may also involve bridging or bone remodeling (Wollstein et al., 2020).

The dental health of the analyzed individuals was notably compromised. In 100% of the individuals with dental remains (12 skeletons) and in the 14 isolated teeth (incisors, canines, premolars, and molars), some type of infectious or physiological process was observed (Figure 7.2). Marked wear on premolars and molars was common, along with severe abrasion on the incisors, leading to complete erosion of the crown and exposure of the pulp cavity, as seen in the dentitions of individuals PM2_E-1, PM2_E-3, and PM2_E-5. Dental calculus was also found at the cemento-enamel junction (PM2_E-1, PM2_E-3, and some isolated incisors), along with caries on the crown and root area (several isolated teeth) and hypercementosis in the root area of an isolated first molar. Some individuals exhibited teeth that had enamel infractions (such as the non-adult PM2_E-6), and there was antemortem tooth loss with alveolar resorption in those areas. In the mandible of the female individual PM2-E2, osteolysis was observed around the second premolar, which

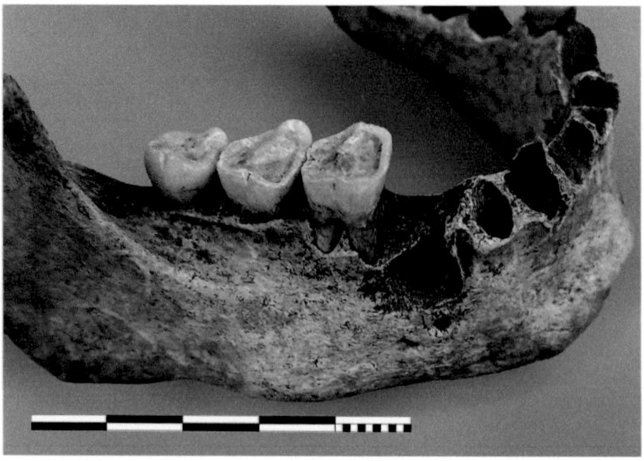

FIGURE 7.2 Oral pathologies at Playa del Mango, Granma, Cuba.

could indicate a dentoalveolar abscess caused by periodontal disease that led to tooth loss during life (Robledo and Trancho, 2007; Assis et al., 2018).

Enthesopathies

The 14 adult individuals analyzed exhibited marked development at the sites of muscle and ligament insertion in the upper limbs. The highest degrees of enthesis development were observed at the insertion of the costoclavicular ligament (proximal clavicle), where 57% (8/14) of the inspected clavicles showed strong development (Figure 7.3). In addition, the insertion of the gluteus maximus in the femur was strong in 53% (11/21) of the femora analyzed (Figure 7.4). The insertion of the brachial muscle in the ulna showed strong development in 33% (7/21) of the ulnae. Other entheses with significant evidence of muscular activity included the insertion of the deltoid muscle in the clavicle, with 5% (1/17) showing strong development and 67% (11/17) showing moderate development. The supinator muscle in the ulna was present at a strong grade of 16% (3/19) and at a moderate grade of 21% (4/19). Other entheses, such as the insertion sites of the conoid ligament in the clavicle, the deltoid and brachioradialis muscles in the humerus, the triceps brachii muscle in the ulna, the vastus medialis in the femur, the quadriceps tendon in the patella and tibia, the soleus muscle

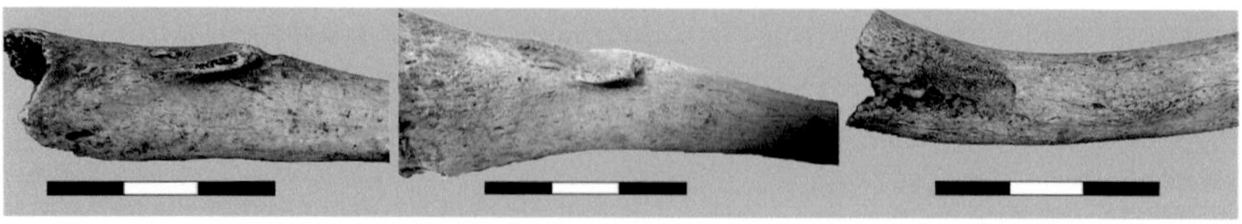

FIGURE 7.3 Enthesopathic changes in the clavicles.

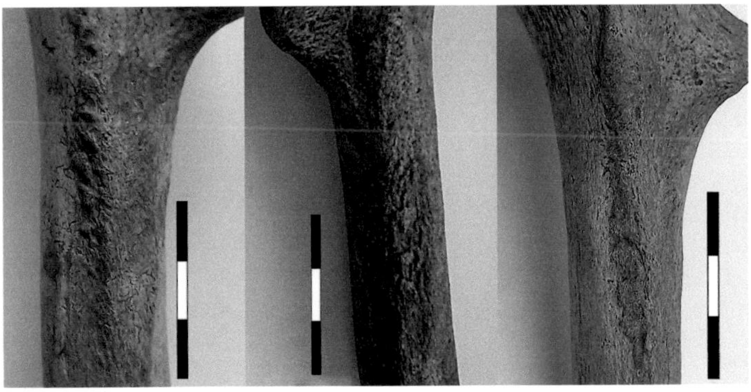

FIGURE 7.4 Enthesopathic changes in the femur.

in the tibia, and the Achilles tendon in the calcaneus, exhibited moderate and strong degrees in this sample, although in fewer bony elements than the aforementioned enthesis changes.

Discussion

The fragmentary nature of Playa del Mango human remains had detrimental effects on the estimation of sex, age, and stature. The four new individuals identified add to the 40 skeletons reported by Utset (1941) at M1. More recently, Morales et al. (2023) estimated 14 individuals for M1. However, these may have been remnants of the remains described by Utset in his field diary following his excavation in 1941, and they would be included with the 40 skeletons—whose locations are unknown—that the specialist claimed to have unearthed. To the number cited by Utset (1941), two other individuals reported at the site can be added: a skull and additional articulated remains recovered during the 2015 campaign and a complete skeleton of an infant whose location is unknown that was discovered "around 1979 by the amateur archaeologist Houghet Arteaga" (González Herrera et al., 2021: 260). In this regard, it can be concluded that at least 42 individuals have been unearthed from M1 to date. This number could be higher if fragmented and disarticulated human remains recovered by archaeologists from the Academy of Sciences between 1980 and 1986 are considered. While their report does not specify the number of individuals, they suggest that the bones—found in plowed fields—could correspond to other individuals not seen by Utset (1941). Therefore, the estimated MNI for the Playa del Mango site, considering both M1 and M2, rises to 67 individuals. It is important to note that the physical boundaries of the cemetery are unknown, and future excavations will likely increase this number.

In contrast to the trend observed in other early burial sites from Cuba, the proportion of non-adults found at Playa del Mango is low in comparison to adults.

TABLE 7.5 Representation of age at early Indigenous precolonial sites from Cuba

Site	*% Non-adults*	*N*	*Non-adults*	*Adults*
Cueva de los Niños	100	13*	13*	0
Cueva del Infierno	86.3	66	57	9
Cueva Calero	80.9	63	51	12
Cueva Marién II	78	50	39	11
Cueva de la Santa	76.5	34	26	8
Cueva del Perico I	73.5	106	78	28
Canímar Abajo	61.0	213	130	83
Playa del Mango	**36**	**39**	**14**	**25**

Taken from García Beatón (1997). Data from Canímar Abajo taken from Roksandic et al. 2015.

* In a later revision, the MNI was raised to 17.

In addition to the infants and children found at M2, Morales et al. (2023) reported seven non-adults at M1. Additionally, two more infants were identified: one excavated by Chinique de Armas in 2015 and the skeleton previously mentioned being reported by Houghet Arteaga (González Herrera et al., 2021). The previous data indicates that of the 39 skeletons with estimated ages for the site, 64% (25/39) reached adulthood, while 36% (14/39) died in non-adult stages. This tendency is opposite to previous observations, where the number of non-adults is typically higher (Table 7.5).

The absence of individuals in the senile category is consistent with other results found in early Cuban sites such as Cueva de la Santa, Cueva del Perico I, Cueva del Infierno, and Canímar Abajo, where very few individuals had reached the mature stage (Garcell, 2006; Luna, 2011; Morales, 2009; Piñón, 2013; Rodríguez, 1998; Torres and Rivero, 1970). This may suggest that, like Mesoamerican hunter-gatherer populations (Hernández Flores and Serrano Sánchez, 2017), ancient groups in Cuba have generally had a low life expectancy, rarely reaching the senile stage.

The high correlation between sex estimations using morphological parameters and those using genetic determinations indicates that the population has a sexual dimorphism that allows osteological characteristics to be distinguishable without a significant margin of error. This supports the sex estimation made for those skeletons whose sex was not genetically determined. This is in high contrast with other sites such as Canímar Abajo, where the correspondence between sex estimations using morphometric parameters and the genetic results was 46% (García et al., 2022).

In general terms, the height of Playa del Mango's male individuals (M: 162.68 cm) was consistent with those calculated by Rivero de La Calle (1969) in Cueva de Los Indios and Bolufé (2015) in Cueva del Perico I (M: 164.2 cm), Cueva Calero (M: 161.2 cm), and Guayabo Blanco (M: 162.0 cm). In contrast, Playa del Mango male stature was higher than the males of Canímar Abajo Old (M: 157.3 cm) and Young cemeteries (M: 154.4 cm), although the misidentification of the sex at that site likely contributed to this difference. In the case of females, the only Playa del Mango woman where stature was estimated (F: 145.06 cm) is significantly smaller than the height of female individuals reported in sites such as Cueva del Perico I (F: 157.4 cm) and the older cemetery of Canímar Abajo (F: 157.3 cm), but is consistent with the values reported in other sites such as the younger component of Canímar Abajo cemetery (F: 146.7 cm), Cueva Calero (F: 147.3 cm), and Guayabo Blanco (F: 150.6 cm). The sexual dimorphism in height observed at Playa del Mango (Δ = 17.62 cm) is significantly greater than the differences in height between sexes in other Indigenous populations on the island: Cueva del Perico I (Δ = 6.8 cm), Canímar Abajo: Old Cemetery (Δ = 0 cm), Young Cemetery (Δ = 7.7 cm), Cueva Calero (Δ = 13.9 cm), and Guayabo Blanco (Δ = 11.4 cm). Although the sample size is small, these variations in height suggest that the differences may have been influenced by both diverse genetic traits and the influence of living in different environments.

Overall, a low number of pathologies in the postcranial skeleton were observed. Although this low occurrence may be influenced by the absence of bone elements and their considerable fragmentation, it seems that the incidence of inflammatory-infectious disease was lower than observed in Canímar Abajo (Campillo, 2009). The identified conditions, such as periosteal reactions and osteomyelitis in a single individual, constitute non-specific indicators dependent on the environment, cultural factors, and individual response (Weston, 2008; Roberts, 2019). Proposed etiologies include infectious diseases, traumatic injuries, and nutritional deficits, which should be considered non-recurring in this population due to their unique morbidity. In M1, Morales et al. (2023) observed other lesions such as *cribra orbitalia*, adding to the paleopathological lesion at the site. Cribra orbitalia is a mild form of hyperostotic osteoporosis that may be associated with inflammation in the orbital region due to infectious processes or with hypertrophy of the bone marrow, which is frequently associated with anemia (Wapler et al. 2004; Brickley, 2018). It can also be caused by osteoporosis and hemorrhagic processes (as seen in scurvy). Additionally, cribra orbitalia is commonly mistaken for post-mortem erosion (Wapler et al. 2004). The fracture pattern observed in the upper limbs of three different individuals may indicate overuse. Fracture patterns allow for the interpretation of the mechanisms and forces applied, especially when contextualized with archaeological data (Gilmour et al., 2015). In particular, fractures in the metacarpals may be indicative of gripping or manipulating heavy objects, related to activities that require intensive use of the hands, such as stone tool manufacture, which are present at the site and demand constant and strong contact with work surfaces (Robb, 1994).

The high frequency of dental pathologies was also observed in M1 by Morales et al. (2023). Oral pathologies are related to the use of the mouth in daily activities, resulting in fractured teeth from pressure or leverage, infractions, and differential occlusal abrasion (Flensborg, 2013). The presence of caries is consistent with a diet rich in carbohydrates where root cultigens are frequently consumed (Chinique de Armas et al., 2022). The high incidence of caries is also linked to poor hygiene, which erodes the enamel and, in some cases, leads to abscesses and infections (Flensborg, 2013). It is noteworthy that antemortem tooth loss is not frequent.

Although muscles do not act in isolation, the results obtained allow for some inferences about individual patterns of mechanical muscle demand and to infer some of the most recurrent movements in the sample (Stirland, 1998). These inferences suggest the performance of certain activities when considering other elements of the archaeological record and ethnographic chronicles. The enthesopathies of the costoclavicular ligament, which limits the elevation of the clavicle, are often observed as depression or bony exostosis and are related to activities that require backward and downward arm movements (Hawkey and Merbs, 1995). In ancient populations, costoclavicular enthesopathies have been associated with activities such as sewing, hunting, transporting heavy loads on the shoulders, and maintaining extended arms (Capasso et al., 1999; Kennedy, 1989). Enthesopathies of this type were recorded in South American populations of the Middle Paraná: the multi-component funerary site La Lechuza—comprised of hunters, fishers, gatherers, and potters—dated to 130 AD to 250 AD (Curetti, 2009).

The deltoid muscle is involved in all movements of the arm and shoulder and actively participates in arm abduction or elevation, as well as in flexion and extension (Capasso et al., 1999). The pronounced development of this area, similar to that of the costoclavicular region, is also linked to activities such as throwing objects (Hawkey and Merbs, 1995) and rowing (Curetti, 2009). Among the materials found at the site, several objects—stone spear and arrow tips (depending on weight and design)—have been recorded, which could support this idea without excluding other possible activities related to horticultural populations (Caro and Fernández, 2006), as suggested by the consumption of maize or sweet potato found in studies of dental calculus and stable isotopes of individuals at Playa del Mango (Chinique de Armas et al., 2022).

Changes in the insertion points on the ulna also show significant development, especially at the insertion point of the brachial muscle located at the base of the coronoid process of this bone, appearing as a depression surrounded by a thick ridge. This muscle is involved in the process of flexing the forearm over the arm and has been linked in other populations to manual activities such as weaving (Caro and Fernández, 2006). Both this muscle and the development of the supinator crest are associated with applying force with flexed arms to drag heavy objects, involving backward pulling movements (López Bueis, 1999). These elbow flexions often leave clear marks at the insertion of the biceps brachii (Acosta, 2012), although this is not observed in this sample.

Unlike elbow flexion, the triceps brachii and the supinator of the ulna act in extension (Capasso et al., 1999). The hyperactivity of these muscles has been associated with throwing different types of spears during hunting, spear fishing, and activities like rowing or swimming; these activities coincide with archaeological evidence from Playa del Mango, around the Las Playas lagoon system, for obtaining marine specimens suitable for diet and others whose vertebrae were transformed into necklace beads, including teeth from shark species turned into pendants (Córdova and Arredondo, 1988; González Herrera, 2020; Pérez, 1981).

The gluteus maximus, a highly active muscle in these individuals, is responsible for the rotation and extension of the thigh. Along with the soleus, it is associated with activities such as walking or running while carrying weight (Niinimäki and Baiges Sotos, 2013). Movements related to the gluteus maximus in the thigh—rising from a seated position, maintaining a standing posture in unstable conditions, or squatting (Al Oumaoui et al., 2004; Kennedy, 1989)—combined with the development of insertion points in the upper limbs suggest activities such as rowing and using canoes for transport along watercourses in the region, whether in the lagoon system or river channels (Figure 7.1).

Changes in the insertion points on the patella and tibia (quadriceps tendon and soleus muscle), common in individuals who spend many hours squatting, could also be related to textile or basketry activities, as well as food processing for later cooking, where the individual spent many hours in the same position (Espinoza, n.d.; Jiménez-Brobeil et al., 2004). These same lesions on the tibia and patella are also associated with activities that require bearing heavy weights, with the knees serving as the balance point; similarly, such lesions on the tibia and calcaneus (involving stress on the adductor muscle of the hallux that inserts into the Achilles tendon) are related to covering long distances (Capasso et al., 1999; Scabuzzo, 2012).

In summary, the traces described at the insertion points of the upper limbs are clearly linked to the manual activities that the population performed, resulting from the crafting of the objects found, as well as their systematic use. The extensive representation of lithic artifacts (perforators, scrapers, projectile points, punches, blades, flakes with notches, and burins) indicates that lithic knapping was frequent. Grinding stones have been found—with used-wear-patterns—whose quantity in the area exceeds the evidence recorded at other archaeological sites with similar socioeconomic affiliations in the archipelago. The archaeological findings indicate the presence of shellworking tradition, evidenced by the presence of scrapers and gouges, some with signs of handling, as well as pieces for weaving, including punches and a needle made from fish bones. It is shown in the archaeological record that there was the manufacture of body adornments and superstructural artifacts (daggers, spheres). This site includes the biconical perforation of shark teeth and small lithic pebbles as earrings. On the other hand, the search and collection of raw materials—such as various tuff rocks located more than 30 km away—impact some enthesopathy markers related to walking; however, muscle fatigue does not indicate significant pressure, possibly due to the typically flat terrain, with no changes in slope and the use of canoes for transportation.

Conclusions

The Playa del Mango site constitutes a significant funerary area in the Río Cauto-Gulf of Guacanayabo region due to the number of human remains exhumed to date. Notably, there is a prevalence of adult individuals over non-adults, which contrasts with the typical preponderance of non-adults in early Indigenous sites in Cuba.

The average height values tend to be in the medium range among male individuals, a factor that cannot be determined for the female population due to the small sample size; however, there is a significant difference of 13.4 cm between the shortest male and the only estimated female height.

Overall, the population did not show a prevalence of diseases with bony manifestations, except for oral pathologies, which are common in most archaeological populations of the Cuban archipelago. The observed enthesopathic changes correspond to individuals who exploited flat areas without abrupt slopes and spent time performing tasks in a squatting position, although the most notable signs of impact are located in the clavicles.

Acknowledgments

Our highest gratitude to all members of the rural community of El Mango that received us as family in their territories. Thanks to all members of our archaeological team, especially to our colleague from Granma, Jose Manuel Yero. Thank you to Madeline Lischka for her assistance in editing the first version of this manuscript. Our gratitude to the direction of the Cuban Institute of Anthropology for granting access to the human remains of Playa del Mango. This research was funded by the Social Science and Humanities Research Grant led by Dr. Yadira Chinique at the University of Winnipeg (SSHRC-IG 435-2021-1144).

References

Acosta, M. A. 2012. Una mirada a los marcadores óseos de actividad: Aproximación al periodo temprano (340 a. C.-440 d. C.) del valle geográfico del río Cauca. *Revista Colombiana de Antropología*, *48*(1), 169–187.

Al Oumaoui, I., Jiménez-Brobeil, S. & du Souich, P. 2004. Markers of activity patterns in some populations of the Iberian Peninsula. *International Journal of Osteoarchaeology*, *14*(5), 343–359.

Assis, S., Casimiro, S. & Cardoso, F. A. 2018. Is it a cystic or a cyst-like condition? Discussing the etiology of an unusual large mandibular lesion in a Roman skeleton from Quinta da Torrinha/Quinta de Santo António (Almada, Portugal). *International Journal of Paleopathology*, *22*, 149–157.

Bolufé, R. 2015. *Variaciones de la estatura y las características morfométricas craneofaciales en poblaciones pescadoras-recolectoras-cazadoras de Cuba*. Tesis en opción al título de licenciada. Universidad de La Habana, Facultad de Biología.

Brickley, M. B. 2018. Cribra orbitalia and porotic hyperostosis: A biological approach to diagnosis. *American Journal of Physical Anthropology*, *167*(4), 896–902.

Buikstra, J. E. & Ubelaker, D. M. 1994. *Standards for data collection from human skeletal remains*. Arkansas Archeological Survey Research Series No. 44.

Campillo, E. 2009. *Paleopatologías en aborígenes cubanos del sitio arqueológico Canímar Abajo, Matanzas, Cuba. Tesis en opción al título Académico de Máster en Antropología*. Universidad de La Habana, Facultad de Biología.

Capasso, L., Kennedy, K. A. & Wilckzak, C. A. (1999). *Atlas of occupational markers on human remains*. Edigrafital S.p.A, 1–184.

Caro, L. & Fernández, M. 2006. *Estudio antropológico de los restos óseos de los sepulcros de los reyes Juan II de Castilla e Isabel de Portugal y del infante Don Alfonso de la Cartuja de Miraflores*. Informe antropológico inédito. Elaborado para la junta de Castilla y León, 6.

Chinique de Armas, Y., Buhay, W. M., Rodríguez Suárez, R., Bestel, S., Smith, D., Mowat, S. D. & Roksandic, M., 2015. Starch analysis and isotopic evidence of consumption of cultigens among fisher–gatherers in Cuba: The archaeological site of Canímar Abajo. Matanzas. *Journal of Archaeological Science, 58*, 121–132.

Chinique de Armas, Y., González Herrera, U. M., Buhay, W. M.; Yero, J. M., Viera, L. M. & Burchell, M. 2020. Chronology of the archaeological site of Playa del Mango, Río Cauto, Granma. Cuba. *Radiocarbon*, *62*(5), 1221–1236.

Chinique de Armas, Y., González Herrera, U. M., Hernández Godoy, S., Garcell Domínguez, J., Viera Sanfiel, L.M., Laffoon, J. 2025. Lifeways Diversity in early Antillean Societies: A Multi-Isotope Approach.

Chinique de Armas, Y., González Herrera, U. M., Reyes Serrano, I., Buhay, W. M., Skelton, S., Rodríguez Suárez, M, Roksandic, M. & Laffoon J. 2022. Multiproxy paleodietary reconstruction using stable isotopes and starch analysis: The case of the archaeological site of Playa del Mango, Granma, Cuba. *Journal of Archaeological Science Reports*, *46*, 103671.

Chinique de Armas, Y., Roksandic, M., Suárez, R. R., Smith, D. G. & Buhay, W. M. 2016. *Isotopic evidence of variations in subsistence strategies and food consumption patterns among 14 "Fisher-gatherer" populations of Western Cuba. Cuban archaeology in the Caribbean.* University Press of Florida, 125–146.

Córdova, A. P. & Arredondo, O. 1988. Análisis de restos dietarios del sitio arqueológico El Mango, Río Cauto, Granma. In: M. E. Zulueta (Ed.), *Anuario de Arqueología.* Editorial Academia, 111–132.

Cruxent, J. M. y Rouse, I. 1961. *Arqueología cronológica de Venezuela. Estudios monográficos VI*. Unión Panamericana.

Curetti, P. 2009. *Evidencias osteológicas de estrés musculo-articular en una población prehistórica del Paraná Medio*. 2nd edition. Editorial Juris..

Del Ángel, A. & Cisneros, B. 2004. Modification of regression equations used to estimate stature in Mesoamerican skeletal remains. *American Journal of Physical Anthropology*, *125*, 264–265.

Espinoza, P. O. H. n.d. La actividad ocupacional se refleja hasta en los huesos. Un aporte de la antropología física a los estudios del trabajo. Unpublished manuscript.

Filippini, J., Pezo-Lanfranco, L. & Eggers, S. 2019. Estudio regional sistemático de treponematosis en conchales (sambaquis) precolombinos de Brasil. *Chungará (Arica)*, *51*(3), 403–425.

Flensborg, G. A. 2013. Paleopatologías bucales y tendencias paleodietarias en grupos cazadores-recolectores de la transición pampeano-patagónica oriental, durante el Holoceno tardío. *Relaciones de la Sociedad Argentina de Antropología*, *XXXVIII*(1), 199–222.

Garcell, J. 2006. *Sitio Bacuranao 1. Testimonio del culto funerario de grupos preagroalfareros*. Tesis en opción al título académico de Master en Arqueología, Instituto Cubano de Antropología, La Habana.

García Beatón, L. 1997. *Estudio antropológico de los restos esqueletales aborígenes procedentes de la Cueva del Infierno, San José de las Lajas, La Habana*. Tesis de Diploma, Facultad de Biología, Universidad de La Habana.

García, N., Chinique de Armas, Y., y Morales, D. 2022. *Comparación de la estimación del sexo obtenida mediante métodos osteoscópicos y morfométricos con la determinación sexual a partir de ADN antiguo en individuos del sitio arqueológico Canímar Abajo, Matanzas, Cuba*. Póster presentado en el Congreso de la International Association for Caribbean Archaeology (IACA), Varadero, Cuba.

Gilmour, R. J., Gowland, R., Roberts, C., Bernert, Z., Kiss, K. K. & Lassányi, G. 2015. Gendered differences in accidental trauma to upper and lower limb bones at Aquincum, Roman Hungary. *International Journal of Paleopathology*, *11*, 75–91.

González Herrera, U. M. 2020. *Informe científico-técnico final del Proyecto Arqueología de prácticas mortuorias en sociedades aborígenes de bajos niveles productivos de Cuba* P.N.C.T. 0439. Departamento de Arqueología, Instituto Cubano de Antropología.

González Herrera, U. M., Chinique de Armas, Y., Viera, L. M.; Morales, D., Godo, P. P. & Yero, J. 2015. Informe parcial de trabajo de campo realizado en el sitio arqueológico aborigen Playa El Mango, municipio Río Cauto, Granma. Archivo del Instituto Cubano de Antropología.

González Herrera, U. M., Morales, D., Fernández R., Hernández I., Leal S., Gutiérrez N. & Chinique de Armas Y. 2021.Relación de sitios aborígenes con presencia de restos humanos. In U. M. González Herrera, D. Morales y R. Fernández (Eds.), *Compendio de sitios arqueológicos aborígenes con presencia de restos humanos en Cuba*. Aspha Ediciones, 365. ISBN 978-987-3851-21-6

Hackett, C. J. 1976. *Diagnostic criteria of syphilis, yaws, and treponarid (treponematoses) and of some other diseases in dry bones (for use in osteo-archaeology)*. Springer-Verlag, 134.

Hawkey, D. E. & Merbs, C. F. 1995. Activity-induced musculoskeletal stress markers (MSM) and subsistence strategy changes among ancient Hudson Bay Eskimos. *International Journal Osteoarchaeology*, *5*, 324–338.

Henderson, C. Y., Mariotti, V., Pany Kucera, D., Villotte, S., Wilczak, C., 2013. Recording specific entheseal changes of fibrocartilaginous entheses: Initial tests using the Coimbra method. *International Journal of Osteoarchaeology*, *23*, 152–162.

Henderson, C.Y., Mariotti, V., Pany-Kucera, D., Villotte, S., Wilczak, C., 2016. The new Coimbra method: A biologically appropriate method for recording specific features of fibrocartilaginous entheseal changes. *International Journal of Osteoarchaeology*, *26*(5), 925–932.

Henderson, C. Y., Mariotti, V., Santos, F., Villotte, S. & Wilczak, C. (2017). The new Coimbra method for recording entheseal changes and the effect of age-at-death. *Bulletins et Mémoires de la Société d'Anthropologie de Paris*, *29*(3–4), 140–149.

Hernández Flores, R. H. & Serrano Sánchez, C. S. 2017. La interrelación entre modo de vida y fenómenos vitales en la población prehistórica de México. Una reconsideración necesaria. *Antropología Americana*, 2(4). 109–136.

Jiménez-Brobeil, S. A., Al Oumaoui, I. & Esquivel, J. A. (2004). Actividad física según sexo en la cultura argárica. Una aproximación desde los restos humanos. *Trabajos de Prehistoria*, *61*(2), 141–153.

Karakostis, F. A. & Harvati, K. (2021). New horizons in reconstructing past human behavior: Introducing the "Tübingen University Validated Entheses-based Reconstruction of Activity" method. *Evolutionary Anthropology: Issues, News, and Reviews*, *30*(3), 185–198.

Karakostis, F. A., Jeffery, N. & Harvati, K. (2019). Experimental proof that multivariate patterns among muscle attachments (entheses) can reflect repetitive muscle use. *Scientific Reports*, *9*(1), 16577.

Kennedy, K. A. R. 1989. Skeletal markers of occupational stress. In M. Iscan & K. A. R. Kennedy (Eds.), *Reconstruction of life from the skeleton*. Alan R Liss, 129–160.

Lambacher, N., Gerdau Radonic, K., Bonthorne, E. & Valle de Tarazaga, F. J. 2016. Evaluating three methods to estimate the number of individuals from a co-mingled context. *Journal of Archaeological Science Reports, 10*. 674–683.

López Bueis, I. 1999. Marcadores de estrés musculoesquelético en los huesos largos de una población española (Wamba, Valladolid). *Biomecánica*, *7*(13). 94–102.

Luna, Y. 2011. *Estudio Antropológico de los restos óseos humanos de aborígenes adultos del sitio arqueológico Canímar Abajo, Matanzas, Cuba. Campaña de excavación del 2007.* Tesis. Facultad de Biología, Universidad de La Habana.

Mariotti, V., Facchini, F. & Belcastro, M. G. 2004. Enthesopathies—Proposal of a standardized scoring method and applications. *Collegium Antropologicum*, *28*, 145–159.

Mariotti, V., Facchini, F. & Beslcastro, M. G. 2007. The study of entheses: Proposal of a standardised scoring method for twenty-three entheses of the postcranial skeleton. *Collegium antropologicum*, *31*(1), 291–313.

Morales, D. 2009. *Caracterización osteológica de aborígenes adultos exhumados en Canímar Abajo, durante la campaña de excavación del 2006*. Tesis en opción al título académico de máster en Antropología. Universidad de La Habana. Facultad de Filosofía e Historia.

Morales, D., González Herrera, U. M., Yero, J. & Chinique de Armas, Y. 2023. Contribución antropológica al estudio de restos humanos recuperados en el yacimiento arqueológico aborigen de playa del Mango, Suboriente de Cuba. *Sociales*, 12, 223–247.

Nägele, K., Posth, C., Orbegozo, M., Chinique de Armas, Y., Hernández Godoy, S., González Herrera, U., Nieves Colón, M., Sandoval Velasco, M., Mylopotamitaki, D., Radzeviciute, R., Laffoon, J., Pestle, W., Ramos Madrigal, J., Lamnidis, T., Schaffer, W., Carr, R., Day, J., Arredondo, C., Rangel, A., Martínez Fuentes, A., Crespo Torres, E., Roksandic, I., Stone, A., Lalueza Fox, C., Hoogland, M., Roksandic, M., Hofman, C. J. & Schroeder, H. 2020. Genomic insights into the early peopling of the Caribbean. *Science*, *369*(6502), 456–460.

Niinimäki, S. & Baiges Sotos, L. 2013. The relationship between intensity of physical activity and entheseal changes on the lower limb. *International Journal of Osteoarchaeology*, *23*(2), 221–228.

Ortner, D. J. & Putschar, W. G. J. 1985. *Identification of pathological conditions in human skeletal remains*. Smithsonian Institution Press.

Pérez, P. 1981. *Carta Informativa No. 18*, Época II, Departamento de Arqueología, Instituto de Ciencias Sociales, Academia de Ciencias de Cuba, La Habana, Cuba.

Piñón, A. 2013. *Osteobiografía de los individuos adultos exhumados en las campañas de excavación de 2010 y 2011, en Canímar Abajo, Matanzas, Cuba*. Tesis de licenciatura. Universidad de la Habana, La Habana.

Rivero de La Calle, M. 1969. La estatura en los aborígenes de Cuba del grupo no ceramista. Datos métricos y morfológicos de sus huesos largos. *Revista de la Universidad de La Habana, 194*, 35–49.

Robb, J. (1994). Skeletal signs of activity in the Italian Metal Ages: Methodological and interpretative notes. *Human Evolution, 9*, 215–229.

Roberts, C. A. 2019. Infectious disease: Introduction, periostosis, periostitis, osteomyelitis, and septic arthritis. In J. E. Buikstra (Ed.), *Ortner's identification of pathological conditions in human skeletal remains*. Academic Press, 285–319.

Robledo, B. & Trancho, G. 2007. Las huellas de la enfermedad en los restos humanos. Dpto. Zoología y Antropología Física. Universidad Complutense de Madrid. https://www.researchgate.net/publication/238751698_Las_huellas_de_la_enfermedad_en_los_restos_humanos

Rodríguez, D. 1998. *Estudio antropológico de los restos óseos procedentes de la Cueva del Perico I, Bahía Honda, Pinar del Río. Trabajonde Diploma*. Facultad de Biología, Universidad de La Habana.

Roksandic, M. & Armstrong, S. D. 2011. Using the life history model to set the stage(s) of growth and senescence in bioarchaeology and paleodemography. *American Journal of Physical Anthropology, 145*, 337–347.

Roksandic, M., Buhay, W. M., de Armas, Y. C., Suárez, R. R., Peros, M. C., Roksandic, I., Mowat, S., Viera, L. M., Arredondo, C., Martínez, A. & Smith, D. G. 2015. Radiocarbon and stratigraphic chronology of Canímar Abajo, Matanzas, Cuba. *Radiocarbon, 57*(5), 755–763.

Scabuzzo, C. 2012. Estudios bioarqueológios de marcadores de estrés ocupacional en cazadores recolectores pampeanos del holoceno temprano-medio: Análisis de la serie Esqueletal de arroyo seco 2. *Revista Argentina de Antropología Biológica, 14*(1), 17–31.

Stirland A. 1998. Musculoskeletal evidence for activity: Problems of evaluation. *International Journal of Osteoarchaeology, 8,* 354–362.

Suárez, R. G. & Forero, C. C. 2012. Fracturas de metacarpianos: actualización de conceptos. *Junio de 2012,* 56.

Tabío, E. & Rey, E. 1979. *Prehistoria de Cuba*. Editorial Científico Técnica.

Torres, P. & Rivero de la Calle, M. (1970). *La Cueva de la Santa. En Simposium XXX Aniversario de la Sociedad Espeleológica de Cuba*. Serie Espeleológica y Carsológica No. 13. Academia de Ciencias de Cuba.

Travieso, R., Rodríguez, D., Rivero de La Calle, M. y Marques, S. 1999. Estudio de los restos óseos humanos aborígenes encontrados en la Cueva del Perico I, Pinar del Río, Cuba. *Revista de Biología, 13*(2), 88–94.

Utset, M. B. 1941. *Notas de campo sobre exploración arqueológica de 1941. Subfondo de expedientes de sitios arqueológicos de Cuba*. Cuban Institute of Anthropology.

Vento, E. 2004. La treponematosis precolombina en Cuba. *Revista Medica Electrónica, 26*(6). http://www.revmedicaelectronica.sld.cu/index.php /rme/article/view/183.

Wapler, U., Crubézy, E., Schultz, M. (2004). Is cribra orbitalia synonymous with anemia? Analysis and interpretation of cranial pathology in Sudan. *American Journal of Physical Anthropology*, 123(4), 333–339.

Weston, D. A. 2008. Investigating the specificity of periosteal reactions in pathology museum specimens. *American Journal of Physical Anthropology, 137*(1), 48–59.

Weston, D. A. 2012. Nonspecific infection in paleopathology: Interpreting periosteal reactions. In A. Grauer (Ed.), *A companion to paleopathology*. Blackwell, 492–512.

White, T. & Folkens, P. 2005. *The human bone manual*. Elsevier Academic Press.

White, T., Black, M. & Folkens A. 2012. *Human osteology* (3rd ed.). Elsevier Academic Press.

Wollstein, R., Trouw, A., Carlson, L., Staff, I., Mastella, D. J. & Ashmead, D. 2020. The effect of age on fracture healing time in metacarpal fractures. *Hand*, *15*(4), 542–546.

8

THE CARIBBEAN SLAVE CEMETERIES. EXAMPLE OF THE FRENCH WEST INDIES

Patrice Courtaud and Thomas Romon

Since the 2000s, the number of colonial funeral complexes excavated in the French overseas departments has been in the order of 15. Their size varies from a few burials to several hundred. They were mainly studied in Guadeloupe and Martinique (Figure 8.1), the islands on which this chapter will focus. On the other hand, in other French West Indies locations, archaeological research has been less active and human remains less well preserved or totally destroyed, as in French Guiana, where the excavation of the Loyola Jesuit convent is an example (Croteau and Auger 1998; Le Roux et al. 2009). The amount of attention paid to colonial era funerary sites in the French West Indies can be contrasted to other island of the Lesser Antilles, where considerable work has been done (Courtaud, 2010; Handler & Lange, 1978; Haviser, 1999; Orser, 1998; Vergès, 2008; Watters, 1987), however, such sites have also remained overlooked in other Caribbean islands.

Of particular interest is the cemetery of Anse Sainte Marguerite (Guadeloupe), which remains one of the most important, not only because of the number of burials excavated, but also because it is the first for which the question of the archaeological recognition of a slave cemetery was raised. A second example is the cemetery of Anse Bellay in Martinique, where excavation is ongoing.

French West Indies during the Colonial Period

The colonization of Guadeloupe and Martinique did not occur until the beginning of the seventeenth century, when French settlers took possession of the island. Since discovery to colonization, these islands were occupied by Amerindians, then the Spanish, French, British, and Dutch, who collected wood and water and traded with the Indigenous inhabitants.

DOI: 10.4324/9780429444449-11

Since 1635, French colonization and incessant conflicts with the British and Dutch brought about many upheavals – the elimination of Amerindian populations primarily during the seventeenth century, the massive influx of people transported from West Africa, leading to the establishment of a hierarchical, segregationist, and racial society, dominated by a few white individuals, and the creation of a religious framework for cultural practices, with the imposition of Catholicism. These upheavals constituted a new form of organization, colonial society, which included mainly European, Caribbean (Creole[1]) and African individuals. These groups (outlined below), with very distinct characteristics, were each buried in their own cemeteries.

Planters – these are mainly whites. They marked their social position by erecting tombs, often monumental, in their family cemeteries. If located on a plantation, these cemeteries are called "master" cemeteries. Planters could also be buried in the parish church or in privileged areas of the parish cemetery (Bégot 1998).

Craftsmen and merchants – white or free of colour.[2] They lived mainly in the city and were most often buried in the parish cemetery (Pérotin-Dumon 2000; Verrand 2001). Prior to the French Revolution, they were not permitted burial inside the church. After the Revolution and the consideration of hygienic measures, these cemeteries, which have become communal, were moved outside the city.

Religious people – of European origin and may be similar to planters. They were buried in their own cemeteries on their plantations or in their religious houses. They could also be buried in the parish church.

Foreigners and seafarers – if they died in Guadeloupe, they were buried a hospital cemetery, frequently after a hospital stay (Paya and Romon 2001; Verrand 1999). Until the nineteenth century, these cemeteries were located at the hospital, then similar to the communal cemeteries, they moved outside the city walls and disappeared at the end of the nineteenth century. After this time, the deceased from the hospital were buried in district cemetery.

The military – they are largely metropolitan. Members of the troop are placed in the cemetery of the military hospital (Kacki and Romon 2013, 2015; Verrand 1999). The remains of officers are often repatriated to Europe. Otherwise, they may be buried in the cemetery of the military fort or parish cemetery.

Slaves – most commonly of African origin, they could be Creole or African. According to the *Code Noir*, they could be baptized, which was often the case, and were buried in the "Negro Cemetery", which could be specific to a single plantation or common to several. In the city, the "Negro Cemetery" was either totally isolated or adjacent to the parish cemetery, being separated from it by a wall. In all cases, the slave cemetery was always separate from the other burial sites, often located in marginal areas unsuitable for agriculture or the economic development of the island.

History of Archaeological Research on Colonial Cemeteries

The first publication on colonial cemeteries in Guadeloupe was by M. Barbotin (1978) on human remains discovered at Anse Nogent in Sainte-Rose. These remains were interpreted as coming from the cemetery of the first French settlers who lived at Pointe Allègre in 1635. However, the information concerning these remains is very incomplete, and the site was destroyed. At the beginning of the 1980's some rescue operations included colonial graves at Anse-des-corps (Petit-Canal), Anse Sainte-Marguerite (Le Moule), and Plage des Raisins-Clairs (Saint-François). Ten years later, at the first true archaeological funerary excavation in Guadeloupe, colonial graves were excavated on the Pointe-des-Pies site (Saint-François; Richard 1994). However, the real objective of this operation was the Amerindian strata into which the tombs were inserted. In 1994, an archaeological survey was carried out on the pre-Columbian site of Anse Sainte-Marguerite (Brokke and Knippenberg 1994), recognizing the presence of burials from the colonial period.

The year 1995 and hurricanes Luis and Marilyn marked a turning point in the history of the archaeology of colonial cemeteries in Guadeloupe. Human bones were discovered as a result of a swell on the Plage du Vieux Fort in Clugny (Sainte-Rose), as well as on the pre-Columbian site of Morel au Moule. A landslide revealed graves on the Morne Dauphine in Matouba (Saint-Claude). Reflection on the status of these funeral complexes, then unknown from historical sources, was initiated and led to the development of a research program involving colonial period funerary sites. It is with this perspective that archaeological research was undertaken, starting in 1997, on the Anse Sainte-Marguerite slave cemetery (Courtaud et al. 1999; Courtaud and Romon 2004).

The Cemeteries of Basse-Terre

In the early 2000s, excavations in Basse-Terre of the well-known cemeteries of Saint-François and the Charité Hospital, provided new data on parish and hospital cemeteries, respectively, in Guadeloupe at the end of the Old Regime, prior to the French Revolution. These results provide a basis for comparison when studying funerary complexes not documented by historical sources.

The Parish Cemetery in the Saint-François District

Until the beginning of the nineteenth century, the parish cemetery, which became "communal" in 1837, was located near the parish church. Metropolitan hygienic laws were then applied in the French colonies, with extra-mural displacement of cemeteries. However, these rules were not applied in the same way to all municipalities of Guadeloupe. Today, some municipalities, such as Anse Bertrand or Vieux-Habitant, still have their cemeteries in their original locations, alongside the church. Others moved their cemeteries, but they have been overtaken by urbanization. Finally, some, such as Port-Louis or Sainte-Anne, still have their cemeteries in peripheral areas.

The current city of Basse-Terre is the result of the merger, during the Revolution, of two parishes, Basse-Terre and Saint-François. Until the second third of the eighteenth century, each of these two parishes had its own cemetery, adjacent to the church. At the end of the eighteenth century, the cemetery of the parish of Basse-Terre was moved extra-murally and then, at the joining of the two parishes, abandoned in favour of that of Saint-François, which remained in its original location. The parish church of Saint-François was erected as a cathedral in 1850 and its cemetery was moved extra-murally in the first quarter of the nineteenth century (Figure 8.1).

Mitigating archaeological operations carried out around the Basse-Terre Cathedral (Bonnissent and Romon 2004; Romon et al. 2009) made it possible to study the cemetery of the Saint-François parish. Maintained by the Capuchins, the cemetery operated from the last quarter of the seventeenth century to the beginning of the nineteenth century. Study of parish registers (Verrand 2001) indicates that the cemetery was intended for the free population of the district, but freedmen were also registered as being buried. Intramural burials located inside the church were present and a cemetery reserved for slaves is adjacent, separated by a wall.

The opening of a 3m × 4m excavation in a space next to the cathedral revealed 45 graves, including those of seven men, 12 women, three children, and three adolescents. Two phases could be identified. The older phase did not demonstrate any immature individuals, only adults of both sexes. They were buried supine, heads facing southwest. The use of a trapezoidal coffin was split evenly between use in tombs and for burials directly in the ground. The younger phase contained immature individuals, during which coffin use became more systematic, switching to a hexagonal shape. The orientation of the deceased, head southwest, was no longer systematic, as some had the head oriented northeast. Skeletal reductions are common in both phases, with an ossuary appearing during the second, younger phase.

An analysis of individual health status showed many signs of bone degeneration, mainly cervical and lumbar osteoarthritis, which was related to age. Dental wear, as a result of clay tobacco pipe smoking, was frequent. One adult had a consolidated fracture of the right humerus, while another had medial convexity of the tibiae and fibulae that may represent rickets (Brickley and Ives 2008). As this childhood disease, linked to vitamin D deficiency, was relatively rare in Guadeloupe, it is possible that the subject was a first-time European migrant.

The graves were organized in successive rows with deliberate groupings. The shift in the orientation of the deceased with respect to the cardinal points is interpreted as a technical constraint related to the orientation of the church, which follows the urban structure, parallel to the shoreline, itself oriented south-east / north-west. Associated grave goods consist mainly of metal pins and bone buttons, signifying dressed burials. No religious elements (i.e. rosaries, medals, etc.,) were found.

The eighteenth-century parish cemetery, reserved for whites and freedmen, adjoins the church and was characterized by individuals buried in coffins or placed directly in the ground, with the deceased mostly resting their heads to the west. Children had their own reserved areas. Apart from the remains of clothing, there are no associated furniture.

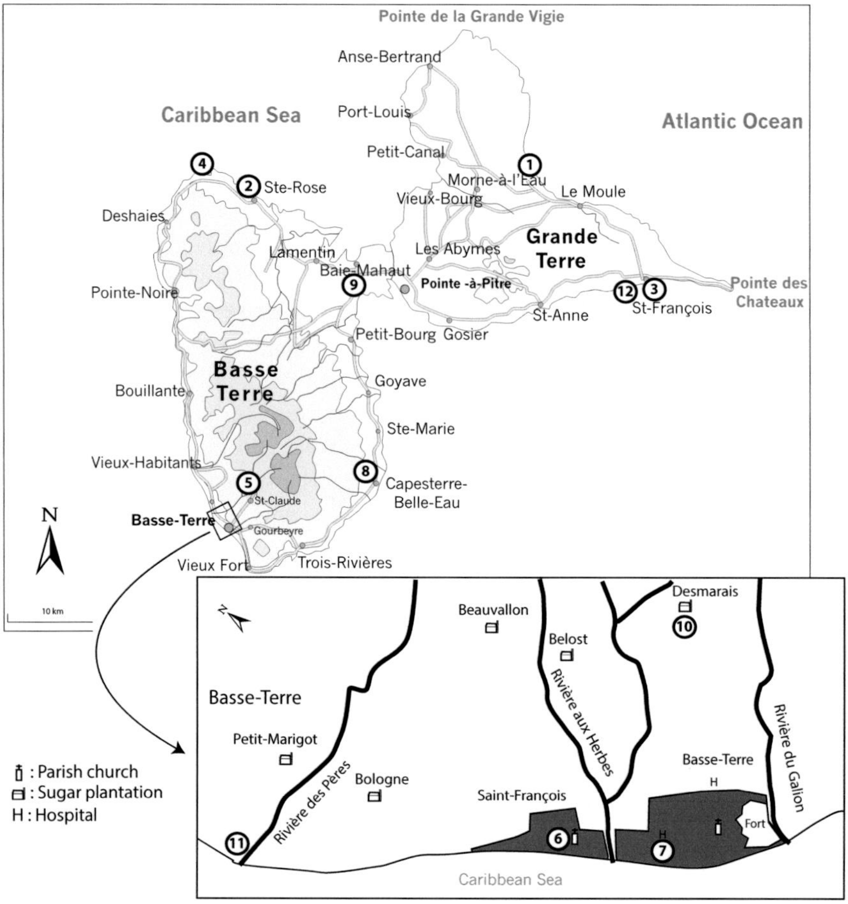

FIGURE 8.1 Map of Guadeloupe with the location of the places mentioned. (1) Slave cemetery of Anse Sainte-Marguerite (Le Moule). (2) Cemetery of the colonial period of the Anse Nogent (Sainte-Rose). (3) Cemetery of the colonial period of Pointe des Pies (Saint François). (4) Probable slave cemetery of Clugny Beach in Sainte-Rose. (5) Probable slave cemetery from Morne Dauphine (Saint-Claude). (6) Saint-François parish cemetery (Basse-Terre). (7) Cemetery of the Charité Hospital (Basse-Terre). (8) Probable slave cemetery from Doyon (Capesterre-Belle-Eau). (9) Probable slave cemetery from Jabrun (Baie-Mahault). (10) Cemetery of the plantation Desmarais. (11) Second cemetery of the military hospital of Basse-Terre at the Pointe des Pères (Baillif). (12) Slave cemetery of the Raisins Clairs beach (Saint-François) (© Th. Romon).

The Cemetery of the Hospital of La Charité

The La Charité Hospital cemetery was located south of the chapel of the Frères de la Charité and was managed by religious congregations. The study of historical sources (Verrand 1999) indicates that, in addition to the most disadvantaged and, in particular, the most committed Europeans, the hospital housed sailors and foreigners. Until the construction of the Basse-Terre military hospital in 1766, it also welcomed soldiers. The La Charité Hospital cemetery operated between 1664 and 1788.

A part of this cemetery was excavated over an area of 59 m^2, as part of a rescue excavation (Paya and Romon 2001). One hundred and nine primary burials were discovered, including 48 men, three women, three children, and five adolescents, showing a clear predominance of adult males. The grave plan shows a completely different organization from that of the cemetery of Basse-Terre Cathedral. Three burial phases were identified, each characterized by a change in the orientation of the graves along an east-west or north-south axis. When the cemetery was saturated, there was a levelling, a cleaning of the surface, and a gathering of the disturbed bones in pits, after which a new burial phase began with an orientation perpendicular to the previous one.

The most frequent graves types are empty pit excavations covered with nailed wooden planks. Individuals were placed on their backs, one after the other. This type of structure allows the grave to be reused and possibly to control the decomposition stage of the bodies. Burials immediately backfilled with soil are rare. These had individuals placed on their stomach or one side and are interpreted as sloppy burials, perhaps indicating a desire not to handle the body. Graves with coffins are also rare; only two have been identified by the presence of nails from trapezoid-shaped coffins. Finally, six rectangular pits contained secondary burials. These are interpreted as ossuaries intended to receive the bone elements disturbed by the management of the cemetery.

As with the parish cemetery, the main type of burial is an individual grave, with most subjects being placed on their backs. However, two types of burials, not found in the parish cemetery, have been identified. The first is double burials (Duday 2009), most probably reflecting the expense of digging two pits for two deaths that occurred close in time. The second is the presence of several burials containing individuals placed on their sides or stomachs. These placements may reflect the rapid burial of individuals who may be contagious. Most of the primary burials were disturbed by the digging of new graves. The displaced parts were either placed in the new grave as a reduction or collected in ossuary pits. Within the latter, a tendency to collect the skulls has been highlighted.

The excavated area contains mainly young male adults. Bone and dental health is satisfactory. Most of the individuals died from diseases that did not leave any traces on the skeleton. Four individuals have consolidated fractures of a section of the leg, and one has vertebral lesions similar to those found in tuberculosis (Baker 1999). No grave goods were directly associated with the burials, and there was no religious material and no indication of clothing.

The Anse Sainte-Marguerite Cemetery

This funerary complex is situated on the northeast coast of Grande-Terre (Figure 8.1), to the north of Le Moule, near the village of Gros Cap. It was situated on a beach, where it occupied the slope of a coastal dune (Figure 8.2). The geographical location of this cemetery, far from any place of worship, in a rural area largely devoted to sugar cane crops, designated it a priori as a funerary complex destined for a slave population (Courtaud 2002, 2010). The graves provide both cultural information relating to early populations, particularly their relationship with death, and biological data about past populations.

Funerary practices

The *Code Noir* (Sala-Molins 2006) was a decree originally created in 1685 under the reign of Louis XIV. It defined the conditions of slavery in the French colonial empire. Seven articles of the *Code Noir* state that Catholicism is the only religion for all slaves, who must be baptized as soon as possible after arrival in the colonies (art. 1 to 7).

As for burials, owners were required (art. 14) to bury their baptized slaves in holy ground, in cemeteries designed for this purpose. As for the others, a simple burial in a field was sufficient. Non-baptized slaves, of whom there were many during the Old Regime (before the French Revolution of 1789), were thus deprived of cemetery burials. Baptized slaves to be deposited in holy ground were not to be buried in the same zones as freemen.

FIGURE 8.2 View from the plateau of Anse Sainte-Marguerite (ASM) beach, where the cemetery was installed on the coastal slope of the dune (© P. Courtaud).

Archaeological observations were carried out for individual burials and for the cemetery as a whole, in order to address questions relating to the organization and management of the sepulchral area. Different questions were raised, including:

- For baptized slaves, were Christian burial practices respected, and which parameters provide answers to this question?
- Conversely, is it possible to identify non-baptized individuals?
- Do certain elements indicate the persistence of African practices? These would involve symbolism and complex interpretations beyond the scope of this paper, given the large size of the African and the multitude of different populations. A priori, we do not expect any specific practices, but rather practices different from Christian rites, like for example, the association of decorative elements or objects from Africa.

In order to discern the organization and management of burial areas, it is imperative to work on an extensive area and excavate a sufficient number of skeletons.

The Biology of the Buried Population

The first aspect considered when studying the biology of the buried population is the study of selection (i.e. who was being buried in the cemetery). This involves estimating age at death and identifying the biological sex of the deceased. The exploitation of these essential biological data enables us to produce a demographic profile of the living population. However, in this case we are not dealing with a natural population, but a population partly made up of transported people, which substantially modifies the demographic parameters used for a "natural" population.

New arrivals regularly succombed to premature death, mainly due to deplorable living conditions. This introduces another aspect, namely health conditions, which are revealed by bone markers indicating trauma, mechanical stress, dietary deficiencies, and infections. This aspect will not be discussed further here, as studies in this domain are ongoing, notably relating to infection diseases (Dutour et al. 2005). Some studies regarding the sanitary state of slaves have been conducted (Khudabux 1999; Rathburn 1987; Owsley et al. 1987).

Results

First, the aim was to recognize the existence of a cemetery from colonial times, to specify its limits, its conservation, and the typology of the burials.

From east to west, the coastal area is divided into four parts (Figure 8.3):

- the current beach, about 30 m wide, is made up of partly consolidated coral sand (beach-rock), with no archaeological remains. The development of a coastal

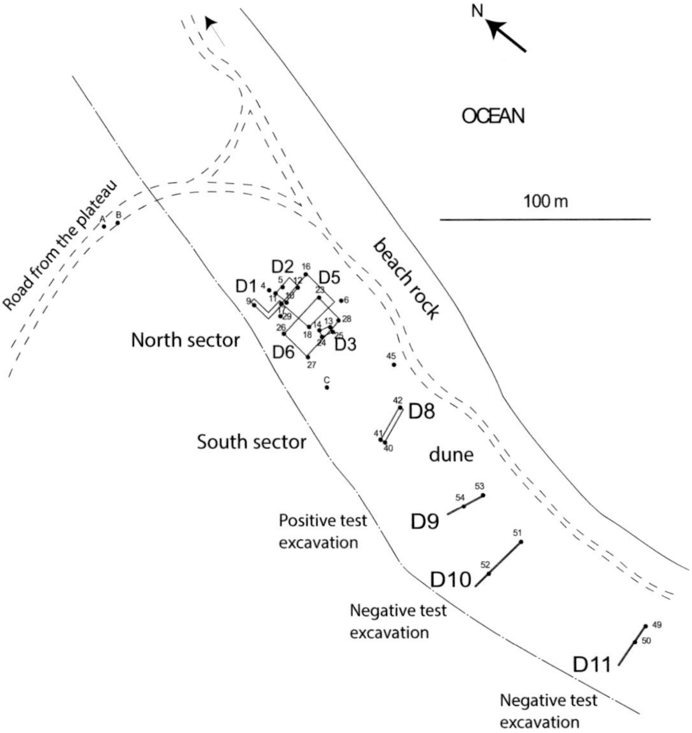

FIGURE 8.3 Plan of the two excavated sectors, north and south (ASM, © P. Courtaud).

path resulted in the disappearance of this part of the coastal dune and the graves situated there;

- behind the beach, a ridge of coastal dunes with a width of 40–60 m and a height ranging from 0.50 m to 2 m. It is only partially conserved as sand retrieval damaged the northern extremity and the western slope. The total length of the dune is over 200 m, but it has been fractioned into five unequal parts by sand plundering, each of which is covered by some vegetation. The preserved part of the funerary complex is located in this sector;
- a zone covered by shrub vegetation. The limestone bedrock, at a depth of a metre, underlies a clay level, overlain by vegetation;
- a cliff leads up to the plateau, which lies at an altitude of 60 m above the coast.

The existence of sea-rolled, fragmented human remains on the current beach is the only material evidence of ancient, destroyed burials. Previously, they were considered to belong to the pre-Columbian period, due to the presence of an underlying Amerindian site, disturbed by the digging of colonial burials and subsequently by the partial destruction of the dune. The identification of significant zones disturbed by the massive recovery of sand in the 1970s and the threat of the progressive disappearance of the dune were decisive in the decision to conduct archaeological operations on this site.

Different surveys showed that the initial cemetery extended over a length of 200 m and must initially have included several hundred deceased individuals, probably close to a thousand, which does not agree with the hypothesis that the deceased were from a single plantation. Indeed, as this part of the island was only colonized at the end of the first half of the eighteenth century, the cemetery would have been used over the course of a century, or slightly more. This cemetery would thus have gathered the dead from several plantations or from a townland. At the end of the eighteenth century, in the neighbouring townland of Petit-Canal, 88% of the population were slaves. Historical data concerning the cemeteries from the colonial period (Boutin 1998) mention small clusters of slave cemeteries installed in habitations, urban cemeteries like in Basse Terre and Pointe-à-Pitre, but there is no mention of funerary complexes gathering the slaves from several plantations.

We distinguished two zones inside the cemetery:

The Northern Sector

The first three excavation seasons focused on the northern sector, where burials were numerous and well preserved. The explored sector yielded close to 200 graves (Figure 8.4). Bone preservation is excellent, which is exceptional in a tropical environment (Figure 8.5).

However, not all skeletons are so well preserved. Firstly, children's graves are smaller and were not buried as deeply and were thus exposed to perturbations. Children are not as well represented as adults. Secondly, many graves were reused (Figure 8.6). They are accompanied by re-organization and reductions and consequently by the dislocation of certain skeletons. The graves were organized in rows and were oriented east/west with the skulls lying to the west in the majority of cases (Figure 8.7).

Inversions of orientation were observed for about 5% of the remains, generally for children. Corpses were buried supine in coffins, even for the youngest individuals. Well-preserved nails attest to the presence of coffins. For adult coffins, 40 nails were used, whereas children's coffins used about 20 nails. Two graves were without coffins, with bodies placed directly in the ground. One of these is a "classical" child's grave from all viewpoints, with a three to four-year-old child in the primary position. Sometimes, osteological indicators clearly show that decomposition occurred in a confined space (Duday 2009), which excludes the hypothesis of an original empty space of decomposition, and therefore a hermetically sealed coffin. The corpses were deposited in an extended position. Simple graves were not the norm, but many graves were reused, and very frequent reductions occurred inside the grave. In this way, the same structure can contain the remains of several individuals, and the remains of the same individual can also be distributed in two different burials. Reuse and reduction are frequent in medieval cemeteries and

FIGURE 8.4 View of the northern sector of the excavation. The graves were only identified on the surface by the limits of the excavation of the grave pit. No architectural elements were preserved, except some coffin nails (ASM, © P. Courtaud).

result from the management of limited areas. Here, the existence of unused space near plural graves appears to demonstrate the will to gather the deceased in the afterlife. The criteria for this gathering appears to be linked to kinship or ethnicity.

A complex of two burials (Figure 8.8) is a good illustration of this, where an adult grave was partially intersected by a child's one. The adult skull was secondarily deposited above the child's skull. These funerary rites suggest that particular care was taken here and that the unusual image of the adult skull "protecting" the child's grave could have symbolic value. Here, we are constrained by the limits of archaeology. We observe a particular disposition, but any explanation must remain a hypothesis.

These numerous groups, the quasi-systematic use of coffins, and the respect for orientation give a very "Christian" image to this funerary complex. No African practices were discerned. Several rare burials contained objects, such as a kaolinite pipe, rosary beads (Figure 8.9), and even a medal of Saint Mary dating to 1852.

This latter discovery is significant as it provides a date, as it is impossible to obtain radiocarbon dates for such recent periods. We thus have proof that the cemetery was used after the definitive abolition of slavery (1848 for French colonies), which raises questions as to the future of these funerary complexes. It appears

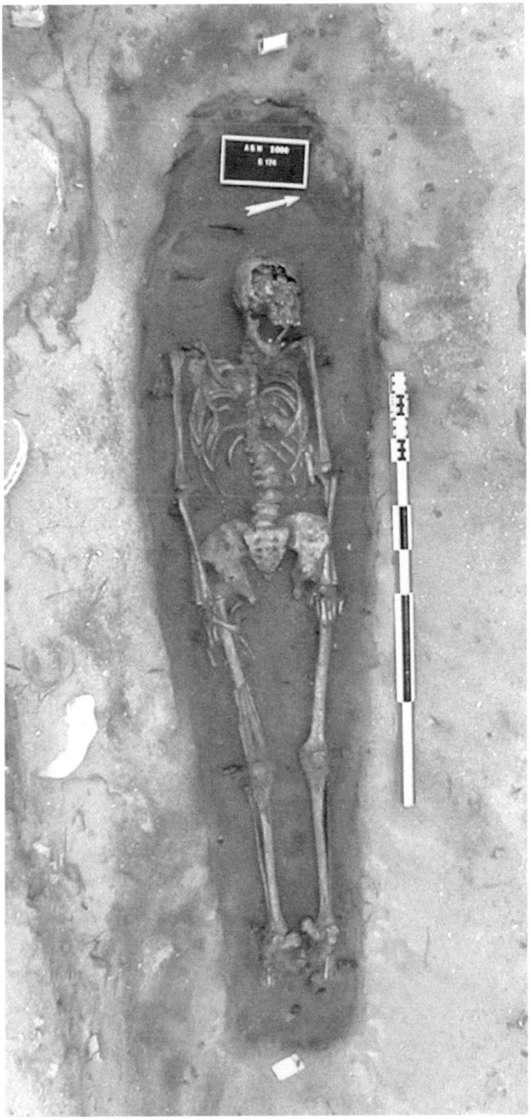

FIGURE 8.5 View of a simple grave of an adult lying on his back with the head towards the west (S.174). The remains of a hexagonal coffin are represented by the nails distributed around the skeleton. (ASM, © P. Courtaud).

that, at Sainte-Marguerite, some of the new free population continued to use the ancient slave cemetery for a while. Was this due to a deliberate choice to be near the graves of their ascendants or because of prohibited access to holy ground? This is one of the main questions raised by this excavation. Adults were buried dressed, as shown by the presence of bone and sometimes mother-of-pearl buttons near the

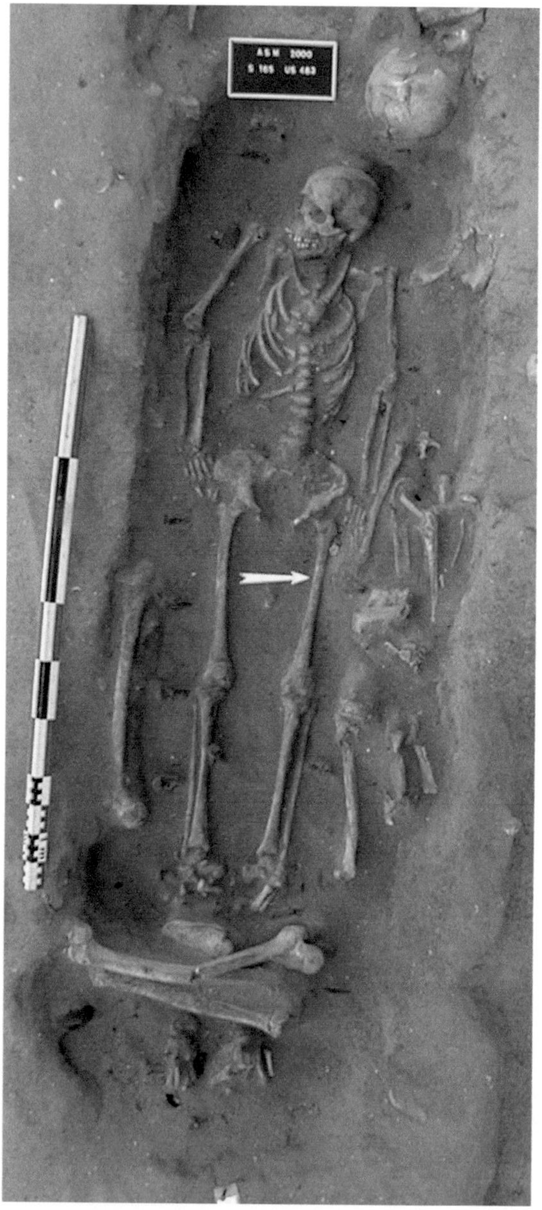

FIGURE 8.6 Anse Sainte-Marguerite – View of a reduced grave following the installation of a new occupant. The bones of the first deceased appear to be dislocated around the grave (S.165).

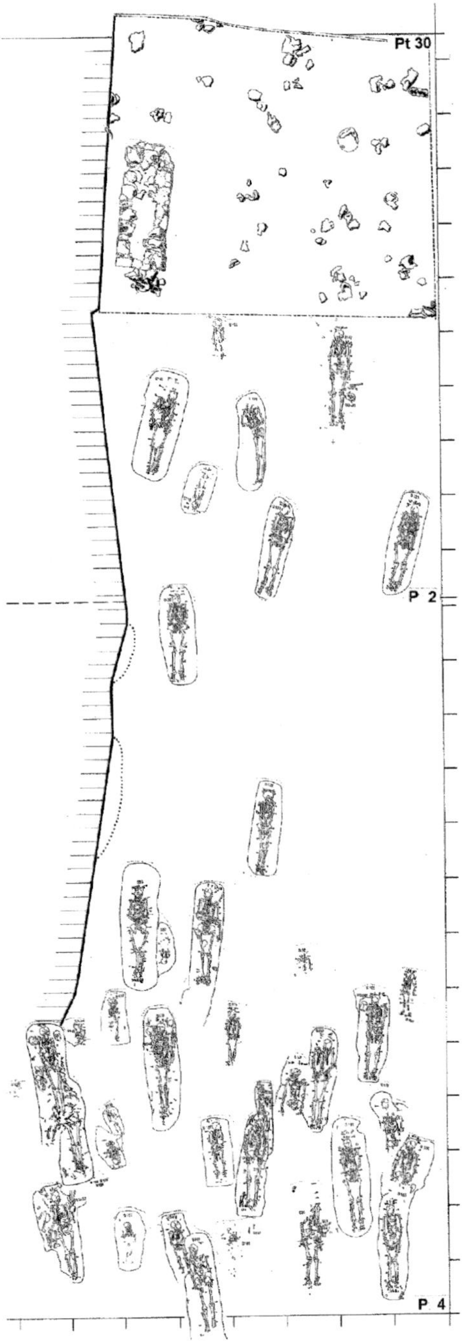

FIGURE 8.7 Plan of part of the northern sector where the tombs follow an east-west axis (ASM, © P. Courtaud).

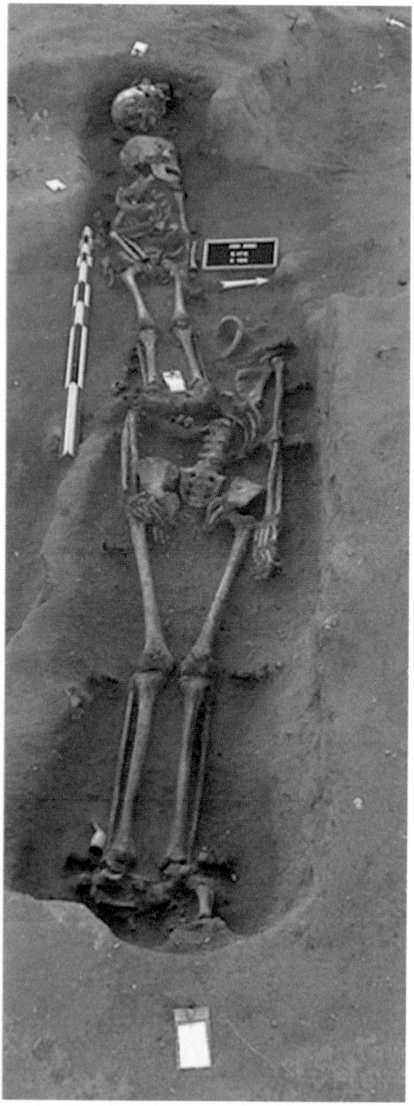

FIGURE 8.8 This example illustrates reductions and redistributions of human remains after grouping remains together. The adult's skull was moved and placed above the child's skull. This appears to be pre-meditated rather than accidental (S.178 – S.185) (ASM, © P. Courtaud).

pelvis area, indicating underpants, or in the thorax area, indicating a shirt. The only known references are archaeological. Similar pieces were found during a rescue operation in a grave in the Anse Bertrand chapel (Pichon 1996), in the north of the island.

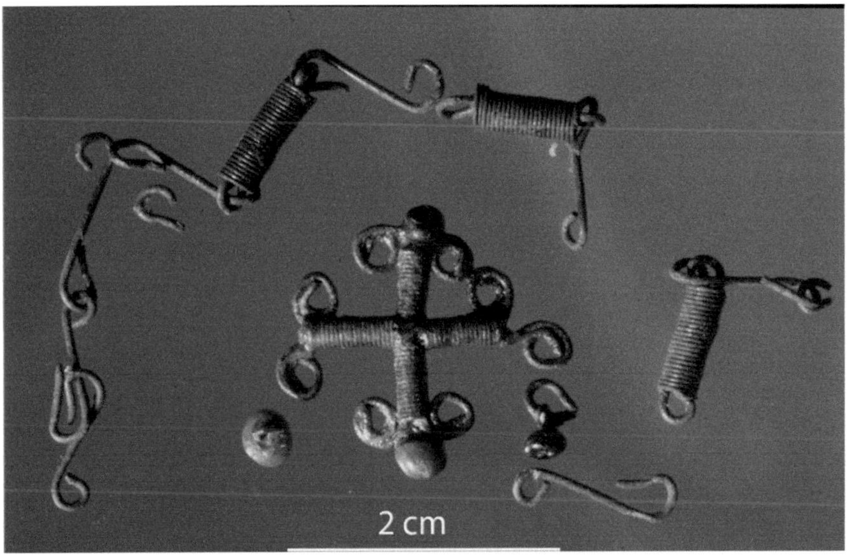

FIGURE 8.9 Detail of rosary beads covered by the hand of a deceased individual in the northern sector (ASM, © P. Courtaud).

Another element indicates the use of the cemetery after 1848. This is the discovery of two partially destroyed brick-built tombs, where the exposed part is missing. The foundation is made up of crude or roughly squared limestone blocks, of small coral bricks assembled with mortar surrounding the coffin. One of them contains the skeleton of an adult with Pott's disease (Ortner and Putschar 1985), spinal tuberculosis, which may be the cause of death. The cranial morphology of this individual evokes an African origin. The architecture of the tomb is not like settlers' tombs (Bégot 1998) but looks like the Delanoé cemetery (Boutin 1998), a small funerary complex on the plateau. Although no specific funerary practices point to an African origin, there is evidence of dental mutilation, consisting of filed front teeth, for three individuals (Figure 8.10). As mentioned above, good preservation, particularly of skulls, allows for the observation and quantification of certain morphological characteristics that point to an African origin. The African characteristics, mainly of the face, are particularly evident (Howells 1989, 1995).

Two hundred and fifteen individuals were exhumed in all, comprising 126 adults and 89 children and adolescents. The latter are underrepresented, as are newborns. However, this must be interpreted with caution, given the singular renewal modalities of this population. The 82 individuals for whom the coxal bone is well preserved are made up of 39 females and 33 males (10 unknown). This population appears to have been particularly disadvantaged; adult death occurred frequently

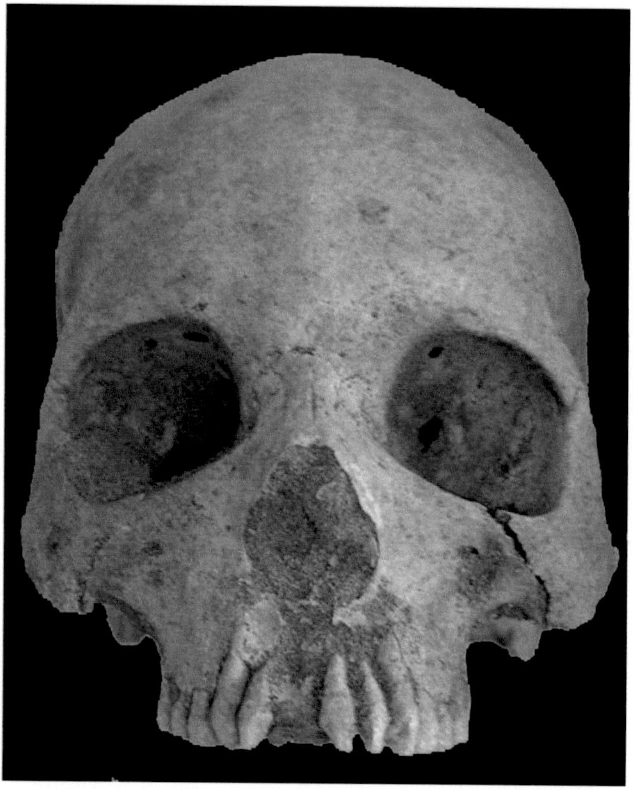

FIGURE 8.10 Anterior view of a skull with filed and pointed upper incisors. This mutilation was carried out on an adolescent or young adult (ASM, © P. Courtaud).

before the age of 30, and certain significant pathological signs point to frequent biological stress (Dutour et al. 2005; Romon and Courtaud 2002). On the basis of the archaeological material from the burials and nearby, this north sector was used during the first half of the nineteenth century and a part after the abolition of slavery. We cannot, however, propose a date for the end of the funerary use of this zone.

Once we attained our initial aim to uncover a large area and gather a consequential sample of exhumed skeletons, we pursued excavations for a supplementary season in a zone further to the south. Our aim was to check the homogeneity of funerary management and selection for burial.

The South Sector

A survey was carried out in a band perpendicular to the dune, located about 50 m to the south, where perturbations did not appear to be too marked (Figure 8.3). We discovered a more anarchic organization, with regard to the organization of space and funerary architecture. This area was clearly different from the northern sector.

It was dated to the second half of the eighteenth century by surface ceramic finds. None of the burials contained material clearly associated with the deceased, apart from an intact bone crucifix found in relation to a grave, but in a secondary position. It was lying between the leg bones but probably decorated the coffin lid. Fifty-three graves were identified, corresponding to the remains of 63 individuals, mainly from primary burials, whereas 16% of the individuals were disturbed. This sector yielded a majority of burials directly in the ground (55%), and the head facing west in half of the cases. The grave area thus appears to be disorganized. Moreover, an absence of bone or mother-of-pearl buttons indicates there were no clothes, just an iron belt buckle. Four more individuals with dental mutilation were uncovered.

Discussion

Since the end of the 1990s, the number and diversity of cemeteries explored in Guadeloupe and the quality of the documentation collected have been a quite exceptional reference point. Because of the partitioning of colonial society, which imposed its own cemetery on each population group, the work carried out on funeral complexes constitutes a privileged field of investigation for questions relating to the status of cemeteries, funeral practices, their composition, and the biological characteristics of the populations they contain.

In cemetery terms, the transposition of metropolitan practices to the colonies is unequivocal. It testifies to the domination and supremacy of Whites. However, there is a certain adaptation of colonial society and its specific composition: enslaved people. This results in the creation of unique funeral spaces. These "Negro" cemeteries are in many ways identical to parish cemeteries in terms of organization, functioning, and sacredness.

Almost all the cemeteries discovered in Guadeloupe over the past 20 years were forgotten in the collective memory, especially those intended for enslaved people. Such graveyards, which were also far from religious, military, or civilian buildings, had until then been interpreted as slave cemeteries. Determining the status of the buried populations, and therefore the identity of these cemeteries, is one of the major challenges for archaeologists and historians. Several types of information must be taken into account:

- Historical sources: Archives and parish registers are very informative documents. However, they may be deficient due to often poor preservation (due to fire, cyclones, etc.). They contain very limited information in reference to burial sites or enslaved people in general. Moreover, let us not forget that all the writings were produced by the free population.
- Funeral archaeology: The management and organization of cemeteries and the nature of the funeral practices provide valuable information.
- The biology of the deceased: Data on age at death, sex, and general health status are extremely valuable.

The information collected from the Anse Sainte-Marguerite site can be extended to those obtained from other burials or cemeteries suspected of containing enslaved people: the Raisins Clairs (Saint-François; Rouquet 2015) located on a beach and three others inland, Jabrun (Baie-Mahaut; Martias 2006), de Doyon (Capesterre-Belle-Eau; Martias 2005), and Morne Dauphine (St-Claude) (Figure 8.1). These rescue archaeology operations have not been detailed here because they have so far been less documented.

The Anse Bellay Burial Site

Anse Bellay is a cobble beach located on the southern edge of Fort de France Bay (Martinique), right at the mouth of the Ravine Maudite, and just in front of the Ilet à Ramier. It is an 80-m-long, east-west-oriented cordon, behind which a mangrove develops. The cove is the Conservatoire du Littoral's property, and, as such, it is protected from anthropogenic activities. In August 2007, the swell that followed Hurricane Dean brought to light human bones and pre-Columbian ceramics. Two evaluation campaigns commissioned by the Direction des Affaires Culturelles (DAC) of Martinique and carried out by the Institut National de Recherches Archéologiques Préventives (INRAP) revealed pre-Columbian occupations and a cemetery from the colonial period. The latter is subject to an excavation programme, in progress and in collaboration with the UMR CNRS 5199, the University of the French Antilles, INRAP, and the DAC's Archaeological Service.

The cemetery is located on the barrier beach. Its downstream part has been destroyed by marine erosion (Figure 8.11). The remaining surface area is estimated

FIGURE 8.11 View of the cemetery of Anse Bellay (© T. Romon).

to be 70 m^2 and to contain about 80 graves. No surface markers were identified, and there is no trace of enclosure. Graves correspond to single burials, and their pits intersect with the underlying pre-Columbian stratigraphic layers.

The excavation of a 40 m^2 area revealed 48 burials from the colonial period. The density of the graves (i.e., 1.2 individuals per m^2) is in line with the densities of graves in contemporary cemeteries previously excavated in the region. These are single primary burials. The individuals were buried in a supine position, each one in its own pit, with the lower limbs in an extended position. The preferred orientation was east-west, with the head to the west, although there is some variability. Several graves were intercutting (Figure 8.12).

When digging a grave, the gravedigger may have disturbed the remains of previously buried individuals. They were then removed and repositioned in the pit of the new grave. This is a common funerary gesture referred to as "reduction" that is still in use today. As the first body is already decayed at the time of the disturbance, this kind of manipulation implies that the burial space was used for quite a long period. This demonstrates that Anse Bellay is not a cemetery related to a catastrophic event (epidemic, shipwreck, war, etc.), where individuals would have been buried in a relatively short timeframe. Moreover, if not fortuitous, intercutting between the graves may indicate a desire to bring the individuals together. This would therefore imply that the identity of the deceased and the place where they were buried were known (Figure 8.13).

The exhumed population includes young and old adults of both sexes, as well as adolescents and children of all ages, including newborns. It is therefore highly unlikely that this cemetery was that of soldiers associated with the nearby Fort de l'Ilet à Ramiers. Indeed, such a military population would have been mainly composed of young adult males.

In addition, the upper incisors of the individual from grave 51 have been intentionally sharpened (Figure 8.14). In West Africa, it was a common cultural practice related to initiation rites, but it was not practised in Martinique during the colonial period. It is therefore assumed that individuals who carry such stigmata were born in Africa and then transported to be enslaved in the Americas. Other individuals show particular abrasions of their teeth that are characteristic of the use of the so-called "Dutch" clay pipes, which are abundantly recovered during archaeological excavations of sites from the colonial period. These pipes were used by both free and enslaved people.

No artefacts, such as buttons, medals, or rosaries, were found in association with the skeletons, in contrast with what is sometimes observed in other colonial cemeteries. The few artefacts found are all in a secondary position, in the filling of the graves, and correspond mainly to ceramic shards that are dated to the eighteenth century. There is no evidence that coffins were used. The dead were most likely buried in a simple fashion, probably wrapped in shrouds, as indicated by some taphonomic evidence (e.g., position of the feet of some individuals).

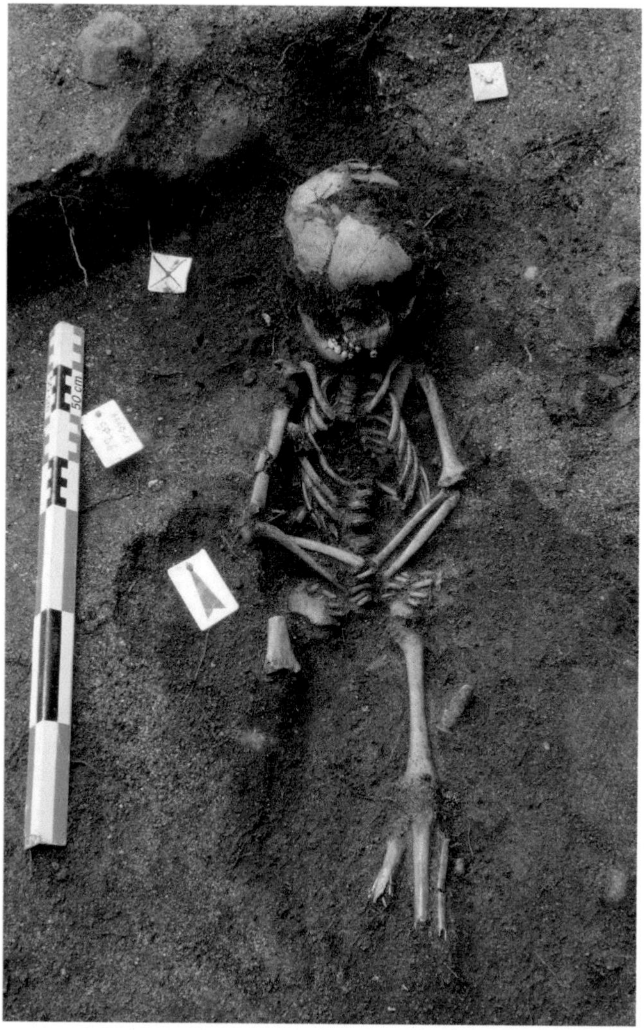

FIGURE 8.12 View of a simple grave of a child lying on his back with the head towards the north-east (S.36) (© T. Romon).

Conclusion

No evidence for funeral practices of African origin have been discovered. On the contrary, funeral rites seem to conform to the Christian rite with an east-west orientation of the graves with the head lied to the west.

At Anse Sainte-Marguerite, the coffins are commonly used in the nineteenth-century sector and only sometimes (one out of two cases) in the eighteenth-century sector, which suggests that at that time fewer subjects were baptized, which results in a more anarchic organization of the funeral space. Like Anse Bellay, the cemetery

FIGURE 8.13 View of a set of graves of adults in different positions with the head towards the west (S.48, S.51, S. 53) (© T. Romon).

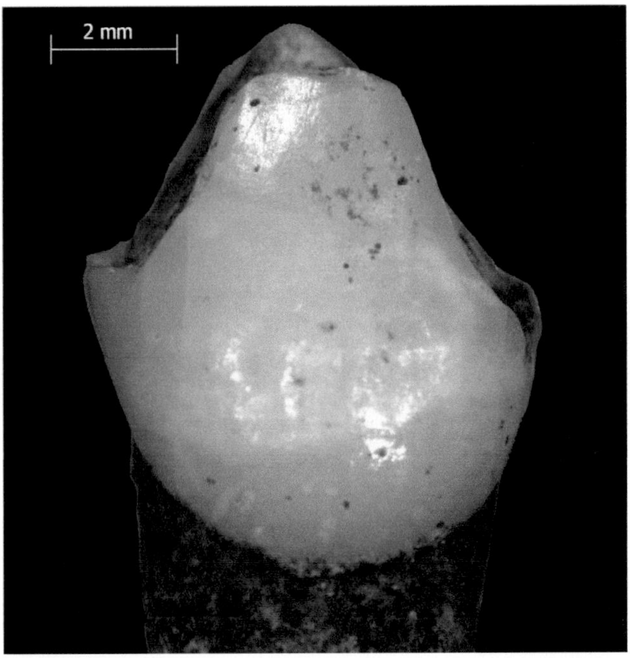

FIGURE 8.14 A sharpened upper incisor (S.51) (© T. Romon).

is located in a remarkable place: in a rural area, which was occupied by settlers since the eighteenth century and where there was a high proportion of slaves.

Anse Bellay cemetery is located at the boundary between the parishes of Anses-d'Arlet (5.5 km) and Trois-Ilets (4.5 km), which are known to have parish cemeteries adjoined to churches, by the sea, i.e., an area of little interest for the plantation economy because it is not well-suited for cultivation, and was owned by the king. In addition, we observed some variations in burial orientation and no evidence of coffins.

The first archaeological results obtained at Anse Bellay cemetery converge towards the same interpretation. It was probably neither a parish cemetery, nor a military cemetery, nor a disaster cemetery, nor a master cemetery. The hypothesis of a slave cemetery appears more likely. However, the first results have to be taken with caution given the small size of the excavated area and the small number of graves studied so far. This work needs to be supplemented by further fieldwork and additional analyses (in particular genetic and biochemical analyses) and by further historical investigations.

The indications that led us to identify a cemetery as containing enslaved individuals are above all elements of dating, attributing them to the colonial period. The geographical location away from towns and religious places, in marginal areas such as a beach here, are decisive factors. While these arguments indicate the servile status of the buried population, there are also biological characteristics reflecting poor health and the presence of the practice of dental mutilation, which occurs in some African populations (Handler 1994).

Does this mean that we have found a scheme that makes it possible to irrefutably characterize the status of the people buried in these sites? The servile condition of an individual cannot be deduced from any of these facts, whether archaeological or anthropological. It is the combination of multiple strands of information that leads us towards a proposed function of these cemeteries and towards the identity of the buried population. It is therefore not the individuals we identify as enslaved but the funeral complex we assume to be a slave cemetery (Courtaud 2015).

This research is part of a more global question regarding knowledge of past societies, considered from a transdisciplinary perspective, based on collaborations involving archaeologists, anthropologists, and historians. This research theme, which is in progress in French-speaking territories, requires the study of colonial society as a whole, in order to restore the links that existed between each of its components.

Acknowledgements

We are grateful to the archaeological services of the Direction des Affaires Culturelles of Guadeloupe and Martinique and to the Regional Council of Guadeloupe for the support of research. We would like to thank the municipalities of Le Moule

(Guadeloupe) and Les Anses d'Arlet (Martinique). This research benefitted from the scientific framework of the University of Bordeaux's IdEx "Investments for the Future" program / GPR "Human Past".

Notes

1 The term Creole refers to any person born in the West Indies, regardless of their origin and status.

2 Freedmen.

References

Baker, B. M. (1999). Early manifestation of tuberculosis in the skeleton. In: Pálfi et al. (ed.), *Tuberculosis: Past and Present*. Budapest-Szeged: Golden Book and Tuberculosis Foundation, pp. 301–311.

Barbotin, M. (1978). Découverte de crânes, fémurs et autres os… *Bulletin de la Société d'histoire de la Guadeloupe*, 38, pp. 3–37.

Bégot, D. (1998). Les tombes de colons: Un art colonial. In: *Patrimoine de Guadeloupe, Les Cimetières*. Special Issue, Ed. Société d'Histoire de la Guadeloupe, 1, pp. 21–36.

Bonnissent, D. and Romon, T. (2004). *Fouilles de la Cathédrale de Basse-Terre*. Basse-Terre: Document final de synthèse, Service régional de l'archéologie de la Guadeloupe.

Boutin, R. (1998). Le champ de repos In Patrimoine de Guadeloupe, Les Cimetières. Ed. *Société d'Histoire de la Guadeloupe*. Special Issue, Ed. Société d'Histoire de la Guadeloupe, 1, pp. 5–12.

Brickley, M. and Ives, R. (2008). *The Bioarchaeology of Metabolic Bone Disease*. Amsterdam: Academic Press.

Brokke, A. J. and Knippenberg, S. (1994). *Evaluation archéologique du site de l'Anse Sainte-Marguerite, sous la direction de X. ROUSSEAU*. Basse-Terre: Service régional de l'archéologie de la Guadeloupe.

Courtaud, P. (2002). Le site de l'Anse Sainte-Marguerite (Guadeloupe, Grande Terre). Présentation d'un cimetière d'époque coloniale. In: A. Delpuech, J.- P. Giraud, and A. Hesse (ed.), *Archéologie précolombienne et coloniale des Caraïbes*. Fort-de-France: Actes du CTHS, pp. 283–294.

Courtaud, P. (2010). Approche archéologique des populations serviles. L'archéologie funéraire aux Antilles. In: M. Cottias, E. Cunin, and A. de Almeida Mendes (ed.), *Les traites et les Esclavages. Perspectives historiques et contemporaines*. Karthala, pp. 301–308.

Courtaud, P. (2015). Le cimetière, miroir de l'esclavage. In: P. Courtaud, S. Kachi, and Th. Romon (ed.), *Cimetières et Identités,* Thanat'Os, Ausonius Ed., pp. 103–114. Bordeaux: Ausonius Éditions – Maison des Sciences de l'Homme d'Aquitaine.

Courtaud, P. and Romon, Th. (2004). Le site de l'Anse Sainte-Marguerite (Guadeloupe, Grande-Terre. Présentation d'un cimetière d'époque coloniale. *Journal of Caribbean Archaeology*, Special issue 1: pp. 58–67.

Courtaud, P., Delpuech, A., and Romon, Th. (1999). Archeological investigations at colonial cemeteries on Guadeloupe: African slave burial sites or not? In: J. B. Haviser (ed.), *African Sites: Archaeology in the Carribbean*. Princeton: Markus Wiener Publishers, pp. 277–290.

Croteau N. and Auger, R. (1998). La mise en valeur du site de Loyola à Rémire, Guyane: communique l'archéologie coloniale. In: A. Delpuech, J.-P. Giraud, and A. Hesse (ed.),

Archéologie précolombienne et coloniale des Caraïbes. Fort-de-France: Actes du CTHS, pp. 315–312.

Duday, H. (2009). *The archaeology of the dead: Lectures in archaeothanatology (Vol. 3)*. Oxford and Philadelphia: Oxbow books.

Dutour, O., Romon, Th., Ardagna, Y., Tatillon, C., and Courtaud, P. (2005). Paléoépidémiologie de la tuberculose en Guadeloupe: Le cimetière d'esclaves de l'Anse Sainte-Marguerite.. In: UMR 6568 Marseille (ed.), *L'Homme et ses images, Mesures, Représentations, Constructions, Actes du XXVe colloque du Groupement des Anthropologues de Langue Française*, Marseille: Compact Disk, pp. 355–362.

Handler, J. S. (1994). Determining African Birth from skeletal Remains: A note on Tooth Mutilation. *Historical Archaeology*, 28(3), pp. 113–119.

Handler, J. S. and Lange, F. W. (1978). *Plantation Slavery in Barbados: An Archaeological and Historical Investigation*. Cambridge: Harvard University Press, 368 p.

Haviser, J. B. (1999). *African Sites Archaeology in the Caribbean*. Princeton: Markus Wiener Publishers.

Howells, W. W. (1989). *Skull shapes and the map: Craniometric analyses in the dispersion of modern Homo (Vol. 79)*. Papers of the Peabody Museum of Archaeology and Ethnology 79. Cambridge, MA: Harvard University Press.

Howells, W. W. (1995). *Who's Who in Skulls, Ethnic Identification of Crania from Measurements. Papers of the Peabody Museum of Archaeology and Ethnology (Vol. 82). Antiquity*. Peabody Museum.

Kacki, S. and Romon, Th. (2013). From burials to population identity: Archaeological appraisal of the status of a Lesser Antilles colonial cemetery (Baillif, Guadeloupe). *Antiquity*, 87, pp.829–839

Kacki, S. and Romon, Th. (2015). Le cimetière de la Rivières des Pères (Baillif, Guadeloupe): Identification du statut des défunts par l'approche archéo-anthropologique. In: P. Courtaud, S. Kachi, Th. Romon (ed.), *Cimetières et Identités, Thanat'Os.* Ausonius Ed., pp. 117–132. Bordeaux: Ausonius Éditions – Maison des Sciences de l'Homme d'Aquitaine.

Khudabux, M. R. (1999). Effects of life conditions on the health of a negro slave community in Suriname. In: J. B. Haviser (ed.), *African sites: Archaeology in the Caribbean.* Princeton: Markus Wiener Publishers, pp. 291–312.

Le Roux, Y., Auger, R. and Cazelles N. (2009). *Les Jésuites et l'esclavage Loyola: L'habitation des Jésuites de Rémire en Guyane française*. Quebec: Presses de l'Université du Québec.

Martias, R. (2005). *Doyon, Capesterre-Belle-Eau, Guadeloupe. Rapport de diagnostic archéologique*. Basse-Terre: Service régional de l'archéologie de la Guadeloupe.

Martias, R. (2006). *Jabrun, Baie-Mahault, Guadeloupe. Rapport de diagnostic archéologique*. Basse-Terre: Service régional de l'archéologie de la Guadeloupe.

Orser, C., E. (1998). The archaeology of African diaspora. *Annual Review of Anthropology*, 27, pp. 63–82.

Ortner, D. J. and Putschar, W. G. J. (1985). *Identification of pathological conditions in human skeletal remains*. Washington, DC: Smithsonian Institution Press.

Owsley, D. W., Orser, C. E., Mann, R. W., Moore-Jansen, P. H., and Montgomery, R. L. (1987). Demography and pathology of an urban slave population from New Orleans. *American Journal of Physical Anthropology*, 74(2), pp. 185–197.

Paya, D. and Romon, Th. (2001). *Le Cimetière de l'hôpital de la Charité, Palais de Justice (Basse-Terre, Guadeloupe). Document Final de Synthèse*. Basse-Terre: Service régional de l'archéologie de la Guadeloupe.

Pérotin-Dumon, A. (2000). *La ville aux îles, la ville dans l'île*. Basse-Terre et Pointe-à-Pitre, Guadeloupe, 1650–1820, Paris: Editions Karthala.

Pichon, M. (1996). *Rapport de fouille archéologique à la chapelle d'Anse Bertrand.* Basse-Terre: Service régional de l'archéologie de la Guadeloupe.

Rathbun, T. A. (1987). Health and disease at a South Carolina plantation: 1840–1870. *American Journal of Physical Anthropology*, 74(2), pp. 239–253.

Richard, G. (1994). Premier indice d'une occupation précéramique en Guadeloupe continentale. *Journal de la Société des Américanistes*, 80(1), pp. 241–242

Romon, Th. and Courtaud, P. (2002). Etude biologique des squelettes du cimetière d'époque coloniale de l'Anse Sainte-Marguerite (Le Moule, Guadeloupe). In: A. Delpuech, J.- P. Giraud, and A. Hesse (ed.), *Archéologie précolombienne et coloniale des Caraïbes*, Fort-de-France: Actes du CTHS, pp. 295–299.

Romon, Th., Courtaud, P., Paya, D., Bonnissent, D., and Verrand, L. (2009). La place des esclaves dans les cimetières coloniaux. *Trois exemples guadeloupéens. Archéopages*, 25, pp. 46–51.

Rouquet, J. (2015). *Plage des Raisins Clairs, Rapport Final d'opération, Fouille de sauvetage archéologique urgent.* Basse-Terre: Service régional de l'archéologie de la Guadeloupe.

Sala-Molins, Louis 2006. *Le Code Noir ou le calvaire de Canaan.* Ed. Paris: PUF.

Vergès, F. (2008). Archéologie de l'esclavage, archéologie de l'absence. In: J.P. Demoule and B. Stieger (eds.), *L'avenir du passé*. Modernité de l'archéologie, La Découverte, pp. 105–113. Paris: La Decouverte.

Verrand, L. (1999). *Etude du cimetière de l'hôpital Saint Jean-Baptiste de la Charité.* Basse-Terre: Guadeloupe – Volet de recherche CAOM 1999, Service régional de l'archéologie de la Guadeloupe

Verrand, L. (2001). *Le cimetière de la cathédrale de Basse-Terre. Rapport d'étude d'archives au centre des archives d'outre-mer*. Basse-Terre: Service régional de l'archéologie de la Guadeloupe.

Watters, D. R. (1987). Excavations at the Harney site slave cemetery, Montserrat, West Indies. *Annals of Carnegie Museum*, 56(18), pp. 289–318.

9

BIOARCHAEOLOGY OF THE NEWTON PLANTATION BURIAL GROUND, BARBADOS

Slavery, Sugar, and the Embodiment of Violence

Kristrina A. Shuler

Introduction

During the mid-seventeenth century, the Caribbean Island of Barbados in the Lesser Antilles emerged as the economic and political epicenter of a burgeoning global sugar industry built through the trafficking and enslavement of human laborers from Africa (Figure 9.1). In total, 2.7 million Africans were shipped to the British Caribbean between 1625 and 1807 (Eltis and Engerman 2011, 392), with an estimated 350,000 to Barbados alone (Handler and Lange 1978, 21). Caribbean sugar plantations were among the most brutal labor systems in the colonial Atlantic world (Higman 1995; Kiple 2002; Kiple and Higgins 1992; Roberts 2013). Enslaved laborers worked long hours in intense climates with exposure to dangerous conditions at every stage of the production process (Mintz 1986; Roberts 2013; Sheridan 1989). Even with policies aimed at amelioration in the late eighteenth century, the daily lives of enslaved workers on Barbadian plantations were little improved, and disciplinary tactics may have increased as planters aimed to maximize labor while resources such as land, food, and fuel were becoming increasingly scarce (Roberts 2013, 58). A late eighteenth-century observer claimed "…the great body of the slaves, the field-people on sugar-plantations, are generally treated more like beasts of burden than like human creatures" (William Dickson 1789, 6; Handler and Lange 1978, 77). Roberts (2013, 14) has argued that the resistance paradigm that has shaped the history of slavery since the mid-twentieth century has undermined the significant power differentials and broad spectrum of violence endured by enslaved populations in the Atlantic world. While such narratives have undeniably placed the enslaved as active agents in history, there has been an unintended consequence of limiting discourse surrounding the violence and suffering experiences of the Caribbean enslaved who were commodified and exploited for

DOI: 10.4324/9780429444449-12

their labor. Despite the wealth of documentation that is available about the institution of chattel slavery, direct first-hand accounts of the treatment, health, and lives of the enslaved in Barbados and the Caribbean are rare.

Bioarchaeology offers a unique and complementary means of exploring the life histories of past populations within a biocultural framework (Agarwal and Glencross 2011, Larsen 2015, Martin et al. 2013), including variable experiences of human trauma (Martin et al. 2012). Through individual, community-wide, and even global datasets, skeletal and dental evidence provide insight into demography (e.g., age, sex, and ancestry), growth and nutritional stress, diseases, traumatic injuries, and a lifetime of activities, all of which leave their signatures on the body. Bioarchaeological data are thus not only essential for properly reconstructing life histories but also for revealing evidence of pain and suffering (Martin and Harrod 2016). For groups who remain historically invisible, including underrepresented and misrepresented populations of the African Diaspora and enslaved colonial Africans, textual biohistories can be supplemented, and in some cases contested, through biological data from archaeological sites (Tiesler and Zalaba 2010, 70; Shuler 2005a, 2011; Shuler et al. 2019). More than 50 years of collaborative research from the Newton Plantation Burying Ground in Barbados has contributed insight into the health, nutrition, and life histories of enslaved Africans at the epicenter of industrialized sugar production. This chapter offers a synthesis and critical reappraisal of bioarchaeological research from the site, including the need for greater engagement with descendant communities, which is essential to understanding the past and addressing the pervasive legacy of colonialism in archaeology.

Newton Plantation Excavations and Bioarchaeology: 1970s–1980s

Anthropologist Jerome Handler began exploring cultural practices and ethnohistory in Barbados in 1963 (see Armstrong 2019 for a detailed chronology of archaeological research and history of the island). By the early 1970s, Handler began a collaborative archaeological project with Fred Lange that utilized the abundant archival materials from early colonial plantations to search for evidence of "intermediary cultures between Africa and the present" (Handler et al. 1978, 1; cf. Jamieson 1995). They set out to identify the living quarters of enslaved Africans on 14 plantations, and, while their search for architectural structures was unsuccessful, oral history from the local community led to the identification of an unmarked cemetery (Farmer 2021) on the grounds of the former Newton Plantation. They renegotiated the project in the field and excavated at the cemetery for two months in 1972 and six weeks in 1973 (Handler et al. 1989, 4), exposing remains from 104 skeletons from a relatively small area. Burials were encountered in five low earthen mounds and four non-mound burial complexes (Handler and Lange 1978, 108–132). Mound 2 contained the highest concentration of burials (n = 47), while the largest mound (Mound 1) contained only a single interment (Handler and Lange

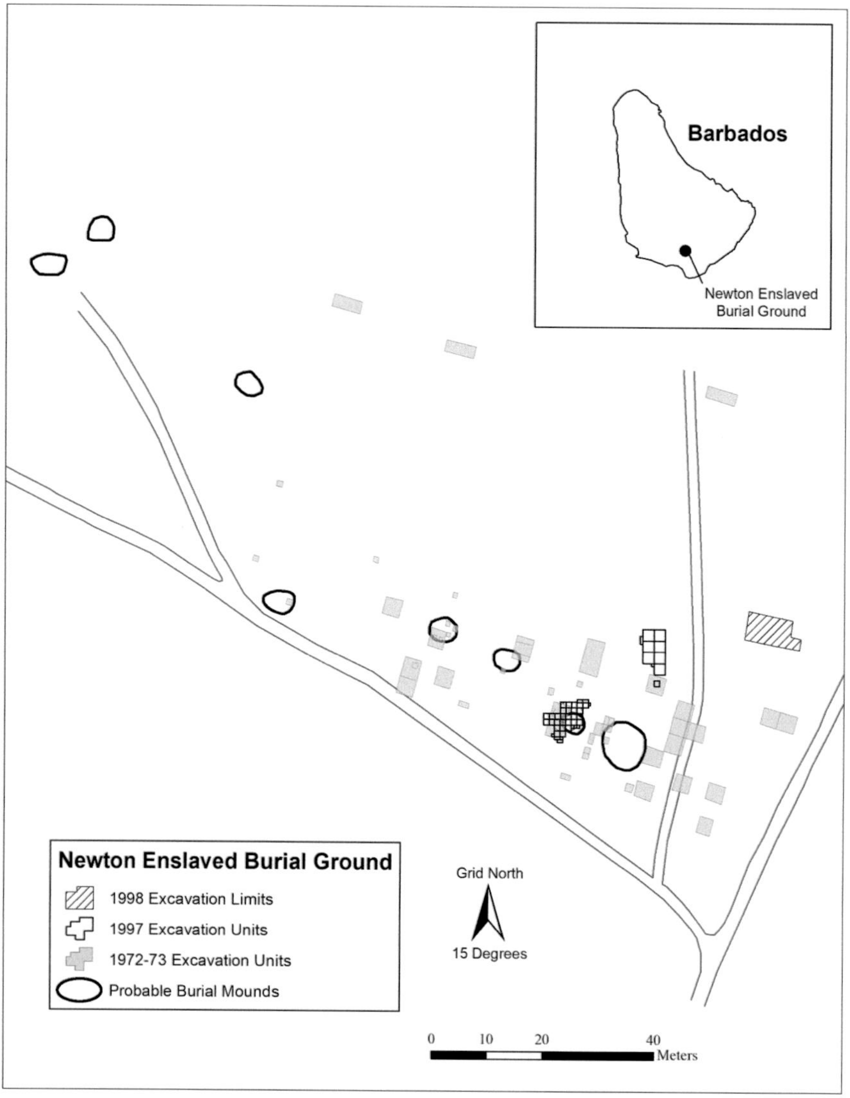

FIGURE 9.1 Map showing the location of Newton Plantation Cemetery, Barbados (Map by Eric Sipes).

1978, 110–112). The latter was found in the prone position, head facing west, with no evidence of a coffin or artifacts. Most burials were aligned on an east-west axis, with west-headed coffin burials being the most common. According to Handler and Lange (1978, 162–165), these represent later graves from the site based on relative proveniences and associated artifact dates. Grave goods included white clay pipes, locally produced earthen wares, imported Buckley ware, porcelain, glass beads, metal knives, jewelry, and buttons. Two burials were distinctive, revealing potential evidence of differential status, including Burial 9 (female and sole burial in Mound 1), who was hypothesized to have been negatively viewed within her community, and Burial 72, who was suggested to have been a possible healer of high status based on grave accouterments (Handler and Lange 1978, 111, 125–132; Handler 1996, 1997). Following the 1970s excavations:

> *Many of the bones recovered at Newton during the early 1970s were poorly preserved owing to burial near the surface and the effects of a tropical environment; in addition, there were many multiple interments with disturbance and commingling in the graves, resulting in many disarticulated bones. These bones were analyzed in the field as they were excavated, but the analysis was largely limited to aging and sexing. Attempts were made to deposit the skeletal materials at the Barbados Museum, but the Director at the time was uninterested, for he lacked storage facilities and trained curatorial personnel; and storage space could not be obtained elsewhere on the island. With budgetary limitations preventing shipping all of these materials to the United States, most of the bones were* carefully reburied *at the Newton site*
>
> *(Handler et al. 1989, 6, emphasis added)*

Permissions and ethical considerations for the original excavations and subsequent relocation of human remains are not well documented in scholarly literature aside from a 1989 survey report where a letter from the Barbados Prime Minister is included that shows official governmental support for the archaeological research at that time (Handler et al. 1989). Armstrong et al. (2019, 27) have recently pointed to a series of heated newspaper articles between Handler and distinguished Barbadian historian Sir Hilary Beckles. This exchange was published in Barbados newspapers, The Advocate and the Nation in March and April 2000:

> *Despite the protests of the director of the [Barbados] Museum, Ms. Alissandra Cummins, Handler who had exhumed the mortal remains of enslaved ancestors from their resting place at Newton Plantation, and took them out of the country without informing the Museum of the inventory, continues to be a problem. He has returned some bones though in a deplorable condition. The leadership of the Museum is not pleased. Its official position is that Handler continues to hold on to more of the mortal remains of our ancestors. The University of South Illinois where Handler was based has been keen to assist by returning what was in its possession. They have handled Handler's case internally to my satisfaction." In part 2 of this exchange goes on to note that "Kevin Farmer, who the University of the West Indies and Government has trained at great cost, and who is now the Museum's specialist in the field of historical archaeology, has not been able to gain access to the field documents relevant to the excavation that took place at Newton*
>
> *(https://glc.yale.edu/sites/default/files/files/mpi/2011/1816-Revolt-2000.pdf; pages 19 and 23)*

In total, the dental and limited cranial remains of 104 individuals unearthed during 1970s excavations were transported to Southern Illinois University Carbondale, where they were extensively studied by physical anthropologist Robert Corruccini, colleagues, and students throughout the 1980s and early 1990s leading to some of the earliest and most influential bioarchaeological research into the lives of enslaved Africans to date.

These studies included assessment of temporomandibular joints and dental malocclusion, which suggested decreased mastication and reliance on an increasingly processed diet over time (Corruccini and Handler 1980; Handler and Corruccini 1983), while the ubiquity of dental caries (20%; $n = 19/94$ occlusal and 57.4% $n = 54/94$ interproximal surfaces) was argued to be consistent with frequent consumption of soft carbohydrates and poor dental care (Corruccini et al. 1985). Hypercementosis was found to be exceptionally high (89.4% of individuals), which Corruccini et al. (1987b) associated with high rates of periodontal disease. Enamel hypoplasia was present in most individuals, with deeply palpable hypoplasias, or "major growth arrest lines", noted in 20% of individuals (Corruccini et al. 1985, 702; Corruccini et al. 1982; Handler and Corruccini 1983). Overall, they argued that the frequency and severity of dental defects were consistent with poor diet and episodic starvation among the enslaved laborers on the plantation (Corruccini et al. 1985; Corruccini et al. 1987b). In a reanalysis of the dental remains, Ritter (1991, 101) found consistent rates of clearly palpable hypoplastic grooves (19%) but noted that hypoplastic frequency was not in line with expected stress levels, unlike the higher rates of cusp asymmetry, which, like historical accounts, suggested that slaves at Newton may have been severely stressed, dying of acute diseases, leaving dentitions unaltered. The small number of cranial remains observed from the 1970s excavations displayed no evidence of anemia and low rates of non-specific infection, although this was possibly due to sampling (Corruccini et al. 1982, 450; Handler and Corruccini 1983). However, a later study by Jacobi et al. (1992, 156) revealed dental stigmata indicative of congenital syphilis in three individuals, potentially affecting as many as 10% of the plantation. The 1970s remains documented previously unknown evidence of lead poisoning with an average bone lead content of 117.6 +/−94.9 ppm (Corruccini et al. 1987a), which is nearly four times higher than in U.S. slaves (Aufderheide et al. 1985). Clinical symptoms would have been mild to severe intestinal and neurologic distress, and Handler (1986) drew upon ethnohistorical sources to suggest the association of bone lead values to the consumption of contaminated rum by the enslaved. Lower lead values (44.7 ppm) were reported for individuals displaying cultural tooth modifications than those without (126.2 ppm), suggesting a potential pattern of co-occurring dental filing, low skeletal lead, slight hypercementosis, and burial practices (east and north-headed orientation, and absence of coffins) that distinguished African from Caribbean-born individuals in the sample (Corruccini et al. 1987a; Handler 1986).

Bioarchaeological research at Newton has also provided a means of comparing historical and skeletal data for assessing accuracy in paleodemography and expectations for health indicators. Contrary to predictions, Corruccini et al. (1989) found that infant mortality was low and a life expectancy at birth was 29 years, which is inconsistent with the archivally documented life expectancy of 20 years at birth and evidence of very high infant mortality. Infants and children were seriously underestimated in the 1970s sample, while young and middle-aged categories were overrepresented, which may be the result of sampling, aging biases, differential preservation, and/or non-stationarity as many of the enslaved laborers were forcibly imported at Newton (Corruccini et al. 1989). While skeletal demography proved less reliable from the 1970s sample, most of the skeletal and dental patterns from the earlier studies appeared to be consistent with extensive historical evidence for poor health, periodic starvation, and high levels of protein-calorie malnutrition (Corruccini et al. 1982; Handler and Corruccini 1983). Along with the extensive archaeological research that has been published on the material culture from the site (Handler and Lange 1978; Handler 1996, 1997, 1998), bioarchaeological analyses published in the 1980–1990s generated substantial information on the lives of the Caribbean enslaved population.

Newton Plantation Excavations and Bioarchaeology: 1990s–present

The Barbados Museum and Historical Society (BMHS) acquired the Newton Plantation Slave Burying Ground in 1993 and leads in the burying ground's "protection, preservation, and interpretation as a site of memory for generations of enslaved people and as a legacy for their descendants" (The Colonial Williamsburg Foundation, n.d.). With the partnership and guidance of the BMHS, excavations were allowed in 1997 by the author (then a Ph.D. candidate supervised by R. Corruccini at SIUC) and in 1998 by Ray Pasquariello (a Ph.D. candidate in historical archaeology supervised by D. Armstrong at Syracuse University). Prior to the start of the project, the BMHS hosted a press conference where the museum and archaeologists addressed community questions about the project.

Fieldwork in 1997 was conducted in May and June with graduate students from SIUC and Syracuse University, members of the local community, and regular site visits by Ms. Phillipa King, who was the Deputy Director of the Barbados Museum at that time. Excavations focused on previously excavated areas of the site where human remains had been uncovered by researchers and left in situ (Figure 9.2; Handler and Lange 1978; Handler et al. 1989). The site grid was reestablished based on personal communications between R. Riordan, F. Lange, and R. Corruccini (1995), field notes from Riordan, and information provided by Handler and Lange (1978). Notably, the absence of a permanent site datum or geographical benchmark made the realignment of the grid tenuous. A total of 32 square meters was re-excavated in the location of Mound 2 and 30 square meters northeast of Mound 2, near the documented location of Unit 30N9W Burial Complex (Handler and Lange 1978, 117–125). Excavations and documentation followed standard

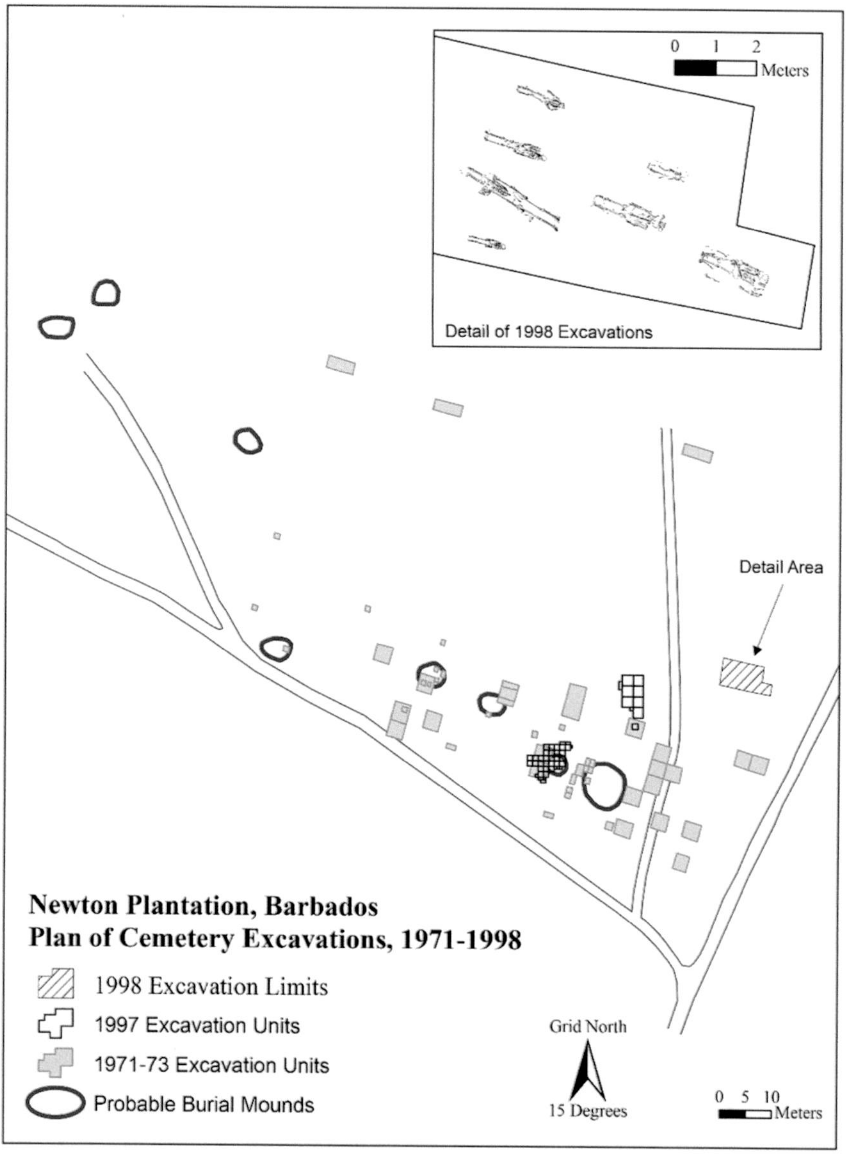

FIGURE 9.2 Map showing the location of excavation areas at Newton Plantation (Map by Eric Sipes).

protocol (e.g., Bass 1987; Buikstra and Ubelaker 1994; Ubelaker 1989), with each burial exposed, recorded, mapped, and photographed in plain view. Skeletal elements were carefully removed, packaged, and labeled with provenience. Surrounding soil was screened through 1/8" mesh to ensure that each individual's remains were kept together for later reburial. In all, 35 articulated/semi-articulated burials were excavated in 1997 with dense clusters of disturbed and commingled human remains found close to the surface (10–20 cm) in the southeastern and southwestern portions of Mound 2. Evidence of prior disturbance included the presence of a modern glass bottle and a 1965 British coin near masses of commingled skeletal remains that were largely grouped by element. Fragmentary long bones had been taped together, while others had candy wrappers adhering to human remains (Shuler 1997, unpublished field notes on file with the Barbados Museum). Upon encountering this, work at the site was halted, and the museum was contacted immediately to assess the situation. Ms. King visited the following day, after which we documented the condition of the remains before carefully removing them. The following summer, R. Pasquariello returned to the site and excavated a previously undisturbed area northeast of Mound 1 with assistance from the Barbados Museum and Dr. Karl Watson and students from the University of the West Indies. Researchers excavated seven distinct burial shafts and recovered the remains of 14 individuals across 44 square meters. In all, 49 individuals and an estimated MNI of 24 commingled individuals were excavated in the 1990s. Individuals ranged from very complete (20%), moderately complete (37%), extremely incomplete (39%), and dental remains only (4%). Fieldnotes were deposited with the museum at the conclusion of each season, and, with the permission of the Barbados Museum, human remains from each of the two field seasons were officially loaned and transported by freight to SIUC and later relocated to Auburn University with approval, where researchers from numerous institutions around the world have collaborated to offer new insight into issues of systemic health, migration, and the identity of the Barbadian past.

All bioarchaeological research conducted with human remains from Newton Plantation since 1997 has been with approvals from the BMHS directors and the museum board. Investigation of the 45 most complete skeletons revealed a life expectancy of 19.95 years, which is nearer to the archival sources than the 29 years reported from dental studies from the cemetery (Corruccini et al. 1989; Shuler 2011). Dental analyses were generally consistent with earlier findings, though Shuler (2005a, b) reports a far higher rate of caries (75%) that evenly impacted occlusal and interproximal tooth surfaces. The average caries rate of 1.09 per individual seems low, but high antemortem tooth loss (AMTL) (7.35 per individual) was observed in both females (7.17) and males (8.8). Rates of loss per individual were not noted for the earlier sample but were reported as high overall (51%) (Corruccini et al. 1982, 451). Handler and Corruccini (1983) suggest that frequent, non-random AMTL in molars indicates the practice of dentistry (Corruccini et al. 1987b). Hypercementosis in the 1990s sample (37%) was lower than previously reported in the 1970s sample (52%), and only adolescents and adults were affected, though it does appear to have been age progressive and affecting males more than females (Corruccini et al. 1987b; Shuler 2005a). Rates of linear enamel hypoplasia (LEH) were highly consistent across studies of the two samples at 20% (Corruccini et al. 1985), 19% (Ritter 1991), and 17% (Shuler 2005b).

Archival records for Barbados and elsewhere in the Caribbean suggest that pervasive malnutrition plagued the enslaved community (Handler 2006 a, b, 177), and inadequate macro and micronutrients likely caused widespread nutritional disorders such as marasmus, kwashiorkor, pellagra, beriberi, anemias, and scurvy as well as eye disorders caused by vitamin A and/or riboflavin deficiencies (Handler 2006 a, b; Kiple 2002). Metabolic diseases affecting the skeleton were generally low at Newton in contrast to many colonial populations (Shuler 2005a). Porotic hyperostosis was absent and cribra orbitalia was infrequent, though equally expressed in males and females (13%). Geophagy, or "pica", potentially provided some minerals and iron (Handler 2006a; Handler and Corruccini 1983), though Shuler (2005a) argues that any advantages would have been offset by an increased parasite load (e.g., hookworms) resulting in, rather than preventing, anemias (Layrisse et al. 1964). Dietary iron and calcium were available to the enslaved population through black strap molasses that was widely available, as well as lentils, pigeon peas, and iron cookware (Kiple 2002). Vault and orbital lesions that were previously attributed to iron, lysine, and B3 (niacin) deficiency, especially in corn-based diets (Stuart-Macadam 1985; 1987), have more recently been attributed to acquired megaloblastic anemia from inadequate B12 intake, folate deficiency, unsanitary living conditions, and possibly weaning stress (Walker et al. 2009; Oxenham and Cavill 2010). At Newton, the availability of local fish and seafood along with salted meats may have sufficiently mitigated B12 deficiency. Osteomalacia was observed in only one adult (N26), and possible rickets were seen in two subadults (N49 and N53), though these changes appear alongside other modifications that may have been congenital. Vitamin D deficiency would be less likely to have impacted the Creole-born individuals at Newton who would have likely spent substantial time outside contrary to the enslaved in industrial settings in the U.S. (Kelley and Angel 1987), although individuals who survived the Middle Passage may have experienced long periods of time without adequate nutrition or sunlight, leading to diseases such as rickets and osteomalacia. Stature estimates from the site offer additional insight into growth and nutritional status, providing a window into the stressful experiences of childhood, which are largely invisible for enslaved children (Teelucksingh 2006). Though two adults, one male (Burial 22–182.5 cm) and one female (Burial 52–167.5 cm), were taller than anticipated from archival records on Barbados, stature estimates for the remaining nine individuals who could be measured fall close to historically documented stature for nineteenth-century Barbadian Creole males (169.73 cm; s.d. 4.61) and females (159.3 cm; s.d.1.06), and are consistent with the relatively high stress experienced by sugar producers relative to other economies in the Caribbean and U.S. (Shuler 2005a, 2011). Still, the small sample size warrants caution when making comparisons.

Archival records from the Caribbean and Barbados describe symptoms associated with numerous communicable diseases (Alden and Miller 1987; Handler 2006 a, b; Handler and Corruccini 1983; Higman 1995; Kiple 1987, 2002; Kiple and

Higgins 1992; Sheridan 1985), though acute infections and epidemics (e.g., yellow fever, cholera, smallpox, and tetanus) are less likely to have affected the skeleton. The most common chronic infections known to affect Barbadian enslaved individuals (e.g., leprosy, tuberculosis, and treponemal infections) are generally known to produce osseous changes in fewer than 5% of cases (Shuler 2011). However, it was somewhat surprising that the more complete sample showed no signs of treponemal diseases among children or adults since dental stigmata indicative of congenital syphilis were previously observed in three individuals, which Jacobi et al. (1992) argue may have affected as many as 10% of individuals. Despite the absence of sequelae attributable to specific diseases, 41% of individuals from the site displayed inflammatory changes to the periosteal surfaces, especially in the long bones. Periosteal inflammation is widely understood to be a strong indicator of generalized infection, though it can occur from a wide variety of insults and activities (Weston 2008; 2012). The frequency and severity of lesions that were most often seen on the tibiae and femora of males, including an adult male individual with severe bilateral lesions (Burial 22), are highly consistent with injuries sustained from sharp instruments used during fieldwork to harvest sugarcane, though surprisingly this individual had no clearly visible signs of trauma based on macroscopic observations (Shuler 2011). Physical evidence of trauma was remarkably less visible on the skeletal remains of individuals in the cemetery at Newton Plantation, although systematic radiographs, CT scans, and surface scans of the sample have not been conducted, which may reveal greater evidence of healed traumas. Visible non-lethal injuries were observed in single bone elements from one male and two females who also had malalignment during healing. While the absence of intentional physical trauma related to abuse may be indicative of the decline in corporal punishment during the eighteenth century (Roberts 2013, 290–291), extreme work-related stress appears to have been common at Newton. In an initial assessment of activity stressors, Shuler (2005a) observed osteophytic lipping/eburnation of joints in 59% of individuals and vertebral herniations in 25% of individuals. Entheseal changes in the form of roughened, resorbed, and/or pitting and lipping at the muscle attachment sites on any major long bones were reported in 76% of individuals, although revised assessment of entheseal modifications and activity are in progress by K. Williams (Ph.D. student, SUNY Binghamton) using recent standardized scoring techniques (e.g., Henderson et al. 2016). Similarly, other markers such as osteoarthritis and trauma are being reevaluated using current standards. According to historical accounts of Barbadian plantations, strenuous labor would have impacted both men and women, though the previously collected skeletal evidence at Newton seems to suggest that the toll of stress in women resulted in decreased longevity and greater activity-related impacts to much of the body (Shuler 2005a, 320; 2011). In a comparison of biomechanics in commingled elements from enslaved rice producers from Hagley Plantation and at Newton, Stevens (2016) examined cross-sectional geometry of upper and lower limbs and found that impacts to upper bodies from Newton were consistent with archival predictions, though lower bodies of individuals from Hagley represent differences due to terrain, mobility, and/or load bearing in this group (Shuler et al. 2019).

Multi-isotopic and elemental Pb investigations of 26 individuals have provided insight into the natal origins of individuals buried within the cemetery (Schroeder et al. 2009, 2013) and an invaluable avenue for comparing the health of African and Creole-born individuals from the site. Evidence of high lead burden in the dental enamel demonstrates lifetime exposure to this heavy metal starting in early life for children born on the island (Schroeder et al. 2013). As lead burden in the enslaved population was arguably tied to the consumption of rum (Corruccini et al. 1987a; Handler 1986; Laffoon et al. 2020), Shuler and Schroeder (2013) used enamel lead as a proxy for potential exposure to ethanol during early childhood. Analyses of 26 individuals found evidence in one individual (N49) who displayed elevated enamel lead (3.8 g/g) with congenital defects (e.g., co-occurring vertebral and mandibular anomalies) consistent with clinical and experimental descriptions of alcohol-related birth defects (ARBD). Combining skeletal and isotopic data shows that dental filing ($n = 2$), as expected, was associated with African isotopic signatures, but no differences were found for birthplace and mortuary treatment (burial location, direction, or coffins). Barbadian-born individuals displayed periosteal inflammation on any long bone more frequently than non-Barbadian-born individuals. Lesions on long bones were assessed as active or healed at the time of death following guidelines by Mensforth et al. (1978) and descriptions in Ortner (2003). Most health indicators did not show clear patterns by location in the cemetery or by social markers, including gender or place of birth. Barbadian-born individuals demonstrated more chronic inflammation near the end of life with more frequent active lesions at the time of death. This may lend some support to Roberts' (2013) argument about increased workload relative to tracked hours during the eighteenth century, though without adequate methods for dating the interments, this remains unclear. The assessment of the more complete remains has allowed for systemic studies of growth, infectious disease, trauma, and activity-related stress from the site (Shuler 2005a, 2011; Shuler and Schroeder 2013; Shuler et al. 2019; Stevens 2016) and the first clear evidence of Barbadian ($n = 18$) versus African birth ($n = 7$) using geochemical analyses of isotopes ($\delta^{13}C$, $\delta^{15}N$, $\delta^{18}O$ and $^{87/86}Sr$) (Schroeder et al. 2009). Genetic distance studies of dental non-metric traits, odontometrics, and osteometrics found that European admixture was low (5.38–10.25%) (Munson 2012), similar to what was reported (5–10%) from the earlier sample from the site (Corruccini et al. 1982; Ritter 1991).

Newton Plantation and African Diaspora Bioarchaeology: Past and Future

For more than 50 years, researchers have engaged in multidisciplinary research on the Newton Plantation cemetery in Barbados, drawing upon archival, archaeological, and bioarchaeological data to provide insight into the everyday lives and life experiences of enslaved Africans whose labor supported the emergence and globalization of the sugar industry. A biocultural synthesis that positions Newton within the emerging world system (Armstrong 2019; Shuler and Cunningham 2024) is essential to understanding the health and life experiences of the enslaved population (Shuler 2005a, 2011). Since the 1980s, skeletal remains from diasporic sites and anatomical series have been used to explore the life experiences and violence of enslaved Africans (Angel 1980; Barrett and Blakey 2011; Kelley and Angel 1987; Mack and Blakey 2004; Rathbun 1987; Rathbun and Scurry 1991; Smith and Rathbun 1997) and free populations (de la Cova 2012; Rankin-Hill 2016; Rose 1985) in a range of social and economic contexts in the U.S., demonstrating the substantial physical impacts of slavery and persistent health inequalities that continue to impact descendant populations today (Dressler et al. 2005; Farmer 2001; Gravlee 2009). Fewer studies have assessed the health impacts in Caribbean nations, largely because of smaller samples that are often not well preserved (Shuler 2005a; Mathena and Zuckerman 2017).

Generalized comparisons of bioarchaeological studies from the Caribbean, U.S., and South America suggest that growth conforms to archival predictions with a secular increase among free groups, except the severely stressed rural community at Cedar Grove (Martin et al. 1987; Rose 1985; Shuler 2005a). LEH at Newton was substantially higher than observed among the enslaved buried in colonial New York (71%; Null et al. 2009) and the Remley Plantation in Charleston (85%; Rathbun 1987). While this may indicate less exposure to stress during childhood, as noted among first generation Africans at Campeche (Cucina 2010), it may also result from AMTL and hidden frailty in the sample (Wood et al. 1992). Comparative studies are also compounded by variable scoring methods and interobserver error since studies may count fainter lines without requiring them to be palpable (Danforth et al. 1993). Anemia rates from Barbados, Suriname, and Montserrat are reported as generally low to moderate, although higher rates were observed from enslaved individuals on U.S. plantations such as Remley in South Carolina (27% male; 36% female), and less frequently reported among freedmen from Cedar Grove, Arkansas (20% male and 5% female) and Dallas (4% male and 3% female) (Rathbun and Steckel 2002, 217). Infection/inflammatory response is prevalent across diasporic sites, although it is difficult to directly compare due to the wide differences in reporting and highly variable etiologies that demand qualitative contextualized assessment (Weston 2008, 2012). While rates of periosteal inflammation at Newton were lower than reported from other diasporic sites, including Remley Plantation in South Carolina (Rathbun and Steckel 2002), New York (Barrett and Blakey 2011), and even Cedar Grove in Arkansas, where 71% of individuals were affected (Rose 1985), Shuler (2011) has argued that enslaved laborers from Barbados were an especially vulnerable population, as indicated archivally and seen through both skeletal and archival evidence of many individuals dying young.

Venereal disease was rarely documented in Barbados or the Caribbean (Handler 2006 b, 12–13), perhaps due to the highly stigmatized nature of the disease. Congenital syphilis identified from the 1970s sample at Newton by Jacobi et al. (1992) is comparable to the 10% frequency reported for the Freedman Cemetery in Dallas (Condon et al. 1998). However, the rate at Newton is far lower than reported from a colonial Waterloo cotton plantation in Suriname, where 26% of skeletons displayed evidence of treponemal infection and 56% of total cases were congenital (Khudabux 1999). Although the exact form of treponemal infection at Waterloo was often difficult to diagnose skeletally, the author argues that the rates are far too high to have resulted from an endemic treponeme such as yaws, which is typically found at a prevalence of 1% skeletally in hot, tropical climates (Khudabux 1999, 304). Null et al. (2009, 182) report skeletal evidence of treponemal disease at the New York African Burial Ground (ABG) in 16.1% (40 individuals) of individuals over age 15, with more males than females affected. Blakey and Rankin-Hill (2004, 86) argue that visible evidence of venereal disease was potentially more widespread at the ABG due to sexual abuse and concubinage of domestic servants by white owners. They argue that the occurrence of saber shins without stellate scars in this sample suggest yaws, but that congenital syphilis was also observed and was likely related to maternal venereal infections from sexual violence and concubinage of domestic servants by white owners (Blakey 2009, 38). Similar abuses were likely experienced by the enslaved populations on Caribbean plantations (Shuler 2011).

In general, high rates of work-related stress such as degenerative joint disease and Schmorl's nodes have been reported for both sexes in bioarchaeological studies of enslaved groups. In South Carolina, Rathbun (1987) reports high frequencies of degenerative joint changes for both males and females at Remley Plantation. Unlike Newton sugar laborers with higher frequencies in the elbows and wrists, these enslaved rice planters demonstrate most joint changes in the hips and shoulders, although both groups show higher rates in the lower vertebrae (Shuler 2005a). At Remley, males displayed more herniated vertebral disks and ulnar modification than women, but at Newton Plantation there were no differences by sex in Schmorl's nodes. According to Rathbun (1987), the specific activities associated with these changes have not been discerned, as these changes can occur from any number of activities. Likewise, St. Peter's Street Cemetery in New Orleans provided evidence of diverse labor roles with greater differences between enslaved men and women in this urban U.S. setting (Owsley et al. 1987) than reported from bioarchaeological studies of rural Caribbean communities where historic documents attest to strenuous field labor for many individuals of both sexes, in addition to domestic labor for women (Bush 1990; Morrissey 1989). Wilczak et al. (2009) reported that the bodies of the enslaved at the ABG reveal a picture of frequent activity stresses among adults, with the ankles and the lower body more frequently impacted than the upper body, with relatively little shoulder stress. The more frequent cervical than thoracic/lumbar patterning for osteophytosis for both males and females in this sample suggested heavy lifting or potentially carrying/dragging things from the neck, but again, a wide variety of activities could produce such changes (Wilczak et al. 2009). Patterns indicate a variety of activities for the entire group, which is supported by the diverse labor roles of the urban enslaved population in New York (Barrett and Blakey 2011, 238–242). Comparisons of the ABG sample with other urban (St. Peter's) and rural (Remley) enslaved populations support a greater diversity of activity patterns for rural laborers who engaged in farming (Wilczak et al. 2009). At the ABG, the high frequency of traumas of the cranial and postcranial bones, especially the common nature of perimortem injury, argues for interpersonal violence. In a comparison of injuries at the ABG and other enslaved groups, Wilczak et al. (2009, 460) report that the pattern observed in the urban enslaved site in New Orleans similarly suggests violence but that the bodies of the enslaved from the Catoctin Iron Furnace appear to represent accidental (industrial) injuries, which may suggest the potential for differential treatment of these skilled workers. At Newton, the historical evidence points to extremely high rates of activity by men, women, and children working in sugar agriculture, but work stress, trauma, and infectious diagnosable diseases are surprisingly low compared to industrial and urban slaves in the U.S. However, trauma occurs in many forms (Klaus 2012) as a spectrum of violence that impacted the colonial world, including the enslaved at Newton (Roberts 2013).

Bioarchaeologists studying colonial contexts have increasingly applied structural violence theory following Klaus (2012) as a powerful tool for examining the embodiment of historically deep social inequities and persistent multiaxial modes of suffering (Farmer et al. 2006, 1686) that may be obscured and invisible in archival accounts. As described by Galtung (1969), structural violence pertains to harm or limits to individual potential due to unequal political and economic structures such as the large-scale commodity production systems of the Atlantic world, where enslaved Africans were exploited for their labor (Inikori 2020). The overall bioarchaeological picture at Newton is highly consistent with widespread structural violence (Martin and Harrod 2016, 167), which has also been discussed in detail by Mathena and Zuckerman (2017) in their assessment of 26 individuals from Danish St. Croix. At Newton, assessment of individual traits presents a skewed view of the health impacts, as the absence of severe sequelae for infectious diseases that were well documented archivally is likely due to individuals dying early from a syndemic of acute infections, chronic malnutrition, and work-related stress (DeWitte and Stojanowski 2015; Wood et al. 1989). Through a contextualized synthesis of multidisciplinary data emerges a more complex narrative of early mortality, widespread malnutrition that included episodic starvation, poor growth, and frequent localized inflammatory responses on the lower limbs, mainly in males, which are consistent with work-related field injuries and the brutal conditions of enslavement in the former British West Indies (Roberts 2013; Shuler 2011; Shuler et al. 2019).

Bioarchaeology is uniquely positioned to reveal violence and atrocities suffered in the past (Blakey 2001; Martin et al. 2012, 2013, 2015; Martin and Harrod 2016) through material indicators of warfare, interpersonal conflict, and captivity that are often absent from other forms of knowledge. Yet, as Blakey (2020) and others (Dunnavant 2021; Halcrow et al. 2021) have rightly pointed out, white bioarchaeologists have often ignored and been remiss in addressing the pain and suffering experiences of descendent communities in the present and the rights over the bodies of their ancestors. Anthropology as a field has a long and even violent colonial past that includes ties to eugenics (Blakey 2020), a past that biological anthropology is only recently beginning to recon with and acknowledge as a field of science (Fuentes 2021). The ABG project in New York (Blakey and Rankin-Hill 2009) serves as a model for scholarly engagement with descendent communities, demonstrating the vital importance of connecting the archaeological past to the present (Blakey 2020). Today, a new generation of scholars from within BIPOC communities in the U.S. and Caribbean are engaged with studies of archaeological sites and museum collections while calling for action that includes stronger protections for ancestral remains and acknowledgement of the pervasive racism and colonialist practices within archaeology (Dunnavant et al. 2021). This new generation of scholars includes Ph.D. and M.A. students who are working on bioarchaeological research from Newton Plantation with the support of the BMHS using the most up-to-date standards in the field. Among these scholars, Andreana Cunningham (2023) examined the biosocial impacts of the slave trade on individuals from Newton Plantation and other diasporic sites by exploring the complex morphological relationships and the entanglements of phenotype, culture, and migration. Ciele Rosenberg (2022) recently assessed fluctuating asymmetry in dental and skeletal remains at Newton for evidence of early developmental stress. Kortney Williams (Ph.D. student, SUNY Binghamton) is beginning an investigation of comparative evidence of violence, trauma, and activity-related stresses across diasporic sites, including Newton. Researchers are continuing to explore undocumented aspects of social identities, health, and violence in the past to understand the life experiences and inequalities more fully in the world today, but to truly do this requires acknowledging the positions of archaeologists and bioarchaeologists within this narrative. It requires increased presence, collaboration, and engagements with Caribbean scholars and descendant communities who have allowed research to be conducted on their ancestors and their historic and precontact sites (Smith 2019). As archaeologists and bioarchaeologists, it is essential that we recognize and address these issues and, most importantly, to listen and give back to the communities and nations in which we work (Halcrow et al. 2021).

The Newton Plantation Enslaved Burying Ground is the oldest-known cemetery associated with the enslaved on the island and one of the largest in the world that is associated with the colonial history of sugar production. As recently stated during an interview at the Newton Plantation Burying Ground with Deputy Director Kevin Farmer of the BMHS, "What we do with sites like this is seek to bring them alive, so we have a better understanding of what our history is" (Kevin Farmer, July 1, 2021).

Acknowledgements

Research at Newton Plantation has been possible due to the continued support of the Barbados Museum and the nation of Barbados. I am indebted to Barbados Museum Director Alissandra Cummins, Deputy Director Kevin Farmer, and numerous museum staff, past and present. I am grateful for the support of the museum, the community, and the many scholars who have contributed and continue to contribute to understanding the lives of the enslaved men, women, and children who lived at the Newton Plantation. Funding for the 1997 excavation was provided to Kristrina Shuler Herndon and Robert Corruccini through NSF SBR-96-22706. Field notes, documents, and maps from the 1970s excavations were kindly provided by Robert Riordon, and maps for this chapter were produced by Eric D. Sipes who also provided editorial assistance. I want to thank the editors of this volume for the invitation to participate and for their patience. Their suggestions improved this manuscript, though any errors or omissions remain solely my own.

References

Agarwal, S. C., and B. A. Glencross. 2011. *Social Bioarchaeology*. Malden, MA: Wiley-Blackwell Publishing.

Alden, D., and J. C. Miller. 1987. "Unwanted Cargoes." In *The African Exchange*, edited by Kenneth Kiple, 35–109. Durham, NC: Duke University Press.

Angel, J. L. 1980. "Appendix III: Family Skeletons from Clifts Plantation." In *Field Archaeology of the Clifts Plantation Site: Westmoreland County, Virginia*, edited by F.D. Neiman, 221–231. Stratford, VA: Robert E. Lee Memorial Association.

Armstrong, D. V. 2019. "Early-Seventeenth-Century Settlement in Barbados and the Shift to Sugar, Slavery, and Capitalism." In *Power, Political Economy, and Historical Landscapes of the Modern World Interdisciplinary Perspectives*, edited by Christopher DeCorse, 145–167. Albany: SUNY Press.

Aufderheide, A. C., J. L. Angel, J. O. Kelley, A. C. Outlaw, M. A. Outlaw, G. Rapp Jr., and L. E. Wittmers Jr. 1985. "Lead in Bone III. Prediction of Social Correlates from Skeletal Lead Content in Four Colonial American Populations (Catoctin Furnace, College Landing, Governor's Land, and Irene Mound)." *American Journal of Physical Anthropology* 66(4): 353–361.

Barrett, A. R., and M. L. Blakey. 2011. "Life Histories of Enslaved Africans in Colonial New York: A Bioarchaeological Study of the African Burial Ground." In *Social Bioarchaeology*, edited by Sabrina C. Agarwal and Bonnie A. Glencross, 212–151. Malden, MA: Wiley-Blackwell.

Bass, W. M. 1987. *Human Osteology: A Laboratory and Field Manual*. Columbia, MO: Missouri Archaeological Society.

Blakey, M. L. 2001. "Bioarchaeology of the African diaspora in the Americas: Its Origins and Scope." *Annual Review of Anthropology* 30: 387–422.

Blakey, M. L. 2009. "History and Comparison of Bioarchaeological Studies in the African Diaspora." In *The African Burial Ground Project*, edited by Michael L Blakey and Leslie Rankin-Hill, 19–40. Washington, DC: Howard University Press.

Blakey M.L. 2020. Archaeology under the blinding light of race. *Current Anthropology* 61 (22): S182–S197.

Buikstra, J. E., and D. H. Ubelaker. 1994. "Standards for Data Collection from Human Skeletal Remains." *Research Series 44*. Fayetteville, Arkansas: Arkansas Archaeological Survey.

Bush, B. 1990. *Slave Women in Caribbean Society, 1650–1838*. Bloomington: Indiana University Press.

Condon, C. G., J. L. Becker, H. J. H. Edgar, J. M. Davidson, J. R. Hoffman, P. Kalima, D. Kysar, S. Moorehead, V. M. Owens, and K. Condon. 1998. "Freedman's Cemetery: Site 41DL316, Dallas, Texas, Assessments of Sex, Age-at Death, Stature, and Date of Interment for Excavated Burials." *Archaeology Studies Program,* Texas Department of Transportation, Austin.

Corruccini, R. S., E. M. Brandon, and J. S. Handler. 1989. "Inferring Fertility from Relative Mortality in Historically Controlled Cemetery Remains from Barbados." *American Antiquity* 54(3): 609–614.

Corruccini, R. S., and J. S. Handler. 1980. "Temporomandibular Joint Size Decrease in American Blacks: Evidence from Barbados." *Journal of Dental Research* 59(9): 1528.

Corruccini, R. S., J. S. Handler, and K. P. Jacobi. 1985. "Chronological Distribution of Enamel Hypoplasias and Weaning in a Caribbean Slave Population." *Human Biology* 57(4): 699–711.

Corruccini, R. S., J. S. Handler, R. J. Mutaw, and F. W. Lange. 1982. "Osteology of a Slave Burial Population from Barbados, West Indies." *American Journal of Physical Anthropology* 59(4): 443–459.

Corruccini, R. S., A. C. Aufderheide, J. S. Handler, and L. E. Wittmers. 1987a. "Patterning of Skeletal Lead Content in Barbados Slaves." *Archaeometry* 29(2): 233–239.

Corruccini, R. S., K. P. Jacobi, J. S. Handler, and A. C. Aufderheide. 1987b. "Implications of Tooth Root Hypercementosis in a Barbados Slave Skeletal Collection." *American Journal of Physical Anthropology* 74(2): 179–184. https://doi.org/10.1002/ajpa.1330740206.

Cucina, Andrea. 2010. "Social Inequality in the Early Spanish Colony: Oral Pathologies and Dental Enamel Hypoplasias in the Skeletal Sample from Campeche." In *Natives, Europeans, and Africans in Colonial Campeche:* History and Archaeology, edited by Vera Tiesler, Pilar Zabala, and Andrea Cucina, 111–129. Gainesville: University Press of Florida.

Cunningham, A. S. 2019. "Cranial Morphological Distribution of African Descent Peoples as Reflectors of Migration, Variation, and Race Construction." *Paper Presented at the International Association for Caribbean Archaeology Biannual Conference*. Barbados. July 27th, 2019.

Cunningham, A. 2023. "Reframing Diaspora: Southeastern African Contributions to Biosocial Variation in Atlantic Afro-Descendant Groups." PhD diss., University of Florida.

Danforth, M. E., K. S. Herndon, and K. B. Propst. 1993. "A Preliminary Study of Patterns of Replication in Scoring Linear Enamel Hypoplasias." *International Journal of Osteoarchaeology* 3(4): 297–302.

de la Cova, C. 2012. "Patterns of Trauma and Violence in 19th-Century-Born African American and Euro-American Females." *International Journal of Paleopathology* 2(2–3): 61–68.

DeWitte, S. N., and C. M. Stojanowski. 2015. "The Osteological Paradox 20 Years Later: Past Perspectives, Future Directions." *Journal of Archaeological Research* 23(4): 397–450.

Dressler, W. W., K. S. Oths, and C. C. Gravlee. 2005. "Race and Ethnicity in Public Health Research: Models to Explain Health Disparities." *Annual Review of Anthropology* 34(1): 231–252.

Dunnavant, J., D. Justinvil, and C. Colwell. 2021. "Craft an African American Graves Protection and Repatriation Act." *Nature* 593: 337–340.

Eltis, D., and S. L. Engerman, eds. 2011. *The Cambridge world history of slavery: Volume 3, AD 1420–AD 1804*. New York: Cambridge University Press.

Farmer, K. 2021. "Exploring and Understanding Newton Slave Burial Ground." *Visit Barbados* https://www.youtube.com/watch?v=0WmwFcf9veo

Farmer, P. 2001. *Infections and Inequalities: The Modern Plagues*. Barkeley, LA, London: University of California Press.

Farmer, P., B. Nizeye, S. Stulac, and S. Keshavjee. 2006. "Structural Violence and Clinical Medicine." *PLoS Medicine* 3(10): e449.

Fuentes, A. 2021. "Biological anthropology's critical engagement with genomics, evolution, race/racism, and ourselves: Opportunities and challenges to making a difference in the academy and the world." *American Journal of Physical Anthropology* 175(2): 326–338.

Galtung, J. 1969. "Violence, Peace, and Peace Research." *Journal of Peace Research* 6(3): 167–191.

Gershon, L. 2021. "After Breaking Ties with Britain, Barbados Announces Heritage District Tracing Slavery's Toll." *Smithsonian Magazine*. https://www.smithsonianmag.com/smart-news/barbados-announces-new-heritage-district-documenting-slavery-180979175/

Gravlee, C. C. 2009. "How Race Becomes Biology: Embodiment of Social Inequality." *American Journal of Physical Anthropology* 139(1): 47–57.

Halcrow, S., A. Aranui, S. Halmhofer, A. Heppner, N. Johnson, K. Killgrove, and G. R. Schug. 2021. "Moving beyond Weiss and Springer's Repatriation and Erasing the Past: Indigenous Values, Relationships, and Research." *International Journal of Cultural Property* 28(2): 211–220.

Handler, J. S. 1986. "Lead Contact and Poisoning in Barbados Slaves: Historical, Chemical, and Biological Evidence." *Social Science History* 10(4): 399–427.

Handler, J. S. 1996. "A Prone Burial from a Plantation Slave Cemetery in Barbados, West Indies: Possible Evidence for an African-type Witch or other Negatively Viewed Person." *Historical Archaeology* 30(3): 76–86.

Handler, J. S. 1997. "An African-Type Healer/Diviner and His Grave Goods: A Burial from a Plantation Slave Cemetery in Barbados, West Indies." *International Journal of Historical Archaeology* 1(2): 91–130.

Handler, J. S. 1998. "Life Histories of Enslaved Africans in Barbados." *Slavery and Abolition* 19(1): 129–141.

Handler, J. S. 2006a. "Diseases and Medical Disabilities of Enslaved Barbadians, from the Seventeenth Century to Around 1838: Part 2." *The Journal of Caribbean History* 40(2): 177–214.

Handler, J. S. 2006b. "Diseases and Medical Disabilities of Enslaved Barbadians, From the Seventeenth Century to around 1838 (Part I) 1." *The Journal of Caribbean History* 40(1): 1–38.

Handler, J. S., M. D. Conner, and K. P. Jacobi. 1989. "Searching for a Slave Cemetery in Barbados, West Indies: A Bioarchaeological and Ethnohistorical Investigation." *Research Paper 59, Center for Archaeological Investigations*. Carbondale: Southern Illinois University.

Handler, J. S., and R. S. Corruccini. 1983. "Plantation Slave Life in Barbados: A Physical Anthropological Analysis." *The Journal of Interdisciplinary History* 14(1): 65–90.

Handler, J. S., and R. S. Corruccini. 1986. "Weaning among West Indian Slaves: Historical and Bioanthropological Evidence from Barbados." *The William and Mary Quarterly* 43(1): 111–117.

Handler, J. S., and F. W. Lange. 1978. *Plantation Slavery in Barbados: An Archaeological and Historical Investigation.* Cambridge: Harvard University Press.

Henderson, C. Y., V. Mariotti, D. Pany-Kucera, S. Villotte, and C. Wilczak. 2016. "The new 'Coimbra Method': A Biologically Appropriate Method for Recording Specific Features of Fibrocartilaginous Entheseal Changes." *International Journal of Osteoarchaeology* 26(5): 925–932.

Higman, B. W. 1995. *Slave Populations of the British Caribbean, 1807–1834*. Barbados, Jamaica, Trinidad and Tobago: The Press, University of the West Indies.

Inikori, J. E. 2020. "Atlantic Slavery and the Rise of the Capitalist Global Economy." *Current Anthropology* 61(S22): S159–S171.

Jacobi, K. P., D. C. Cook, R. S. Corruccini, and J. S. Handler. 1992. "Congenital Syphilis in the Past: Slaves at Newton Plantation, Barbados, West Indies." *American Journal of Physical Anthropology* 89(2): 145–158.

Jamieson, R. W. 1995. "Material Culture and Social Death: African-American Burial Practices." *Historical Archaeology* 29(4): 39–58.

Kelley, J. O., and J. L. Angel. 1987. "Life Stresses of Slavery." *American Journal of Physical Anthropology* 74(2): 199–211.

Khudabux, M. R. 1999. "Effects of Life Conditions on the Health of a Negro Slave Community in Suriname." In *African Sites: Archaeology in the Caribbean*, edited by Jay B. Haviser, 291–312. Princeton, NJ: Markus Wiener Publishers.

Kiple, K. F. 1987. *The African Exchange: Toward a Biological History of Black People.* Durham, NC: Duke University Press.

Kiple, K. F. 2002. *The Caribbean Slave: A Biological History*. Cambridge: Cambridge University Press.

Kiple, K. F., and B. T. Higgins. 1992. "Yellow Fever and the Africanization of the Caribbean." *Disease and Demography in the Americas*, edited by J. W. Verano and D. H. Ubelaker, 237–248. Washington, DC: Smithsonian Institution Press.

Klaus, H. D. 2012. "The Bioarchaeology of Structural Violence: Theoretical Model and Case Study." In *The Bioarchaeology of Violence*, edited by Debra L. Martin, Ryan P. Harrod, and Ventura R. Perez, 29–62. Gainesville: University Press of Florida.

Laffoon, J. E., K. A. Shuler, A. R. Millard, J. N. Connelly, and H. Schroeder. 2020. "Isotopic Evidence for Anthropogenic Lead Exposure on a 17th/18th Century Barbadian Plantation." *American Journal of Physical Anthropology* 171(3): 529–538.

Larsen, C. S. 2015. *Bioarchaeology: Interpreting Behavior from the Human Skeleton*. Cambridge: Cambridge University Press.

Layrisse, M., M. Roche, E. Rusián, and G. López-Calzón. 1964. "The Relationship between Anemia and Hookworm Infection." *American Journal of Epidemiology* 79(3): 279–301.

Mack, M. E., and M. L. Blakey. 2004. "The New York African Burial Ground Project: Past Biases, Current Dilemmas, and Future Research Opportunities." *Historical Archaeology* 38(1): 10–17.

Martin, D. L., and R. P. Harrod. 2016. "10 The Bioarchaeology of Pain and Suffering: Human Adaptation and Survival during Troubled Times." *Archeological Papers of the American Anthropological Association* 27(1): 161–174.

Martin, D. L., A. L. Magennis, and J. C. Rose. 1987. "Cortical Bone Maintenance in an Historic Afro-American Cemetery Sample from Cedar Grove, Arkansas." *American Journal of Physical Anthropology* 74(2): 255–264.

Martin, D. L., R. P. Harrod, and V. R. Perez. 2012. *The Bioarchaeology of Violence.* Gainesville: University Press of Florida.

Martin, D. L., R. P. Harrod, and V. R. Perez. 2013. *Bioarchaeology: An Integrated Approach to Working with Human Remains.* Gainesville, FL: Springer.

Martin, D. L., R. P. Harrod, and V. R. Perez. 2015. "Bioarchaeological Contributions to the Study of Violence." *American Journal of Physical Anthropology* 156(S59): 116–145.

Mathena, S. A., and M. K. Zuckerman. 2017. "Sugar and Suffering: A Case Study on Injury Recidivism, Overall Health, and Structural Violence from the Holy Trinity Lutheran Church." In *Broken Bones, Broken Bodies: Bioarchaeological and Forensic Approaches for Accumulative Trauma and Violence*, edited by Caryn E. Tegtmeyer and Martin Debra L., 121–140. New York: Lexington Books.

Mensforth, R. P., C. O. Lovejoy, J. W. Lallo, and G. J. Armelagos. 1978. "The Role of Constitutional Factors, Diet, and Infectious Disease in the Etiology of Porotic Hyperostosis and Periosteal Reactions in Prehistoric Infants and Children." *Medical Anthropology* 2: 1–59.

Mintz, S. W. 1986. *Sweetness and Power: The Place of Sugar in Modern History.* New York: Penguin.

Morrissey, M. 1989. *Slave Women in the New World: Gender Stratification in the Caribbean.* Lawrence: University Press of Kansas.

Munson, S. N. 2012. *Genetic Admixture and Tooth Size in an Enslaved Population from Newton Plantation, Barbados.* MA Thesis, Southern Illinois University Carbondale.

Null, C. C., M. L. Blakey, K. J. Shujaa, L. M. Rankin-Hill, and S. H. H. Carrington. 2009. "Osteological Indicators of Infectious Disease and Nutritional Inadequacy." In *The African Burial Ground Project*, edited by M. Blakey, and L. Rankin-Hill, 169–197. Washington, DC: Howard University Press.

Ortner, D. J. 2003. *Identification of Pathological Conditions in Human Skeletal Remains.* 2nd edition. Washington, DC: Smithsonian Institution Press.

Owsley, D. W., C. E. Orser Jr, R. W. Mann, P. H. Moore-Jansen, and R. L. Montgomery. 1987. "Demography and Pathology of an Urban Slave Population from New Orleans." *American Journal of Physical Anthropology* 74(2): 185–197.

Oxenham, M. F., and I. Cavill. 2010. "Porotic Hyperostosis and Cribra Orbitalia: The Erythropoietic Response to Iron-Deficiency Anaemia." *Anthropological Science* 118(3): 199–200.

Rankin-Hill, L. M. 2016. "Identifying the First African Baptist Church: Searching for Historically Invisible People." In *New Directions in Biocultural Anthropology*, edited by Martin D. and M.K. Zuckerman, 133–156. Hoboken, NJ: John Wiley & Sons.

Rathbun, T. A. 1987. "Health and Disease at a South Carolina Plantation: 1840–1870." *American Journal of Physical Anthropology* 74(2): 239–253.

Rathbun, T. A., and J. D. Scurry. 1991. "Status and Health in Colonial South Carolina: Belleview Plantation, 1738–1756." In *What Mean These Bones*, edited by Mary Lucas Powell, Patricia S. Bridges, and Ann Marie Wagner Mires, 148–164. Tuscaloosa: The University of Alabama Press.

Rathbun, T. A., and R. H. Steckel. 2002. "The Health of Slaves and Free Blacks in the East." In *The Backbone of History: Health and Nutrition in the Western Hemisphere*, edited by Richard H. Steckel and Jerome C. Rose, 208–225. Cambridge: Cambridge University Press.

Ritter, E. B. 1991. *Dental Developmental Disruption and Aspects of Afro-American Biohistory*. Ph. D. Dissertation, Southern Illinois University, Carbondale.

Roberts, J. 2013. *Slavery and the Enlightenment in the British Atlantic, 1750–1807*. Cambridge: Cambridge University Press.

Rose, J. C. 1985. *Gone to a Better Land: A Biohistory of a Rural Black Cemetery in the Post-Reconstruction South*. Fayetteville: Arkansas Archaeological Survey.

Rosenberg, C. L. 2022. "Fluctuating Asymmetry and the Embodiment of Maternal Stress at Newton Plantation, Barbados." Master's thesis, North Carolina State University.

Schroeder, H., K. A. Shuler, and S. R. Chenery. 2013. "Childhood Lead Exposure in an Enslaved African Community in Barbados: Implications for Birthplace and Health Status." *American Journal of Physical Anthropology* 150(2): 203–209.

Schroeder, H., T. C. O'Connell, and J. A. Evans. 2009. "Trans-Atlantic Slavery: Isotopic Evidence for Forced Migration to Barbados." *American Journal of Physical Anthropology* 139(4): 547–557.

Seymour, T. 2021. "David Adjaye Plans Slavery Museum in Barbados as New Republic Severs Ties with Britain." *The Art Newspaper*, published online December 2, 2021. https://www.theartnewspaper.com/2021/12/06/barbados-announces-museum-and-memorial-to-victims-of-western-slavery-as-it-cuts-ties-with-british-monarchy

Sheridan, R. B. 1985. *Doctors and Slaves: A Medical and Demographic History of Slavery in the British West Indies*. Cambridge: Cambridge University Press.

Sheridan, R. B. 1989. "Changing Sugar Technology and the Labour Nexus in the British Caribbean, 1750–1900, with Special Reference to Barbados and Jamaica." *New West Indian Guide/Nieuwe West-Indische Gids* 63(1–2): 59–93. https://doi.org/10.1163/13822373-90002033.

Shuler, K. A. 2005a. *Health, History, and Sugar: A Bioarchaeological Study of Enslaved Africans from Newton Plantation, Barbados, West Indies*. Ph.D. Dissertation, Southern Illinois University Carbondale.

Shuler, K. A. 2005b. "Nutritional Health of Enslaved Africans from Newton Plantation, Barbados: New Data." *Journal of the Barbados Museum & Historical Society* LI: 166–186.

Shuler, K. A. 2011. "Life and Death on a Barbadian Sugar Plantation: Historic and Bioarchaeological Views of Infection and Mortality at Newton Plantation." *International Journal of Osteoarchaeology* 21: 66–81.

Shuler, K. A., and A. Cunningham. n.d. "African Diaspora Bioarchaeology: Osteological Evidence from the Americas, Caribbean, Africa, and South Atlantic Regions." Unpublished manuscript in progress.

Shuler, K. A., and A. Cunningham. 2024. "Bioarchaeological Approaches to African Diaspora in the Twenty-First Century: Intercontinental and Global Legacies of Displacement." *Journal of African Diaspora Archaeology and Heritage* 13(3): 215-260. https://doi.org/10.1080/21619441.2024.2406652

Shuler, K. A., and H. Schroeder. 2013. "Evaluating Alcohol Related Birth Defects in the Past: Skeletal and Biochemical Evidence from a Colonial Rum Producing Community in Barbados, West Indies." *International Journal of Paleopathology* 3(4): 235–242.

Shuler, K. A., H. Schroeder, and W. D. Stevens. 2019. "Health and Life Histories of Enslaved Sugar Producers from the Newton Plantation, Barbados: 35 Years of Bioarchaeological

Research." In *Pre-Colonial and Post-Contact Archaeology in Barbados: Past, Present and Future Research Directions*, edited by de Waal, Maaike S., Niall Finneran, and Matthew C. Reilly with Douglass Armstrong and Kevin Farmer, 269–286. Leiden: Sidestone Press.

Smith, F. 2019. "St. Nicholas Abbey. Centering People in Plantation Archaeology in Barbados in the Twenty-First Century." In *Pre-Colonial and Post-Contact Archaeology in Barbados: Past, Present and Future Research Directions*, edited by de Waal, Maaike S., Niall Finneran, and Matthew C. Reilly with Douglass Armstrong and Kevin Farmer, 133–156. Leiden: Sidestone Press.

Smith, S. D., and T. A. Rathbun. 1997. "Folly Island: An African-American Union Brigade Cemetery in South Carolina." In *Remembrance Archaeology and Death*, edited by David A. Poirier and Nicholas F. Bellantoni, 65–77. New York: Bergin & Garvey.

Stevens, W. D. 2016. *Enslaved Labor in the Gang and Task Systems: A Case Study in Comparative Bioarchaeology of Commingled Remains*. Ph.D. Dissertation, University of South Carolina.

Stuart-Macadam, P. 1985. "Porotic Hyperostosis: Representative of a Childhood Condition." *American Journal of Physical Anthropology* 66(4): 391–398.

Stuart-Macadam, P. 1987. "Porotic Hyperostosis: New Evidence to Support the Anemia Theory." *American Journal of Physical Anthropology* 74(4): 521–526.

Teelucksingh, J. 2006. "The 'Invisible Child' in British West Indian Slavery." *Slavery and Abolition* 27(2): 237–250.

The Colonial Williamsburg Foundation. n.d. *Barbados Museum and Historical Society: Bridgetown, Barbados*. Accessed November 22, 2021. https://slaveryandremembrance.org/partners/partner/?id=P0026.

Tiesler, V., and P. Zalaba. 2010. "Dying in the Colonies: Death, Burial, and Mortuary Patterning in Campeche's Main Plaza." In *Natives, Europeans, and Africans in Colonial Campeche: History and Archaeology*, edited by Vera Tiesler, Pilar Zalaba, and Andrea Cucina, 70–74. Gainesville: University Press of Florida.

Ubelaker, D. H. 1989. *Human Skeletal Remains, Excavation, Analysis, Interpretation*. Second Edition. Washington D.C: Taraxacum.

Walker, P. L., R. R. Bathurst, R. Richman, T. Gjerdrum, and V. A. Andrushko. 2009. "The Causes of Porotic Hyperostosis and Cribra Orbitalia: A Reappraisal of the Iron-Deficiency-Anemia Hypothesis." *American Journal of Physical Anthropology* 139(2): 109–125.

Weston, D. A. 2008. "Investigating the Specificity of Periosteal Reactions in Pathology Museum Specimens." *American Journal of Physical Anthropology* 137(1): 48–59.

Weston, D. A. 2012. "Nonspecific Infection in Paleopathology: Interpreting Periosteal Reactions." In *A Companion to Paleopathology*, edited by Anne L. Grauer, 15: 492–512. Oxford: Wiley-Blackwell.

Wilczak, C., R. Watkins, C. Null, and M. L. Blakey. 2009. "Skeletal Indicators of Work: Musculoskeletal, Arthritic and Traumatic Effects." In *The African Burial Ground Project*, edited by M. Blakey, and L. Rankin-Hill, 199–223. Washington, DC: Howard University Press.

Wood, J. W, G. R. Milner, H. C. Harpending, K. M. Weiss, M. N. Cohen, L. E. Eisenberg, D. L. Hutchinson, R. Jankauskas, G. Cesnys, G. Česnys, and M. A. Katzenberg. 1992. "The Osteological Paradox: Problems of Inferring Prehistoric Health from Skeletal Samples [and Comments and Reply]." *Current Anthropology* 33(4): 343–370.

Diet and Population Mobility

10

INFANT AND CHILDREN FEEDING PRACTICES IN THE PRECOLONIAL CARIBBEAN

A Cross-Cultural Analysis

Yadira Chinique de Armas and Jason E. Laffoon

Introduction

Infant and child feeding practices are culturally determined activities with important implications for the survival of humans, as they affect the fertility, morbidity, and mortality patterns of the population (Riordan 2005; Stuart-Macadam 1995; Quinlan 2007). In comparison to other primates, humans have a longer period of growth and development (Bogin 1997; Kaplan et al. 2000; Walker et al. 2006). During the first years of life, they require food provisioning from older individuals and learned behaviour to successfully reach their biological and social maturity (Sellen 2007). The first six months of human life, for instance, is a critical period in which infants require a high-energy diet for their rapid growth and development. Breastmilk contains most of the essential nutrients required for early development and is easier to digest by the underdeveloped gastrointestinal tracts of infants compared to other types of food (Polin et al. 2010). Through breastfeeding, mothers also provide infants with immunity to environmental pathogens, thus increasing their chances of survival (Hanson 2004). Studies on modern populations suggest that by six months of age, infants' nutritional needs require the incorporation of more complex foods (Dewey and Brown 2003; Michaelsen et al. 2000), which marks the beginning of the weaning process. However, while not entirely sufficient, breast milk often remains an important source of energy, nutrients, and anti-infective factors, which lower infant mortality and morbidity (Brown et al. 1990; Naylor and Marrow 2001). While the cost of breastfeeding is mainly at the mother´s expense (Hewlett and Winn 2014), other individuals can provide food to infants and children during the weaning process, dispersing the burden of childcare from mothers to other members of the population (Tsutaya 2017, 2). Although the duration of

DOI: 10.4324/9780429444449-14

breastfeeding in humans is generally shorter in comparison to other primates, population investment in rearing children benefits the group by allowing mothers to invest more time in new offspring, reducing the inter-birth intervals (Jones 2011; Kramer 2005, 2010). The complete cessation of breastfeeding represents the end of the weaning process (Katzenberg et al. 1996).

Weaning and post-weaning food types are highly variable among human populations and depend on numerous factors, including environmental, biological, and cultural constraints such as dietary practices of the population, availability of resources, infant and child preferences, and sex and health status (Chinique de Armas and Roksandic 2018; Tsutaya 2017). With the establishment of stable isotope analysis to reconstruct the dietary practices of ancient populations (Chisholm et al. 1982; van der Merwe and Vogel 1978; Tauber 1981), a new avenue of possibilities was opened to examine not only individual-specific aspects of nutrition but also how dietary practices, parental investment, and cultural perceptions of proper infant nutrition vary among populations and how these affect infant health and survival (Chinique de Armas and Roksandic 2018; Katzenberg et al. 1996; Tsutaya and Yoneda 2015). The use of stable isotopes for dietary reconstructions is based on the premise that the isotopic composition of bones and teeth reflects the diet of an individual (Tieszen et al. 1983; Tsutaya and Yoneda 2015). Combining different stable isotope results ($\delta^{15}N$, $\delta^{13}C$, $\delta^{18}O$, $\delta^{34}S$) from tissues formed at different ages (e.g., bone collagen, teeth enamel) enable longitudinal studies of the dietary practices of individuals and populations.

While several stable isotope studies have examined the diet of adult and non-adult populations in the precolonial Caribbean (Buhay et al. 2013; Chinique de Armas et al. 2015, 2016; Laffoon et al. 2013, 2016; Pestle 2010), research studies that try to understand the biological and cultural determinants of infant and child diets are relatively recent and scarce (Chinique de Armas and Pestle 2018; Chinique de Armas et al. 2013, 2017). Recent findings indicate that while no direct association existed between the length of weaning and subsistence strategies in the precolonial Caribbean, the availability of cultigens may have affected weaning food choices (Chinique de Armas and Pestle 2018). Differences in the carbon source of diet were observed between non-adults and female adults in the horticultural pre-colonial population of Canímar Abajo in Cuba, suggesting that not all adult foods were available to infants and young children (Chinique de Armas et al. 2017). In this chapter, we summarize previous research on breastfeeding and weaning practices (BWPs) of Caribbean populations. In addition, we analyse published stable isotope data derived from bone collagen and teeth enamel in non-adults and adults from Antillean and circum-Caribbean populations to further examine potential dietary differences between non-adults and adults within the same populations and the diversity in weaning and post-weaning food choices among circum-Caribbean groups. This way, we aim to shed light on the biological and cultural determinants of non-adult dietary practices in the Caribbean.

The Use of Stable Isotopes of Nitrogen and Carbon to Reconstruct Infants and Children's Feeding Practices

Bone consists of organic and mineral portions, almost entirely composed of collagen and calcium phosphates, respectively (Aerssens et al. 1998; Fratzl et al. 2004; Hillson 1996; Rho et al. 1998). Stable carbon isotope analyses can be applied to collagen and the mineral portion of bones, while stable nitrogen isotope analyses can be performed on collagen only. A tooth is composed of dentin, enamel, and cementum tissues. Dentin, structurally similar to bone, allows for the same types of isotope analyses. Enamel consists almost entirely of apatite (96% dry weight) and can be used for oxygen and carbon stable isotope analyses (Hillson 1996).

Nitrogen ($\delta^{15}N$) is the most frequently used element in the isotopic reconstruction of BWPs since its isotopic composition enriches progressively through trophic levels (Bocherens and Drucker 2003; Minagawa and Wada 1984; Schoeninger and DeNiro 1984). Given that the only source of nutrients for a nursing infant is breast milk, infants are one trophic level higher than their mothers. After birth, the $\delta^{15}N$ value of bone collagen promptly increases by 2–3‰ for exclusively breastfed human infants (Fogel et al. 1989; Fuller et al. 2006; Millard 2000; Richards et al. 2002). Since typical weaning food proteins are significantly depleted in nitrogen isotopic composition compared to breast milk, weaned non-adult nitrogen isotopic compositions in tissues are also significantly depleted (Schurr 1998; Tsutaya and Yoneda 2013, 2015). By analysing the $\delta^{15}N$ values of bones synthesized during infancy and childhood, we can estimate the ages of weaning food introduction and breastfeeding cessation.

Carbon isotope values ($\delta^{13}C$) have been extensively used as an indicator of the types of foods used during supplementation (Dupras et al. 2001; Fuller et al. 2006; Richards et al. 2002), since different food sources have different $\delta^{13}C$ values. Isotopic distinctions between potential terrestrial plant sources result from differences in their photosynthetic pathways (C_3, C_4, or CAM). The C_3 plants include wheat, rice, trees, montane and wetland grasses, all root crops, legumes, vegetables, nuts, and most fruits and have $\delta^{13}C$ values averaging −25‰; whereas the C_4 plants include sorghum, millets, maize, sugarcane, some amaranths, chenopods, and pasture grass plants and have typical $\delta^{13}C$ values near −12‰ (O'Leary 1981; Schwarcz and Schoeninger 2011). CAM plants are all succulents, and $\delta^{13}C$ values encompass the range of both C_4 and C_3 plants (Williams et al. 2009). Furthermore, $\delta^{13}C$ of marine sources averages around −12‰, compared to −20‰ for terrestrial sources (Chisholm et al. 1982; Fry 2008). Although the $\delta^{13}C$ of bone collagen primarily reflects dietary protein in breast milk, weaning foods, and adult foods, it is also influenced by other diet macronutrients (Froehle et al. 2010; Schwarcz 2000). In contrast, $\delta^{13}C$ values of apatite reflect those of the total diet, including all macronutrients (Krueger and Sullivan 1984; Lee-Thorp 1989).

The quantitative evaluation of contributions from food sources in infant and children's diets requires knowledge of physiological processes, as food components are represented in body tissues in proportions that are not necessarily equal to the proportion of intake (Tsutaya and Yoneda 2015). Studies have detected diet-specific variations in the diet-collagen fractionation factors, suggesting that the exact dietary offsets ($\Delta\delta^{13}C_{diet\text{-}col}$) are influenced by the isotopic compositions of the macronutrients consumed (Froehle et al. 2010), which could lead to the misrepresentation of the food consumed if static isotopic offsets are used. For this reason, there is a recent tendency towards the use of Bayesian models to isotopically reconstruct dietary practices in the Caribbean, which allows us to discuss the results in terms of probabilities (Chinique de Armas and Pestle 2018; Chinique de Armas et al. 2017; Pestle and Laffoon 2018). This is especially useful in a region such as the Caribbean, where the isotopic values of some marine and terrestrial plants and animals significantly overlap (Pestle 2010, 2013).

Finally, there is a bias associated with the use of cross-sectional samples of skeletal populations (concentrating on non-adults of different ages) to reconstruct dietary practices at early life stages. Such approaches are potentially problematic in light of the "osteological paradox", which postulates that the skeletal series represents only dead individuals who did not survive past the age at which they are examined and thus are not necessarily representative of the actual health status and life history of the population (Wood et al. 1992). Recent studies are trying to overcome the individual heterogeneity in the risk of death and disease (Czermak et al. 2018; Sandberg et al. 2014) when reconstructing non-adult feeding practices by focusing on tissues in adult specimens that formed during growth and development (infancy, childhood, adolescence).

Dataset

The Sample

Papers on paleodietary reconstruction using stable isotope analyses ($\delta^{15}N$, $\delta^{13}C_{co}$, $\delta^{13}C_{en}$) in the Antilles and circum-Caribbean region were searched using Google Scholar. This resulted in a total of 24 populations from 19 publications (Figure 10.1, Table 10.1). Assigned subsistence strategies, culture, and age at death of the populations were based on each paper's description. For $\delta^{15}N$ and $\delta^{13}C_{co}$, the age at death was considered to be representative of the stable isotope concentration of the sampled bone tissues for infants, children, juveniles, and adults (Buikstra and Ubelaker 1994), while for $\delta^{13}C_{en}$ we considered the age of tooth (enamel) formation (Hillson 1996) (Table 10.1).

Bayesian Mixing Model and Statistics

An open-source Bayesian probability mixing model: Weaning Age Reconstruction with Nitrogen isotopes (WARN) (Tsutaya and Yoneda 2013) was used to determine the length of BWPs and assess the $\delta^{15}N$ values of the weaning foods used during the

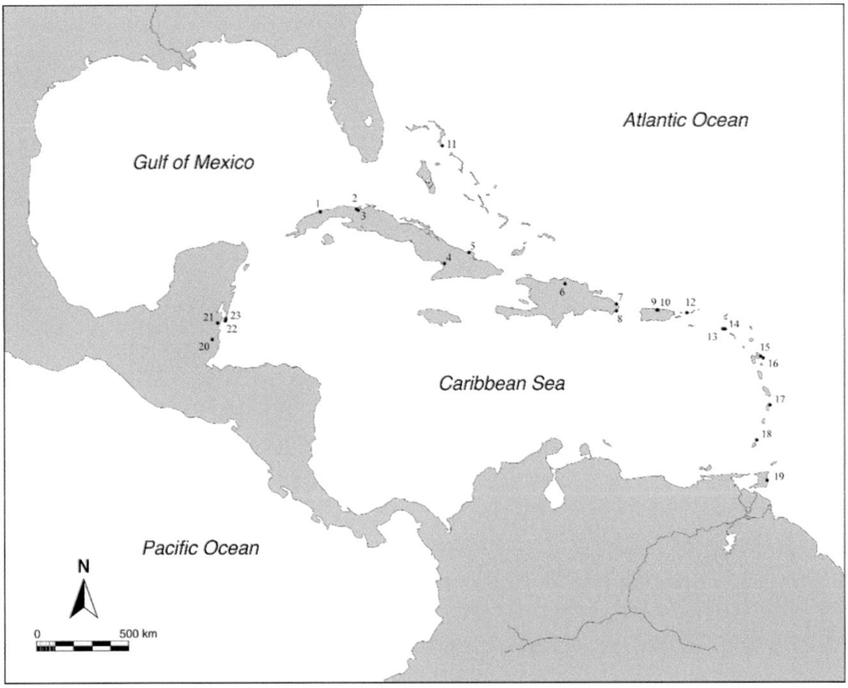

FIGURE 10.1 Location of sites included in this study. Cuba: Cueva del Perico I (1), Canímar Abajo (2), Cueva Calero (3), Playa del Mango (4), Chorro de Maita (5); Dominican Republic: El Flaco (6), Punta Macao (7), El Cabo (8); Puerto Rico: Paso del Indio (9), Maisabel (10); Bahamas: Imperial Lighthouse (11); Virgin Islands: Tutu (12); Saba: Spring Bay (13), Kelbey's Ridge (14); Guadalupe: Morel (15), Anse a la Gourde (16); St. Lucia: Lavoutte (17); Grenada: Carriacou (18); Trinidad: Manzanilla (19); Belize: Pactibun (20), Altun Ha (21), Marco González (22), San Pedro (23).

weaning process. From this model, the weaning parameters, such as the age at the start of weaning (t_1), which represents the age when foods other than breastmilk were introduced into infants' diets, the age at the end of weaning or when breastfeeding ceased (t_2), the nitrogen isotopic enrichment of the juvenile from his/her mother (E), and the $\delta^{15}N$ value of the weaning foods ($\delta^{15}N_{wnfood}$), were determined along with their posterior probabilities. This model accounts for changes in the turnover rates of bone collagen, which are higher during the first years of life. In this model, the non-adult $\delta^{15}N$ values are compared against the nitrogen isotopic enrichment between them and females (Tsutaya and Yoneda 2013). In this study, only sites with more than five infants/children available were processed using the WARN model. Female $\delta^{15}N$ isotopic composition was available for all sites except for Cueva del Perico I and Cueva Calero, where the average values of the population were used instead.

TABLE 10.1 Summary of the samples included in this study

Site/Region	*Subsistence & Culture*	*Infants*			*Children*			*Juveniles*			*Adults*		*Sources*
		$\delta^{15}N$	$\delta^{13}C_{co}$	$\delta^{13}C_{en}$	$\delta^{15}N$	$\delta^{13}C_{co}$	$\delta^{13}C_{en}$	$\delta^{15}N$	$\delta^{13}C_{co}$	$\delta^{13}C_{en}$	$\delta_{15}N$	$\delta^{13}C_{co}$	
Cueva del Perico I, Cuba	Fisher-gatherers	10		*	2		*	n/a		*	11		1, 2
Canimar Abajo OC, Cuba	Low level food producers	2		9	n/a		7	n/a		5	11		3, 4, 5
Canimar Abajo YC, Cuba	Low level food producers	24		6	5		10	n/a		3	23		3, 4, 5
Cueva Calero, Cuba	Fisher-gatherers	8		n/a	1		n/a	n/a		n/a	12		1, 2
Playa del Mango, Cuba	Low level food producers	3		13	n/a		4	n/a		9	17		6
El Chorro de Maita, Cuba	Agriculturalists	n/a		2	n/a		10	n/a		n/a	n/a		7
El Flaco, Dominican Republic	Agriculturalists	n/a		2	n/a		3	n/a		n/a	n/a		8
Punta Macao, Dominican Republic	Agriculturalists	n/a		1	n/a		3	n/a			n/a		9
El Cabo, Dominican Republic	Agriculturalists	n/a		n/a	n/a		3	n/a		n/a	n/a		9
Paso del Indio, Puerto Rico	Agriculturalists	6		n/a	13		n/a	1		n/a	57		10, 2
Maisabel, Puerto Rico	Agriculturalists	n/a		3	1		2	n/a		n/a	17		11, 7

Imperial Lighthouse, Bahamas	Agriculturalists	n/a	n/a	1	n/a	n/a	n/a	n/a	11
Tutu, Virgin Islands	Agriculturalists	n/a	n/a	4	5	2	n/a	18	7, 8, 12, 13
Spring Bay, Saba	Agriculturalists	n/a	n/a	1	n/a	n/a	n/a	n/a	11
Kelbey's Ridge, Saba	Agriculturalists	n/a	1	1	4	1	n/a	3	7, 13
Morel, Guadalupe	Agriculturalists	n/a	3	n/a	4	n/a	n/a	n/a	8
Anse a la Gourde	Agriculturalists	n/a	7	1	48	n/a	2	40	8
Lavoutte, St. Lucia	Agriculturalists	n/a	1	2	4	1	n/a	25	14
Carriacou, Grenada	Agriculturalists	n/a		n/a		3		11	15
Manzanilla, Trinidad	Agriculturalists	n/a	n/a	n/a	5	n/a	n/a	n/a	7
Marco Gonzalez, Belize	Agriculturalists Mayas	7	n/a	3	n/a	n/a	n/a	18	16
San Pedro, Belize	Agriculturalists Mayas	2	n/a	4	n/a	2	n/a	30	16
Pactibun, Belize	Agriculturalists Mayas	n/a	n/a	2	n/a	n/a	n/a	20	17
Altun Ha Belize	Agriculturalists Mayas	7	n/a	5	n/a	3	n/a	37	18

[1]Chinique de Armas et al. (2016), [2]Chinique de Armas and Pestle (2018), [3]Chinique de Armas et al. (2015), [4]Chinique de Armas et al. (2017), [5]Chinique de Armas et al. (2025), [6]Chinique de Armas et al. (2022), [7]Laffoon et al. (2013), [8]Laffoon et al. (2019), [9]Laffoon (2012), [10]Pestle (2010), [11]Stokes (1998), [12]Norr (2002), [13]Laffoon and Hoogland (2012), [14]Laffoon et al. (2016), [15]Krigbaum et al. (2013), [16]Williams et al. (2005), [17]Braun (1987), [18]White et al. (2001). * Information available in Laffoon and Chinique de Armas (this volume). n/a: not available.

The Bayesian mixing model: Stable Isotope Analysis in R (SIAR) (Jackson et al. 2009; Parnell et al. 2010; Phillips and Koch 2002) was used to estimate possible food sources in infants' and children's diets and probabilistic changes in source importance over time. The isotopic composition of Caribbean food sources (flora and fauna) and their concentration dependencies were taken from Pestle (2010) and cited literature (Table 10.2). In order to estimate the importance of breastmilk in non-adult diets, we estimated the carbon and nitrogen isotopic compositions of female individuals for all sites following the procedure described in Chinique de Armas et al. (2017). Inferences were based on the carbon and nitrogen variations between collagen and milk by combining the relationships between mother's milk and hair (De Luca et al. 2012; Romek et al. 2013) and hair to collagen (O'Connell and Hedges 1999 a, b). Breast milk carbon and nitrogen concentration dependencies were taken from Romek et al. (2013) (Table 10.2).

Differences between diet and collagen were estimated based on the type of resource. Differences between milk and collagen were estimated from the differences between mother milk and infant's hair (Romek et al. 2013) and hair to collagen (O'Connell and Hedges 1999 a, b), following Chinique de Armas et al. (2017). The other sources were adjusted in $\delta^{15}N$, by using the value of $3.6 \pm 1.2‰$ reported by Ambrose (2000). In the case of $\delta^{13}C$, the difference between protein sources of diet and collagen was calculated based on the difference between collagen-protein ($\Delta^{13}C_{col\text{-}prot} = 4.77 \pm 1.7‰$), using values gathered by Froehle et al. (2010). By using the same database, the difference between sources rich in carbohydrates and collagen was calculated from the difference in diet-collagen ($\Delta^{13}C_{col\text{-}diet} = 3.69 \pm 2.64‰$). Data coming from animal diets (animal feeding experiments) with low protein values were excluded since they were observed to be out of the regression line reported by Kellner and Schoeninger (2007). $\delta^{13}C_{en}$ values were adjusted according to the estimate of the offset between whole diet and bioapatite for humans ($\Delta^{13}C_{diet} - _{ba} = 10.1 \pm 0.4‰$) of Fernandes et al. (2012).

Statistical analysis was performed in populations with at least three individuals within the categories used for comparison. When the samples were normally distributed, comparisons between two groups of samples were performed by using a T-test. To compare more than two groups of samples, we used a one-way ANOVA (F) with a Tukey–Kramer post-hoc test. For comparing more than two groups that did not fit the normal distribution, we used a Kruskral–Wallis (H) test with a Dunn test. Statistical significance was set at $\alpha=0.05$ for all tests performed.

A Time to Wean? Variability in Breastfeeding and Weaning Practices among Antillean and Mesoamerican Populations

Based on the WARN Model parameters, the age at the start of weaning (t_1), and the duration of the process (t_2-t_1) were variable among circum-Caribbean populations (Table 10.3). Maximum density estimators (MDE) values suggested that the non-adults from Cueva del Perico I began the weaning process (t_1) the earliest

TABLE 10.2 Isotopic values and standard deviations (‰ AIR for nitrogen; VPDP for carbon) for probable source diet components used in the SIAR calculations for all populations included in this study (Pestle 2010 and cited literature); estimated source diet concentration dependences (%) and standard deviations used in the SIAR calculations; estimated diet-collagen fractionation offsets (‰ AIR for nitrogen; VPDP for carbon) and standard deviations used in the SIAR calculations

Source	*Source Isotopic Compositions (‰)*				*Concentration Dependences (%)*				*Diet-Collagen Fractionations (‰)*			
	$\delta^{15}N_{source}$	*Sd*	$\delta^{13}C_{source}$	*Sd*	*Nitrogen*	*Sd*	*Carbon*	*Sd*	$\Delta\delta^{15}N_{diet\text{-}col}$	*Sd*	$\Delta\delta^{13}C_{diet\text{-}col}$	*Sd*
Breastmilk Antilles	6.22	0.93	−23.77	3.37	1.61	0.27	49.13	3.63	4.42	0.11	5.46	0.67
Breastmilk Mayas	6.25	0.47	−14.79	2.36	1.61	0.27	49.13	3.63	4.42	0.11	5.46	0.67
Marine Fish[a]	8.39	2.94	−11.25	2.91	15.01	1.02	45.03	1.03	3.60	1.20	4.77	1.72
Estuarine Riverine Fish	9.35	1.70	−16.29	2.63	13.28	1.24	47.82	2.53	3.60	1.20	4.77	1.72
Terrestrial Animals	5.92	2.44	−21.58	2.59	13.16	0.82	44.89	0.59	3.60	1.20	4.77	1.72
Maize	3.31	1.63	−9.77	0.68	2.12	0.24	41.97	0.43	3.60	1.20	3.69	2.64
Root Cultigens[b]	3.84	1.51	−22.42	1.20	1.09	0.09	41.36	1.34	3.60	1.20	3.69	2.64
Legumes	2.23	1.16	−22.86	0.77	3.82	0.33	41.78	0.41	3.60	1.20	3.69	2.64
Zamia	−0.60	0.24	−25.59	0.92	1.5	0.67	42.64	3.44	3.60	1.20	3.69	2.64
Fruits (C3)	2.93	2.33	−23.77	3.37	1.37	0.61	43.46	4.50	3.60	1.20	3.69	2.64

[a]It includes Fish (2, 3 carivores and reef fish), [b]Root cultigens and legumes were also analyzed in combination by using original average values.

TABLE 10.3 The WARN MDE parameters of the non-adults and the average $\delta^{15}N$ values of the adult/females in the Antilles and Mesoamerica

Site	*Region*	*n*	t_1	t_2	t_2-t_1	*E*	$\delta^{15}N_{wnfood}$ (‰)	$\delta^{15}N_{female}$ *Average/SD (‰)*	$\delta^{15}N_{adults}$ *Average/SD (‰)*
Cueva del Perico I	Greater Antilles Cuba	12	0.8	2.6	1.8	2.3	11.3	-	11.5/2.3
Canimar Abajo	Greater Antilles Cuba	29	1.8	2.8	0.8	1.0	7.5	10.9/1.2	11.2/1.4
Cueva Calero	Greater Antilles Cuba	9	2.0	4.3	2.3	1.2	8.2	-	10.0/2.2
Paso del Indio	Greater Antilles Puerto Rico	20	2.7	5.9	3.2	1.4	8.5	9.5/0.7	9.6/0.7
Marco González	Mesoamerica Belize	10	1.1	2.5	1.4	1.6	9.0	10.2/0.9	10.5/0.8
San Pedro	Mesoamerica Belize	8	2.2	3.7	1.5	3.2	9.2	9.6/0.6	9.8/0.8
Altun Ha	Mesoamerica Belize	15	3.4	4.4	1.0	2.4	10.7	10.1/0.5	10.3/0.6

(MDE: 0.8 years; 95% Credibility Interval [CI]: 0–2.9 years), followed by those from Marco Gonzales (MDE: 1.1 years.; 95% CI: 0–2.5 years), Cueva Calero (MDE: 2.0 years; 95% CI: 0–4.3 years), Canímar Abajo (MDE: 1.8 years; 95% CI: 0.2–2.6 years), San Pedro (MDE: 2.2 years; 95% CI: 0.9–3.3 years), Paso del Indio (MDE: 2.7 years; 95% CI: 0.8–5.2 years), and Altun Ha (MDE: 3.4 years; 95% CI: 1.3–4.3 years). However, when credible intervals are included in those comparisons, it is possible to observe a large degree of overlap for all age parameters, making conclusions about sequence more contingent.

The high variability observed in WARN parameters may be caused by the small size of the sample per age range and the variability of weaning trajectories within the populations of interest (Figure 10.2). During the first two years of life, some infants are highly enriched in $\delta^{15}N$ in comparison to the female average, suggesting that breastmilk was the most important source of protein in their diet (indicated in Figure 10.2 as BMilk, dashed ovals). However, there is a great variability between infants of similar age, some of which have nitrogen values even lower

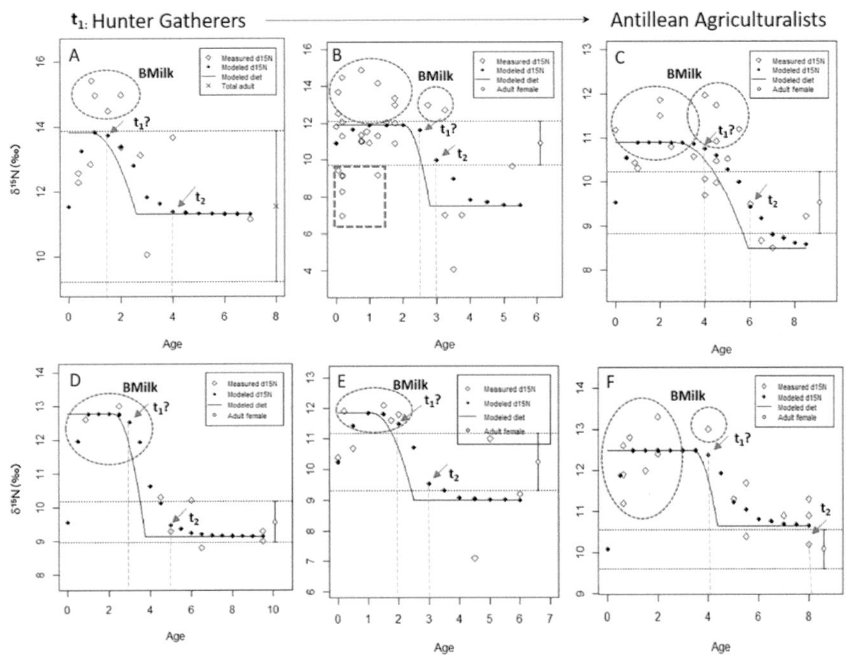

FIGURE 10.2 Transverse comparison of $\delta^{15}N$ (‰ Air) values by non-adult age calculated from reconstructed maximum density estimators (MDEs). Mean and SD range for adult females/adults are indicated with open circles or crosses, respectively. A: Cueva del Perico I, B: Canímar Abajo, C: Paso del Indio, D: Marco González, E: San Pedro, F: Altun Ha. Cueva Calero was removed because of the high variability in adults $\delta^{15}N$ values. t_1: start of weaning, t_2: end of weaning, BMilk: Breastmilk. Ovals/circles indicate breastfeeding infants.

than the female individual values, suggesting that they were weaned at early ages (Figure 10.2: B, dashed rectangle).

While infants have an exclusive consumption of breastmilk, their $\delta^{15}N$ values are 2–3‰ higher than their mothers' due to the trophic level effect (Bocherens and Drucker 2003; Minawaga and Wada 1984; Schoeninger and DeNiro 1984). In terms of the ^{15}N enrichment from breastfeeding (E), the group with the highest ^{15}N enrichment between mothers and their respective breastfeeding children is from San Pedro (MDE: 3.2‰; ≥95% CI: 2.7–3.7‰) followed by those from Altun Ha (MDE: 2.4‰; 95% CI: 2.1–2.7‰), Cueva del Perico I (MDE: 2.3‰; 95% CI: 1.6–3.1‰), Marco Gonzales (MDE: 1.6‰; 95% CI: 1.0–2.4‰), Paso del Indio (MDE: 1.4‰; 95% CI: 1.1–1.8‰), Cueva Calero (MDE: 1.1‰; 95% CI: 0.4–1.9‰), and Canímar Abajo YC (MDE: 1.0‰; 95% CI: 0.3–1.8‰) (Table 10.3). The low nitrogen enrichment observed in Canímar Abajo infants may be influenced by a bimodal distribution in which some infants have very low nitrogen values (Figure 10.2: B), supporting the idea of early supplementation. Low nitrogen values at early ages can also be possible when breastmilk isotopic signals have not been completely fixed in bone collagen tissues, as suggested by the similarity between $\delta^{15}N$ values of those infants and the female individual values (Chinique de Armas et al. 2017). Nevertheless, it is likely that infants ate supplements at early ages through adult food provisioning or self-acquisition, as previously reported by Kramer (2005, 2010) and Greenwald et al. (2016).

The start of weaning has been associated with the moment in which the stable nitrogen values start to decrease (Schurr 1998). Both the WARN parameters and the visual inspection of Figure 10.2 (t_1) suggest that most ancient populations introduced supplementary food to their infants around two years of age, or even after, which contradicts current medical recommendations that support the importance of supplementation in infant diet by six months of age (Dewey and Brown 2003; Michaelsen et al. 2000) and ethnographic reports in which supplementation is observed within the first year of life (Gutiérrez de Pineda and Muirden 1948; Sellen and Smay 2001; Watson 1968). A delayed signal of supplementation in nitrogen isotope values is likely associated with the fact that $\delta^{15}N$ values primarily reflect the protein source of diet (Fernandes et al. 2012), while typical weaning foods are high in carbohydrates and low in protein (Sellen and Smay 2001; WHO 1998). Ethnographic studies of the Island Caribs and the Warao of the Amacuro and Orinoco Deltas have identified the use of herbal teas, fruits, and root cultigens such as sweet potatoes, yautia, and arrowroots to supplement non-adult diets (Du Tertre 1667; Hill and Muirden 1956; Hodge and Taylor 2016; Taylor 1938, 1946, Wilbert 1972). Because these foods are low in protein, we may predict that their consumption does not have an important contribution to infant $\delta^{15}N$ values, leading to a misrepresentation of the age at the start of weaning when the 95% credibility intervals are not considered (Chinique de Armas and Pestle 2018). The use of oxygen isotopes to assess external water sources in diet may improve the resolution by which we can currently detect the age at the start of weaning in archaeological populations.

While there is a strong relationship between $\delta^{15}N$ values and diet (trophic level enrichment), the observed variability in some infant nitrogen values may also be associated with several physiological factors. For instance, the $\delta^{15}N$ values of infants may decrease because of growth and development (Reitsema and Muir 2015; Waters-Rist and Katzenberg 2010) or can increase because of nutritional stress (Fuller et al. 2005; Warinner and Tuross 2010). Given that these data rely on individuals who did not survive to the age at which they are studied, it is possible that some of them may be affected by one or more of these physiological factors, thus not representing the actual weaning trajectory of their source population. Future studies of adult first molar dentine will help to better understand the BWPs of Caribbean populations (Chinique de Armas and Roksandic 2018).

In terms of the duration of the weaning process (t_2-t_1), the non-adults from Paso del Indio had the longest duration, followed by those from Cueva Calero, Cueva del Perico I, San Pedro, Marco Gonzales, Altun Ha, and Canímar Abajo, correspondingly (Table 10.3). The age at the end of weaning (t_2) was the highest in Paso del Indio (MDE: 5.9 years; 95% CI: 2.1–3.5 years) followed by those from Cueva Calero (MDE: 4.8 yrs.; 95% CI: 2.5–7.2 years), Altun Ha (MDE: 4.4 years; 95% CI: 3.8–5.6 years), San Pedro (MDE: 3.7 years; 95% CI: 3.0–4.4 years), Canímar Abajo (MDE: 2.8 years; 95% CI: 2.1–3.5 years) and Cueva del Perico I (MDE: 2.8 years; 95% CI: 1.6–7.6 years), and Marco González (MDE: 2.5 years; 95% CI: 1.8–3.9 years) (Table 10.3). It is possible that this order is influenced by the type of food of the adult population since groups that depend mainly on terrestrial sources, such as Cueva Calero and Paso del Indio, exhibited the highest age at the end of weaning, suggesting that typical diets of this population could mimic the isotopic values of the weaning food, making it difficult to detect the end of the process.

The relationship between the start of the weaning process and a group's subsistence strategy was explored in a previous study on the reconstruction of BWPs in the Caribbean (Chinique de Armas and Pestle 2018). The results indicated that the weaning process began earlier in the "Hunter-Fisher-Gatherer" groups (Cueva del Perico I and Cueva Calero combined) than the horticulturalists from Canímar Abajo and the Antillean and Mayan Agriculturalist groups. This result is consistent with previous archaeological and ethnographic reports (Howcroft 2013; Sellen and Smay 2001). When Cueva Calero is removed from the analysis, because of the high variability of adult nitrogen values, this association between the age at weaning and the subsistence strategy is stronger (Figure 10.2: A, B, and C), although it is important to acknowledge the wide range by the 95% credibility intervals in comparison to the MDE values. No direct association was found between the age at the end of weaning and subsistence (Chinique de Armas and Pestle 2018). Antillean and Mesoamerican agriculturalists had different strategies and food subsistence patterns that are reflected in the weaning process. This further emphasize the important role of cultural perceptions and cultural dietary practices in BWPs.

Infants, Children and Juvenile Diets in the Antilles and Mesoamerica

The Cueva del Perico I population(s) seems to have weaned their children with higher $\delta^{15}N$ food values (MDE: 11.3‰; 95% CI: 9.5–13.1‰), followed by those from Altun Ha (MDE: 10.7‰; 95% CI: 10.3–11.1‰), San Pedro (MDE: 9.2‰; 95% CI: 8.9–9.5‰), Marco Gonzales (MDE: 9.0‰; 95% CI: 8.4–9.6‰), Paso del Indio (MDE: 8.5‰; 95% CI: 7.6–9.4‰), and Cueva Calero (MDE: 8.3‰; 95% CI: 6.4–9.9‰). Lastly, the group from Canímar Abajo has the lowest average weaning food $\delta^{15}N$ values (MDE: 8.3‰; 95% CI: 6.4–9.9‰) (Table 10.3). The $\delta^{15}N$ values of the weaning food are consistently lower than the average value of their respective adults (Table 10.3 and 10.4), supporting the idea that weaning food is typically low in proteins, especially marine sources.

The delayed signal of supplementation in infants at early ages (Figure 10.2) suggests that breast milk was still an important source of protein, at least for some infants of most populations during their first two years of life. The SIAR model supports the idea of breastmilk contribution being high during the first years of life (Table 10.5, Figure 10.3: A1-A2, B1-B2), especially for infants from the Antilles (Figure 10.3: A2, Table 10.5). The SIAR model values showed variability in the importance of several resources within populations (note differences between LCI and HCI in Table 10.5). This variability may be explained by the fact that individuals with ages between 0 and 3 years are grouped under the infant category. In addition, the WARN model indicated that not all infants have the same weaning trajectory, not even within the same population at the same ages, which supports the idea that weaning practices are highly variable and dependent on multifactorial cultural and personal behaviours. As mentioned before, infant dietary variability is associated with different factors, including food provisioning and self-acquisition. Differences in food provisioning are related not only to the dietary practices of the population but also to mothers choosing not to breastfeed or to prolong or truncate the weaning process because of health-related reasons or personal perceptions. At the same time, infants and children are not passive agents since they usually also forage independently (Bird and Bliege 2005; Bock 2005; Eerkens and Bartelink 2013) or select some food over others, according to their likes. This variability of weaning food choices can lead to stable isotope differences. Chronological differences among infants from the same population may be another source of variability since cemeteries typically group individuals from several generations.

In terms of $\delta^{13}C_{co}$, recent research on BWPs in the Caribbean showed that statistically significant differences in diet existed between Antillean and Mesoamerican breastfeeding infants and weanlings (Chinique de Armas and Pestle 2018). This pattern is also evident in the SIAR model (Figure 10.3: A1 vs. B1, Table 10.5) and suggests that Antillean infants had weaning foods consisting mainly of C_3 based resources (terrestrial animals, root cultigens, legumes, and other resources with similar isotopic values), while Mesoamerican infants had a Marine/C_4 plant

TABLE 10.4 Nitrogen and carbon stable isotope values of bone collagen and enamel of Caribbean and circum-Caribbean populations

Site	*Region*	*Infants*			*Children*			*Juveniles*			*Adults*	
		$\delta^{15}N$ *(SD)*	$\delta^{13}C_{co}$ *(SD)*	$\delta^{13}C_{en}$ *(SD)*	$\delta^{15}N$ *(SD)*	$\delta^{13}C_{co}$ *(SD)*	$\delta^{13}C_{en}$ *(SD)*	$\delta^{15}N$ *(SD)*	$\delta^{13}C_{co}$ *(SD)*	$\delta^{13}C_{en}$ *(SD)*	$\delta^{15}N$ *(SD)*	$\delta^{13}C_{co}$ *(SD)*
Cueva del Perico I	Antilles Cuba	13.4 (1.6)	−19.3 (1.4)	–	12.4 (1.8)	−20.2 (0.9)	*	–	–	–	11.5 (2.3)	−20.6 (2.8)
Canimar Abajo OC	Antilles Cuba	11.7 (0.5)	−18.1 (3.3)	−10.4 (1.3)	–	–	−10.3 (2.4)	–	–	−9.6 (1.0)	10.8 (1.9)	−15.0 (2.3)
Canimar Abajo YC	Antilles Cuba	11.5 (2.0)	−19.0 (2.1)	−9.9 (3.2)	8.1 (3.3)	−21.3 (3.5)	−10.6 (2.3)	–	–	−10.2 (1.4)	11.2 (1.4)	−14.5 (2.2)
Cueva Calero	Antilles Cuba	11.7 (1.2)	−22.3 (3.2)	–	8.5 (–)	−18.6 (–)	–	–	–	–	10.0 (2.2)	−22.2 (3.1)
Playa del Mango	Antilles Cuba	12.9 (–)	−18.1 (–)	−12.6 (0.9)	8.10 (–)	−17.0 (–)	−12.2 (1.6)	–	–	−12.5 (1.0)	9.3 (0.5)	−17.9 (0.9)
El Chorro de Maita	Antilles Cuba	–	–	−12.6 (0.9)	–	–	−12.2 (1.6)	–	–	–	–	–
El Flaco	Antilles D. Republic	–	–	−12.4 (0.3)	–	–	−11.9 (0.3)	–	–	–		–
Punta Macao	Antilles D. Republic	–	–	−12.7 (–)	–	–	−13.3 (0.9)	–	–	–	–	–
El Cabo	Antilles D. Republic	–	–	–	–	–	−12.7 (0.2)	–	–	–		–
Paso del Indio	Antilles P. Rico	9.7 (3.2)	−19.0 (3.2)	–	10.3 (1.1)	−19.1 (0.4)	–	9.2	−19.7	–	9.7 (0.8)	−19.1 (0.5)
Maisabel	Antilles P. Rico	–	–	−11.6 (0.9)	11.3 (–)	−18.0 (–)	−12.8 (0.4)	–	–	–	9.5 (0.7)	−18.1 (1.0)
Imperial Lighthouse	Bahamas	–	–	–	8.6 (–)	−15.0 (–)	–	–	–	–		–

(Continued)

TABLE 10.4 (Continued)

Site	*Region*	*Infants*			*Children*			*Juveniles*			*Adults*	
		$\delta^{15}N$ *(SD)*	$\delta^{13}C_{co}$ *(SD)*	$\delta^{13}C_{en}$ *(SD)*	$\delta^{15}N$ *(SD)*	$\delta^{13}C_{co}$ *(SD)*	$\delta^{13}C_{en}$ *(SD)*	$\delta^{15}N$ *(SD)*	$\delta^{13}C_{co}$ *(SD)*	$\delta^{13}C_{en}$ *(SD)*	$\delta^{15}N$ *(SD)*	$\delta^{13}C_{co}$ *(SD)*
Tutu	Virgin Islands	–	–	–	11.9 (0.7)	−15.5 (0.2)	−11.3 (1.1)	11.5 (0.4)	−15.7 (1.0)	–	12.2 (0.9)	−15.4 (1.1)
Spring Bay	Antilles Saba	–	–	–	10.9	−15.8	–	–	–	–	–	–
Kelbey's Ridge	Antilles Saba	–	–	−11.9 (–)	11.1 (0.4)	−15.6 (0.3)	−11.7 (1.36)	10.1	−15.7	–	10.7 (0.5)	−15.7 (0.9)
Morel	Antilles Guadalupe	–	–	−11.9 (0.7)	–	–	−12.1 (0.6)	–	–	–	–	–
Lavoutte	Antilles St. Lucia	–	–	−12.0 (–)	11.6(0.2)	−15.2 (0.4)	−10.8 (1.1)	–	–	–	11.6 (0.5)	−16.1 (0.7)
Carriacou	Antilles Granada	–	–	–	–	–	–	11.1 (0.5)	−14.0 (1.2)	–	11.1 (0.5)	−12.4 (0.6)
Marco Gonzalez	Mesoamerica Belize	11.4 (0.6)	−6.4 (0.7)	–	9.1 (2.0)	−6.9 (0.9)	−9.1 (2.1)	–	–	–	10.5 (0.9)	−7.8 (1.0)
San Pedro	Mesoamerica Belize	12.8 (0.3)	−5.5 (0.7)	–	9.7 (0.7)	−6.1 (0.5)	n/a	9.2 (0.2)	−5.8 (0.3)	–	9.8 (0.8)	−7.4 (1.5)
Pactibun	Mesoamerica Belize	–	–	–	9.0 (1.4)	−12.8 (1.3)	–	–	–	–	9.1 (0.8)	−10.4 (1.8)
Altun Ha	Mesoamerica Belize	12.3 (0.7)	−10.9 (2.7)	–	11.4 (1.0)	−12.1 (2.5)	–	10.8 (0.6)	−12.7 (1.3)	–	10.3 (0.5)	−11.8 (2.0)

* Information available in Laffoon and Chinique de Armas (Chapter 11).

signal (Figure 10.3: A1 vs. B1, Table 10.5). This would appear, at least in part, to be a consequence of lower maize consumption (Figure 10.3: A3 vs. B3), and higher use of terrestrial C_3 plants as weaning foods (Figure 10.3: A4 vs. B4) among the Antillean infants in comparison to the Mesoamerican infants. This supports Chinique de Armas and Pestle's (2018) statement pointing to weaning food choices for non-adults being affected by the subsistence strategy of the adults within various populations within the Caribbean. It is also possible that the maize contribution is being underestimated in populations where nixtamalization was not practised (Sefa-Dedeh et al., 2004). However, the comparison between $\delta^{13}C_{ap}$ of Antillean agriculturalists and Mayan populations supports that such dietary differences are

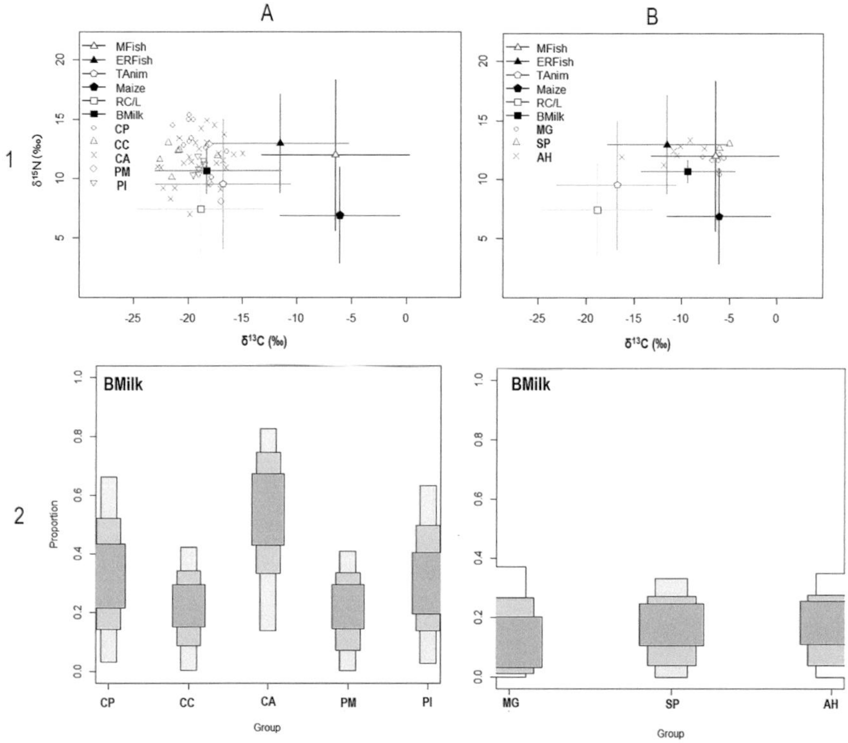

FIGURE 10.3 SIAR model isotopic distributions of infant bone nitrogen and carbon stable isotope values with respect to possible food sources for Antillean (A1) and Mesoamerican (B1) infants; SIAR model % mode estimates and accompanying high and low 95% credibility intervals (error bars) for breast milk (BMilk), maize and root cultigens and legumes (RC/L) for Antillean (A2–4) and Mesoamerican infants (B2–4). Cueva del Perico I (CP), Cueva Calero (CC), Canímar Abajo (CA), Playa del Mango (PM), Paso del Indio (PI), Marco González (MG), San Pedro (SP), and Altun Ha (AH).

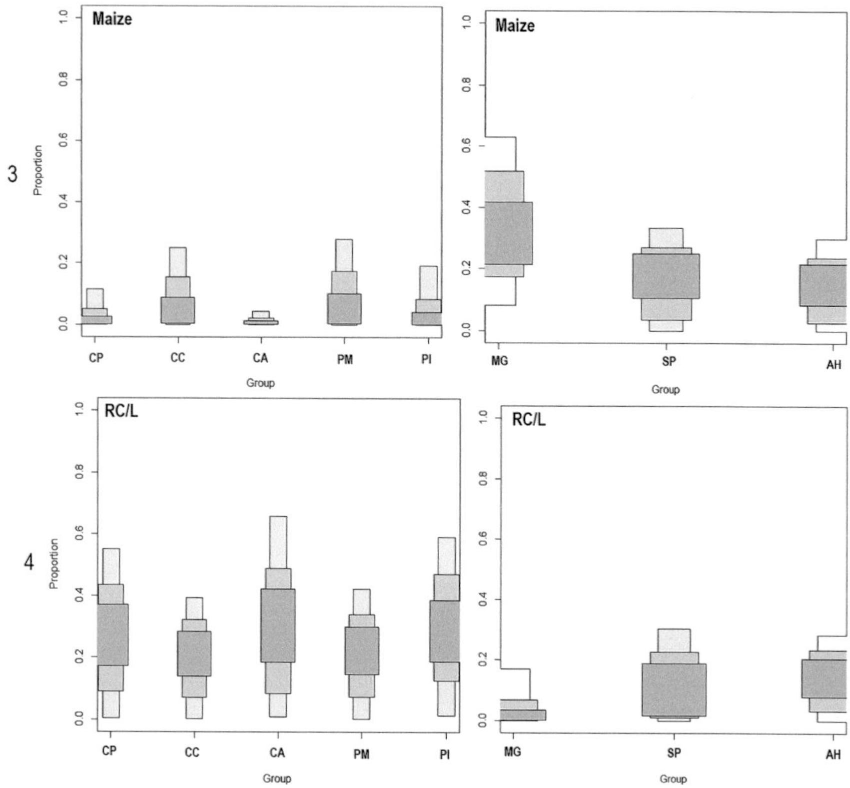

FIGURE 10.3 Continued

not related only to the protein source of diet but also the general dietary practices of the populations (Chinique de Armas and Pestle 2018).

Maize consumption versus root cultigens/legumes consumption is variable among populations and inversely related (Figure 10.3: A3, B3, A4, B4, Table 10.5). While the SIAR model parameters support that there is a low probability for maize being used as a supplement to Antillean infants' diets at Cueva del Perico I, Cueva Calero, Canímar Abajo, Playa del Mango, and Paso del Indio during their first three years of life (note in Table 10.5 that LCI for maize is zero and the mode is typically low), there is a higher probability for the use of root cultigens and legumes as supplements (note in Table 10.5 that LCI for root cultigens/legumes is higher than zero and the mode is between 21 and 30%). In contrast, infants from Mesoamerica showed values (Table 10.5, Figure 10.3) that support the use of maize as supplements versus root cultigens/legumes for most populations. These values may also be influenced by the isotopic contribution of breast milk since its values for Antillean populations are close to the values of the root cultigens and legumes, while Mayas have more positive $\delta^{13}C_{co}$ values. Ethnographic studies support the use of

TABLE 10.5 SIAR mixing model ($\delta^{15}N$ and $\delta^{13}C_{co}$) mode estimates and accompanying high (HCI) and low (LCI) 95% credibility intervals (CI) for potential diet protein sources for Cueva del Perico (CP), Canímar Abajo (CA), Cueva Calero (CC), Playa del Mango (PM), Paso del Indio (PI), Maisabel (Ma), Imperial Lighthouse Cave (ILC), Tutu (Tt), Spring Bay (SB), Kelbey's Ridge (KR), Anse la Gourde (AG), Lavoutte (La), Marco Gonzalez (MG), San Pedro (SP) and Altun Ha (AH).

Sources	*Marine Fish (95% CI)*			*Riverine/Estuarine Fish (95% CI)*			*Terrestrial Animals (95% CI)*			*Maize (95% CI)*			*Root cultigens and legumes (95% CI)*			*Breast milk (95% CI)*		
Infants	*Mode*	*LCI*	*HCI*	*Mode*	*LCI*	*HCI*	*Mode*	*LCI*	*HCI*	*Mode*	*LCI*	*HCI*	*Mode*	*LCI*	*HCI*	*Mode*	*LCI*	*HCI*
CP	1.1	0.0	17.8	11.3	0.0	38.3	1.5	0.0	27.4	0.8	0.0	11.5	26.9	0.2	55.1	29.0	3.2	66.1
CA	0.7	0.0	6.7	7.0	0.4	15.4	1.0	0.0	14.8	0.3	0.0	4.3	29.8	0.7	65.6	58.8	13.9	82.5
CC	5.6	0.0	25.4	21.6	3.9	36.2	12.6	0.0	29.5	1.9	0.0	25.0	21.2	0.1	39.4	22.9	0.3	42.2
PM	2.1	0.0	26.8	3.1	0.0	29.2	22.4	0.3	39.4	2.2	0.0	27.9	23.4	0.1	42.2	22.4	0.4	41.0
PI	1.4	0.0	16.8	7.2	0.0	25.3	3.9	0.0	30.7	1.1	0.0	19.3	27.6	1.3	59.1	27.4	2.7	63.4
MG	17.7	0.9	42.2	13.9	0.2	30.1	1.3	0.0	18.0	30.2	8.0	62.9	1.0	0.0	16.9	16.1	0.0	37.2
SP	18.4	0.1	34.8	18.6	0.0	34.8	14.0	0.0	30.0	17.3	0.0	33.4	14.7	0.0	30.6	17.1	0.0	33.4
AH	18.7	0.6	31.6	24.6	6.5	42.5	11.0	0.0	24.6	15.1	0.0	29.9	14.9	0.0	28.4	18.6	0.0	34.9
Children																		
CP	17.8	0.0	32.0	20.1	0.0	33.8	18.5	0.0	32.6	8.9	0.0	30.8	18.8	0.0	33.6	21.2	0.0	18.8
CA	2.1	0.0	27.4	2.2	0.0	28.6	22.9	0.2	40.0	2.4	0.0	30.6	21.8	0.0	39.5	22.6	0.1	38.4
CC	18.6	0.0	32.1	19.1	0.0	32.4	19.9	0.0	33.2	16.4	0.0	32.5	18.2	0.0	33.2	20.2	0.0	33.1
PI	0.8	0.0	5.7	3.7	0.0	9.1	4.8	0.0	19.3	0.7	0.0	5.9	46.5	0.0	74.1	35.8	0.0	62.3
Ma	18.9	0.0	32.4	19.9	0.0	32.6	20.9	0.0	32.8	19.7	0.0	32.2	19.9	0.0	33.3	20.0	0.0	33.4
ILC	18.8	0.0	32.3	17.4	0.0	32.4	20.2	0.0	33.3	19.1	0.0	32.5	19.9	0.0	33.2	20.0	0.0	32.8
Tt	5.7	0.0	22.9	21.2	1.3	34.4	19.8	0.2	32.7	1.9	0.0	21.4	23.7	1.6	40.5	23.6	1.8	39.5
SB	20.1	0.0	32.2	19.3	0.0	32.4	19.5	0.0	32.7	19.4	0.0	32.5	20.3	0.0	33.1	19.6	0.0	33.1
KR	19.0	0.0	32.0	20.5	0.0	32.7	20.6	0.0	32.8	18.7	0.0	32.0	18.2	0.0	33.3	20.0	0.0	33.2
AG	19.3	0.0	32.4	19.9	0.0	32.9	20.0	0.0	32.3	20.3	0.0	32.7	20.3	0.0	32.1	19.2	0.0	32.4
La	10.5	0.0	28.5	21.1	0.2	32.8	21.9	0.1	33.2	4.5	0.0	28.7	20.6	0.2	35.4	21.8	0.2	35.3

(Continued)

TABLE 10.5 (Continued)

Sources	*Marine Fish (95% CI)*			*Riverine/Estuarine Fish (95% CI)*			*Terrestrial Animals (95% CI)*			*Maize (95% CI)*			*Root cultigens and legumes (95% CI)*			*Breast milk (95% CI)*		
Infants	*Mode*	*LCI*	*HCI*	*Mode*	*LCI*	*HCI*	*Mode*	*LCI*	*HCI*	*Mode*	*LCI*	*HCI*	*Mode*	*LCI*	*HCI*	*Mode*	*LCI*	*HCI*
MG	22.0	0.0	38.0	2.4	0.0	29.7	2.6	0.0	31.5	22.7	0.1	43.6	2.2	0.0	29.1	20.5	0.0	35.2
SP	5.8	0.0	36.1	3.2	0.0	26.6	3.9	0.0	30.1	23.5	0.5	56.8	1.9	0.0	28.5	20.4	0.0	37.6
Pa	14.7	0.0	28.7	19.7	1.6	35.1	16.7	0.4	31.5	14.7	0.0	29.1	18.0	0.5	32.7	16.1	0.0	31.6
AH	13.0	0.0	29.9	11.5	0.0	30.7	21.1	0.5	35.4	19.4	0.0	31.7	21.1	0.4	34.9	20.5	0.0	32.1
Adults																		
CP	2.0	0.0	29.0	23.7	0.4	40.3	27.1	0.0	49.0	1.5	0.0	23.8	30.7	4.4	64.9	–	–	–
CA	1.2	0.0	7.7	6.2	12.5	12.0	10.5	0.9	26.5	1.4	0.0	9.8	74.6	12.5	84.8	–	–	–
CC	1.4	0.0	20.9	4.1	0.0	25.2	32.1	2.4	62.4	1.4	0.0	23.5	38.1	9.7	75.2	–	–	–
PM	1.2	0.0	7.7	6.2	0.6	12.0	10.5	0.9	26.5	1.4	0.0	9.9	74.6	60.0	84.8	–	–	–
PI	0.9	0.0	3.8	8.9	5.4	11.9	9.1	4.0	15.7	0.2	0.0	2.1	80.0	73.7	84.4			
Ma	1.3	0.0	7.9	8.1	1.0	13.8	10.4	0.8	28.5	1.0	0.0	7.8	74.6	58.2	84.4	–	–	–
Tt	2.7	0.0	13.3	36.0	22.4	47.2	7.6	0.0	30.0	0.9	0.0	8.5	45.6	26.0	58.2	–	–	–
KR	8.5	0.0	14.8	20.9	0.2	28.3	14.9	1.1	26.6	9.5	0.0	16.1	46.3	2.3	54.8	–	–	–
AG	8.5	2.3	14.8	20.9	14.6	28.3	14.9	6.4	26.6	9.5	13.2	16.1	46.3	34.8	54.8	–	–	–
La	8.0	1.3	14.8	28.5	20.4	38.6	8.9	0.8	21.7	1.2	0.0	9.6	48.1	34.6	59.3	–	–	–
MG	9.4	0.7	22.7	13.5	5.2	22.2	6.4	0.1	13.7	64.6	50.3	74.5	1.5	0.0	11.5	–	–	–
SP	5.7	0.6	12.1	8.9	3.6	13.9	5.0	0.2	10.9	74.6	67.7	81.2	2.7	0.0	10.1	–	–	–
Pa	5.1	0.9	10.8	14.1	8.9	19.0	6.2	0.5	14.7	40.2	32.3	47.0	33.1	24.2	41.5	–	–	–
AH	5.4	0.0	9.0	5.4	0.5	9.5	5.1	0.0	15.6	57.4	49.0	64.8	27.4	16.2	36.5	–	–	–

atole (a maize-based beverage) to supplement Mesoamerican infants' diets at early ages (Villa Rojas 1945), while the use of C_3 plants such as fruits and root cultigens to wean infants has been reported for other groups in the Orinoco valley (Du Tertre 1667; Taylor 1938, 1946; Hill and Muirden 1956; Wilbert 1972).

According to the SIAR model parameters when combining $\delta^{13}C_{co}$ and $\delta^{15}N$, higher reliance on root cultigens versus maize is also apparent in Antillean children's diets from two populations described as "fisher-gatherers" or "low level food producers" from Cuba: Cueva del Perico I and Canímar Abajo (see modes and HCI in Table 10.5, Figure 10.4). The same pattern was observed for some agriculturalist populations such as Paso del Indio, Tutu, and Lavoutte, while a mixed diet with similar probabilities for the contribution of both maize and root cultigens/legumes was observed for children from Cueva Calero, Maisabel, Imperial Lighthouse Cave, Anse à la Gourde, Kelbey's Ridge, and Spring Bay. Within Mesoamerican groups, a higher reliance on maize was observed relative to root cultigens/legumes for Marco Gonzalez and San Pedro, while for children from Pactibun and Altun Ha, the SIAR model suggests that both maize and root cultigens/legumes were an important source in their diets (Figure 10.4, Table 10.5). In contrast, adult values indicated that most Antillean populations had a diet where C_3 resources such as root cultigens and legumes had a major contribution to the diet, while Mayan populations had a higher reliance on maize or other C_4-based resources (Figure 10.4, Table 10.5).

Variability between adults and non-adults is not fully supported by the statistical analyses, probably as a consequence of the high variability that exists among individuals grouped within the same age ranges. For $\delta^{15}N$, no statistically significant variations by age range were observed for Cueva del Perico I (CP: $F = 2.238$, $df = 2$, $p = 0.133$), Cueva Calero (CC: $F=2.7$, $df=2$, $p=0.0943$), Playa del Mango (PM: $t=-0.68$, $df=2.04$, p-value $=0.57$), Tutu (Tt: $t=0.52$, $df=5.97$, p-value $=0.62$), Carriacou (Ca: $t = 0.14$, $df = 3.54$, p-value $= 0.90$), or San Pedro (SP: $t = 0.43$, $df = 4.24$, p-value $= 0.69$). Statistically significant variations were observed for Canímar Abajo (CA: $F = 6.67$, $df = 2$, $p = 0.00$), Paso del Indio (PI: $F = 9.33$, $df = 3$, $p = 0.00$), Marco González (MG: $H = 7.566$, $df = 2$, $p = 0.00269$) and Altun Ha (AH: $F = 23.9$, $df = 3$, $p = 0.00$). Variations are related to statistically significant differences between children's diets and both infants and adults for the populations of Canímar Abajo and Marco González, while for Paso del Indio adult and non-adult isotope values (infants and children) differ, which is consistent with a prolonged weaning process. In the case of Altun Ha, adult values are different from infants probably as a consequence of breastfeeding. No statistically significant variations in $\delta^{13}C_{co}$ were observed between adults and non-adults for most populations (CP: $F = 0.891$, $df = 2$, $p = 0.426$; CC: $F = 0.641$, $df = 2$, $p = 0.538$; CA: $F = 31.74$, $df = 2$, $p = 0.00$; PM: $t = -0.35$, $df = 4.07$, p-value $= 0.75$; PI: $H = 3.17$, $df = 3$, $p = 0.37$; Tt: $t = 0.56$, $df = 19.04$, $p = 0.58$; Ca: $t = 2.22$, $df = 2.25$, $p = 0.14$; MG: $F = 6.21$, $df = 2$, $p = 0.00$; AH: $F = 0.645$, $df = 3$, $p = 0.59$). In the case of Canímar Abajo, variation was due to statistically significant differences between adults and both infants and children, as had been previously reported (Chinique de Armas and Pestle 2018; Chinique de Armas et al. 2017), while in the case of Marco González, statistically significant differences were found between adults and infants.

In terms of dietary protein sources, most Antillean children had a mixed diet where terrestrial, marine, especially riverine/estuarine resources, were exploited, but with intra- (Table 10.4) and inter-population variations (Figure 10.4 and 10.5,

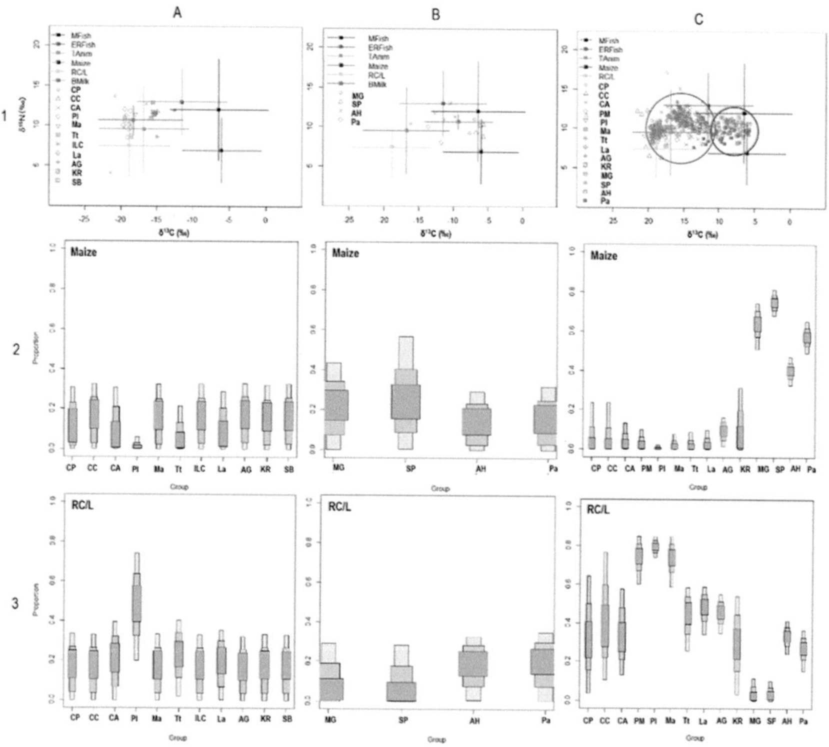

FIGURE 10.4 SIAR model isotopic distributions of children's (columns A and B) and adults' (column C) bone nitrogen and carbon stable isotope values with respect to possible food sources for Antillean (A1 and C1) and Mesoamerican (B1 and C1) sites; SIAR model % mode estimates and accompanying high and low 95% credibility intervals (error bars) for maize, root cultigens and legumes (RC/L), marine fish (MFish), riverine/estuarine fish (REFish) and terrestrial animals (TAnim) for Antillean (A2–6) and Mesoamerican children (B2–6) and adults (C2–6). Cueva del Perico I (CP), Cueva Calero (CC), Canímar Abajo (CA), Paso del Indio (PI), Maisabel (Ma), Tutu (Tt), Imperial Lighthouse Cave (ILC), Lavoutte (La), Anse e la Gourde (AG), Kelbey's Ridge (KR), Spring Bay (SB), Marco González (MG), San Pedro (SP), Altun Ha (AH), Pactibun (Pa), and Playa del Mango (PM).

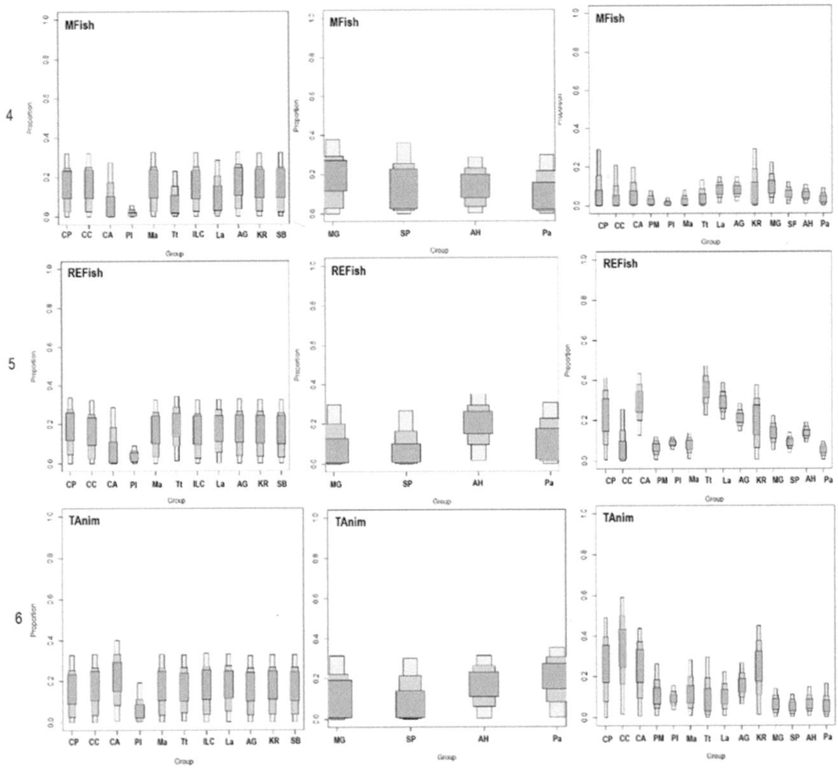

FIGURE 10.4 Continued.

Table 10.5). Statistically significant variations were found between Caribbean children's bone collagen for both carbon (H = 28.47, df = 5, p = 0.00) and nitrogen (F = 3.72, df = 5, p = 0.01). Differences in nitrogen isotope values are due to statistically significant differences between Canímar Abajo and both Altun Ha and Tutu, whose $\delta^{15}N$ are higher (Figure 10.5: A1). Most children available for comparison had values consistent with the consumption of terrestrial C_3-based resources (terrestrial animals and C_3 plants) and riverine/estuarine animals. Please note that all values covered by the shadow are consistent with C_3-based resources when adjusted (Figure 10.5: A1), although marine resources in the Caribbean tend to have lower nitrogen values (Keegan and DeNiro 1988).

For $\delta^{13}C_{co}$, statistically significant differences were found between Mayan sites and both Canímar Abajo and Paso del Indio. It suggests that Maya children had a higher contribution of C_4/Marine resources in their diet than Canímar Abajo and Paso del Indio. In addition, statistically significant differences existed between the Tutu and Canímar Abajo children, supporting previous observations in adults that suggested variability in dietary practices between Antillean populations (Chinique

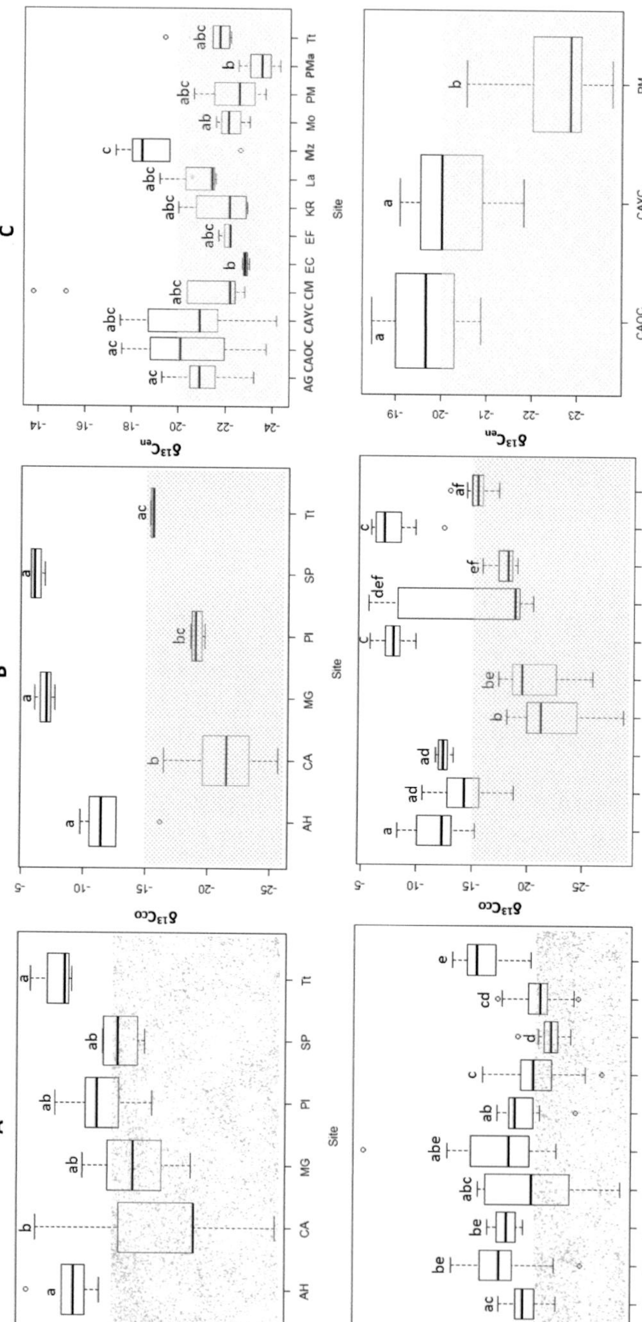

FIGURE 10.5 $\delta^{15}N$, $\delta^{13}C_{co}$, $\delta^{13}C_{en}$ isotopic variations among children (row 1) and adults (row 2) from the Caribbean. Different letters mean statistically significant differences.

de Armas et al. 2016). The Tutu $\delta^{13}C_{co}$ are consistent with a mixed diet where both C_4/Marine and C_3-based resources were present (values under the shadow, when adjusted, are consistent with a C_3 terrestrial/riverine-based diet). Higher variability in stable isotope values of collagen was found at Canímar Abajo, the only "low-level food producer/fisher-gatherer" population included in the comparison. It may support previous observations suggesting that groups with less predictable foods tend to exploit a wider range of resources, leading to greater isotopic differences (Chinique de Armas and Pestle 2018).

Carbon enamel values ($\delta^{13}C_{en}$) suggest greater variability among populations regarding the consumption of maize. Children from Canímar Abajo (CA), Chorro de Maita (CM), Anse à la Gourde, Tutu (Tt), Lavoutte (La), and Manzanilla (Mz) showed a mixed pattern in which both C_3 and C_4 based resources were important (Table 10.6, Figure 10.6). In contrast, low consumption of maize in comparison to root cultigens, and legumes was found in children from Playa del Mango (PM), Punta Macao (PMa), El Cabo (EC), Maisabel (Ma), Kelbey's Ridge (KR), El Flaco (EF), and Morel (Mo) (Table 10.6, Figure 10.6). The results from Canimar Abajo and Lavoutte contradict the SIAR model estimations when combining carbon collagen and nitrogen stable isotope values (Table 10.5), suggesting that SIAR model estimates, using $\delta^{13}C_{co}$ and $\delta^{15}N$, may be influenced by the consumption of marine resources (Figure 10.5).

No statistically significant variability was found for $\delta^{13}C_{en}$ between age ranges per site (CAYC: $H = 1.96$, $df = 2$, $p = 0.38$; CAOC: $F = 0.123$, $df = 2$, $p = 0.885$; PM: $F = 0.032$, $df = 2$, $p = 0.968$; AG: $F = 0.087$, $df = 3$, $p = 0.967$; Mo: $t = -0.48$, $df = 4.10$, p-value $= 0.66$), indicating that whole diet for those sites did not change significantly between infants, children, and juveniles. This result, along with a lower variability among SIAR values for different food sources in the same population when using only $\delta^{13}C_{en}$, supports the idea that while plant consumption was relatively stable throughout individuals' lives, protein sources of diet highly influenced differences between non-adults and adults (Figure 10.5). It is also possible there was a superposition of isotopic information among different teeth used for comparison since each of these teeth was formed over a long period of time.

Statistically significant variations between Antillean sites were found for teeth mainly formed during infancy (1st molar or similar: $H = 15.61$, $df = 5$, $p = 0.01$), childhood (2nd molar or similar: $H = 29.90$, $df = 12$, $p = 0.00$) and adolescence (third molar: $F = 13.62$, $df = 2$, $p = 0.00$). Variations between infants (AG, CAOC, CAYC, Ma, and Mo) are driven by the statistically significant differences between Playa del Mango infants and both Anse à la Gourde and Canímar Abajo (older and younger cemetery) infants. These differences may be influenced by the diet of non-adults and cultural perceptions during the weaning process (Figure 10.5).

There are statistically significant variations in $\delta^{13}C_{en}$ for Antillean children. In agreement with the SIAR model, most populations had a high dependency on C_3-based resources, with the exception of Manzanilla, where the effect of a

TABLE 10.6 SIAR model ($\delta^{13}C_{en}$) percent mode estimates and accompanying high (HCI) and low (LCI) 95% credibility intervals (CI) for potential diet protein sources for Canímar Abajo (CA1: Young cemetery, CA2: Old cemetery), Playa del Mango (PM), Chorro de Maita (CM), Punta Macao (PMa), El Cabo (EC), Maisabel (Ma), Kelbey's Ridge (KR), Anse la Gourde (AG), Tutu (Tt), Lavoutte (La), Manzanilla (Mz), El Flaco (EF) and Morel (Mo)

Sources	*Maize (95% CI)*			*Root cultigens (95% CI)*			*Legumes (95% CI)*			*Zamia (95% CI)*			*Breast milk (95% CI)*		
Infants	*Mode*	*LCI*	*HCI*	*Mode*	*LCI*	*HCI*	*Mode*	*LCI*	*HCI*	*Mode*	*LCI*	*HCI*	*Mode*	*LCI*	*HCI*
CA1	25.1	5.7	39.1	23.3	0.0	38.0	22.4	0.0	38.0	21.9	0.0	35.8	21.8	0.0	38.0
CA2	23.5	15.3	31.3	21.9	0.0	37.6	22.0	0.0	36.6	20.8	0.2	35.2	23.4	0.5	36.4
PM	4.6	0.0	14.0	25.2	0.0	42.1	25.1	0.0	42.6	26.8	0.7	40.2	27.0	0.1	49.1
CM	12.4	0.0	30.8	23.6	0.0	40.7	24.2	0.0	40.6	25.1	0.2	40.7	22.4	0.1	39.0
PMa	16.1	0.0	37.0	22.9	0.0	39.5	23.2	0.0	39.6	22.7	0.0	40.1	24.5	0.0	39.1
Ma	15.3	1.8	28.0	23.2	0.0	40.9	23.1	0.0	40.9	24.0	0.3	39.8	23.3	0.2	38.2
KR	18.6	0.0	37.2	23.9	0.0	39.3	24.9	0.0	39.3	23.8	0.0	39.3	22.7	0.0	39.1
AG	21.3	14.0	28.2	21.8	0.1	39.3	23.8	0.2	39.6	23.1	0.8	39.1	15.5	0.0	30.2
La	22.5	0.0	38.2	23.7	0.0	39.7	23.9	0.0	39.4	23.6	0.0	39.4	21.1	0.0	39.4
EF	10.0	0.0	29.3	25.1	0.0	41.7	25.3	0.0	42.1	26.1	0.4	42.4	22.2	0.0	38.6
Mo	13.0	1.0	25.6	23.0	0.0	42.0	24.2	0.1	41.9	24.5	0.5	41.3	23.6	0.0	37.0
Children															
CA1	22.5	0.1	32.8	22.5	0.0	37.9	21.0	0.0	37.5	22.6	0.0	35.6	22.1	0.0	39.5
CA2	23.7	8.9	36.2	20.6	0.0	37.8	22.0	0.0	37.5	20.8	0.0	35.8	23.0	0.0	38.4
PM	10.6	0.1	23.1	24.5	0.0	42.1	24.1	0.0	41.9	25.6	0.3	41.1	25.1	0.4	40.7
CM	23.2	7.29	35.8	22.9	0.0	38.1	22.4	0.0	37.8	21.9	0.0	36.1	21.5	0.0	37.6
PMa	4.1	0.0	21.3	23.5	0.0	41.9	24.3	0.0	42.8	28.3	0.8	47.4	24.5	0.1	40.8
EC	7.2	0.0	18.9	25.4	0.1	43.3	26.7	0.0	45.6	27.1	1.4	46.2	19.0	0.0	37.0
Ma	6.4	0.0	29.5	24.0	0.0	41.8	24.6	0.0	42.2	25.4	0.3	43.5	23.5	0.0	39.1

KR	14.9	1.9	27.3	23.6	0.0	40.9	22.6	0.0	40.4	23.5	0.2	39.4	24.5	0.2	38.8
AG	20.2	14.7	25.0	22.3	0.0	42.1	22.6	0.1	41.1	23.2	1.1	41.1	17.5	0.7	23.8
Tt	17.3	6.6	27.9	24.0	0.0	39.8	22.9	0.0	39.5	23.8	0.3	38.3	23.3	0.4	37.5
La	20.2	8.6	32.1	23.0	0.0	38.6	23.1	0.0	38.1	22.4	0.2	36.9	22.6	0.2	36.5
Mz	28.5	13.2	42.8	20.4	0.0	37.3	19.8	0.0	37.0	4.8	0.0	34.4	19.6	0.0	37.4
EF	12.1	1.1	23.8	24.6	0.0	42.0	25.3	0.3	43.3	25.5	0.9	42.5	21.7	0.0	36.2
Mo	11.7	1.6	21.0	24.7	0.0	43.5	25.2	0.2	44.2	26.0	0.9	42.8	16.7	0.0	35.4
Juveniles															
CA1	25.0	7.2	37.0	23.1	0.0	37.9	22.3	0.0	37.2	22.3	0.0	35.9	24.0	0.0	37.0
CA2	28.8	19.2	37.4	24.1	0.0	37.8	20.8	0.0	37.1	15.0	0.0	34.5	17.0	0.0	33.1
PM	9.3	0.7	16.3	26.1	0.0	44.1	25.1	0.2	43.6	25.8	1.0	43.9	23.9	0.4	37.7
AG	20.6	1.5	34.3	24.0	0.0	39.0	23.6	0.0	38.8	22.5	0.0	38.1	23.7	0.0	37.7

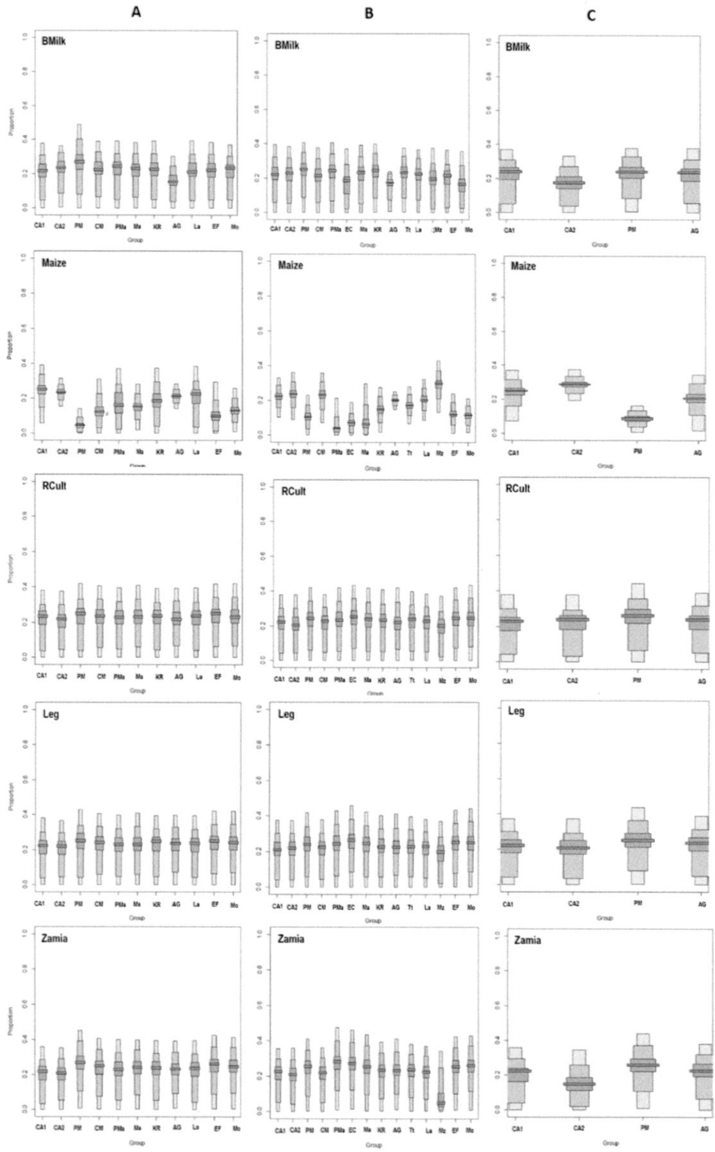

FIGURE 10.6 SIAR mixing model ($\delta^{13}C_{en}$) percent mode estimates and accompanying high (HCI) and low (LCI) 95% credibility intervals (error bars) for maize, root cultigens/legumes (RC/L), and zamia in the diet of infants (first molars or equivalent – row A), children (second molar or equivalent – row B), and juveniles (third molar – row C) of Antillean populations. Canímar Abajo (CA1: Young cemetery, CA2: Old cemetery), Playa del Mango (PM), Chorro de Maita (CM), Punta Macao (PMa), El Cabo (EC), Maisabel (Ma), Kelbey's Ridge (KR), Anse la Gourde (AG), Tutu (Tt), Lavoutte (La), Manzanilla (Mz), El Flaco (EF) and Morel (Mo).

mixed diet on stable isotope values is more apparent (Figure 10.5: C1). Within the negative $\delta^{13}C_{en}$ values that characterize the consumption of C_3 plants, it is possible to distinguish differences. There are populations with very low $\delta^{13}C_{en}$ values, such as El Cabo (EC) or Punta Macao (PMa) while most populations exhibited variable values, going from very low values comparable to those of the sites previously mentioned to higher values typical of a mixed diet where some C_4 plants are present, such as Manzanilla (Figure 10.5: C1). This pattern of variability is also evident at older ages (adolescents), at least for the sites of Canímar Abajo and Playa del Mango (Figure 10.5: C2). Although not all sites have more than three children available for comparison, it is possible to observe, in general terms, that children's diets seem to be lower in both nitrogen and carbon than adults (Figure 10.5: A1 vs. A2, B1 vs. B2, C1 vs. C2), suggesting that resources such as C_3 plants may have played an important role as weaning and transitional foods (Chinique de Armas and Pestle 2018; Chinique de Armas et al. 2017; Laffoon and Chinique de Armas this volume). It was possible to observe a marked difference in the consumption of C_4/marine resources between non-adults and adults in some populations, as is the case for Canímar Abajo and Anse à la Gourde. For Canímar Abajo, this was interpreted as differential consumption of maize (Chinique de Armas et al. 2017), a topic that needs to be re-evaluated in light of the new evidence and the variability of stable isotope values observed at the site (Chinique de Armas et al. forthcoming b). Most populations included in this study are the result of several generations that lived in the Caribbean from the middle Holocene, a period highly variable in terms of climate (Peros et al. forthcoming) and characterized by movement and interactions of culturally and biologically diverse populations (Nägele et al. 2020). It suggests that combining different lines of evidence within the framework of absolute chronologies is necessary to understand variations in dietary practices in the circum-Caribbean region.

Final Remarks

Feeding practices of non-adults in the Caribbean were variable and likely influenced by cultural and personal perceptions about breastfeeding, weaning and food preferences. During the first years of life, most infants seem to have had a diet lower in proteins in comparison to adults, with breastmilk sources potentially having an important contribution to diet, especially for Antillean populations. In general terms, for both infants and children, weaning, and post-weaning foods in the Antilles were mixed and variable, with a predominance of C_3-based resources for most populations (plants and animals) and probably more riverine/estuarine oriented. There are statistically significant intra- and inter-population variations in diet across the region, especially between Mesoamerican and Antillean populations, demonstrating the role of cultural differences in dietary practices. The marked differences between stable isotope values of these areas are likely due to the greater importance of maize for the diet of Mayan populations in comparison to the role that the plant had within the varied horticulture that characterized most Antillean subsistence economies.

Within the Antilles, there were statistically significant variations as well (within and between populations), indicating that a more refined approach, considering different lines of archaeological and bioarchaeological evidence and recognizing the role of individual variability, is necessary to understand cross-regional differences in subsistence strategies and food consumption patterns. While some populations are more C_3-based, others have a pattern more consistent with a mixed C_3-based – C_4/marine signal. Importantly, there are differences in diet between adults and non-adults, at least for some groups, suggesting that not all adult foods, especially protein sources, were considered appropriate to wean infants. The nature of the fragmented archaeological record and the lack of absolute chronologies for all individuals for whom stable isotopes are available make it challenging to draw conclusions related to dietary practices and population-/individual-based variations. Future stable isotope analysis of non-adults in the Caribbean may further contribute to a better understanding of the different strategies that allowed the survival of infants and children in the region, including aspects related to parental investment, possible gender-oriented behaviours, and health-related issues.

Acknowledgements

The authors would like to thank the editors for their comments. We would also like to thank Luis Manuel Viera Sanfiel for his help with the elaboration of Figure 10.1 and Jilliane Celize Rebustillo for her contribution to data gathering and reference editing. Part of this research was funded by the Social Sciences and Humanities Research Council of Canada (Insight Development Grant SSHRC 430-2020-01179 and SSHRC Postdoctoral Fellowship 756–2016–0180) and several University of Winnipeg Internal Grants (Major Grant 2017 and 2018, Partnership Development Grant 2017 and 2018).

References

Aerssens J, Boonen S, Lowet G, and Dequeker J. 1998. Interspecies differences in bone composition, density, and quality: potential implications for in vivo bone research. *Endocrinology* 139: 663–670.

Ambrose SH. 2000. Controlled diet and climate experiments on nitrogen isotope ratios of rats. In: Ambrose SH and Katzenberg MA (Eds.), *Biogeochemical approaches to paleodietary analysis*. New York: Kluwer Academic, 243–259.

Bird DW, and Bliege R. 2005. Human hunting seasonality. In: Brockman D and van Shaik C (Eds.), *Seasonality in primates: Studies of living and extinct human and non-human primates*. Cambridge: Cambridge University Press, 243–266.

Bocherens H, and Drucker D. 2003. Trophic level isotopic enrichment of carbon and nitrogen in bone collagen: Case studies from recent and ancient terrestrial ecosystems. *International Journal of Osteoarchaeology* 13: 46–53.

Bock J. 2005. Farming, foraging, and children's play in the Okavango Delta, Botswana. In: Pellegrini AD and Smith PK (Eds.), *The nature of play: Great apes and humans*. New York: The Gulford Press, 254–281.

Bogin B. 1997. Evolutionary hypotheses for human childhood. *Yearbook of Physical Anthropology* 40: 63–89.

Brown KH, Stallings RY, de Kanashiro HC, Lopez de Romaña G, and Black RE. 1990. Effects of common illnesses on infants' energy intakes from breast milk and other foods during longitudinal community-based studies in Huascar (Lima), Peru. *American Journal of Clinical Nutrition* 52: 1005–1013.

Buhay WM, Chinique de Armas Y, Rodríguez Suárez R, Arredondo C, Smith DG, Armstrong SD, and Roksandic M. 2013. A preliminary carbon and nitrogen collagen isotopic investigation on skeletal remains recovered from a pre-Columbian burial site, Matanzas Province, Cuba. *Applied Geochemistry* 32: 76–84.

Buikstra JE, and Ubelaker DH. 1994. Standards for data collection from human skeletal remains. In: Buikstra JE and Ubelaker DH (Eds.), *Proceedings of a seminar at the field, Museum of Natural History*. Arkansas Archaeological Survey research series 44, 1–206. Fayetteville: Arkansas Archaeological Survey.

Chinique de Armas Y, and Pestle W. 2018. Assessing the association between subsistence strategies and the timing of weaning among indigenous archaeological populations of the Caribbean. *International Journal of Osteoarchaeology* 28: 492–509.

Chinique de Armas Y, Buhay WM, Rodríguez Suárez R, Bestel S, Smith D, Mowat SD, and Roksandic M. 2015. Starch analysis and isotopic evidence of consumption of cultigens among fisher-gatherers in Cuba: the archaeological site of Canímar Abajo, Matanzas. *Journal of Archaeological Science* 58: 121–132.

Chinique de Armas Y, Buhay WM, Rodríguez Suárez R, and Roksandic M. 2013. Relación entre el proceso de ablactación y la mortalidad infantil de los individuos subadultos del sitio arqueológico Canímar Abajo, Matanzas, Cuba. *Cuba Arqueológica. Revista digital de Arqueologia de Cuba y el Caribe* 6: 27–36.

Chinique de Armas Y., González Herrera U.M., Reyes I., Buhay W.M., Skelton S., Rodríguez Suárez R., Roksandic M., Laffoon J. 2022. Multiproxy paleodietary reconstruction using stable isotopes and starch analysis: The case of the archaeological site of Playa del Mango, Granma, Cuba. *Journal of Archaeological Science Reports* 46: 103671. https://doi.org/10.1016/j.jasrep.2022.103671.

Chinique de Armas Y., Hernández Godoy S.T., Viera L.M., Buhay W.M., Laffoon J. 2025. Earliest evidence of sedentism in the Antilles: Multiple isotope data from Canímar Abajo, Cuba. *Proceedings of the National Academy of Sciences* 1 (122): e2413963121. https://doi.org/10.1073/pnas.2413963121.

Chinique de Armas Y, and Roksandic M. 2018. Breastfeeding and weaning in ancient populations: Challenges and future directions. *International Journal of Osteoarchaeology* 28: 477–478.

Chinique de Armas Y, Roksandic M, Nikitović D, Rodríguez Suárez R, Smith DG, García Jordá D, and Buhay WM. 2017. Isotopic reconstruction of the weaning process in the archaeological population of Canímar Abajo, Cuba: A Bayesian probabilities mixing models approach. *PLOS One* 12: e0176065.

Chinique de Armas Y, Roksandic M, Rodríguez Suárez R, Smith D, and Buhay WM. 2016. Isotopic evidence of variations in subsistence strategies and food consumption patterns among fisher-gatherer populations from western Cuba. In: Roksandic I (Ed.), *Cuban archaeology in the circum-Caribbean context*. Gainesville: University Press of Florida, 125–146.

Chisholm BS, Nelson DE, and Schwarcz HP. 1982. Stable-carbon isotope ratios as a measure of marine versus terrestrial protein in ancient diets. *Science* 216: 1131–1132.

Czermak A, Schermelleh L, and Lee-Thorp J. 2018. Tracing weaning by imaging assisted fine-resolution tooth dentine microsampling. *International Journal of Osteoarchaeology* 28 (5): 535–541.

De Luca A, Bolsseau N, Tea I, Louvet I, Robins RJ. Forhan A, Charles MA, and Regis H. 2012. d^{15}N and d^{13}C in hair from newborn infants and their mothers: A cohort study. *Pediatric Research* 71: 598–604.

Dewey KG, and Brown KH. 2003. Update on technical issues concerning complementary feeding of young children in developing countries and implications for intervention programs. *Food and Nutrition Bull* 24: 5–28.

Du Tertre JB. 1667. Concerning the native of the Antilles. *General History of the Antilles Occupied by the French* 2: 360–419.

Dupras TL, Schwarcz HP, and Fairgrieve SI. 2001. Infant feeding and weaning practices in Roman Egypt. *American Journal of Physical Anthropology* 115: 204–212.

Eerkens J, and Bartelink EJ. 2013. Sex-biased weaning and early childhood diet among middle Holocene Hunter-Gatherers in Central California. *American Journal of Physical Anthropology* 152: 471–483.

Fernandes R, Nadeau MJ, and Grootes PM. 2012. Macronutrient-based model for dietary carbon routing in bone collagen and bioapatite. *Archaeological and Anthropological Sciences* 4: 291–301.

Fogel ML, Tuross N, and Owsley DW. 1989. Nitrogen isotope tracers of human lactation in modern and archaeological populations. In: *Annual Report of the Director of the Geophysical Laboratory* 88. Washington, DC: Carnegie Institution: 111–117.

Fratzl P, Gupta HS, Paschalis EP, and Roschger P. 2004. Structure and mechanical quality of the collagen-mineral nano-composite in bone. *Journal of Material Chemistry* 14: 2115–2123.

Froehle AW, Kellner CM, and Schoeninger MJ. 2010. FOCUS: Effect of diet and protein source on carbon stable isotope ratios in collagen: Follow up to Warinner and Tuross (2009). *Journal of Archaeological Science* 37: 2662–2670.

Froehle CM, Kellner CM, and Schoeninger MJ. 2010. FOCUS: Effect of diet and protein source on carbon stable isotope ratios in collagen: follow up to Warinner and Tuross (2009). *Journal of Archaeological Science* 37: 2662–2670.

Fry B. 2008. *Stable Isotope Ecology*. United States: Springer.

Fuller BT, Fuller JL, Harris DA, and Hedges REM. 2006. Detection of breastfeeding and weaning in modern human infants with carbon and nitrogen stable isotope ratios. *American Journal of Physical Anthropology* 129: 279–293.

Fuller BT, Fuller JL, Sage NE, Harris DA, O'Connell TC, and Hedges REM. 2005. Nitrogen balance and d^{15}N: Why you're not what you eat during nutritional stress. *Rapid Communications in Mass Spectrometry* 19: 2497–2506.

Greenwald AM, Eerkens JW, and Bartelink EJ. 2016. Stable isotope evidence of juvenile foraging in prehistoric Central California. *Journal of Archaeological Science: Reports* 7: 146–154.

Gutiérrez de Pineda V, and Muirden SJ. 1948. Social organization in La Guajira. *Revista Del Instituto Etnológico Nacional*. Bogotá, https://ehrafworldcultures.yale.edu/document?id=sc13-001.

Hanson LA. 2004. *Immunobiology of human milk: How breastfeeding protects babies*. Texas, USA: Pharmasoft Publishing.

Hewlett BS, and Winn S. 2014. Allomaternal nursing in humans. *Current Anthropology* 55: 200–229.

Hill GW, and Muirden S. 1956. *Warao of the Amacuro Delta.* Caracas: Universidad Central de Venezuela, https://ehrafworldcultures.yale.edu/document?id=ss18-004.
Hillson S. 1996. *Dental Anthropology*. Cambridge: Cambridge University Press.
Hodge WH, and Taylor D. 2016. The ethnobotany of the Island Caribs of Dominica. *Webbia* 12: 513–644.
Howcroft R. 2013. *Weaned upon a time: Studies of the infant diet in prehistory*. [Unpublished doctoral dissertation]. Stockholm University.
Jackson AL, Inger R, Bearhop S, and Parnell A. 2009. Erroneous behavior of Mix SIR, a recently published Bayesian isotope mixing model: A discussion of Moore and Semmens (2008). *Ecology Letters* 12: E1–E5.
Jones JH. 2011. Primates and the evolution of long, slow life histories. *Current Biology* 21: R708–R717.
Kaplan H, Hill K, Lancaster J, and Magdalena Hurtado A. 2000. A theory of human life history: Diet, intelligence, and longevity. *Evolutionary Anthropology: Issues, News, and Reviews* 9: 156–185.
Katzenberg MA, Herring DA, and Saunders SR. 1996. Weaning and infant mortality: Evaluating the skeletal evidence. *Yearbook of Physical Anthropology* 39: 177–199.
Keegan WF, and DeNiro MJ. 1988. Stable carbon and nitrogen isotope ratios of bone collagen used to study coral-reef and terrestrial components of prehistoric Bahamian diet. *American Antiquity* 53: 320e336.
Kellner CM, and Schoeninger MJ. 2007. A simple carbon isotope model for reconstructing prehistoric human diet. *American Journal of Physical Anthropology* 133: 1112e1127.
Kramer KL. 2005. Children's help and the pace of reproduction: Cooperative breeding in humans. *Evolutionary Anthropology* 14: 224–237.
Kramer KL. 2010. Cooperative breeding and its significance to the demographic success of humans. *Annual Review of Anthropology* 39: 414–436.
Krigbaum J, Fitzpatrick SM, and Bankaitis J. 2013. Human paleodiet at grand bay, Carriacou, Lesser Antilles. *The Journal of Island and Coastal Archaeology* 8: 210–227.
Krueger HW, and Sullivan CH. 1984. Models for carbon isotope fractionation between diet and bone. In: Turnland JR and Johnson PE (Eds.), *Stable isotopes in nutrition*. Washington DC: American Chemical Society, 205–220.
Laffoon J, and Chinique de Armas Y. 2025. Isotopic analysis of indigenous paleodiet and paleomobility in the western Caribbean: enamel strontium and carbon isotope results from Cueva del Perico I, Cuba (this volume). In: Hoogland M, Weston DA and Chinique de Armas Y (Eds.), *The Routledge handbook of bioarchaeology in the Caribbean*. Abingdon: Routledge.
Laffoon JE. 2012. *Patterns of paleomobility in the ancient Antilles: an isotopic approach*. PhD dissertation, Leiden University.
Laffoon JE, and Hoogland ML. 2012. Migration and mobility in the circum-Caribbean: integrating archaeology and isotopic analysis. In: Kaiser, K., Burger, J. and Schier, W. (Eds.), *Population dynamics in prehistory and early history: new approaches using stable isotopes and genetics*, 337–353. Berlin: De Gruyter.
Laffoon JE, Hoogland ML, Davies GR, and Hofman CL. 2019. A multi-isotope investigation of human and dog mobility and diet in the pre-colonial Antilles. *Environmental Archaeology* 24: 132–148.
Laffoon JE, Hoogland MLP, Davies GR, and Hofman CL. 2016 Human dietary assessment in the Pre-colonial Lesser Antilles: New stable isotope evidence from Lavoutte, Saint Lucia. *Journal of Archaeological Science: Reports* 5: 168–180.
Laffoon JE, Valcárcel Rojas R, and Hofman CL. 2013. Oxygen and carbon isotope analysis of human dental enamel from the Caribbean: Implications for investigating individual origins. *Archaeometry* 55: 742–765.

Lee-Thorp JA. 1989. *Stable carbon isotopes in deep time: The diets of fossil fauna and hominids*. [Unpublished doctoral dissertation].University of Cape Town.
Michaelsen KF, Weaver L, Branca F, and Robertson A. 2000. *Feeding and nutrition of infants and young children: Guidelines for the WHO European Region, with emphasis on the former Soviet countries*. WHO Regional Publications. European Series 87. Copenhagen: World Health Organization.
Millard AR. 2000. A model for the effect of weaning on nitrogen isotope ratios in humans. In: Goodfriend GA, Collins MJ, Macko SA, and Wehmiller JF (Eds.), *Perspectives in amino acid and protein geochemistry*. New York: Oxford University Press, 51–59.
Minawaga M, and Wada E. 1984. Stepwise enrichment of^{15}N along food chains: further evidence and the relation between d^{15}N and animal age. *Geochimica et Cosmochimica Acta* 48: 1135–1140.
Nägele K, Posth C, Iraeta M, Chinique de Armas Y, Hernández ST, González UM, Nieves-Colón MA, Sandoval-Velasco M, Mylopotamitaki D, Radzeviciute R, Laffoon J, Pestle WJ, Ramos-Madrigal J, Lamnidis TC, Schaffer WC, Carr RS, Day JS, Arredondo C, Rangel A, Martínez-Fuentes AJ, and Crespo-Torres E, et al. 2020. Genomic insights into the early peopling of the Caribbean. *Science* 369 (6502): 456–460. DOI: 10.1126/science.aba8697.
Naylor AJ, and Marrow AL. 2001. *Developmental readiness of normal full term infants to progress from exclusive breastfeeding to the introduction of complementary foods*. Washington DC: LINKAGES/Wellstart International.
Norr L. 2002. Bone isotopic analysis and prehistoric diet at the Tutu site. In: Righter E (Ed.), *The Tutu archaeological village site*. London: Routledge, 263e273.
O'Connell TC, and Hedges REM. 1999a. Isotopic comparison of hair and bone: archaeological analyses. *Journal of Archaeological Science* 26: 661–665.
O'Connell TC, and Hedges REM. 1999b. Isotopic comparison of hair, nail and bone: modern analyses. *Journal of Archaeological Science* 28: 1247–1255.
O'Leary M. 1981. Carbon isotope fractionation in plants. *Phytochemistry* 20: 553–567.
Parnell AC, Inger R, Bearhop S, and Jackson AL. 2010. Source partitioning using stable isotopes: Coping with too much variation. *PLoS One* 5: e9672.
Pestle WJ. 2010. *Diet and society in prehistoric Puerto Rico*. [Unpublished doctoral dissertation]. University of Illinois.
Pestle WJ. 2013. Stable isotope analysis of Paleodiet in the Caribbean. In: Keegan W, Hofman CL, and Rodríguez Ramos R (Eds.), *The Oxford handbook of Caribbean archaeology*. Oxford: Oxford University Press, 407–417.
Pestle W, Laffoon J. 2018. Quantitative paleodietary reconstruction with complex food webs: An isotopic case study from the Caribbean. *Journal of Archaeological Science: Reports* 17: 393–403.
Phillips DL, and Koch PL. 2002. Incorporating concentration dependence in stable isotope mixing models. *Oecologia* 130: 114–125.
Polin RA, Fox WW, and Abman SH. 2010. *Fetal and Neonatal Physiology*. 4th ed. Philadelphia: Elsevier.
Quinlan RJ. 2007. Human parental effort and environmental risk. *Proceedings of the Royal Society of London* 274: 121–125.
Reitsema LJ, and Muir AB. 2015. Brief Communication: Growth velocity and weaning d^{15}N "dips" during ontogeny in *Macaca mulatta*. *American Journal of Physical Anthropology* 157: 347–357.
Rho J- Y, Kuhn-Spearing L, and Zioupos P. 1998. Mechanical properties and the hierarchical structure of bone. *Medical Engineering and Physics* 20: 92–102.

Richards MP, Mays S, and Fuller BT. 2002. Stable carbon and nitrogen isotope values of bone and teeth reflect weaning age at the Medieval Wharram Percy Site, Yorkshire, UK. *American Journal of Physical Anthropology* 119: 205–210.

Riordan J. 2005. *Breastfeeding and human lactation*. Sudbury: Jones and Bartlet.

Romek KM, Julien M, Frasquet-Darrieux M, Tea I, Antheaume I, and Hankard R, et al. 2013. Human baby hair amino acid natural abundance 15N-isotope values are not related to the^{15}N-isotope values of amino acids in mother's breast milk protein. *Amino Acids* 45: 1365–1372.

Sandberg PA, Sponheimer M, Lee-Thorp J, and Van Gerven D. 2014. Intra-tooth stable isotope analysis of dentine: A step toward addressing selective mortality in the reconstruction of life history in the archaeological record. *American Journal of Physical Anthropology* 155: 281–293.

Schoeninger MJ, and DeNiro MJ. 1984. Nitrogen and carbon isotopic composition of bone collagen from marine and terrestrial animals. *Geochimica et Cosmochimica Acta* 48: 625–639.

Schurr MR. 1998. Using stable nitrogen-isotopes to study weaning behavior in past populations. *World Archaeology* 30: 327–342.

Schwarcz HP. 2000. Some biochemical aspects of carbon isotopic paleodiet studies. In: Ambrose SH and Katzenberg MA (Eds.), *Biogeochemical approaches to paleodietary analysis*. New York: Kluwer Academic/Plenum, 189–209.

Schwarcz HP, and Schoeninger MJ. 2011. Stable isotopes of carbon and nitrogen as tracers for paleodiet reconstruction. In: Baskaran M (Ed.), *Handbook of environmental isotope geochemistry* 1. Berlin: Springer, 725–742.

Sefa-Dedeh S, Cornelius B, Sakyi-Dawson E, and Afoakwa EO. 2004. Effect of nixtamalization on the chemical and functional properties of maize. *Food Chemistry* 86 (3): 317–324.

Sellen DW. 2007. Evolution of infant and young child feeding: implications for contemporary public health. *Annual Review of Nutrition* 27: 123–148.

Sellen DW, and Smay DB. 2001. Relationship between subsistence and age at weaning in "preindustrial" societies. *Human Nature* 12: 47–87.

Stokes AV. 1998. *A biogeographic survey of prehistoric human diet in the West Indies using stable isotopes*. [Unpublished doctoral dissertation]. Gainesville: University of Florida.

Stuart-Macadam P. 1995. Breastfeeding in prehistory. In: Stuart-Macadam P and Dettwyler KA (Eds.), *Breastfeeding: biocultural perspectives*. Foundations of Human Behavior. New York: Aldine De Gruyter, 75–99.

Tauber H. 1981. 13C evidence for dietary habits of prehistoric man in Denmark. *Nature* 292: 332–333.

Taylor DM. 1938. Caribs of Dominica. *Bulletin Anthropological Papers. Washington* 103: 59.

Taylor DM. 1946. Kinship and social structure of the Island Caribs. *Southwestern Journal of Anthropology* 2: 180–212.

Tieszen LL, Boutton TW, Tesdahl KG, and Slade NA. 1983. Fractionation and turnover of stable carbon isotopes in animal tissues: implications for $\delta^{13}C$ analysis of diet. *Oecologia* 57: 32–37.

Tsutaya T. 2017. Post-weaning diet in archaeological human populations: A meta-analysis of carbon and nitrogen stable isotope ratios of child skeletons. *American journal of physical anthropology* 164 (3): 546–557. https://doi.org/10.1002/ajpa.23295.

Tsutaya T, and Yoneda M. 2013. Quantitative reconstruction of weaning ages in archaeological human populations using bone collagen nitrogen isotope ratios and approximate Bayesian computation. *PLoS ONE* 8: e72327.

Tsutaya T, and Yoneda M. 2015. Reconstruction of breastfeeding and weaning practices using stable isotope and trace element analyses: A review. *Yearbook of Physical Anthropology* 156: 2–21.

Van der Merwe NJ, and Vogel JC. 1978. 13C content of human collagen as a measure of prehistoric diet in woodland North America. *Nature* 276: 815–816.

Villa Rojas A. 1945. *The Maya of east central Quintana Roo*. Carnegie Institution of Washington 559, Washington, DC: University of Texas.

Walker R, Hill K, Burger O, and Hartado AM. 2006. Life in the slow lane revisited: Ontogenetic separation between chimpanzees and humans. *American Journal of Physical Anthropology* 129: 577–583.

Warinner C, and Tuross N. 2010. Brief communication: tissue isotopic enrichment associated with growth depression in a pig: implications for archaeology and ecology. *American Journal of Physical Anthropology* 141: 486e493.

Waters-Rist AL, and Katzenberg MA. 2010. The effect of growth on stable nitrogen isotope ratios in subadult bone collagen. *International Journal of Osteoarchaeology* 20: 172–191.

Watson LC. 1968. Guajiro personality and urbanization. *Latin American Studies*. Los Angeles: Latin American Center, University of California.

White CD, Pendergast DM, Longstaffe FJ, and Law KR. 2001. Social complexity and food systems at Altun Ha, Belize: the isotopic evidence. *Latin American Antiquity* 12: 371–393.

WHO. 1998. *Complementary feeding of young children in developing countries: a review of current scientific knowledge*. Geneva: World Health Organization.

Wilbert J. 1972. *The Fishermen: the Warao of the Orinoco Delta. Survivors of El Dorado; four Indian cultures of South America*. New York: Praeger Publishers.

Williams JS, White CD, and Longstaffe FJ. 2009. Maya marine subsistence: isotopic evidence from Marco González and San Pedro, Belize. *Latin American Antiquity* 20: 37–56.

Wood JW, Milner GR, Harpending HC, and Weiss KM. 1992. The osteological paradox: problems of inferring prehistoric health from skeletal samples. *Current Anthropology* 33: 343–370.

11

ISOTOPIC ANALYSIS OF INDIGENOUS PALEODIET AND PALEOMOBILITY IN THE WESTERN CARIBBEAN

Enamel Strontium and Carbon Isotope Results from Cueva del Perico I, Cuba

Jason E. Laffoon and Yadira Chinique de Armas

Introduction

Traditional approaches to the study of paleodiet in archaeological research have often focused on the analyses of food refuse (archaeobotany and archaeozoology) (Twiss 2007), while traditional approaches to the study of paleomobility have generally focused on large-scale changes in material cultural distributions (Rouse 1986). Roughly coinciding with the founding of the sub-field of bioarchaeology (Buikstra 1977), isotope analyses of human skeletal remains have been increasingly applied to paleodietary and, subsequently, paleomobility reconstructions (Makarewicz and Sealy 2015). Isotopic bioarchaeology offers several major advantages to these more traditional archaeological approaches. The most important of these advantages concerns the individual unit of analysis in isotopic approaches, which permits not only highly detailed reconstructions of individual behaviors and life experiences but also the explicit application of social archaeological inquiries and the exploration of variation within, and between, different social groups. Using a bottom-up approach, individual stable isotopic data also afford the opportunity to make broader inductive generalizations about various facets of the movements and foodways of ancient societies.

According to traditional culture historical models for the pre-colonial Caribbean, the region was occupied by two very distinct cultural groups during the Late Ceramic Age (ca. AD 600–1492) (Rouse and Allaire 1978). Although masking much cultural diversity, these two groups are often characterized as (1) sedentary, ceramic-using agriculturalists residing in the Bahamian archipelago, the Lesser Antilles, and most of the Greater Antilles (including Puerto Rico, Hispaniola, Jamaica, and eastern Cuba) and (2) mobile, aceramic foragers (hunter-gatherer-fishers) who resided in the western portion of Cuba and are referred to alternatively

DOI: 10.4324/9780429444449-15

as Preagroalfereros (Tabío 1984), Ciboney (Rouse 1992, Tabío and Rey 1979), or appropriators (Guarch Delmonte 1990). In recent decades, numerous lines of evidence concerning the so-called 'Archaic' peoples of the broader Caribbean and of Cuban Archaic populations in particular have raised serious questions about this simplistic dichotomy, the validity of the term Archaic as a categorical label (Hofman and Antczak 2019), and the utility of Caribbean cultural history classification systems more broadly (e.g., Pestle et al. 2013). For example, the widespread presence of ceramics (Godo 1999; Rodríguez Ramos et al. 2008), large burial populations at some sites (Chinique de Armas et al. 2016), and the early appearance and persistent use of numerous cultivated plants by Cuban Archaic groups (Chinique de Armas et al. 2015; Mickleburgh and Pagán-Jiménez 2012; Pagán-Jiménez et al. 2015) clearly contradict the main diagnostic criteria used to define them. Recent research has also revealed substantial biological and cultural variation between different archaic Cuban populations, indicating that the concept of archaic may mask more than it reveals (Chinique de Armas and Roksandic 2019).

Principles of Isotope Studies

Stable Carbon and Nitrogen Isotope Analysis of Paleodiet

Isotopic analyses are useful for dietary studies because different foods (plant and animal) vary in their isotope composition for carbon and/or nitrogen (DeNiro and Epstein 1978, 1981). In plants, differences in carbon stable isotope values ($\delta^{13}C$) can derive from differences in the main source of carbon between terrestrial and marine plants, namely atmospheric carbon and dissolved inorganic carbonate, respectively (Chisholm et al. 1982; DeNiro and Epstein 1978; Schoeninger and DeNiro 1984; Schoeninger et al. 1983). Among terrestrial plants, carbon stable isotope values also vary primarily owing to differences in photosynthetic pathways (Bender 1971; Smith and Epstein 1971). The vast majority of plants use the C_3 pathway and have low carbon isotope values (Smith and Epstein 1971), while a much smaller proportion use the C_4 pathway and have much higher values (Sage and Monson 1998). A third type of photosynthetic pathway, known as CAM, produces carbon isotope values that are generally intermediate between C_3 and C_4 plants (Bender et al. 1973), although in the Caribbean CAM and C_4 plants have overlapping carbon isotope values (Pestle 2010). Furthermore, both C_4 and CAM plants (such as pineapple) were quite rare in the pre-colonial Caribbean, with maize representing the only economically important C_4 crop documented to have been consumed by Indigenous peoples in substantial quantities (Newsom and Wing 2004).

Nitrogen isotope values ($\delta^{15}N$) also vary between aquatic and terrestrial ecosystems, with higher $\delta^{15}N$ values generally found in both marine and freshwater settings (DeNiro and Epstein 1978; Schoeninger et al. 1983). Another major source of variation in nitrogen isotope values results from the trophic level effect, whereby a stepwise enrichment (increase) in nitrogen isotope values is observed between trophic levels of an ecosystem such that plants generally possess the lowest values, followed sequentially by herbivores and carnivores (DeNiro and Epstein 1978; Schoeninger and DeNiro 1984). It is important to note that although the trophic level effect exists within all ecosystems, it is a relative effect and thus highly influenced by the nitrogen isotope values of primary producers (plants) at the bottom of the food chain. As such, absolute nitrogen isotope values of animal and plant tissues can vary widely between different regions, even among similar taxa. Additionally, aquatic ecosystems generally possess higher trophic levels than terrestrial ones, leading to higher nitrogen isotope values among high trophic-level aquatic carnivores and terrestrial animals that consume them, including some human populations (DeNiro and Epstein 1978; Schoeninger et al. 1983).

The isotopic ecology of the Caribbean region is highly complex with the presence of marine, freshwater, and terrestrial plants and animals and C_3, C_4, and CAM plants (Pestle 2010; 2013; Stokes 1998). Additional complications also arise from the presence of legumes (beans) and zamia (a widely distributed edible plant also known as *coontie*) with very low $\delta^{15}N$ values, in addition to certain shellfish and various reef fish with combined $\delta^{13}C$ and $\delta^{15}N$ values that overlap with other food groups such as C_4 plants (Keegan and DeNiro 1988; Pestle 2010). Furthermore, certain broad food categories (e.g. birds, terrestrial mammals, freshwater, and marine fish) do not display clear discrete clustering of combined carbon and nitrogen isotope values but rather tend to overlap substantially (Pestle 2013). While these factors complicate human dietary reconstructions in the Caribbean based on stable isotope analyses, they do not preclude the successful application of this method as demonstrated by numerous studies (e.g. Buhay et al. 2013; Chinique de Armas et al. 2015; Krigbaum et al. 2013; Laffoon and de Vos 2011; Laffoon et al. 2016; Norr 2002; Pestle 2010; Stokes 1998). It is important to keep in mind, however, that stable isotope analysis can only provide insights into the relative proportions of different food sources (e.g. C_3 plants, C_4 plants, CAM plants, freshwater animals, terrestrial animals, and marine animals) and cannot be used to definitively identify the presence of specific foods.

The isotope composition of the diet is faithfully recorded in various biological tissues of humans and non-human animals (DeNiro and Epstein 1978; Schoeninger et al. 1983). However, in order to estimate the isotopic composition of diet from the isotope values measured in tissues, it is necessary to consider fractionation processes that enrich isotope values as elements are ingested and incorporated into various tissues (e.g., Fernandes et al. 2012; Pestle et al. 2015). In practice, the main skeletal sampling materials used for human paleodiet reconstruction are collagen and hydroxylapatite (also known as bioapatite). Collagen tissues are highly biased to the protein component of diet, with circa 75% of collagen carbon obtained directly from ingested protein (Fernandes et al. 2012; Froehle et al. 2010). By contrast, bioapatite carbon isotope values are highly correlated with the average carbon isotope values of the whole diet (Ambrose and Norr 1993; Tieszen and Fagre 1993), and thus among omnivores they are more sensitive to lipid and carbohydrate (plant) food sources (Harrison and Katzenberg 2003). Collagen samples can be extracted from bones that, owing to remodeling, continually change throughout life (reflecting diet over months, years, or even decades) or from dentine that reflects dietary intake over specific intervals from in *utero* through childhood and adolescence, depending on the tooth type (Hillson 1996). Collagen tends to be fairly resistant to taphonomic processes, and various quality control indicators are available to provide indications of the integrity of the sample material and the reliability of the measured isotope signals (Ambrose 1990; DeNiro 1985). By contrast, bone bioapatite is much more susceptible to diagenetic contamination, and there is less consensus concerning which, if any, quality control indicators are effective for identifying or mediating against potential post-depositional alteration of apatite isotopic signatures (Pestle and Colvard 2012). Enamel, however, is generally regarded as an extremely robust tissue and highly resistant to diagenesis (Koch et al. 1997), and thus one common strategy is to analyze the carbon isotope signal of bioapatite in dental enamel with the caveat that the isotopic signal reflects non-adult as opposed to adult diet.

Strontium Isotope Analysis of Paleomobility

The principle of using strontium isotope analysis of human skeletal tissues to study mobility is that strontium occurs as a trace element in consumed food and water and is incorporated into skeletal tissues owing to chemical similarities with the major element calcium (Bentley 2006). Strontium is a ubiquitous element in the lithosphere and enters the food chain primarily from the chemical weathering of bedrock minerals into the soil, from where it is taken up by plants via groundwater. As ^{87}Sr is a product of the decay of ^{87}Rb with a half-life of 4.88×1010 years, $^{87}Sr/^{86}Sr$ ratios vary in rocks based on age and initial rubidium concentration (Faure and Mensing 2005). Spatial variation of $^{87}Sr/^{86}Sr$ in the biosphere is predominately controlled by variation in the underlying lithosphere but can also be strongly influenced by non-geological sources of strontium from the hydrosphere and atmosphere (Laffoon et al. 2012; Whipkey et al. 2000).

Strontium isotope analysis was initially developed in the field of isotope geochemistry and widely applied as a dating tool. Ericson (1985) first proposed and demonstrated the utility of the method for archaeological research into paleomobility. Although not initially treated with much enthusiasm by archaeologists, changes in the dominant theoretical paradigm within the field combined with technical advances led to increased interest in this method in the mid-1990s (Ezzo et al. 1997; Grupe et al. 1997; Price et al. 1994). In the last two decades, subsequent methodical developments and improved understandings of the strontium cycle, as well as potential sources of variation, have contributed to the further increased popularity of the strontium isotope method in archaeology, and today, it is arguably the most commonly employed analytical tool for investigating patterns of ancient human mobility and migration from the archaeological record (Slovak and Paytan 2012).

Most applications involve the measurement of $^{87}Sr/^{86}Sr$ ratios in human tooth enamel. Enamel is the preferred sample material as it forms in relatively well-documented intervals in an individual's youth, does not undergo any subsequent remodeling once enamel mineralization is completed (Hillson 1996), and has been shown to be highly resistant to post-mortem alteration of biogenic isotopic signals (diagenesis) in the burial environment (Koch et al. 1997). Unlike light stable isotopes, which undergo fractionation during various biochemical processes, strontium isotopes do not measurably fractionate as they move through the food web and into human or animal tissues (Blum et al. 2000). Because of this, enamel $^{87}Sr/^{86}Sr$ signatures can be directly compared to a range of (measured or estimated) $^{87}Sr/^{86}Sr$ variation for a local ecosystem or population (Price et al. 2002). Individuals with $^{87}Sr/^{86}Sr$ ratios that fall within the local range are deemed to be locals, whereas individuals with $^{87}Sr/^{86}Sr$ ratios that fall outside the local range can be identified as nonlocals.

There are several major interpretive caveats to this approach. Nonlocal enamel $^{87}Sr/^{86}Sr$ ratios could result from having spent one's childhood in an isotopically distinct location, justifying the inference of subsequent migration to the location of burial. Conversely, it could also occur if the individual simply consumed large proportions of nonlocal (imported) food as a child (Wright 2005). For isotopic locals, the main interpretive issue is equifinality (meaning the same result), which refers to the fact an individual may have an enamel $^{87}Sr/^{86}Sr$ ratio similar to the local range of isotope values because he/she originated from a location with similar $^{87}Sr/^{86}Sr$ ratios (Price et al. 2007). As such, there is always the possibility that individuals who appear to be local, based on their enamel $^{87}Sr/^{86}Sr$ signal, are actually nonlocals. The main interpretive consequence is that estimates of the number of nonlocals, and thereby the degree of mobility, based on the strontium isotope method alone can be best considered as minimum estimates (Laffoon and Leppard 2019).

In the Caribbean region, the three major sources of strontium in the biosphere derive from bedrock geology, precipitation, and dryfall, such as Saharan dust transported by the wind across the Atlantic (Bataille et al. 2012). Substantial efforts have been made to map the spatial variation of strontium isotopes available to biological organisms, also known as bioavailable strontium isotope ratios, in the Caribbean region (Laffoon et al. 2012). Laffoon and colleagues (2012) measured $^{87}Sr/^{86}Sr$ in a wide range of biogenic sample materials from across the region and found that although bedrock was the major source of strontium in most locations, substantial inputs of non-geological strontium could also be observed. Bataille and colleagues (2012) subsequently developed a multi-source predictive model to map spatial strontium isotope variation across the landscape (isoscape) for the circum-Caribbean region. Subsequent sampling and analysis of bioavailable $^{87}Sr/^{86}Sr$ have validated the general predictive power of their model, but with some exceptions. For example, geographically and geologically representative sampling of Trinidad and Tobago and the Bahamas archipelago (Ostapkowicz et al. 2017; Schulting et al. 2018) for $^{87}Sr/^{86}Sr$ analysis found no influence of Saharan dust on bioavailable $^{87}Sr/^{86}Sr$. These results indicate that the Caribbean strontium isoscape of Bataille and colleagues (2012) likely overestimates the influence of atmospheric dryfall and suggest that bioavailable $^{87}Sr/^{86}Sr$ in this region can be directly estimated using a simple two-source model with bedrock and sea-derived precipitation as the two main sources of strontium for terrestrial ecosystems.

Background

Site Context

The Cueva del Perico I cemetery is located inside a cave in the northern part of Artemisa province in Cuba (83° 16′ 29.87″ W and 22° 57′ 29.43″ N) (Figure 11.1). It is one kilometer from the Mani-Mani River and 25 km from Bahia Honda (Pino and Alonso 1973). Excavations conducted between 1970 and 1997 at Cueva del Perico I recovered evidence for a minimum of 162 individuals interred at the cave site (Travieso Ruíz et al. 1999). The population was classified as fisher-gatherers (foragers) or Ciboney Guayabo Blanco type (Pino and Alonso 1973). Some artifacts, such as gouges and grinding stones, were found on the site, as well as different

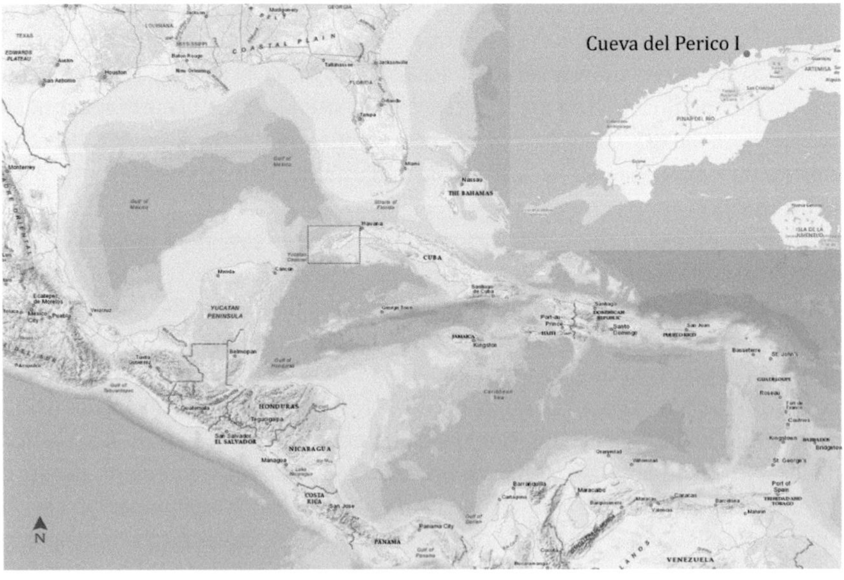

FIGURE 11.1 Map of the Caribbean region with inset indicating the location of the Cueva del Perico I site, Cuba.

animal remains outside of the cave (Brito 1996). Conventional dates indicate that the site was used as a cemetery at least from 100 cal B.C. to 220 cal A.D and until cal A.D. 600–800 (Pino 1995; Cooper 2010). Recent AMS radiocarbon dates obtained directly from human bone collagen samples from this cemetery indicate that it was in use from 2750 to 1190 cal BP (Nägele et al. 2020; Chinique de Armas et al. 2020).

Previous Research

In terms of dietary reconstruction, a previous stable isotope study of paleodiet of this site (Chinique de Armas et al. 2016) generated bone collagen carbon and nitrogen isotope data (n = 11) and concluded that this population relied almost exclusively on terrestrial C_3 foods. However, as collagen is heavily biased toward the protein component of the diet, it is not very sensitive to changes in carbohydrate intake, especially since plants are generally much lower in protein content than animal foods. Therefore, collagen is not an ideal proxy for detecting C_4 inputs unless they form a very significant component of overall food consumption. In contrast, bioapatite carbon isotope values are highly correlated with the whole diet and thus are better suited for detecting minor variations in the relative contributions of C_3 vs. C_4 plants. This is of particular relevance in the Caribbean, where maize represents one of the only major C_4 crops consumed by Indigenous peoples of the region, and because many important issues concerning maize (such as the date of introduction, vectors of dispersal, spatial-temporal distributions, and socio-economic value) remain contested and unresolved (e.g., Mickleburgh and Pagán-Jiménez 2012; Pagán-Jiménez 2011; Pagán-Jiménez et al. 2015; Pestle 2010; Rodríguez Ramos 2010). To further assess dietary patterns for the population of Cueva del Perico I, we applied carbon isotope analysis of enamel samples to more reliably determine the plant contributions to individual diets. These data are also explored relative to the mobility results determined by strontium isotope analysis and in the context of a larger corpus of enamel carbon isotope data for Indigenous Caribbean populations. Because we focus exclusively on dental enamel as a sampling material, the results presented here only apply to non-adult diets, and this work represents the first large-scale study to explore regional and cultural variation in non-adult dietary patterns for the pre-colonial Caribbean region (see also Chinique de Armas and Laffoon, this volume).

Previous paleodiet research at Cueva del Perico I focused exclusively on carbon and nitrogen isotope analysis of bone collagen (Chinique de Armas et al. 2016). The mean collagen $\delta^{13}C$ is −20.6 ± 2.8‰ (1σ), varying from a low of −22.7‰ to a high of −17.4‰, and the mean $\delta^{15}N$ is 11.8 ± 2.3‰ (1σ), varying from 9.0‰ to 17.1‰. These carbon isotope results are very low and are some of the lowest recorded to date for Caribbean populations, with only two other sites having similar values (Cueva Calero and Guayabo Blanco), both of which are also Archaic-related Cuban populations. The mean nitrogen isotope value of Cueva del

Perico I is moderately high but within the range reported for most other Caribbean sites. The high degree of variance in the nitrogen isotope values is notable, possibly indicating a high degree of inter-individual variation in protein sources. The cause(s) of this variation are not known, but it could possibly relate to differences in sex, age, chronology, or other factors. Chinique de Armas and colleagues (2016) also calculated estimates of proportions of potential protein sources using the mixing model SIAR (Stable Isotope Analysis in R). The model outputs indicate a very low average input of marine proteins (<5%) to diet at Cueva del Perico I. Overall, the collagen stable isotope results are consistent with terrestrial C_3 diets with little or no input from C_4 plants and considerable variation in the relative proportions of different protein sources.

Materials and Methods

For the present study, sampled skeletal materials include one tooth each from 17 separate individuals. Sample selection was based on individuals that had been previously sampled for aDNA analysis (Nägele et al. 2020). Samples derive from diverse dental elements and thus do not all represent the same period of formation, however, for individuals, the same tooth was used for strontium and carbon isotope analysis. Sample preparation was conducted in the Laboratory for Archaeological Chemistry at the Faculty of Archaeology, Leiden University, and isotope measurements were conducted at the Department of Earth Sciences at the Faculty of Science, Vrije Universiteit Amsterdam. Enamel (3–5 mg) was extracted using a hand-held dental drill with a diamond-tipped drill tip. Powdered samples were subjected to chemical pre-treatment following the protocol outlined in Bocherens et al. (2011) in order to remove any possible surface contamination or exogenous carbonates.

Afterward, a sub-sample (0.5 mg) of pre-treated enamel was weighed into pre-cleaned glass vials for carbon isotope analysis of the structural carbonate fraction of the enamel bioapatite. For acid dissolution, samples were first flushed with helium gas and then digested in 100% orthophosphoric acid for at least 24 h at 45°C. Carbon isotopes were measured using a ThermoFinnigan GasBench II interfaced with a ThermoFinnigan Delta + isotope ratio mass spectrometer (IRMS). Typical analytical uncertainties for $\delta^{13}C$ are 0.15‰, as determined by long-term analyses of an internal carbonate standard (VICS) normalized to NBS18, NBS19, and NBS20. Carbon isotope values are reported in the delta notation (δ) in permil units (‰) relative to the Vienna PeeDee Belemnite (VPDB) standard.

For strontium isotope analysis, samples (2 mg) were first dissolved in 3 N HNO_3 and loaded onto columns for chemical separation of strontium using Sr-specific resin. Purified Sr was then loaded onto degassed rhenium filaments, and strontium isotope compositions ($^{87}Sr/^{86}Sr$) were measured on a ThermoScientific Triton Plus thermal ionization mass spectrometer (TIMS). $^{87}Sr/^{86}Sr$ ratios were corrected for mass bias relative to the accepted $^{86}Sr/^{88}Sr$ ratio of 0.1194 using the exponential law. The NBS987 international strontium carbonate standard was measured during the same analytical session with long-term average $^{87}Sr/^{86}Sr$ ratios of 0.710247 ±0. 00001 (1σ), while typical procedural blanks are negligible (<40 pg).

Results and Discussion

Sampling information and isotope results are reported in Table 11.1 and plotted in Figure 11.2. The mean $^{87}Sr/^{86}Sr$ ratio is 0.70826 ±.0008 (2σ), and the measured sample $^{87}Sr/^{86}Sr$ ratios vary from 0.70705 to 0.70873. No faunal or floral samples are available in order to independently estimate the local range of bioavailable $^{87}Sr/^{86}Sr$ for this site. Assessments of the expected range of bioavailable $^{87}Sr/^{86}Sr$ therefore rely on comparisons with the bioavailable strontium isoscape of the Caribbean region (Bataille et al. 2012). This isoscape suggests a range of approximately 0.707–0.709 for the site's location, completely encompassing the range of ratios recorded for this population. An alternative approach for identifying outliers (nonlocals) is to calculate the mean ± two standard deviations (±2SD) with the assumption that the majority of analyzed individuals are local. Based on this latter approach, all individuals fall within the mean ±2 SD range and are thus isotopically local, with the exception of one outlier (7152). The outlier's $^{87}Sr/^{86}Sr$ ratio may indicate a nonlocal origin at the site level, although this $^{87}Sr/^{86}Sr$ signal falls

TABLE 11.1 Sampling information and enamel isotope results from the site of Cueva del Perico I, Cuba

Sample ID		*$^{87}Sr/^{86}Sr$*	*$^{87}Sr/^{86}Sr$*	*$\delta^{13}C$*	*$\delta^{13}C$*
	Context		*2SE*	*VPDB*	*SD*
2002	–	0.70766	0.00001	−13.5	0.1
2009	–	0.70835	0.00001	−12.4	0.1
2009.1	–	0.70836	0.00001	−12.2	0.1
2435	–	0.70836	0.00001	−11.9	0.1
2486	E-6, Saco 18	0.70821	0.00001	−12.6	0.1
2493	E-13, Saco 32	0.70832	0.00001	−11.6	0.1
2513	E-17, Saco 91	0.70832	0.00001	−13.4	0.1
7137	–	0.70873	0.00001	−10.4	0.1
7140	–	0.70871	0.00001	−12.8	0.2
7141	–	0.70822	0.00001	−13.0	0.2
7143	–	0.70816	0.00001	−14.1	0.1
7143.2	–	0.70835	0.00001	−12.8	0.2
7146	–	0.70832	0.00001	−13.8	0.1
7148	–	0.70834	0.00001	−12.7	0.4
7149.4	–	0.70826	0.00001	−14.3	0.2
7151	–	0.70873	0.00001	−11.5	0.1
7152	–	0.70705	0.00001	−14.8	0.2

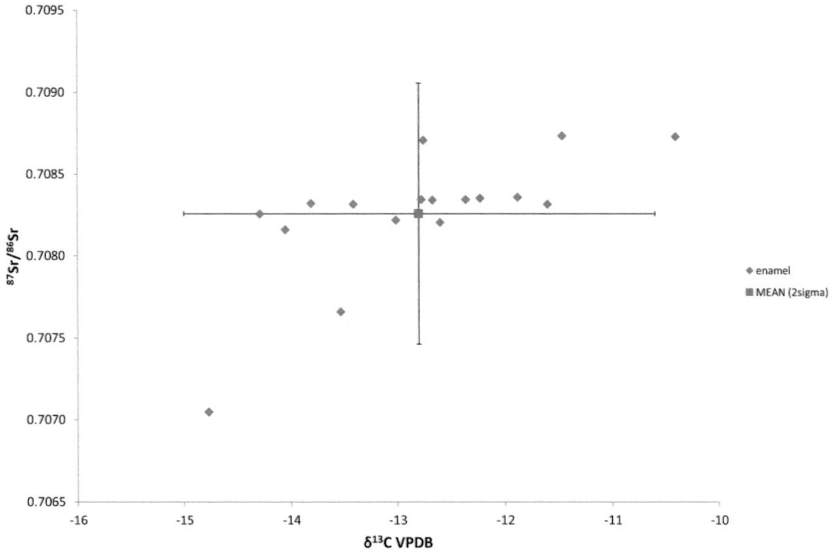

FIGURE 11.2 Chart of enamel strontium and stable carbon isotope data from Cueva del Perico I, Cuba.

within the range of bioavailable $^{87}Sr/^{86}Sr$ for the island of Cuba and numerous other islands throughout the Caribbean. It is not possible to further constrain this individual's origin based on the strontium isotope data alone.

The mean enamel $\delta^{13}C$ is -12.8 ± 2.2‰ (2σ) with a minimum of -14.8‰ and a maximum of -10.4‰. Notably, the single nonlocal outlier (7152) with the lowest $^{87}Sr/^{86}Sr$ ratio also has the lowest $\delta^{13}C$ value in the dataset. This could indicate a possible relationship between mobility and diet, with the nonlocal individual consuming slightly different proportions of food sources at his/her place of childhood origin relative to the local population, as previously observed for other populations in the Caribbean (Laffoon 2016). An alternative explanation is that the enamel $^{87}Sr/^{86}Sr$ ratio has been influenced by the direct consumption of marine resources, which uniformly possess $^{87}Sr/^{86}Sr$ ratios of 0.7092. To test this, we checked the correlation between the $^{87}Sr/^{86}Sr$ and $\delta^{13}C$ values and found a moderately strong correlation between the two (R2 = 0.4709). As such, it is possible that the enamel $^{87}Sr/^{86}Sr$ ratios are correlated with marine resource consumption, a pattern that has not previously been observed among pre-colonial Caribbean populations but has been reported among other insular populations documented to have consumed large proportions of seafood, for example on Iceland (Walser et al. 2020). This explanation seems unlikely to have resulted from the consumption of marine fish considering the very low estimated marine protein input for this population based on the collagen isotope data and associated SIAR model estimates (Chinique de Armas et al. 2016). Alternative factors that could account for the observed correlation are the consumption of marine animal lipids, marine plants, or sea salt; the latter has been reported to have altered enamel $^{87}Sr/^{86}Sr$ ratios of enamel samples in the Mayan region (Fenner and Wright 2014).

In terms of identifying potential C_4 plant food consumption, it is necessary to account for the offset (or fractionation) in carbon isotope values between diet and tissue. The most recent estimate of the offset between the whole diet and bioapatite for humans is +10.1 ±0.4‰ (Fernandes et al. 2012). Pestle (2010) generated a large dataset of floral stable isotope values for the Caribbean and reports that in this region C_3 plants have mean $\delta^{13}C$ of −25.5 ±1.7‰ (from −30.2 to −21.1‰) and C_4 plants have mean $\delta^{13}C$ of −9.6 ±0.9‰ (from −12.9‰ to −8.6‰). Applying the tissue to diet fractionation (offset) factor (−10.1 ±0.4‰) to the enamel $\delta^{13}C$ values from Cueva del Perico I provides an estimate of average whole diet $\delta^{13}C$ of −22.9‰ (from −24.9‰ to −20.5‰). Based on this estimate, all individual whole diet $\delta^{13}C$ values, except one (7137), fall entirely within the range of values for C_3 plants, and even the single exception also falls within this range of variation if the combined uncertainties (of the analyses and the correction) are taken into account. Therefore, the enamel carbon isotope results confirm the conclusion that C_4 plants, including maize, made little or no contribution to diets among this sample population. We also compared these whole diet $\delta^{13}C$ ($\delta^{13}C_{whole\ diet}$) estimates for Cueva del Perico I with $\delta^{13}C_{whole\ diet}$ calculated from enamel carbon isotope values ($\delta^{13}C_{enamel}$) for numerous other Indigenous sites in the insular Caribbean (Figure 11.3). This comparison highlights that the Cueva del Perico I population was one of the most reliant on C_3 plants in the entire Caribbean, with most other sites having intermediate $\delta^{13}C_{whole\ diet}$ values indicating variable mixes of C_3 and C_4/CAM plants. Based on these data, most other Indigenous Caribbean populations consumed minor to moderate quantities of C_4 and/or CAM plants, with the exception of Aruban populations (both Archaic and Ceramic), who appear to have relied much more intensively on C_4 plants and/or marine resources (Mickleburgh and Laffoon 2018) (Table 11.2).

Conclusions

Only one individual at this site can be confidently identified as a nonlocal. Better characterization of the local range of bioavailable strontium isotope variation, preferably via the analysis of local faunal remains, is needed to provide more secure identifications of locals/nonlocals. The low variance in $^{87}Sr/^{86}Sr$ observed for this site could result from mobility within an isotopically homogeneous landscape or conversely reflect low rates of mobility for this population. The former case might be particularly applicable in Cuba, relative to other smaller islands of the Caribbean, since large areas of this island possess broadly similar bioavailable $^{87}Sr/^{86}Sr$ ratios. This highlights one of the main limitations of using the strontium isotope method to quantify mobility, namely the issue of equifinality and false negatives (individuals who were actually mobile but not identified as such based on their $^{87}Sr/^{86}Sr$ signature) (Price et al. 2007). As such, calculations of the proportions of non-local individuals

TABLE 11.2 Statistical summary of enamel carbon isotope results and estimated whole diet carbon isotope values from Cueva del Perico I, Cuba and other Indigenous Caribbean sites

Island/ Region	*Sites*	$\delta^{13}C_{enamel}$		$\delta^{13}C_{whole\ diet}$	
		Mean	*2SD*	*Mean*	*n*
Cuba	Cueva del Perico I	−12.8	2.2	−22.9	17
Cuba	El Chorro de Maíta	−11.4	2.0	−21.5	53
Jamaica	White Marl	−12.2	1.7	−22.3	3
Dominican Rep.	El Flaco & Punta Macao	−11.6	2.6	−21.7	37
Puerto Rico	Maisabel	−11.4	2.8	−21.5	29
Puerto Rico	Maruca	−10.9	3.5	−21.0	2
Bahamas	Preacher's Cave	−9.7	4.2	−19.8	7
St. Thomas	Tutu	−11.0	1.4	−21.1	31
Saba	Kelbey's Ridge & Spring Bay	−11.6	2.2	−21.7	7
St. Kitts	Bloody Point	−11.6	0.9	−21.7	3
Guadeloupe	Anse à la Gourde & Morel	−11.0	1.6	−21.1	79
St. Lucia	Anse Lavoutte	−10.1	1.9	−20.2	33
St. Vincent	Argyle & Escape	−11.6	1.3	−21.7	32
Barbados	Heywoods	−8.6	0.9	−18.7	3
Trinidad	Manzanilla	−10.6	4.1	−20.7	17
Aruba (Ceramic)	Savaneta & Tanki Flip	−3.9	3.7	−14.0	3
Aruba (Archaic)	Malmok	−4.7	1.1	−14.8	4

within a sampled population are always minimum estimates, and comparisons of the proportions of nonlocals across populations should be done with caution. Nevertheless, the proportion of nonlocals at Cueva del Perico I (6% of the sample) is in fact one of the lowest recorded to date for Caribbean populations using strontium isotope analysis (Booden et al. 2008; Hoogland et al. 2010; Laffoon 2012; Laffoon and Leppard 2019), which is particularly notable considering that nearly all of the other Caribbean groups studied to date are considered to have been sedentary. If these results are an accurate reflection of low rates of mobility (or migration) for this population, this would imply a relatively high degree of sedentism and conflict with traditional notions of Archaic Cuban populations as being highly mobile. In light of the long-term human occupation of Cuba, in addition to the various other aforementioned lines of evidence (e.g. large burial populations, cultivation of domesticated plants), these results seem to provide further support for the proposition that not all Cuban Archaic groups fit neatly into either the subsistence divide (between foragers and farmers) or the mobility divide (between nomadic bands and sedentary villagers) (see also, Mickleburgh and Laffoon 2018; Smith et al. 2018).

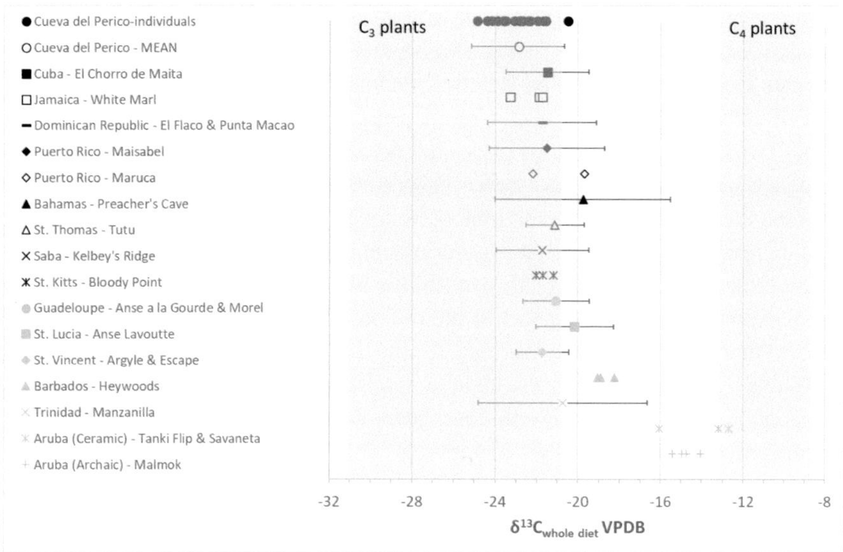

FIGURE 11.3 Cueva del Perico I estimated whole diet carbon isotope values ($\delta^{13}C$ whole diet) compared with other Caribbean Indigenous sites and the range of $\delta^{13}C$ values for C_3 and C_4 plants. Groups with <5 samples are plotted as individual data points; all others are plotted as mean ±2 SD. Data sources: (Laffoon et al. 2013, 2019, 2020; Mickleburgh and Laffoon 2018; Mickleburgh et al. 2019; Pestle 2010; Pestle and Rodríguez Ramos 2018; Schroeder et al. 2018).

In terms of diet, the relatively low enamel $\delta^{13}C$ values are consistent with diets dominated by terrestrial C_3 resources. In agreement with previous studies (Chinique de Armas et al. 2016), we detect no significant contribution of C_4 plants (such as maize) among this population. As previously noted, this food consumption pattern contrasts sharply with that observed for the site of Canímar Abajo, where both stable isotope and starch grain evidence (Buhay et al. 2013; Chinique de Armas et al. 2015) indicates that maize was likely consumed at a very early date and remained an important dietary component throughout the site's long occupation(s). Given the location of both sites near the northern coast of western Cuba, the presumed ease of travel via the coastal areas, and their contemporaneity (between Cueva del Perico I and the younger cemetery at Canímar Abajo), one might expect some degree of interaction between these populations or at least knowledge of each other's cultural practices. Additionally, since maize possesses a high degree of adaptability to various environmental and climatic conditions, there is no apparent ecological constraint that would explain the general lack of maize cultivation/consumption at Cueva del Perico I. As such, the most likely explanation is that the absence of maize in the foodways of this group reflects different cultural traditions, as previously suggested by Chinique de Armas and colleagues (2016). Considering the other evidence for clear cultural differences between various Archaic Cuban populations, such as dental modification practices (Roksandic et al. 2016), it is possible that food represented another medium for the negotiation and expression of cultural or ethnic identity in this region (González Herrera 2016; Pestle 2010; Twiss 2007). The results presented herein provide further evidence for the existence of variation in subsistence economies and food consumption practices between Archaic Cuban populations (Chinique de Armas and Roksandic 2019). This increasingly evident cultural variation not only calls into question the validity of the 'archaic' as a cultural category in this region (Hofman and Antczak 2019) but also provides further support to the notion of the pre-colonial Caribbean as a cultural mosaic (Keegan and Hofman 2016).

While clearly providing new insights into the lifeways of understudied Indigenous populations of the western Caribbean, this study should not be considered definitive but rather hopefully as a stimulus for further investigations of Caribbean mobility and dietary patterns. Several avenues for future research can be proposed. First, there is an obvious need to better characterize strontium isotope variation at multiple scales (site, island, region). In fact, although the current regional bioavailable strontium isotope data set (Giovas et al. 2016; 2019; Laffoon et al. 2012; Pestle et al. 2013; Ostapkowicz et al. 2017; Schulting et al. 2018) is fairly large (n = 800), it is also heavily biased to the eastern half of the Caribbean, and despite the fact that Cuba is the largest island, it is poorly represented in terms of sampling density. Current efforts are focused on improving this situation, but owing to its massive size (nearly 110,000 km^2), there will remain a need to rely on predictive spatial models or isoscapes (Bataille et al. 2012) to assess large-scale variation across Cuba for the foreseeable future.

There is also a clear opportunity for expanding the types of analyses utilized for this study not only to a larger number of individuals from Cueva del Perico I but also to other contemporaneous populations from Cuba and the Greater Antilles more generally. It is notable that despite the fact that stable isotope studies have been applied to the archaeological record of the Caribbean almost since their inception over four decades ago, very little stable isotope analysis of human diet has been conducted to date on Hispaniola or Jamaica (but see Stokes 1998), and very few strontium isotope studies of human mobility have been conducted in the Greater Antilles to date (for exceptions, see Laffoon et al. 2019; 2020; Mickleburgh et al. 2019). Fortunately, the recent trend is very promising, with several important integrated isotopic studies of various Cuban populations conducted in recent years (e.g. Buhay et al. 2013; Chinique de Armas et al. 2015; 2016; Laffoon et al. 2020). There is also much untapped potential for more intensive and extensive isotope bioarchaeology applications in the Caribbean using alternative or unconventional isotope proxies such as sulfur (S) (Sparks and Crowley 2018) or lead (Laffoon et al. 2019b), compound-specific isotope analysis (CSIA) of individual amino acids to better distinguish different food sources (e.g. Honch et al. 2012), coupled serial sampling of incremental dentine layers and isotope analysis to produce high-temporal resolution reconstructions of childhood diet and nutrition (Beaumont et al. 2013; Laffoon et al. 2018), and Bayesian multi-source mixing models to generate quantitative dietary estimates (Chinique de Armas et al. 2016; Chinique de Armas and Pestle 2018; Pestle and Laffoon 2018). Last, but not least, as always, there is much work to be done to better integrate isotopic and other types of bioarchaeological data (e.g. calculus micro-botanicals; skeletal morphology; and proteomics and genomics) both with each other and with other lines of archaeological evidence. Currently, a multi-disciplinary research collaboration, including osteoarchaeological, genomic, and isotopic analysis (of collagen, bone apatite, and dental enamel), is being conducted on the Cueva del Perico I population as well as several other contemporaneous Cuban sites. We are hopeful that this research will provide new insights into the dietary and mobility patterns of Indigenous Cuban groups and contribute to several important ongoing debates about the past lifeways of Caribbean people.

Acknowledgements

The study was funded by the ERC under the European Union's Seventh Framework Programme (FP7/2007–2013)/ERC Grant Agreement No. 319209. This study was conducted as part of the NEXUS 1492: New World Encounters in a Globalising World Synergy grant project directed by Corinne L. Hofman, within the sub-project 'Human Mobility and the Circulation of Materials and Objects' directed by Gareth R. Davies. Thanks to Suzan Verdegaal-Warmerdam for conducting the carbon isotope analysis. Special thanks are also due to Antonio Martinez Fuentes, Carlos Lalueza-Fox, and Hannes Schroeder for facilitating access to the enamel samples. We would like to thank Luis Manuel Viera Sanfiel for providing Figure 11.1.

References

Ambrose, S.H. 1990. Preparation and characterization of bone and tooth collagen for isotopic analysis. *Journal of Archaeological Science*, 17(4), pp. 431–451.

Ambrose, S.H. and Norr, L. 1993. Experimental evidence for the relationship of the carbon isotope ratios of whole diet and dietary protein to those of bone collagen and carbonate. In *Prehistoric Human Bone,* ed. J. Lambert and G. Grupe (pp. 1–37). Berlin/Heidelberg: Springer.

Bataille, C.P., Laffoon, J. and Bowen, G.J. 2012. Mapping multiple source effects on the strontium isotopic signatures of ecosystems from the circum-Caribbean region. *Ecosphere*, 3(12), pp. 1–24.

Beaumont, J., Gledhill, A., Lee-Thorp, J. and Montgomery, J. 2013. Childhood diet: A closer examination of the evidence from dental tissues using stable isotope analysis of incremental human dentine. *Archaeometry*, 55(2), pp. 277–295.

Bender, M.M. 1971. Variations in the $^{13}C/^{12}C$ ratios of plants in relation to the pathway of photosynthetic carbon dioxide fixation. *Phytochemistry*, 10(6), pp. 1239–1244.

Bender, M.M., Rouhani, I., Vines, H.M. and Black, C. 1973.13C/12C ratio changes in Crassulacean acid metabolism plants. *Plant Physiology*, 52(5), pp. 427–430.

Bentley, R.A. 2006. Strontium isotopes from the earth to the archaeological skeleton: A review. *Journal of Archaeological Method and Theory*, 13(3), pp. 135–187.

Blum, J.D., Taliaferro, E.H., Weisse, M.T. and Holmes, R.T. 2000. Changes in Sr/Ca, Ba/Ca and $^{87}Sr/^{86}Sr$ ratios between trophic levels in two forest ecosystems in the northeastern USA. *Biogeochemistry*, 49(1), pp. 87–101.

Bocherens, H., Sandrock, O., Kullmer, O. and Schrenk, F. 2011. Hominin palaeoecology in Late Pliocene Malawi: First insights from isotopes (^{13}C, ^{18}O) in mammal teeth. *South African Journal of Science*, 107(3–4), pp. 1–6.

Booden, M.A., Panhuysen, R.G.A.M., Hoogland, M., de Jong, H., Davies, G.R. and Hofman, C.L. 2008. Tracing human mobility with $^{87}Sr/^{86}Sr$ at Anse à la Gourde, Guadeloupe. In *Crossing disciplinary boundaries and national borders. New methods and techniques in the study of archaeological materials from the Caribbean*, ed. C.L. Hofman and A. van Duijvenbode, pp. 214–225. Leiden: Sidestone Press.

Brito, M.E. 1996. *Arqueozología del sitio Cueva del Perico I, Bahía Honda, Pinar del Río. Bachelor degree*. Faculty of Biology. University of Havana. 67. Unpublished.

Buhay, W.M., de Armas, Y.C., Suárez, R.R., Arredondo, C., Smith, D.G., Armstrong, S.D. and Roksandic, M. 2013. A preliminary carbon and nitrogen isotopic investigation of bone collagen from skeletal remains recovered from a Pre-Columbian burial site, Matanzas Province, Cuba. *Applied Geochemistry*, 32, pp. 76–84.

Buikstra, J.E. 1977. Biocultural dimensions of archaeological study: A regional perspective. In *Biocultural Adaptation in Prehistoric America*, ed. Robert L. Blakely, pp. 67–84. Athens: University of Georgia Press.

Chinique de Armas, Y. and Pestle, W. 2018. Assessing the association between subsistence strategies and the timing of weaning among indigenous archaeological populations of the Caribbean. *International Journal of Osteoarchaeology*, 28(5), pp. 492–509.

Chinique de Armas, Y., Buhay, W.M., Suárez, R.R., Bestel, S., Smith, D., Mowat, S.D., and Roksandic, M. 2015. Starch analysis and isotopic evidence of consumption of cultigens among fisher–gatherers in Cuba: The archaeological site of Canímar Abajo, Matanzas. *Journal of Archaeological Science*, 58, pp. 121–132.

Chinique de Armas, Y., Herrera, U.M.G., Buhay, W.M., Masdeu, J.M.Y., Sanfiel, L.M.V., Burchell, M., Crann, C., González-Quevedo, E.R.G. and Roksandic, M. 2020. Chronology of the archaeological site of Playa del Mango, río Cauto, Granma, Cuba. *Radiocarbon*, 62(5), pp. 1221–1236.

Chinique de Armas, Y. and Roksandic, M. 2019. Biological and cultural diversity of Archaic Age populations from western Cuba. In *Biological Anthropology of Latin America. Historical development and recent advances,* ed. D.H. Ubelaker and S. Colantonio, pp. 161–171. Washington: Smithsonian Contributions to Anthropology, 51

Chinique de Armas, Y., Roksandic, M., Suárez, R.R., Smith, D.G. and Buhay, W.M. 2016. Isotopic evidence of variations in subsistence strategies and food consumption patterns among "fisher-gatherer" populations of western Cuba. In *Cuba Archaeology in the Caribbean*, ed. I. Roksandic, pp. 125–146. Gainesville: University of Florida Press.

Chisholm, B.S., Nelson, D.E. and Schwarcz, H.P. 1982. Stable-carbon isotope ratios as a measure of marine versus terrestrial protein in ancient diets. *Science*, 216(4550), pp. 1131–1132.

Cooper, J. 2010. Pre-Columbian archaeology of Cuba: A study of site distribution patterns and radiocarbon chronologies. In *Island shores, distant pasts: Archaeological and biological approaches to the pre-Columbian settlement of the Caribbean*, ed. S.M. Fitzpatrick and A.H. Ross, pp. 81–107. Gainesville: University of Florida Press.

DeNiro, M.J. 1985. Postmortem preservation and alteration of in vivo bone collagen isotope ratios in relation to palaeodietary reconstruction. *Nature*, 317(6040), pp. 806–809.

DeNiro, M.J. and Epstein, S. 1978. Influence of diet on the distribution of carbon isotopes in animals. *Geochimica et Cosmochimica Acta,* 42(5), pp. 495–506.

DeNiro, M.J. and Epstein, S. 1981. Influence of diet on the distribution of nitrogen isotopes in animals. *Geochimica et Cosmochimica Acta*, 45(3), pp. 341–351.

Ericson, J.E. 1985. Strontium isotope characterization in the study of prehistoric human ecology. *Journal of Human Evolution*, 14(5), pp. 503–514.

Ezzo, J.A., Johnson, C.M., and Price, T.D. 1997. Analytical perspectives on prehistoric migration: A case study from east-central Arizona. *Journal of Archaeological Science*, 24(5), pp. 447–466.

Faure, G. and Mensing, T.M. 2005. *Isotopes: Principles and applications.* 3rd Ed. Hoboken: John Wiley and Sons.

Fenner, J.N. and Wright, L.E. 2014. Revisiting the strontium contribution of sea salt in the human diet. *Journal of Archaeological Science*, 44, pp. 99–103.

Fernandes, R., Nadeau, M.J., and Grootes, P.M. 2012. Macronutrient-based model for dietary carbon routing in bone collagen and bioapatite. *Archaeological and Anthropological Sciences*, 4(4), pp. 291–301.

Froehle, A.W., Kellner, C.M., and Schoeninger, M.J. (2010). FOCUS: Effect of diet and protein source on carbon stable isotope ratios in collagen: Follow up to. *Journal of Archaeological Science*, 37(10), 2662–2670.

Giovas, C.M., Kamenov, G.D., Fitzpatrick, S.M., and Krigbaum, J. 2016. Sr and Pb isotopic investigation of mammal introductions: Pre-Columbian zoogeographic records from the Lesser Antilles, West Indies. *Journal of Archaeological Science*, 69, pp. 39–53.

Giovas, C.M., Kamenov, G.D., and Krigbaum, J. 2019. $^{87}Sr/^{86}Sr$ and ^{14}C evidence for peccary (Tayassuidae) introduction challenges accepted historical interpretation of the 1657 Ligon map of Barbados. *PloS One*, 14(5)): e0216458. https://doi.org/10.1371/journal.pone.0216458.

Godo, P.P. 1999. El problema del Protoagricola de Cuba. Discusión y perspectivas. *El Caribe Arqueológico*, 2, pp. 19–30.

González Herrera, U.M. 2016. Food preparation and dietary preferences among the Arawak Aboriginal Communities of Cuba. In *Cuban Archaeology in the Caribbean*, ed. I. Roksandic, pp. 1168–1184. Gainesville: University of Florida Press.

Grupe, G., Price, T.D., Schröter, P., Söllner, F., Johnson, C.M., and Beard, B.L. 1997. Mobility of Bell Beaker people revealed by strontium isotope ratios of tooth and bone: A study of southern Bavarian skeletal remains. *Applied Geochemistry*, 12(4), pp. 517–525.

Guarch Delmonte, J.M. 1990. *Estructura para las comunidades aborígenes de Cuba*. Holguín: Ediciones Holguín.

Harrison, R.G. and Katzenberg, M.A. 2003. Paleodiet studies using stable carbon isotopes from bone apatite and collagen: Examples from Southern Ontario and San Nicolas Island, California. *Journal of Anthropological Archaeology*, 22(3), pp. 227–244.

Hillson, S. 1996. *Dental Anthropology*. Cambridge: Cambridge University Press.

Hofman, C.L. and Antczak, A.T. eds. 2019. *Early Settlers of the insular Caribbean: Dearchaizing the Archaic*. Leiden: Sidestone Press.

Honch, N.V., McCullagh, J.S., and Hedges, R.E. 2012. Variation of bone collagen amino acid δ13C values in archaeological humans and fauna with different dietary regimes: Developing frameworks of dietary discrimination. *American Journal of Physical Anthropology*, 148(4), pp. 495–511.

Hoogland, M.L., Hofman, C.L., and Panhuysen, R.G.A.M. 2010. Interisland dynamics: Evidence for human mobility at the site of Anse à la Gourde, Guadeloupe. In *Island shores, distant pasts: Archaeological and biological approaches to the pre-Columbian settlement of the Caribbean,* ed. S.M. Fitzpatrick and A.H. Ross, pp. 148–162. Gainesville: University of Florida Press.

Keegan, W.F., and DeNiro, M.J. 1988. Stable carbon-and nitrogen-isotope ratios of bone collagen used to study coral-reef and terrestrial components of prehistoric Bahamian diet. *American Antiquity*, 53(2), 320–336.

Keegan, W.F. and Hofman, C.L. 2016. *The Caribbean before Columbus*. Oxford: Oxford University Press.

Koch, P.L., Tuross, N., and Fogel, M.L. 1997. The effects of sample treatment and diagenesis on the isotopic integrity of carbonate in biogenic hydroxylapatite. *Journal of Archaeological Science*, 24(5), pp. 417–429.

Krigbaum, J., Fitzpatrick, S.M., and Bankaitis, J. 2013. Human Paleodiet at grand bay, Carriacou, Lesser Antilles. *The Journal of Island and Coastal Archaeology*, 8(2), pp. 210–227.

Laffoon, J.E. 2012. Patterns of Paleomobility in the ancient antilles: An isotopic approach. PhD dissertation, Leiden University.

Laffoon, J.E. 2016. Human Mobility and Dietary Patterns in Precolonial Puerto Rico: Integrating multiple isotope data. In *Cuban Archaeology in the Caribbean*, ed. I. Roksandic, pp. 147–167. Gainesville: University of Florida Press.

Laffoon, J.E. and de Vos, B. 2011. Diverse origins, similar diets: An integrated isotopic perspective from Anse à la Gourde, Guadeloupe. In *Communities in Contact: Essays in Archaeology, Ethnohistory & Ethnography of the Amerindian Circum-Caribbean*, ed. C.L. Hofman and A. van Duijvenbode, pp. 187–204. Leiden: Sidestone Press.

Laffoon, J.E., Davies, G.R., Hoogland, M.L., and Hofman, C.L. 2012. Spatial variation of biologically available strontium isotopes ($^{87}Sr/^{86}Sr$) in an archipelagic setting: A case study from the Caribbean. *Journal of Archaeological Science*, 39(7), pp. 2371–2384.

Laffoon, J.E., Espersen, R., and Mickleburgh, H.L. 2018. The life history of an enslaved African: Multiple isotope evidence for forced childhood migration from Africa to the Caribbean and associated dietary change. *Archaeometry*, 60(2), pp. 350–365.

Laffoon, J.E., Hoogland, M.L., Davies, G.R., and Hofman, C.L. 2016. Human dietary assessment in the Pre-colonial Lesser Antilles: New stable isotope evidence from Lavoutte, Saint Lucia. *Journal of Archaeological Science: Reports*, 5, pp. 168–180.

Laffoon, J.E., Hoogland, M.L., Davies, G.R., and Hofman, C.L. 2019. A multi-isotope investigation of human and dog mobility and diet in the pre-colonial Antilles. *Environmental Archaeology*, 24(2), pp. 132–148.

Laffoon, J.E. and Leppard, T.P. 2019.$^{87}Sr/^{86}Sr$ data indicate human post-juvenile residence mobility decreases over time-elapsed since initial Holocene island colonization in the Pacific and Caribbean. *Archaeological and Anthropological Sciences*, 11(5), pp. 1757–1768.

Laffoon, J.E., Rojas, R.V., Weston, D.A., Hoogland, M.L., Davies, G.R., and Hofman, C.L. 2020. Diverse and dynamic dietary patterns in early colonial Cuba: New insights from multiple isotope analyses. *Latin American Antiquity*, 31(1), pp. 103–121.

Laffoon, J.E., Shuler, K.A., Millard, A.R., Connelly, J.N., and Schroeder, H. 2020. Isotopic evidence for anthropogenic lead exposure on a 17th/18th century Barbadian plantation. *American Journal of Physical Anthropology*, 171(3), pp. 529–538.

Laffoon, J.E., Valcárcel Rojas, R., and Hofman, C.L. 2013. Oxygen and carbon isotope analysis of human dental enamel from the Caribbean: Implications for investigating individual origins. *Archaeometry*, 55(4), pp. 742–765.

Makarewicz, C.A. and Sealy, J. 2015. Dietary reconstruction, mobility, and the analysis of ancient skeletal tissues: Expanding the prospects of stable isotope research in archaeology. *Journal of Archaeological Science*, 56, pp. 146–158.

Mickleburgh, H.L. and Laffoon, J.E. 2018. Assessing dietary and subsistence transitions on prehistoric Aruba: Preliminary bioarchaeological evidence. In *The Archaeology of Caribbean and Circum-Caribbean Farmers (6000 BC–AD 1500)*, ed. B. Reid, pp. 288–306. New York: Routledge.

Mickleburgh, H.L., Laffoon, J.E., Pagán Jiménez, J.R., Mol, A.A., Walters, S., Beier, Z.J., and Hofman, C.L. 2019. Precolonial/early colonial human burials from the site of White Marl, Jamaica: New findings from recent rescue excavations. *International Journal of Osteoarchaeology*, 29(1), pp. 155–161.

Mickleburgh, H.L. and Pagán-Jiménez, J.R. 2012. New insights into the consumption of maize and other food plants in the pre-Columbian Caribbean from starch grains trapped in human dental calculus. *Journal of Archaeological Science*, 39(7), pp. 2468–2478.

Nägele, K., Posth, C., Orbegozo, M.I., De Armas, Y.C., Godoy, S.T.H., Herrera, U.M.G., Nieves-Colón, M.A., Sandoval-Velasco, M., Mylopotamitaki, D., Radzeviciute, R., and Laffoon, J. 2020. Genomic insights into the early peopling of the Caribbean. *Science*, 369(6502), pp. 456–460.

Newsom, L.A. and Wing, E.S. 2004. *On land and sea: Native American uses of biological resources in the West Indies*. Tuscaloosa: University of Alabama Press.

Norr, L. 2002. Bone isotopic analysis and prehistoric diet at the Tutu site. In *The Tutu archaeological village site: A multidisciplinary case study in human adaptation*, ed. E. Righter, pp. 263–273. New York: Routledge

Ostapkowicz, J., Brock, F., Wiedenhoeft, A.C., Snoeck, C., Pouncett, J., Baksh-Comeau, Y., Schulting, R., Claeys, P., Mattielli, N., Richards, M., and Boomert, A. 2017. Black pitch, carved histories: Radiocarbon dating, wood species identification and strontium isotope analysis of prehistoric wood carvings from Trinidad's Pitch Lake. *Journal of Archaeological Science: Reports*, 16, pp. 341–358.

Pagán Jiménez, J.R. 2011. Early phytocultural processes in the pre-colonial Antilles. In *Communities in contact: Essays in archaeology, ethnohistory, and ethnography of the Amerindian circum-Caribbean*, ed. C.L. Hofman and A. van Duijvenbode, pp. 87–116. Leiden: Sidestone Press.

Pagán-Jiménez, J.R., Rodríguez-Ramos, R., Reid, B.A., van den Bel, M., and Hofman, C.L. 2015. Early dispersals of maize and other food plants into the Southern Caribbean and Northeastern South America. *Quaternary Science Reviews*, 123, pp. 231–246.

Pestle, W.J. 2010. *Diet and society in Prehistoric Puerto Rico an Isotopic approach*. PhD dissertation, University of Illinois at Chicago.

Pestle, W.J. 2013. Stable isotope analysis of Paleodiet in the Caribbean. In *The Oxford handbook of caribbean archaeology*, ed. W. Keegan, C.L. Hofman, and R. Rodríguez Ramos, pp. 407–417. Oxford: Oxford University Press.

Pestle, W.J. and Colvard, M. 2012. Bone collagen preservation in the tropics: A case study from ancient Puerto Rico. *Journal of Archaeological Science*, 39(7), pp. 2079–2090.

Pestle, W.J., Curet, L.A., Rodríguez Ramos, R., and Rodríguez López, M. 2013. New questions and old paradigms: Reexamining Caribbean culture history. *Latin American Antiquity*, 24(3), pp. 243–261.

Pestle, W.J., Hubbe, M., Smith, E.K., and Stevenson, J.M. 2015. A linear model for predicting δ^{13}Cprotein. *American Journal of Physical Anthropology*, 157(4), pp. 694–703.

Pestle, W.J. and Laffoon, J. 2018. Quantitative paleodietary reconstruction with complex foodwebs: An isotopic case study from the Caribbean. *Journal of Archaeological Science: Reports*, 17, pp. 393–403.

Pestle, W.J. and Rodriguez Ramos, R. 2018. Isotopic challenges and categorical stumbling blocks in Caribbean archaeology. In *The Archaeology of Caribbean and Circum-Caribbean Farmers (6000 BC–AD 1500)*, ed. B. Reid, pp. 273–278. New York: Routledge.

Pestle, W.J., Simonetti, A., and Curet, L.A. 2013.87Sr/86Sr variability in Puerto Rico: Geological complexity and the study of paleomobility. *Journal of Archaeological Science*, 40(5), pp. 2561–2569.

Pino M. 1995. *Actualización de fechados radiocarbónicos de sitios arqueológicos de Cuba hasta diciembre de 1993*. Havana: Editorial Academia.

Pino, M. and Alonso, E. 1973. Excavaciones en la cueva del Perico I. *Serie Espeleológica y Carsológica*, 45, pp. 5–32.

Price, T.D., Burton, J.H., and Bentley, R.A. 2002. The characterization of biologically available strontium isotope ratios for the study of prehistoric migration. *Archaeometry*, 44(1), pp. 117–135.

Price, T.D., Burton, J.H., and Stoltman, J.B. 2007. Place of origin of prehistoric inhabitants of Aztalan, Jefferson Co., Wisconsin. *American Antiquity*, 72(3), pp. 524–538.

Price, T.D., Johnson, C.M., Ezzo, J.A., Burton, J.H., and Ericson, J.A. 1994. Residential mobility in the prehistoric southwest United States: A preliminary study using strontium isotope analysis. *Journal of Archaeological Science*, 21, pp. 315–315.

Rodríguez Ramos, R. 2010. *Rethinking Puerto Rican Precolonial history.* Tuscaloosa: University of Alabama Press.

Rodríguez Ramos, R.R., Babilonia, E., Curet, L.A., and Ulloa, J. 2008. The pre-Arawak pottery horizon in the Antilles: A new approximation. *Latin American Antiquity*, 19(1), pp. 47–63.

Roksandic, M., Alarie, K., Suárez, R.R., Huebner, E., and Roksandic, I. 2016. Not of African descent: Dental modification among indigenous Caribbean people from Canímar Abajo, Cuba. *PloS One*, 11(4)): e0153536. https://doi.org/10.1371/journal.pone.0153536.

Rouse, I. 1986. *Migrations in prehistory: Inferring population movement from cultural remains*. Yale, NH: Yale University Press.

Rouse, I. 1992. *The Tainos: Rise and decline of the people who greeted Columbus*. Yale, NH: Yale University Press.

Rouse, I. and Allaire, L. 1978. *Caribbean chronology. Chronologies in new world archaeology.* New York: Academic Press (Studies in Archaeology Series).

Sage, R.F. and Monson, R.K. (1998). *C4 plant biology*. San Diego: Academic Press.

Schoeninger, M.J. and DeNiro, M.J. 1984. Nitrogen and carbon isotopic composition of bone collagen from marine and terrestrial animals. *Geochimica et Cosmochimica Acta*, 48(4), pp. 625–639.

Schoeninger, M.J., DeNiro, M.J., and Tauber, H. 1983. Stable nitrogen isotope ratios of bone collagen reflect marine and terrestrial components of prehistoric human diet. *Science*, 220(4604), pp. 1381–1383.

Schroeder, H., Sikora, M., Gopalakrishnan, S., Cassidy, L.M., Delser, P.M., Velasco, M.S., Schraiber, J.G., Rasmussen, S., Homburger, J.R., Ávila-Arcos, M.C., and Allentoft, M.E. 2018. Origins and genetic legacies of the Caribbean Taino. *Proceedings of the National Academy of Sciences*, 115(10), pp. 2341–2346.

Schulting, R., Richards, M., Pouncett, J., Manco, B.N., Freid, E., and Ostapkowicz, J. 2018. Absence of Saharan dust influence on the strontium isotope ratios on modern trees from the Bahamas and Turks and Caicos Islands. *Quaternary Research*, 89(2), pp. 394–412.

Slovak, N.M. and Paytan, A. 2012. Applications of Sr isotopes in archaeology. In *Handbook of Environmental Isotope Geochemistry*, ed. Mark Baskaran, pp. 743–768. Berlin: Springer.

Smith, B.N. and Epstein, S. 1971. Two categories of $^{13}C/^{12}C$ ratios for higher plants. *Plant Physiology*, 47(3), pp. 380–384.

Smith, D.G., Chinique de Armas, Y., Suárez, R.R., Buhay, W.M., Roksandic, I., Peros, M.C., and Roksandic, M. 2018. Straddling the subsistence divide: The case of Canímar Abajo and contemporaneous sites in Northwestern Cuba. In *The archaeology of Caribbean and Circum-Caribbean farmers (6000 BC–AD 1500)*, ed. B. Reid, pp. 98–112. New York: Routledge.

Sparks, J.M. and Crowley, B.E. 2018. Where did people forage in prehistoric Trinidad? Testing the utility of a multi-isotope approach for tracking the origins of terrestrial prey. *Journal of Archaeological Science: Reports*, 19, pp. 968–978.

Stokes, A.V. 1998. *A biogeographic survey of prehistoric human diet in the West Indies using stable isotopes*. PhD dissertation, University of Florida.

Tabío, E. 1984. Nueva periodización para el estudio de las comunidades aborígenes de Cuba. *Islas*, 78, pp. 37–52.

Tabío, E. and Rey, E. 1979. *Prehistoria de Cuba.* La Habana: Editorial Ciencias Sociales.

Tieszen, L.L. and Fagre, T. 1993. Effect of diet quality and composition on the isotopic composition of respiratory CO^2, bone collagen, bioapatite, and soft tissues. In *Prehistoric human bone*, ed. J. Lambert and G. Grupe, pp. 121–155. Berlin, Heidelberg: Springer.

Travieso Ruíz, R., Rodríguez D., Rivero de la Calle, M., and Marques, S. 1999. Estudio de los restos óseos humanos aborígenes encontrados en la Cueva del Perico I, Pinar del Río, Cuba. *Revista Biología*, 13, pp. 88–94.

Twiss, K.C. ed. 2007. *The Archaeology of food and identity.* Carbondale: Southern Illinois University.

Walser III, J.W., Kristjánsdóttir, S., Gröcke, D.R., Gowland, R.L., Jakob, T., Nowell, G.M., Ottley, C.J., and Montgomery, J. 2020. At the world's edge: Reconstructing diet and geographic origins in medieval Iceland using isotope and trace element analyses. *American Journal of Physical Anthropology*, 171(1), pp. 142–163.

Whipkey, C.E., Capo, R.C., Chadwick, O.A., and Stewart, B.W. 2000. The importance of sea spray to the cation budget of a coastal Hawaiian soil: A strontium isotope approach. *Chemical Geology*, 168(1–2), pp. 37–48.

Wright, L.E. 2005. Identifying immigrants to Tikal, Guatemala: Defining local variability in strontium isotope ratios of human tooth enamel. *Journal of Archaeological Science*, 32(4), pp. 555–566.

12

EATING LOCAL

Modeling Paleodiet in Later Ceramic Age Puerto Rico

William J. Pestle, Harper Dine and Ali Pourmand

Introduction

Over the past three decades, several attempts have been made to characterize the diet of Later Ceramic Age Puerto Rican populations by means of stable isotope analysis (Pestle 2010a, b; Stokes 1998, 2005). These works have succeeded in providing otherwise unobtainable insights into individual-level diet in the island's prehistory. Ultimately, however, they have been limited in their ability to quantitatively reconstruct dietary composition with any specificity due to the complexity of local foodwebs, which include both a large number of isotopically variable food groups and significant overlap between/among groups (Pestle 2013). Issues of foodweb complexity previously have been compounded by the dearth of appropriate model-based tools for parsing such complexity, as many previous isotope models were incapable of handling undetermined systems (where the number of possible food sources outnumbers the number of isotopic proxies). Fortunately, in the past decade, a number of mixture models for isotopic analysis of paleodiet (e.g., Fernandes et al. 2014) have been developed, which, as a recent pilot study focusing on ancient individuals from St. Thomas, USVI, has shown, can be used probabilistically to determine specific dietary composition even in the face of the extremely complex foodwebs of the insular Caribbean (Pestle and Laffoon 2018).

In this chapter, we present a model-based reconstruction of the paleodiet of over 200 individuals from five Later Ceramic Age Puerto Rican sites representing the period ca. 300–1400 cal. AD. Using these modeled results, we attempt to identify the principal drivers of dietary variability on an insular scale. The results of Bayesian models of dietary composition are here compared with a variety of data, including site locations/attributes, the results of a rigorous program of radiocarbon dating, and the limited results of strontium isotope analysis. Ultimately, on the

DOI: 10.4324/9780429444449-16

basis of these comparisons, it is argued that while chronology and biogeography can explain a large portion of the variability observed, there are residual differences that we attribute to a cultural (dietary) localism (eating local) that manifested in significant dietary differences among and between individuals from different sites. In closing, we examine these dietary data in light of the notion of "community", as presented in the work of Torres (2012), as a useful scale/unit of analysis for studying and understanding cultural diversity in the Later Ceramic Age of Puerto Rico.

The Sample

The individuals considered here are derived from five Puerto Rican Ceramic Age sites: Los Indios, Maisabel, Paso del Indio, Punta Candelero, and Tibes (Table 12.1). While the Maisabel stable isotope data were published by Stokes (1998) and the isotopic data on the individuals from Paso del Indio, Punta Candelero, and Tibes were first presented in Pestle (2010b), full data for the Los Indios individuals are presented here for the first time (Pestle et al. 2024). Archaeological, mortuary, and osteological data for the Maisabel individuals are available in Seigel (1992) and Budinoff (1991), for Paso del Indio and Punta Candelero in Crespo Torres (1991, 2000), for Tibes in Crespo Torres (1998, 2010), and for Los Indios in Crespo-Torres and Muñoz Guevara (2008). The total *n* of isotopically analyzed individuals stands at 229, representing the largest corpus of prehistoric human isotope data for any Caribbean island.

As seen in Figure 12.1, the five sites considered here are located in distinct parts of Puerto Rico, which, like all the Greater Antilles, presents highly differentiated local ecological conditions that may have served to promote or inhibit different dietary behaviors (Newsom and Wing 2004). In broad terms, the north coast sites (Maisabel and Paso del Indio) are characterized by higher rainfall and access to higher energy, rockier, colder, and deeper-water coastal environments, whereas the southern and eastern sites (Los Indios, Punta Candelero, and Tibes) are typified by generally drier conditions and by their proximity to marine environments that are lower energy, with sandy/seagrass bottoms, and which are characterized by warmer waters and more

TABLE 12.1 Detail of sample analyzed in the course of the present study

Site	*Municipio*	*Distance to Coast (km)*	*Stable Isotope n*	*Dated n*	*$Sr^{86/87}$ n*	*Isotopic Data Source*
Los Indios	Santa Isabel	1.0	30	10	-	This study
Maisabel	Vega Baja	0.2	18	18	16	Stokes 1998
Paso del Indio	Vega Baja	6.0	85	85	13	Pestle 2010
Punta Candelero	Humacao	0.1	50	50	7	Pestle 2010
Tibes	Ponce	8.0	46	46	6	Pestle 2010

FIGURE 12.1 Map of Puerto Rico showing location of sites discussed in text.

gradually sloping coastal shelves. One might thus predict *a priori* that diets in these two regions might be distinct by dint of these biogeographic differences.

While there is significant temporal overlap among them, individuals from these five sites nonetheless lived during distinct portions of the Puerto Rican Later Ceramic Age. Figure 12.2 depicts the calibrated 2-sigma ranges for direct radiocarbon dates obtained from some 209 of the individuals for whom we possess isotopic data (Parnell 2014; Reimer et al. 2013). The utility of these direct chronological determinations is difficult to overstate, as it permits us to move beyond discussions framed by cultural historical constructs and instead build interpretations referential to absolute chronology (Pestle et al. 2013). On the basis of the broad temporal sweep these individuals represent, one might posit that dietary composition may have varied diachronically across the Later Ceramic Age and that, for instance, contemporary individuals from different sites might have diets more similar to one another than to individuals from different periods at the same site.

In the following pages, we test both the biogeographical and temporal dimensions of dietary differences in Later Ceramic Age Puerto Rico before proposing a framework by which to consider how this aspect of Indigenous culture may have been structured and maintained.

Methods

Stable Isotope Analysis

We refer readers to detailed discussions of biomolecular extraction and isotopic analysis methods presented in Stokes (1998) and Pestle (2010b). For the present purposes, however, we note that the collagen extraction methods used in both studies were functionally equivalent: ground bone samples were demineralized in HCl, rinsed with NaOH to remove humic, gelatinized in dilute HCl, filtered, condensed, frozen, and freeze-dried. The similarity of the procedures employed should lend itself to comparable results for the various sites under consideration.

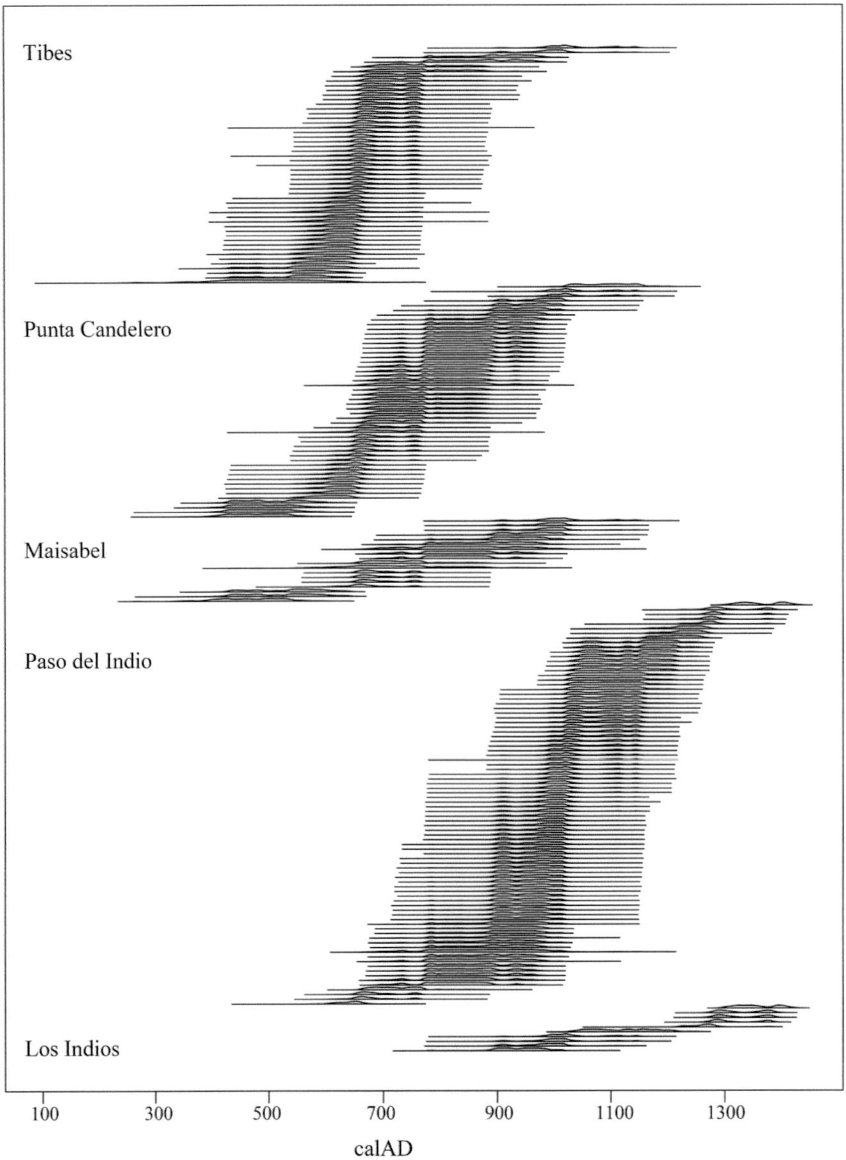

FIGURE 12.2 Calibrated direct radiocarbon dates of individuals analyzed in present study, by site.

Hydroxyapatite isolation also followed similar procedures for organic removal, in that oxidation of organics was performed using NaOCl, but whereas the Maisabel samples were treated for adsorbed carbonates using 1.0 M acetic acid for several days (Stokes 1998, 149), samples from the other four sites were treated with 0.1 M acetic for only four hours (Pestle 2010b, 177–178). This treatment difference can introduce variability in the resulting carbon and oxygen stable isotope values (Garvie-Lok et al. 2004; Koch et al. 1997; Pellegrini and Snoeck 2016; Snoeck and Pellegrini 2015; Yoder and Bartelink 2010). Similarly, as three different laboratories were involved in the analysis of the samples from these five sites, there is further potential for systematic inter-laboratory differences in resulting isotope value stemming from differences in instrumentation and/or calibration (Pestle et al. 2014; Pestle and Szpak, in Press; Szpak et al. 2017). Lamentably, there is little that can be done to mitigate potential differences post hoc, and the reader is instead advised here of their potential existence.

Sample preservation quality was determined using both chemical (collagen yield) and elemental (carbon and nitrogen yield, atomic C/N ratio) data. Only well-preserved samples/individuals (collagen yield >0.5 wt%, carbon yield >4.5 wt%, nitrogen yield >0.9 wt%, atomic C/N ratio between 2.9 and 3.6) were included in subsequent FRUITS modeling. Ultimately, this resulted in a sample of some 229 individuals from the five sites.

Modeling

Individual isotopic data were analyzed using the Bayesian multi-source mixture model known as FRUITS v2.1.1 (Food Reconstruction Using Isotopic Transferred Signals) (Fernandes et al. 2014), which allows for probabilistic and uncertainty-integrated quantification of dietary inputs. FRUITS was chosen because of its capability for incorporating food macronutrient, elemental, and isotopic composition data, as well as source and consumer uncertainty, in its calculations. Prior to modeling, human isotope data were first converted to account for fractionation, with the offset for $\delta^{13}C_{ap}$ stipulated as 10.1 ± 0.4‰, following Fernandes and colleagues (2012). For $\delta^{15}N_{co}$, we employed a trophic fractionation value of 3.6 ± 1.2‰, as recommended by experimental studies of omnivorous animals (Ambrose 2000; DeNiro and Epstein 1981; Hare et al. 1991; Howland et al. 2003; Sponheimer et al. 2003; Warinner and Tuross 2009). Finally, the consumer-foodstuff offset (and error) for $\delta^{13}C_{co}$ was calculated using the linear regression method described by Pestle and colleagues (2015). Instrumental uncertainties of 0.1‰ were additionally stipulated for all isotope systems. Elemental routing was specified as follows: all nitrogen in bone collagen was stipulated as coming from dietary protein, the only macronutrient containing nitrogen, carbon in hydroxyapatite was stipulated to be a weighted average of all dietary carbon, and the carbon in bone collagen was stipulated as reflecting roughly a 3:1 ratio (74 ± 4%:26%) of dietary protein to energy (Fernandes et al. 2012).

Isotopic data on potential faunal and floral foodstuffs ($n = 611$) were garnered from a variety of previous regional studies (Burton et al. 2001; Cardona et al. 2009; Coltrain et al. 2004; Fry et al. 1982; Godley et al. 1998; Keegan and DeNiro 1988; March and Pringle 2003; Moncreiff and Sullivan 2001; Norr 2002; Pestle 2010b; Reich and Worthy 2006; Schimmelmann 1985; Schoeninger and DeNiro 1984; Schwarcz et al. 1985; Stokes 1998; Stoner and Waite 1991; van Klinken

1991; von Fischer and Tieszen 1995). All modern data had their $\delta^{13}C$ values corrected by +1.5‰ to account for the Suess/fossil fuel burning effect (Keeling et al. 1979; Marino and McElroy 1991). Additional tissue corrections (between measured and edible tissues) are fully detailed in Pestle (2010b, Table 38).

The FRUITS model requires the a priori stipulation of food groups. For flora, this division was made on the well-established basis of isotopic differences in photosynthetic pathways, resulting in two groups, C_3 and C_4/CAM plants. For fauna, this determination was more complex and included consideration of chemical composition (both isotopic and macronutrient), ecological niche, and trophic level. Due to isotopic and macronutrient similarities, it was determined that freshwater and terrestrial fauna would be combined for use in subsequent modeling. Conversely, given underlying isotopic differences among the marine taxa analyzed, two marine groupings were stipulated, one of which (Marine 1) consisted primarily of higher trophic level and/or pelagic taxa (with higher $\delta^{15}N$ and lower $\delta^{13}C$ values), and the other (Marine 2) composed of reef/seagrass taxa, with lower $\delta^{15}N$ and higher $\delta^{13}C$ signatures. The macronutrient composition for each food group was determined by reference to the USDA National Nutrient Database for Standard Reference (United States Department of Agriculture 2013). Elemental composition (particularly %C) of each foodstuff/macronutrient group was based on formulae provided by Morrison et al. (2000). Digestibility was determined following Hopkins (1981). Carbon isotope offsets between measured bulk food isotope values and the isotopic values of a foodstuff's fats (bulk-6‰) and carbohydrates (bulk+0.5‰) were based on data from Tieszen (1981). The carbon isotope signature of a measured foodstuff's protein was determined using a mass-balance equation, such that a proportional/weighted average of the $\delta^{13}C$ of protein and energy (fats and carbohydrates) would equal the measured $\delta^{13}C$ bulk value (corrected for the concentration of carbon in each macronutrient and foodstuff-appropriate macronutrient concentration).

The final food group isotope and macronutrient composition as used in the FRUITS simulations are presented in Table 12.2. All FRUITS simulations were performed using 10,000 iterations, as recommended by its developers, and any averages (for site, region, etc.) were based on individual-level modeling results.

AMS Dating

Direct radiocarbon determinations from bone collagen were obtained for 209 of the 229 individuals (91.3%) considered here. These dates have previously been published in Siegel (1992) and Pestle (2008; 2010b). AMS analyses were performed at the University of Arizona's NSF-AMS facility, following their standard pretreatment and analytical protocols for the dating of archaeological bone (acid-base-acid pretreatment, gelatinization, filtration, and graphitization). All Bayesian calibrations used here were performed using BChron (Parnell 2014) and the IntCal13 calibration curve (Reimer et al. 2013). No marine reservoir corrections were applied.

TABLE 12.2 Foodweb macronutrient and isotope data used in FRUITS modeling

		Macronutrient Concentration (%)		Tissue $\delta^{13}C$ (‰)			Tissue $\delta^{15}N$ (‰)
Food Grouping	*Group n*	*Protein*	*Energy*	*Bulk*	*Protein*	*Energy*	*Protein*
C_3 plants	67	14 ± 12	86 ± 14	−25.5 ± 1.7	−26.5 ± 1.7	−25.3 ± 1.7	3.3 ± 2.1
C_4/CAM plants	20	14 ± 12	86 ± 14	−10.4 ± 1.3	−11.4 ± 1.3	−10.2 ± 1.3	5.1 ± 2.9
Freshwater/terrestrial fauna	115	73 ± 14	27 ± 15	−19.8 ± 2.9	−17.6 ± 2.9	−25.8 ± 2.9	6.2 ± 2.7
Marine 1	131	76 ± 15	24 ± 17	−13.9 ± 2.1	−12.7 ± 2.1	−17.8 ± 2.1	13.0 ± 2.7
Marine 2	278	76 ± 15	24 ± 17	−10.7 ± 3.3	−9.5 ± 3.3	−14.6 ± 3.3	5.9 ± 3.0

Sr Isotope Analysis

$^{87/86}$Sr determinations of dental enamel are available for 42 individuals for whom isotopic data are presented here. This includes 16 individuals from Maisabel, for whom strontium isotope data are presented in Laffoon (2012), to which we refer readers for details on sampling and analytical methods. Data for a further 13 individuals from Paso del Indio, seven from Punta Candelero, and six from Tibes are presented here for the first time and were generated using the following procedures.

Enamel samples were collected using a handheld drill from the buccal surface of molars, and an extraction chromatography protocol detailed in Pourmand and colleagues (2010) was used to separate Sr from other matrix elements. The radiogenic ^{87}Sr/^{86}Sr isotope ratios of the samples were measured on a Neptune Plus (Thermo Fisher Scientific, Germany) multi-collector ICP mass spectrometer at the University of Miami (Pourmand et al. 2012). Sample measurements were bracketed with the analysis of two SRM987 standard solutions of matching Sr concentration and matrix to the samples. The measured ^{87}Sr/^{86}Sr ratios in the samples were corrected for mass bias and isobaric interferences, and the final ratio was further adjusted relative to the accepted value of 0.710248 for SRM 987 to allow comparison with literature measurements of radiogenic Sr isotopes. More details on the analytical methodology can be found in Pourmand et al. (2014). The resulting human ^{87}Sr/^{86}Sr values were then compared with local Puerto Rican bedrock and bioavailable ^{87}Sr/^{86}Sr values (Laffoon 2012; Laffoon et al. 2012; Pestle et al. 2013) to ascertain the degree of similarity of the measured enamel values with local strontium signatures and to thereby infer information as to patterns of migration/mobility.

Statistical Analysis

Pairwise comparisons of simulated foodstuff contributions were made using Student's T-tests, with Levene's correction for those comparisons for which subsample variances were unequal. Comparisons among more than two samples were made using one-way ANOVA with Tukey's HSD post-hoc comparison of subsample pairs. Correlations between model results, as well as between model results and other data (e.g., radiocarbon dates), were assessed using Pearson's *r*. An alpha of 0.01 was used for all statistical tests.

Results

At the insular scale (considering all 229 individuals together), the single largest modeled dietary contributor was C_3 plants, which accounted for an average of 43.6 ± 17.7% (range 16.0–63.0%) of a given individual's dietary intake (calories) (Table 12.3). C_4 and CAM plants, by comparison, were far less important to the island diet, forming an average of only 11.3 ± 8.7% (range 6.2–16.3%) of the

TABLE 12.3 Island-wide, regional, and site level results of 229 Ceramic Age individuals using FRUITS modeling (values represent mean ± standard deviation by grouping)

Site	*Parameter*	*C_3 Plants (%)*	*C_4/CAM Plants (%)*	*Freshwater/Terrestrial Fauna (%)*	*High Trophic Marine (%)*	*Reef/Seagrass Marine (%)*
All individuals	Average	43.6 ± 17.7	11.3 ± 8.7	27.8 ± 21.3	9.1 ± 7.6	8.2 ± 7.2
Northern sites	Average	48.1 ± 18.1	10.2 ± 7.9	26.1 ± 20.4	8.4 ± 7.0	7.3 ± 6.5
Southern sites	Average	40.7 ± 17.4	12.1 ± 9.2	28.7 ± 20.5	9.5 ± 7.9	8.9 ± 7.8
Eastern site	Average	38.7 ± 17.4	12.4 ± 9.4	30.0 ± 24.6	10.0 ± 8.2	9.0 ± 8.0
Los Indios	Minimum	23.1 ± 15.6	9.2 ± 7.2	19.0 ± 16.5	7.2 ± 6.0	6.8 ± 5.8
	Average	40.2 ± 17.0	12.2 ± 9.2	28.8 ± 20.3	9.9 ± 8.2	9.0 ± 7.8
	Maximum	57.7 ± 15.4	14.4 ± 10.6	44.0 ± 22.9	13.9 ± 10.3	11.0 ± 9.1
Maisabel	Minimum	38.6 ± 16.0	8.1 ± 6.4	18.6 ± 17.0	6.8 ± 6.2	6.5 ± 5.8
	Average	50.0 ± 17.0	10.3 ± 7.9	23.0 ± 18.8	8.9 ± 7.2	7.9 ± 6.8
	Maximum	58.2 ± 15.7	13.4 ± 10.0	26.2 ± 20.4	13.4 ± 10.1	11.6 ± 9.5
Paso del Indio	Minimum	27.1 ± 19.1	6.2 ± 5.2	17.6 ± 15.6	5.6 ± 5.0	5.2 ± 4.7
	Average	47.7 ± 18.3	10.2 ± 8.0	26.7 ± 20.7	7.1 ± 6.4	8.3 ± 7.0
	Maximum	63.0 ± 14.9	12.8 ± 9.5	45.0 ± 24.4	11.4 ± 8.5	9.2 ± 7.9
Punta Candelero	Minimum	16.0 ± 13.0	9.4 ± 7.5	21.8 ± 18.2	6.5 ± 5.9	7.3 ± 7.0
	Average	38.7 ± 17.4	12.4 ± 9.4	30.0 ± 24.6	10.0 ± 8.2	9.0 ± 8.0
	Maximum	51.8 ± 16.5	16.3 ± 12.1	50.7 ± 23.0	15.5 ± 10.9	12.3 ± 10.4
Tibes	Minimum	29.0 ± 18.5	9.2 ± 7.2	21.9 ± 17.5	6.8 ± 6.7	7.0 ± 6.1
	Average	41.1 ± 17.5	12.2 ± 9.2	28.6 ± 20.7	9.2 ± 7.8	8.8 ± 7.8
	Maximum	53.6 ± 16.7	15.3 ± 10.1	41.8 ± 24.1	11.5 ± 9.4	10.3 ± 9.0

individual diet. The C_3 to C_4/CAM ratio of an average Later Ceramic Age Puerto Rican was thus nearly 4:1. The second largest food source, and thus the primary source of dietary protein at an insular scale, were terrestrial and freshwater animals, which provided an average of 27.8 ± 21.3% (range 17.6–50.7%) of dietary calories. In comparison, marine foods made up a far smaller portion of the average individual's diet. Higher trophic level marine foods accounted for 9.1 ± 7.6% (range 5.6–15.5%), whereas reef and seagrass species comprised just 8.2 ± 7.2% (range 5.2–12.3%) of the average diet. An overall marine-terrestrial protein ratio for the island is thus a surprising average of 0.6:1. The differences among these various sources are significant overall ($F(5,1368)=2281.7$, $p<0.01$), and all pairwise comparisons, with the exception of the two marine sources, show significant difference from one another.

When these data are disaggregated by site, an apparent spatial structure emerges (Table 12.3). While most of the general insular-level trends identified above persist, there are significant differences among the five sites for the mean modeled contribution of all five food categories ($F(4,224)=7.1–53.2$, $p<0.01$). At all sites, C_3 plants are the primary food source, but their contribution ranged from a low of 38.7 ± 17.4% at Punta Candelero to a high of 50.0 ± 17.0% at Maisabel. In contrast, C_4/CAM plants varied far less (but still significantly) from a low of 10.2 ± 8.0% at Paso del Indio to 12.4 ± 9.4% at Punta Candelero. Turning to protein sources, freshwater and terrestrial animal consumption varied substantially among the sites under consideration here, contributing a low of 23.0 ± 18.8% of dietary calories at Maisabel and a high of 30.0 ± 24.6% at Punta Candelero. Higher trophic level marine fauna made up as little as 8.3 ± 7.0% of the diet at Paso del Indio and as much as 9.9 ± 8.2% at Punta Candelero, while reef and seagrass taxa followed a similar pattern, forming just 7.1 ± 6.4% of the average diet at Paso del Indio versus 9.0 ± 8.0% at Punta Candelero.

Pairwise post-hoc comparisons of site pairs make more evident the spatial logic of the inter-site dietary differences. Certain sites show remarkable similarity: three site pairs (Los Indios-Punta Candelero, Los Indios-Tibes, and Punta Candelero-Tibes) are statistically indistinct from one another for all five modeled food groups, while Maisabel and Paso del Indio are significantly different for only one of the five categories (reef/seagrass marine fauna). Conversely, five site pairs (Los Indios-Maisabel, Los Indios-Paso del Indio, Maisabel-Punta Candelero, Maisabel-Tibes, and Paso del Indio-Tibes) are significantly different for four of five modeled food groupings, and Paso del Indio and Punta Candelero are statistically distinct for all five modeled food categories. Dimension reduction of these data using Principal Component Analysis yields two components that, combined, explain 89.7% of the observed variability. Visualization of these components (Figure 12.3) makes obvious the overall similarity of, in particular, Los Indios, Punta Candelero, and Tibes, on the one hand, and Maisabel and Paso del Indio on the other. The differences between these two groups of sites also are evident in that figure.

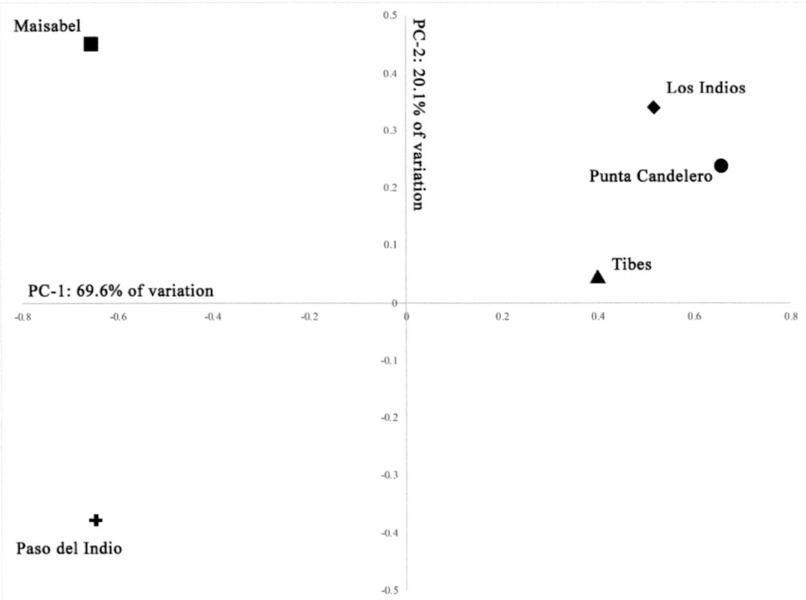

FIGURE 12.3 Results of Principal Component Analysis of five modeled foodweb categories, by site.

Biogeography would thus appear to play a significant role in shaping diet (either directly or indirectly). Indeed, several biogeographically derived statistical comparisons show the degree to which this is the case. First, and as would be expected based on the pairwise comparisons and PCA results presented above, ANOVA analysis of the sites grouped by region/coast (North, South, and East) found significant differences among the three regionally based site groupings for all five modeled food categories ($F(2,226) = 10.4–97.8$, $p < 0.01$). When examining the pairwise comparisons, however, it is clear that only a portion of the observed regional differences are, in fact, significant. Specifically, while the differences between the northern (Maisabel and Paso del Indio) and southern (Los Indios and Tibes) sites, on the one hand, and the northern and eastern (Punta Candelero) sites, on the other, are significant for all five modeled food groupings, there are no significant differences between the eastern and southern site groups.

Second, distance to the coast would also appear to be related to site-level dietary practices. Indeed, when grouped into coastal (Los Indios, Maisabel, and Punta Candelero) and non-coastal (Paso del Indio and Tibes) sites, individuals from the coastal sites would appear to have consumed significantly fewer C_3 plants ($t = -3.9$, $df = 227$, $p < 0.01$), and significantly greater amounts of C4/CAM plants ($t = 4.6$, $df = 227$, $p < 0.01$), high tropic level ($t = 5.0$, $df = 227$, $p < 0.01$), and low trophic level ($t = 7.1$, $df = 227$, $p < 0.01$) marine animals than individuals from non-coastal locations. There were no significant differences in the consumption of freshwater/terrestrial-derived protein ($t = 1.3$, df = 227, $p = 0.2$).

Pearson correlations analysis found a significant positive correlation between distance-to-coast (measured linearly and indicated per site in Table 12.1) and C_3 plant consumption ($r = 0.18$, $n = 229$, $p < 0.01$), and significant negative correlations between distance-to-coast and the contributions of C4/CAM plants ($r = -0.19$, $n = 229$, $p < 0.01$), high trophic marine fauna ($r = -0.27$, $n = 229$, $p < 0.01$), and reef/seagrass taxa ($r = -0.3$, $n = 229$, $p < 0.01$). No significant correlation was observed for freshwater/terrestrial fauna and distance to the coast. Even in those cases where significant correlations were observed, however, the r-squared values involved were small (less than 0.1 in all cases), such that a large portion of the variance in patterns of consumption remains unexplained.

Time is the other obvious vector by which diet might be thought to have differed. Indeed, Pearson correlation analysis of modeled dietary inputs and median probability calibrated radiocarbon dates found significant temporal correlations for all food groupings except freshwater and terrestrial fauna ($r = -0.04$, $n = 209$, $p < 0.01$). C_3 plant consumption and date were significantly positively correlated (r=0.26, n=209, p<0.01), whereas calibrated radiocarbon dates and C_4/CAM (−0.34, $n = 209$, $p < 0.01$), high trophic level marine ($r = -0.44$, $n = 209$, $p < 0.01$), and reef/seagrass taxa ($r = -0.49$, $n = 209$, $p < 0.01$) were all significantly negatively correlated. The r-squared values for the significant temporal correlations are somewhat higher than for distance-to-coast (above), ranging from 0.07 for C_3 plants to 0.24 for reef/seagrass marine animals, but still leave substantial residual variation that can, in turn, be reexamined in biogeographical terms. Furthermore, the chronological shifts towards lower marine foodstuff consumption remain, even when controlled for using partial correlation analysis, which accounted for the diachronic shift in isotopically similar C_4/CAM consumption ($r = -0.38$, $n = 206$, $p < 0.01$ for high trophic level marine fauna, $r = -0.41$, $n = 206$, $p < 0.01$ for reef/seagrass).

Indeed, the broad biogeographic structure discussed above would thus seem to largely hold, even when differences in chronology are fully accounted for. ANOVA analysis of chronologically regressed residuals for the sites grouped by region/coast confirms this, finding significant variation for all five modeled food categories ($F(2,206) = 5.1$–35.1, $p < 0.01$). Pairwise comparisons found the south (Los Indios, Tibes) and east (Punta Candelero) sites to be statistically similar for all five food groups, whereas the northern sites (Maisabel and Paso del Indio) differed from the southern sites for four of five food categories and from Punta Candelero for all five modeled food groupings.

ANOVA of the date-controlled residuals also still finds significant variation among the sites for all five modeled food categories ($F(4,204) = 6.4$–29.3, $p < 0.01$). Pairwise comparisons of site pairs roughly parallel those presented above, in that Maisabel-Paso del Indio and Punta Candelero-Tibes were still statistically indistinguishable for four of the five food categories, and Los Indios-Punta

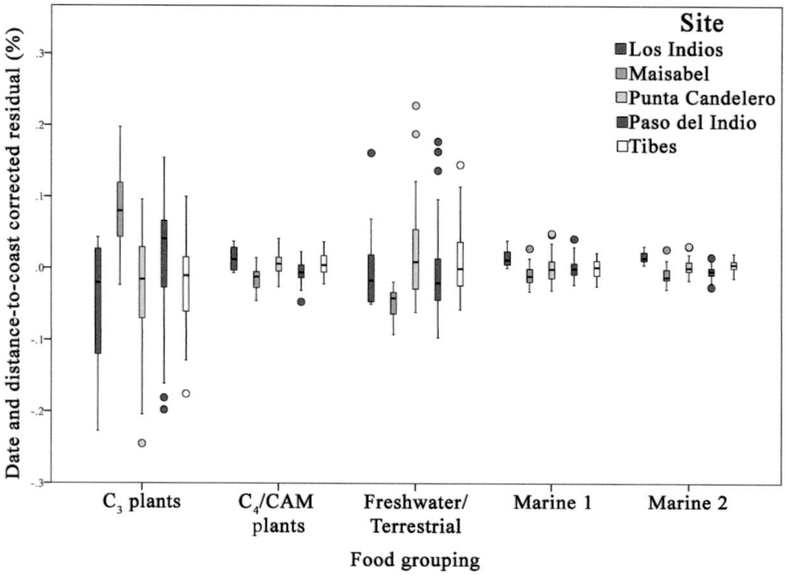

FIGURE 12.4 Date and distance-to-coast corrected residuals of five modeled foodweb categories, by site.

Candelero and Los Indios-Tibes were statistically similar for three of the five food categories. Intriguingly, once controlling for time, Paso del Indio and Tibes were also (similarly) statistically indistinct for three of five modeled food groupings. At the other extreme, four site pairs (Los Indios-Maisabel, Los Indios-Paso del Indio, Maisabel-Punta Candelero, and Maisabel-Tibes) were only similar for one of the five food sources, and Paso del Indio and Punta Candelero were significantly different for all five food categories.

The only means of examining what variability might remain once time and biogeography are both controlled for is to examine the residuals left after regressing modeled dietary contributions to both date and distance-to-coast (Figure 12.4). Even after these other drivers of variation have been controlled for, there are still significant differences among the five sites for all five modeled food groupings (ANOVA, $F(4,204) = 4.3–23.7$, $p < 0.01$). At this level of granularity, it is only sensible to make pairwise comparisons of sites within rather than between regions.

Of the two northern sites (Maisabel and Paso del Indio), there are still significant differences for two of the modeled food groups (C_3 plants and terrestrial/freshwater animals). Of the three southern/eastern sites, Punta Candelero and Tibes are statistically indistinguishable for all five sources, whereas the Los Indios individuals differ from their counterparts at Punta Candelero and Tibes for two food groups (both marine faunal groups). Thus, even after controlling for time and space (in the guise of distance to coast), some small degree of site-level dietary difference persists as

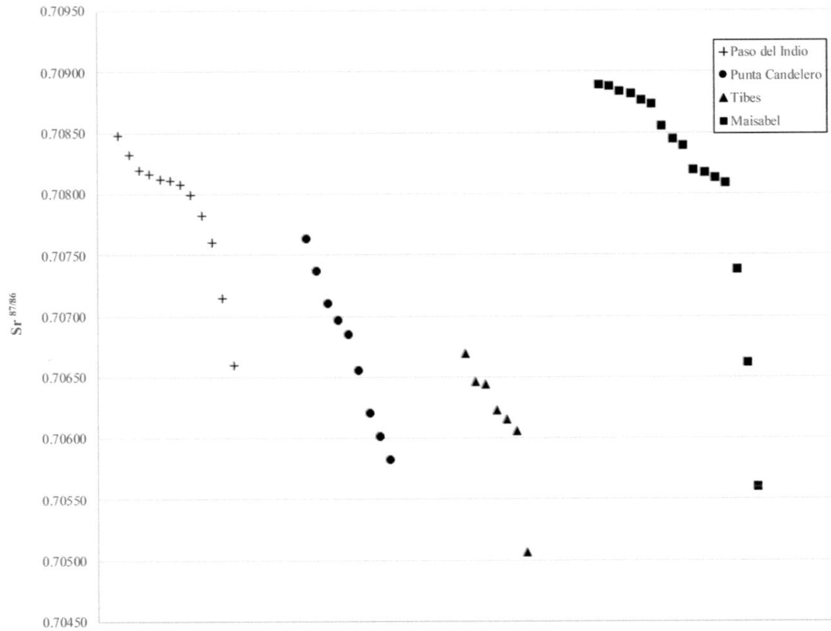

FIGURE 12.5 $Sr^{86/87}$ results for 42 individuals included in the present study, by site. Maisabel data from Laffoon (2012).

a testament to the tendency of individuals buried at these sites to, in some small regard, eat local.

To contextualize and understand these remaining locavore tendencies, we can look to the (admittedly limited) results of strontium isotope analysis of a subset of the individuals for whom we possess modeled dietary data. As seen in Figure 12.5, $Sr^{86/87}$ values for measured individuals from Maisabel ranged from 0.70559 to 0.708879 (Laffoon 2012), at Paso del Indio from 0.70659 to 0.70849, at Punta Candelero from 0.70581–0.70762, and at Tibes from 0.70506 to 0.70670. While the ranges of values observed at all four sites are substantial, it should be noted that in none of these cases do any of the human values fall outside of the range of bioavailable and/or geological $Sr^{86/87}$ signatures recorded in the environs immediately surrounding each site (Laffoon 2012; Pestle et al. 2013). Indeed, in the cases of Punta Candelero and Tibes, one need to go no farther than 5 km from the site to find bioavailable or bedrock strontium values that fully encompass the range of human values observed at each site, while for Paso del Indio this equivalent range is 15 km, and at Maisabel, the required range is just under 20 km. While equivalent values can also be found at other locations farther afield on the island, a testament to Puerto Rico's geological complexity, there is nothing in these strontium isotope results that necessitates that individuals buried at each site need to have been born and raised any farther than 5–20 km away.

Discussion

Bayesian modeling of paleodiet for a large sample of isotopically analyzed individuals from the Later Ceramic Age of Puerto Rico reveals general cultural inertia that shaped diet on an insular scale as well as a host of trends and tendencies that are attributable to chronology and/or biogeography. C_3 plants dominated average dietary carbohydrate intake and only increased in importance over time and with increasing distance from the coast. In comparison, C_4/CAM plants formed a far smaller portion of the diet (averaging roughly one-quarter of the caloric input of C_3 plants) and decreased in importance over time. Surprisingly, on an insular scale, freshwater and terrestrial protein sources contributed nearly two times as many calories as marine animals, although marine protein sources were significantly more important at coastal than interior sites. The importance of marine animals to diet did, however, decrease over time. Multiple regional comparisons of diet found, perhaps unsurprisingly, that region/coast was a powerful predictor of variation, leading us to conclude that biogeographical factors (rainfall and/or the characteristics and composition of nearby marine ecosystems) powerfully shaped dietary options and informed consumer choices, confirming earlier island-scale conclusions reached by Stokes (1998). Chronology is also a powerful explanation for some of the observed dietary differences, revealing, for instance, a steadily increasing reliance on C_3 plants as a mainstay of the Puerto Rican Ceramic Age diet.

There remains, however, residual variation in modeled contributions that we attribute to behavioral (in this case dietary) differences that manifested on a sub-regional, or "community", scale. Josh Torres, speaking of archaeological manifestations of community in prehistoric Puerto Rico, defines community as,

> a group of people who live in proximity to one another within a geographically limited area, who have face-to-face interaction on a regular basis and who share access to social and natural resources…Spatially, the proximity of places of day-to-day dwelling influences the degree to which groups share forms of meaning and behavior as individuals and as members of a group. These frames of reference and social propinquity inherent in consistent face-to-face interaction can produce inclusive communal relationships above individuals and households.
>
> *(Torres 2012, 415)*

While we do not propose that any of the sites/communities in question were isolated from those around them (indeed, at most we would say that they were insulated from one another), we do suggest that residual dietary idiosyncrasies that we observe after controlling for both time and biogeography can be seen as a behavioral consequence of communities that functioned at scale smaller than region or shared coastal orientation. While our data are spatially discontinuous, and in this approximation reflect patterns of diet at discrete individual sites (thus rendering dietary variation in spaces between the tested sites invisible), these dietary data, in combination with the results of strontium isotope analysis, can be taken to speak to shared community-based dietary proclivities that result from the social linkage of people who hail from, and participate in, social networks that coalesced in different settings and at different scales. In other words, one can see in these sub-regional variations a cultural centripetal force functioning to produce and perpetuate small-scale cultural variations between and among sites at a regional scale.

Torres argues that (one of) the fabrics of these communities are the person-to-person interactions facilitated by physical and social proximity, the effective radius of which he sets, based on a variety of other works, at somewhere between 2.5 km (the radius around a site that agriculturalists tend to cultivate) and 25 km (one day's walking distance) (Torres 2012, 135–138). The coincidence of these values and the upper limits of geographic origin (5–20 km) suggested by the strontium isotope results presented here is noteworthy. Based on this coincidence, our proposed rationale by which to explain these differences is that people born and raised in the environs of each site in this study ate like their kin and partners in the community that knitted them together, and when they died they were buried at Tibes, Paso del Indio, or any of the other sites discussed herein. Put most simply, the analyzed individuals ate local (at least in part) because they were, ultimately, locals, and their habits were the product of cultural factors shared in common among members of their community, factors that might not be consonant with others in different communities.

This explanation of the observed variation in modeled diet is consistent with the broader narrative of cultural differentiation seen in post AD 600 Puerto Rico. This Late/Later Ceramic Age on the island saw a diversification of ceramic styles (Curet et al. 2004; Petersen et al. 2004), the emergence of varied social, political, and cultural forms (Curet et al. 2004, 65), as well as shifts in mortuary and other ritual practices (Curet and Oliver 1998; Curet and Torres 2010; Keegan 2009; Oliver 1998). Combined, these changes led Torres to suggest that, during this period,

> while social networks may have been expanding regionally, they were contracting locally… This is not to say that these groups did not have interaction with one another, or perhaps, more distant social groups outside the region, but that these social conditions would have fostered the consolidation of proximally related social collectives, an increased focus on local affairs, and establishing identities in relation to 'others',.
>
> *(Torres 2012, 413–414)*

The modeled dietary data presented here fully support and amplify this conclusion.

In the present case, as in most, if not all, other places and times, human food choices would seem to reflect a variety of influences and decisions. In these modeled data, we get a general picture of what the typical island diet of the period was, but we also can detect in them both regional and temporal variations on that island-wide theme. At the smallest scale, we also find differences that speak to local flavors, to local culinary traditions and ideals, that, in turn, seemed to have been circulated and maintained by the differentiated societies and cultures of Later Ceramic Age Puerto Rico.

Acknowledgements

The authors wish to acknowledge the curatorial institutions of the collections studied in the course of this research: the *Instituto de Cultura Puertorriqueña and Consejo para la Protección del Patrimonio Arqueológico Terrestre*, the *Centro Ceremonial Indigena de Tibes*, the *Autoridad de Carreteras y Transportación*, the *Universidad del Turabo*, and the *Universidad de Puerto Rico – Recinto de Río Piedras*. This work was also supported, in multiple senses of the word, by L. Antonio Curet, Edwin Crespo-Torres, and Luz Veronica Muñoz Guevara. Portions of this work were supported financially by the National Science Foundation under Grant No. BCS-0612727. Versions of this paper were presented at the 15th Congress of the Latin American Association for Biological Anthropology in Mayagüez, Puerto Rico, and the *XIV Conferencia Científica Internacional "Antropología 2018"* in Havana, Cuba.

References

Ambrose, Stanley H. 2000. Controlled Diet and Climate Experiments on Nitrogen Isotope Ratios of Rats. In *Biogeochemical Approaches to Paleodietary Analysis*, edited by S. H. Ambrose and M. A. Katzenberg, pp. 243–259. New York: Kluwer Academic/Plenum Publishers.

Budinoff, Linda C. 1991. An Osteological Analysis of the Human Burials Recovered from Maisabel: An Early Ceramic Site on the North Coast of Puerto Rico. In *Proceedings of the Twelfth Congress of the International Association for Caribbean Archaeology*, edited by L. S. Robinson, pp. 117–134. Martinique: A.I.A.C.

Burton, Robert K., J. Josh Snodgrass, Diane Gifford-Gonzalez, Tom P. Guilderson, Tom Brown, and Paul L. Koch. 2001. Holocene Changes in the Ecology of Northern fur Seals: Insights from Stable Isotopes and Archaeofauna. *Oecologia* 128(1):107–115.

Cardona, Luis, A. Aguilar, and L. Pazos. 2009. Delayed Ontogenic Dietary Shift and High Levels of Omnivory in Green Turtles (Chelonia mydas) from the NW Coast of Africa. *Marine Biology* 156: 1487–1495.

Coltrain, Joan Brenner, M. Geoffrey Hayes, and Dennis H. O'Rourke. 2004. Sealing, Whaling and Caribou: The Skeletal Isotope Chemistry of Eastern Arctic Foragers. *Journal of Archaeological Science* 31: 39–57.

Crespo Torres, Edwin F. 1991. Informe Preliminar sobre Los Enterramientos Humanos en el Yacimiento do Punta Candelero, Puerto Rico. In *Proceedings of the 13th Congress of the International Association for Caribbean Archaeology*, edited by E. N. Ayubi and J. B. Haviser, pp. 73–85, Curaçao.

Crespo Torres, Edwin F. 1998. *Osteobiografia de los Restos Humanos Procedentes del Centro Ceremonial Indigena de Tibes*. Ponce, Puerto Rico: Submitted to Report submitted to the Municipo Autónomo de Ponce.

Crespo Torres, Edwin F. 2000. *Estudio Comparativo Biocultural entre dos Poblaciones Prehistóricas en la Isla de Puerto Rico: Punta Candelero y Paso del Indio, Intstituto de Investigaciones Arqueológicas*. Mexico City, Mexico: Universidad Nacional Autónoma de México.

Crespo Torres, Edwin F. 2010. Ancient Bones Tell Stories: Osteobiography of Human Remains from Tibes Archaeological Site. In *Tibes: People, Power, and Ritual at the Center of the Cosmos*, edited by L. A. Curet and L. M. Stringer, pp. 209–229. Tuscaloosa, AL: University of Alabama Press.

Crespo Torres, Edwin F. and Luz Veronica Muñoz Guevara. 2008. Reporte Final Técnico de Campo de Los Entierros Humanos Procedenetes del Sitio Los Indios (Santa Isabel, Puerto Rico). In *Informe Final Mitigacion Arqueologica Fase III, Conector Sur Santa Isabel [AC-016104]*, edited by Santa Isabel, Puerto Rico, pp. 646–865. Guaynabo: Historic Preservation Consultants.

Curet, L. Antonio and Jose R. Oliver. 1998. Mortuary Practices, Social Development, and Ideology in Precolumbian Puerto Rico. *Latin American Antiquity* 9(3): 217–239.

Curet, L. Antonio and Joshua M. Torres. 2010. Plazas, Bateys, and Ceremonial Centers: The Social and Cultural Context of Tibes in the Ancient History of Puerto Rico. In *At the Center of the Cosmos: People, Power, and Sacred Place at the Ceremonial Center of Tibes, Puerto Rico*, edited by L. A. Curet and L. M. Stringer, pp. 261–286. Tuscaloosa, AL: University of Alabama Press.

Curet, L. Antonio, Joshua M. Torres, and Miguel Rodriguez Lopez. 2004. Political and Social History of Eastern Puerto Rico: The Ceramic Age. In *Late Ceramic Age Societies in the Eastern Caribbean*, edited by A. Delpuech and C. Hofman, pp. 59–85. BAR International Series. vol. 1273. Oxford: Archaeopress.

DeNiro, Michael J. and Samuel Epstein. 1981. Influence of Diet on the Distribution of Nitrogen Isotopes in Animals. *Geochimica et Chosmochimica Acta* 45: 341–351.

Fernandes, Ricardo., Andrew R. Millard, Marek Brabec, Marie-Josée Nadeau, and Pieter Grootes. 2014. Food Reconstruction Using Isotopic Transferred Signals (FRUITS): A Bayesian Model for Dietary Reconstruction. *PLoS One* 9(2): e87436.

Fernandes, Ricardo, Marie-Josée Nadeau, and Pieter M. Grootes. 2012. Macronutrient-based Model for Dietary Carbon Routing in Bone Collagen and Bioapatite. *Archaeological and Anthropological Sciences* 4: 291–301.

Fry, Brian, Ruth Lutes, Mark Northam and Patrick L. Parker. 1982. A(13)C/(12)C comparison of Food Webs in Caribbean Seagrass Meadows and Coral Reefs. *Aquatic Botany* 14(4):389–398.

Garvie-Lok, Sandra J., Tamara L. Varney, and M. Anne Katzenberg. 2004. Preparation of Bone Carbonate for Stable Isotope Analysis: The Effects of Treatment Time and Acid Concentration. *Journal of Archaeological Science* 31: 763–776.

Godley, Brendan J., David R. Thompson, Susan Waldron, and Robert W. Furness. 1998. The Trophic Status of Marine Turtles as Determined by Stable Isotope Analysis. *Marine Ecology Progress Series* 166: 277–284.

Hare, P. Edgar, Marilyn L. Fogel, Thomas W. Stafford, Alva D. Mitchell and Thomas C. Hoering. 1991. The Isotopic Composition of Carbon and Nitrogen in Individual Amino Acids Isolated from Modern and Fossil Proteins. *Journal of Archaeological Science* 18(3): 277–292.

Hopkins, D. T. 1981. Effects of Variations in Protein Digestibility. In *Protein Quality in Humans: Assessment and In Vitro Estimation*, edited by C. E. Bodwell, J. S. Adkins, and D. T. Hopkins, pp. 178–181. Westport, CT: AVI Publishing.

Howland, M. R., L. T. Corr, S. M. M. Young, V. Jones, S. Jim, N. J. Van Der Merwe, A. D. Mitchell, and R. P. Evershed. 2003. Expression of the Dietary Isotope Signal in the Compound-Specific 13C Values of Pig Bone Lipids and Amino Acids. *International Journal of Osteoarchaeology* 13: 54–65.

Keegan, William F. 2009. Central Plaza Burials in Saladoid Puerto Rico: An Alternative Perspective. *Latin American Antiquity* 20(2): 375–385.

Keegan, William F. and Michael J. DeNiro. 1988. Stable Carbon-and Nitrogen-Isotope Ratios of Bone Collagen Used to Study Coral-Reef and Terrestrial Components of Prehistoric Bahamian Diet. *American Antiquity* 53(2): 320–336.

Keeling, Charles D., Wim G. Mook, and Pieter P. Tans. 1979. Recent Trends in the $^{13}C/^{12}$ C Ratio of Atmospheric Carbon Dioxide. *Nature* 277(5692): 121–123.

Koch, P. L., N. Tuross and M. L. Fogel. 1997. The Effects of Sample Treatment and Diagenesis on the Isotopic Integrity of Carbonate in Biogenic Hydroxylapatite. *Journal of Archaeological Science* 24: 417–429.

Laffoon, Jason E. 2012. *Patterns of Paleomobility in the Ancient Antilles: An Isotopic Approach*. Unpublished PhD Dissertation, University of Leiden.

Laffoon, Jason E., G. R. Davies, Menno L. P. Hoogland, and Corinne L. Hofman. 2012. Spatial Variation of Biologically Available Strontium Isotopes ($^{87}Sr/^{86}Sr$) in an Archipelagic Setting: A Case Study from the Caribbean. *Journal of Archaeological Science* 39(7): 2371–2384.

March, James G. and Catherine M. Pringle. 2003. Food Web Structure and Basal Resource Utilization along a Tropical Island Stream Continuum, Puerto Rico. *BIOTROPICA* 35(1): 84–93.

Marino, Bruno D. and Michael B. McElroy. 1991. Isotopic Composition of Atmospheric CO2 Inferred from Carbon in C4 Plant Cellulose. *Nature* 349: 127–131.

Moncreiff, Cynthia A. and Michael J. Sullivan. 2001. Trophic Importance of Epiphytic Algae in Subtropical Seagrass Beds: Evidence from Multiple Stable Isotope Analyses. *Marine Ecology Progress Series* 215: 93–106.

Morrison, D. J., B. Dodson, C. Slater, and T. Preston. 2000. ^{13}C Natural Abundance in the British Diet: Implications for ^{13}C Breathe Tests. *Rapid Communications in Mass Spectrometry* 14: 1312–1324.

Newsom, Lee A. and Elizabeth S. Wing. 2004. *On Land and Sea*. Tuscaloosa: University of Alabama Press.

Norr, Lynette. 2002. Bone Isotopic Analysis and Prehistoric Diet at the Tutu Site. In *The Tutu Archaeological Village Site: A Multidisciplinary Case Study in Human Adaptation*, edited by E. Righter, pp. 263–273. New York: Routledge.

Oliver, José R. 1998. *El centro ceremonial de Caguana, Puerto Rico*. Oxford: BAR International Series. Archaeopress.

Parnell, Andrew. 2014. Bchron: Radiocarbon Dating, Age-Depth Modeling, Relative Sea Level Rate Estimation, and non-Parametric Phase Modeling. *R package version* 4(1): 1–43.

Pellegrini, Maura and Christophe Snoeck. 2016. Comparing Bioapatite Carbonate Pre-Treatments for Isotopic Measurements: Part 2—Impact on Carbon and Oxygen Isotope Compositions. *Chemical Geology* 420: 88–96.

Pestle, W. J., M. Hubbe, E. K. Smith, and J. M. Stevenson. 2015. A Linear Model for Predicting $\delta^{13}C_{protein}$. *American Journal of Physical Anthropology* 157(4): 694–703.

Pestle, William J. 2008. Informe de Laboratori-Fechas de Radiocarbon. In *Informe Final Mitigacion Arqueologica Fase III, Conector Sur Santa Isabel [AC-016104]*, edited by Santa Isabel, Puerto Rico, pp. 866–869. Guaynabo: Historic Preservation Consultants.

Pestle, William J. 2010a. Bone Chemistry and Paleodiet at the Centro Ceremonial Indígena de Tibes, Municipo Autonomo de Ponce, Puerto Rico. In *Tibes: People, Power, and Ritual at the Center of the Cosmos*, edited by L. A. Curet and L. Stringer, pp. 209–230. Tuscaloosa, AL: University of Alabama Press.

Pestle, William J. 2010b. Diet and Society in Prehistoric Puerto Rico, an Isotopic Approach. Unpublished Ph.D. Dissertation, Department of Anthropology, University of Illinois at Chicago.

Pestle, William J. 2013. Stable Isotope Analysis of Paleodiet in the Caribbean. In *The Oxford Handbook of Caribbean Archaeology*, edited by W. Keegan, R. Rodriguez-Ramos and C. Hofman, pp. 407–417. Oxford: Oxford University Press.

Pestle, William J., Antonio Curet, Reniel Rodríguez Ramos, and Miguel Rodríguez López. 2013. New Questions and Old Paradigms: Reexamining Caribbean Culture History. *Latin American Antiquity* 24(3): 243–261.

Pestle, William J, Brooke E Crowley, and Matthew T Weirauch. 2014. Quantifying Inter-laboratory Variability in Stable Isotope Analysis of Ancient Skeletal Remains. *PLoS One* 9(7): e102844.

Pestle, William J. and Jason Laffoon. 2018. Quantitative Paleodietary Reconstruction with Complex Foodwebs: An Isotopic Case Study from the Caribbean. *Journal of Archaeological Science: Reports* 17: 393–403.

Pestle, William J. and Paul Szpak. In Press. Interlaboratory Error in Archaeological Stable Isotope Analysis. In *Bioarchaeology and Isotopes (working title)*, edited by D. Hamilton, H. Bocherens and K. J. Knudson. New York: Routledge.

Pestlc, William J., Maria Nieves-Colón, Harper Dine, and Ali Pourmand. 2024. The Tibes Burials: New Perspectives on Ancient Lives. In *The Archaeology of Tibes: Life, Death, and Memory at an Early Ceremonial Center in the Caribbean*, edited by A. Curet and L. Stringer, pp. 152–175. Tuscaloosa: University of Alabama Press.

Pestle, William J, A. Simonetti, and L. Antonio Curet. 2013. $^{87}Sr/^{86}Sr$ Variability in Puerto Rico: Geological Complexity and the Study of Paleomobility. *Journal of Archaeological Science* 40(5): 2561–2569.

Petersen, James B., Corinne L. Hofman, and L. Antonio Curet. 2004. Time and Culture: Chronology and Taxonomy in the Eastern Caribbean and the Guianas. In *Late Ceramic Age Societies in the Eastern Caribbean*, edited by A. Delpuech and C. L. Hofman, pp. 17–32. BAR International Series. vol. 1273. Oxford: Archaeopress.

Pourmand, Ali and N. Dauphas. 2010. Distribution Coefficients of 60 Elements on TODGA Resin: Application to Ca, Lu, Hf, U and Th Isotope Geochemistry. *Talanta* 81(3): 741–753.

Pourmand, Ali, N. Dauphas, and Y. J. Ireland. 2012. A Novel Extraction Chromatography and MC-ICP-MS Technique for Rapid Analysis of REE, Sc and Y: Revising CI-chondrite and Post-Archean Australian Shale (PAAS) Abundances. *Chemical Geology* 291: 38–54.

Pourmand, Ali, J. M. Prospero, and Arash Sharifi. 2014. Geochemical Fingerprinting of Trans-Atlantic African Dust Based on Radiogenic Sr-Nd-Hf Isotopes and Rare Earth Element Anomalies. *Geology* 42(8): 675–678.

Reich, Kimberly J. and Graham A. J. Worthy. 2006. An Isotopic Assessment of the Feeding Habits of Free-ranging Manatees. *Marine Ecology Progress Series* 322: 303–309.

Reimer, Paula J., Edouard Bard, Alex Bayliss, J Warren Beck, Paul G. Blackwell, Christopher Bronk Ramsey, Caitlin E. Buck, Hai Cheng, R. Lawrence Edwards, Michael Friedrich, Pieter M. Grootes, Thomas P. Guilderson, Haflidi Haflidason, Irka Hajdas,

Christine Hatté, Timothy J. Heaton, Dirk L. Hoffmann, Alan G. Hogg, Konrad A. Hughen, K. Felix Kaiser, Bernd Kromer, Sturt W. Manning, Mu Niu, Ron W. Reimer, David A. Richards, E. Marian Scott, John R. Southon, Richard A. Staff, Christian S. M. Turney, and Johannes van der Plicht 2013 IntCal13 and Marine13 Radiocarbon Age Calibration Curves 0–50,000 years cal BP. *Radiocarbon* 55(4): 1869–1887.

Schimmelmann, Arndt. 1985. *Stable Isotopic Studies on Chitin.* Ph.D. Dissertation, Los Angeles: University of California.

Schoeninger, Margaret J. and Michael J. DeNiro. 1984. Nitrogen and Carbon Isotopic Composition of Bone Collagen from Marine and Terrestrial Animals. *Geochimica et Cosmochimica Acta* 48(4): 625–639.

Schwarcz, H. P., J. Melbye, M. Anne Katzenberg, and M. Knyf. 1985. Stable Isotopes in Human Skeletons of Southern Ontario: Reconstructing Paleodiet. *Journal of Archaeological Science* 12: 187–206.

Siegel, Peter E. 1992. *Ideology, Power, and Social Complexity in Prehistoric Puerto Rico, Anthropology*. Binghamton, NY: SUNY-Binghamton.

Snoeck, Christophe and Maura Pellegrini. 2015. Comparing Bioapatite Carbonate Pre-Treatments for Isotopic Measurements: Part 1—Impact on Structure and Chemical Composition. *Chemical Geology* 417: 394–403.

Sponheimer, M., T. Robinson, L. Ayliffe, B. Roeder, J. Hammer, B. Passey, A. West, T. Cerling, D. Dearing, and J. Ehleringer. 2003. Nitrogen Isotopes in Mammalian Herbivores: Hair δ15 N Values from a Controlled Feeding Study. *International Journal of Osteoarchaeology* 13(1–2): 80–87.

Stokes, Anne V. 1998. *A Biogeographic Survey of Prehistoric Human Diet in the West Indies Using Stable Isotopes*. Gainesville: Department of Anthropology, University of Florida.

Stokes, Anne V. 2005. Ceramic-Age Dietary Patterns in Puerto Rico: Stable Isotopes and Island Biogeography. In *Ancient Borinquen*, edited by P. E. Siegel, pp. 185–201. Tuscaloosa: University of Alabama Press.

Stoner, A. W. and J. M. Waite. 1991. Trophic Biology of Strombus Gigas in Nursery Habitats: Diets and Food Sources in Seagrass Meadows. *The Journal of Molluscan Studies* 57: 451–460.

Szpak, Paul, Jessica Z Metcalfe, and Rebecca A Macdonald. 2017. Best Practices for Calibrating and Reporting Stable Isotope Measurements in Archaeology. *Journal of Archaeological Science: Reports* 13: 609–616.

Tieszen, L. L. 1981. Natural Variation in the Carbon Isotope Values of Plants: Implications for Archaeology, Ecology, and Paleoecology. *Journal of Archaeological Science* 18: 227–248.

Torres, Josuha M. 2012. *The Social Contruction of Community, Polity, and Place in Ancient Puerto Rico (AD 600–AD 1200)*. PhD dissertation, Department of Anthropology, University of Florida.

United States Department of Agriculture. 2013. National Nutrient Database for Standard Reference, *Release* 27.. 2015.

von Fischer, Joseph C. and Larry L. Tieszen. 1995. Carbon Isotope Characterization of Vegetation and Soil Organic Matter in Subtropical Forests in Luquillo, Puerto Rico. *BIOTROPICA* 27(2): 138–148.

van Klinken, Gert Jaap. 1991. Dating and Dietary Reconstruction by Isotopic Analysis of Amino Acids of Fossil Bone Collagen—With Special Reference to the Caribbean. *Publications Foundation for Scientific Research in the Caribbean Region*, Amsterdam.

Warinner, C. and N. Tuross. 2009. Alkaline Cooking and Stable Isotope Tissue-Diet Spacing in Swine: Archaeological Implications. *Journal of Archaeological Science* 36: 1690–1697.

Yoder, C. J. and Eric J. Bartelink. 2010. Effects of Different Sample Preparation Methods on Stable Carbon and Oxygen Isotope Values of Bone Apatite: A Comparison of Two Pre-Treatment Protocols. *Archaeometry* 52(1): 115–130.

13

INVESTIGATIONS AT THE ROYAL NAVAL HOSPITAL CEMETERY (C. 1763–1834) IN ENGLISH HARBOUR, ANTIGUA

Tamara L. Varney, Christopher K. Waters and Reginald Murphy

Introduction

Atop a hill overlooking the former British Royal Naval (RN) Dockyard at English Harbour, Antigua, from the northeast is the site of a cemetery that has been the subject of ongoing research for over two decades. That the cemetery is adjacent to what was once the site of British Royal Naval Hospital facilities is widely known in the area; indeed, the hill is locally known as Hospital Hill. The entire area that the hospital and cemetery are thought to have occupied is now a residential neighbourhood. Due to this development, large portions of the cemetery have been subject to considerable disturbance. Local residents have pointed out areas where such disturbance has destroyed ruins of associated structures, grave markers, and graves over the past few decades. Development is continuing in this area, and the undisturbed portions of the site are in imminent danger of destruction. However, regulations by the National Parks Authority, the managing body, stipulate that recovery is required prior to destruction and more recovery is expected over the next several years.

The RN Dockyard at English Harbour functioned from 1725 to 1889 (Nicholson 1991) and is located on the south shore of Antigua. It was renowned as much for being a safe, sheltered harbour and careening station as its being a hot disease infested graveyard for the British military (ADM 1/306, Francis Holborne to the Admiralty, 18 July 1749; Crewe 1993, 241). Dr. William Turnbull, a surgeon who had served in the RN and author of The Naval Surgeon, asserted that 'the West India station' was 'avowedly the most unhealthy' for the RN (Turnbull 1806, 152). English Harbour was particularly notorious at this time for being an unhealthy place, and it was known as a 'Grave of Englishman' (Nicholson 1993). At the time, the West Indies garrison as a whole shared this reputation as being a death sentence to British servicemen (Buckley 1998; Duffy 1987, 326), as it was thought to have

DOI: 10.4324/9780429444449-17

had one of the highest mortality rates from disease in the British Empire (Buckley 1979, 1998; Curtin 1989; Duffy 1987).

Bioarchaeological research at the RN hospital cemetery site involving analysis of the excavated human skeletal remains and related historical records has opened up a new window into life and death for those serving the RN at the English Harbour Dockyard. Much of the work related to this site has focused on the larger question of the identity of the interred. Methodologies employed in this effort have included standard gross morphological analyses, but much effort has concentrated on the application of chemical analyses, specifically stable isotope and trace elemental analyses of bones and teeth. The work reported here is the first to employ such techniques on colonial-era non-Indigenous populations of the Caribbean and has added to our knowledge of the RN in the West Indies and the lived experiences of African and Afro-descendant people in colonial Antigua.

Site History

The first archaeological activity related to the hospital site was the 1980 salvage excavation of a midden discovered during the construction of a hotel (Nicholson 1993). The site is located on a hilltop overlooking a former RN dockyard at English Harbour from the north. The site of the hospital was initially thought to have been in use from approximately AD 1793 to 1822 based on the types of ceramics found in the midden and a limited search of historical documents (Nicholson 1995). However, that midden may have dated to a relatively narrow time frame, as more recent archival research has revealed that the site atop the hill overlooking the RN Dockyard had been used as the site for RN hospital facilities since the construction of a hospital in the 1760s (ADM 140/1175, General Plans of English Harbour; Convertito 2016, 516). This first iteration of the hospital, owned and run by a contractor, was built between 1763 and 1768 but was destroyed by a hurricane in 1772 (ADM 140/1218, Elevation of Hospital 1763; Convertito 2016, 517–518). Subsequently, a temporary hospital was established in a derelict house belonging to the RN at the base of the hill closer to the Dockyard. In addition, from 1772 to 1790, temporary shelters were set up all over English Harbour to accommodate the sick, including at the site of the ruins of the hilltop hospital (Convertito 2016, 520–521). From 1788 to 1790, hospital facilities that were to be owned by the Navy itself were reconstructed at the original hilltop site (ADM 140/1220, Plan of Hospital 1779–1788; Convertito 2016, 266, 522). The cemetery may very well have been used through the entire timespan that hospital facilities existed in the vicinity of the hill, extending the usage of the cemetery from the 1760s to sometime in the early mid-1800s.

As the Caribbean became strategically less important after the Napoleonic Wars, military operations were consolidated across branches. By 1820, the Admiralty stopped funding the RN Hospital, turning it over to the Board of Ordnance, and it was renamed the General Detachment Hospital (NAAB B00308, St. Philip Marriage Banns, 16 May 1820; ADM 140/1198, Sir Anderson's Report with drawings,

1853–1858). This consolidation effort also included abandoning the nearby Middle Ground Barracks in 1833 and the removal of the West India Regiment troops (black soldiers) stationed there to the site of the RN Hospital (NAAB F00935, Record of Emolument, R.N. Crosse to E.J. Murray MacGregor, 14 November 1833). Nicholson (1993) cites correspondence between the Admiralty and the Antigua Dockyard that noted the last RN burial on the site was in 1822, when Tom Spanker, a sailmaker and enslaved man who worked in the Dockyard, was interred. However, the St. Paul's Church parish burial records list 18 individuals being interred there from 1826 to 1834, with no additional burials being noted after 1834 (NAAB B000293, St. Paul's Burial Records 1816–1825/1826–1840). The last burial occurred on 29 October 1834, when Mary Ann, recorded as a Black 38-year-old woman and likely recently emancipated, was buried (emancipation occurred on August 1, 1834) (NAAB B000293, St. Paul's Burial Records 1816–1825/1826–1840, pg. 58, entry 463). By 1853, the RN hospital ruins were reported to have been entirely removed with the exception of a water tank (ADM 140/1198, Record of Emolument, R.N. Crosse to E.J. Murray MacGregor, 14 November 1833).

Excavation History at the Cemetery

The disturbance of human remains during the grading of a road atop Hospital Hill was reported to local archaeologist Dr. Reginald Murphy, who then conducted a survey and preliminary excavations in the summer of 1997. These activities revealed several tombstones, and test-pitting led to the excavation of one burial to assess the conditions of the remains and the potential depth of the graves. A second grave shaft was noticed in the south profile of the excavated burial, but it was not uncovered.

More intensive excavations followed from 1998 to 2001, directed by co-authors Murphy and Varney. The tombstones noted in 1997 were no longer present. Few grave markers of any kind remained in place at the site, with the remains of four rectangular-shaped horizontal tombstone foundations being clearly visible in a single row at the ground surface. These structures no doubt once provided the bases of more elaborate structures that were similar to those found at other contemporaneous cemeteries on the island. A single grave marker had been in place over the western-most foundation up until a few years previous, but it had been salvaged and taken to the Dockyard Museum, where it remains on display. The stone bears the name of an Assistant Ship's Surgeon named Alexander Bernard who served on the H.M.S. Pyramus and states that he died at the age of 27 years in October of 1821 (Nicholson 1993).

The remains of more than 30 individuals were removed from 26 grave features from 1997 to 2001 (See Table 13.1).

In January 2017, human remains were encountered during the construction of a new housing complex near the original 1997 excavations and again reported to Dr. Reginald Murphy. The remains (B 32) were removed from the site by

TABLE 13.1 Naval Hospital cemetery burials

Burial #	*Age*		*Sex*	*Ancestry*
	Stage	*Years*		
1	Adult	45–49	M	Black
2	Adult	25–29	M	White
3	Adult	30–34	M	White
4a	Adult	50–59	M	Black
4b	Infant[1]	Newborn	?	Indeterminate
5	Adolescent[3]	16–18	M?	White
6	Adult	25–29	M	White
8a	Adolescent[3]	14–16	?	Indeterminate
8b	Adult	?	?	Indeterminate
8c	Adult	?	?	Indeterminate
9a	Adult	18–20	M	Indeterminate
9b	Adult	?	?	Indeterminate
12a	Adult	35–39	M	White
12b	Child[2]	2–4	?	Indeterminate
13	Adult	45–49	M	Black
14	Adult	35–39	M	Indeterminate
15a	Adult	35–39	M	White
15b	Infant	Newborn	?	Indeterminate
16	Adolescent[3]	16–18	M	White
17	Adult	35–39	M	Indeterminate
18	Adult	30–35	M	White
19a	Adult	40–45	M	White
19b	Adult	20–24	M	Indeterminate
20	Infant[1]	1–1.5	?	Indeterminate
21	Infant	0–0.5	?	Indeterminate
22	Adult	20–29	M	Indeterminate
23	Adult	35–39	M	Black
24	Adult	25–29	M	Black
25	Adult	30–35	M	Black
26	Adolescent[3]	14–18	M?	Indeterminate
31	Adult	>14	M	Black
32	Adult	>30	F	Black

[1]infant = 0–2 years;[2] child = os coxa not fused; [3]adolescent – os coxa fused.

Drs. Matthew Brown and Murphy before construction activities proceeded. In the summer of 2019, excavations under the direction of Brown and Murphy were again undertaken in response to the planned construction of new housing adjacent to the area of the 1998–2001 excavations. The remains of approximately 15 individuals were removed from the site during these most recent excavations, many of which were from disturbed contexts rather than discrete graves. In total, the remains of over 47 individuals have been disinterred from the site. Data are still being compiled and interpreted from the most recent excavations, and this chapter focuses on the prior investigations. All archaeological activities were carried out with the permission of local landowners and/or occupants of the land, as well as the permission of the National Parks Authority within whose boundaries the site is located.

Excavations of 1998–2001

The aforementioned row of tombstone foundations (B 2; B 3; B 5; B 6) provided a logical place to begin larger-scale excavations. Two 'unmarked' grave shaft features (B 4 and B 7) were found in the space between two of the tombstone foundations. B 7 turned out to be a very small, shallow shaft feature that did not contain any human remains; however, several skeletal elements of an infant (B 4b) were found scattered in the fill of an adjacent grave (B 4a), and it is likely that these disturbed remains once occupied the small empty grave shaft (B 7). Most of the burials were single interments; however, there was a sole purposeful double burial of an adult male and a newborn infant (B 15a and B 15b). Although it is tempting to believe the two individuals were related, it may be that they were buried together as an expedient measure. One grave shaft contained multiple individuals that appeared to be intentionally interred atop one another, perhaps to maximize space and minimize digging (B 19a and B 19b), while others overlapped one another with an obvious pattern of succession with only the topmost individual being undisturbed and therefore the last to be interred (B 4a, b; 8a, b, c; 9a, b; 12a, b). The very poorly preserved remains of a grave marker were found at the head of B 12. It appeared to have been composed of wood covered with ferrous metal with plain-cut stone at its base. It is not known which of the two occupants of that grave it might have been erected for. Numerous isolated human bones and coffin nails were found in the fill of many of the grave shafts, providing more evidence of disturbance during the digging of graves. The overall pattern of disturbance indicates extensive use of the burial ground, with possible reuse of some grave shafts. These patterns were found again during the 2019 excavations but to an even greater extent. Modern disturbance also had a substantial effect on preservation over time, as half of two of the graves (B 8 and B 9) excavated in 1998 were obliterated by the modern construction of a septic tank.

Although many artifacts related to the hospital were found at the midden site, aside from the odd ceramic fragment found in the grave fill, few items were found at the cemetery. The burials themselves also contained very few artifacts. All graves were very similar to one another; most individuals had been buried in six-sided wooden coffins, which were often evident only by the pattern of nails, almost all made of ferrous material, and/or a brown stain in the soil where the wood had decayed. Aside from the row of burials with tombstone foundations, all of which included coffins, the only other discernible row of graves (B 22–B 26) did not contain any nails or buttons, suggesting that these individuals had been interred without a coffin or clothing. These two rows show the only notable evidence of systematic differential treatment of interment.

Only one individual (B 1) showed obvious signs of preparation for burial aside from placement in a coffin, being covered with the remnants of a cloth shroud, and his feet having been tied together with rope. Eleven of the burials had mainly bone buttons found overlying the pelvic bones in a configuration similar to that seen on trousers favoured by many sailors (Haythornwaite and Younghusband, 1993). Burials 4 and 13 also had a few shell buttons in the mid-thorax and wrist regions. Brass buckles were found in B 5 and B 17, with only the latter having buttons. Two burials (B 5, B 21) contained one to three brass nails, respectively, with one nail in each burial bearing a 'Broad Arrow' that, in this context, no doubt indicates it was the property of the RN (Nicholson 1993). Overall, for individuals over the age of 15 years, no internment patterns were apparent aside from that already mentioned in the two rows.

Four of the children were also interred in six-sided wooden coffins. One newborn infant (B 15b) and an infant of approximately 1.5 years (B 20) had small, rolled brass 'beads' in the neck region. A third infant (B 21) about six months old and a child of about three years (B 12b) had single brass pins found with them. The remains of the fifth child, belonging to a newborn infant, were found in the grave shaft fill of B 4, as described earlier.

General Site Demography

A summary of the burials disinterred during various archaeological activities up to 2017 is given in Table 13.1. Analysis of skeletal remains recovered in the 2019 excavations is currently incomplete and is not included in this table. Sex and age were determined using methods by Buikstra and Ubelaker (1994). With the exception of five individuals under the age of three years and individual B 32, all individuals that could be assessed were males ranging in age from about 14 years to over 60 years of age, with a cluster of individuals at the younger end of this range. B 32, the only female individual found interred at this site, was excavated during the salvage activity in 2017. This skeleton belonged to an older adult woman who appeared to have been buried within the hospital compound rather than in the burial ground.

Whenever possible, an assessment of non-metric cranial features of adult individuals was done to determine ancestry following the criteria of Gill (1986) and Rhine (1990). Eight individuals exhibited White features (B 2; B 3; B 5; B 6; B 15a; B 16; B 18; B 19a), and an equal number exhibited Black features (Burials B 1; B 4a; B 13; B 23; B 24; B 25; B 31; B 32). All the individuals (B 2; B 3; B 5; B 6) interred under the three large grave marker foundations exhibited White features and ranged in age from about 17 to 35 years.[1]

Stable isotope analysis and reconstructed diet

Although the assessment of ancestry was possible for 16 individuals, many of the skeletons could not be assessed due to the absence or damage to their craniofacial features. Given the potential to provide additional information on the lived experience of African-descended peoples in the British RN and Antigua, alternative means of identification were sought. In societies with strict social stratification, as was the case in the colonial British West Indies, foodstuffs are often markers of identity and differentiation (Dietler 2007; Twiss 2012). Stable isotope analyses of human skeletal remains can often aid in the determination of dietary differences among groups within a population. These analyses are based on the fact that hard tissues of the body will retain isotopic signatures that reflect the foodstuffs they consume. As dietary reconstruction via stable carbon (C) and nitrogen (N) isotope analysis is commonplace and well established in bioarchaeology, only a brief explanation will be given here (please see Katzenberg and Waters-Rist (2018) for more details). The relative importance of general food categories in the dietary regime of the individuals or population(s) under study can be elucidated via stable C and N isotope analyses of skeletal remains. Analytical results are expressed as delta or δ values in parts per mil (‰), where the given value is the ratio of stable isotopes in the sample relative to that of an international standard (for carbon $^{13}C/^{12}C$; for nitrogen $^{15}N/^{14}N$). Biological tissues generally contain less ^{13}C than the standard, so their $\delta^{13}C$ values are negative numbers. In contrast, $\delta^{15}N$ values of biological tissues are usually positive numbers since they generally contain more ^{15}N than the standard of atmospheric N_2. The variation in $\delta^{13}C$ values of consumers, such as humans, reflects that of the plants that comprise the base staple(s) of their diet; the $\delta^{13}C$ values of these plants are incorporated into the body tissues of consumers. Plants can be divided into two main groups based on the type of photosynthesis they employ. C_4 plants include maize/corn, millets (including sorghums), and sugar cane and have $\delta^{13}C$ values ranging from −21 to −9‰ with a mean of −13‰. C_3 plants include wheat, rye, oats, barley, temperature trees, rice, beans, fruits, and vegetables including root crops and have $\delta^{13}C$ values ranging from −22 to −38‰, with a mean of −27‰. A third group, CAM plants (succulents), has $\delta^{13}C$ values intermediate to the C_3 and C_4 plants but is not thought to have contributed to the diet of the populations under consideration here. Marine resources have $\delta^{13}C$ values that overlap those of C_4 plants, but the use of $\delta^{15}N$ values can elucidate more detailed dietary information, particularly in cases where foods with overlapped or intermediate $\delta^{13}C$ values may be consumed. Stable nitrogen ($\delta^{15}N$) values indicate the overall trophic level at which a consumer is feeding and are higher towards the upper end of the food chain so that carnivores have higher values than herbivores and herbivores have higher values than plants. Marine animals tend to have higher $\delta^{15}N$ values than terrestrial animals due to the greater number of trophic levels in marine environments (Schoeninger and DeNiro 1984). Human groups consuming substantial amounts of marine protein have been demonstrated to follow the same pattern (Schoeninger et al. 1983).

Isotopic analysis of different components of the skeleton has been shown to provide somewhat different information about diet. The $\delta^{13}C$ and $\delta^{15}N$ values of bone collagen (protein) reflect dietary protein, while the $\delta^{13}C$ values of bone carbonate (mineral) reflect the whole diet and are strongly influenced by carbohydrates and fats (Ambrose and Norr 1993; Tieszen and Fagre 1993). Bone collagen and carbonate $\delta^{13}C$ values are approximately +5‰ and +9‰ respectively relative to dietary values, while $\delta^{15}N$ values of bone collagen tend to be on average +3‰ relative to dietary values (Ambrose et al. 1997). Isotopic data obtained from bone collagen and bone carbonate is displayed in Figures 13.1a and 13.1b, respectively.

Interpretation of the isotopic data was guided in part by information on diet found in historical sources ranging from RN victualling accounts to the journals of individual sailors (Duffy 1987; Crewe 1993, 147; Knight and Wilcox 2010; Lloyd and Coulter 1961; Thomas 1968). Taken together, these sources generally describe the primary staples of the enslaved population of African descent on the British West Indian islands as consisting of a variable mixture of C_4 starches (maize and/or sorghum) and C_3 root crops such as cassava, yam, and taro, rounded out with a variety of C_3 vegetables and fruits. Imported salt fish is usually attributed as the main animal protein, with smaller amounts of poultry and meat of both imported and local origin (Abrahams and Szwed 1983; Dunn 1972; Luffman 1789; Higman 1984; Sheridan 1985; Ward 1988). In contrast, European military personnel tended to maintain a diet typical of their countries of origin, and in the case of the RN in particular, much of the diet was provisioned by the Navy. Their primary staples were imported cereal grains such as wheat and oats, rounded out with locally available foodstuffs. Much of their dietary protein was also imported in the form of salted beef and pork, although that was supplemented with local livestock on occasion (Buckley 1998; Duffy 1987; Dyde 1997; Crewe 1993, 147; Knight and Wilcox 2010; Lloyd and Coulter 1961). The RN is known to have had contracts with the Codrington operations on neighbouring Barbuda for both sheep and cattle (Lowe 1968; Nicholson 1991, 2002; Thomas 1968), and there is some documentary evidence that access to fresh meat, at least while in port at the Dockyard, was not uncommon (ADM 51/3350, ADM 55/1013, Captain's and Ship's logs: Pyramus). The bulk of the RN diet consisted of foodstuffs with C_3 at their base, and the meals of the rank-and-file personnel were likely largely limited to the supplied victuals.

Taking these historical sketches of diet into account, it was expected that individuals consuming a 'European' diet would have bone collagen $\delta^{13}C$ values ranging from about −21 to −17‰ and $\delta^{15}N$ values ranging from 9 to 12‰. In contrast, it was expected that individuals consuming an 'African or enslaved' diet would have bone collagen $\delta^{13}C$ values ranging from about −16 to −12‰ and $\delta^{15}N$ values ranging from 13 to 15‰. As can be seen in Figure 13.1a, the isotopic ratios for bone collagen indeed do show that the individuals of different ancestry did consume different diets, although not as drastically or in precisely the pattern expected. As predicted, individuals of European ancestry have $\delta^{13}C$ values typical of what is expected from consuming a primarily European-style diet. The range of their $\delta^{13}C$ values is also relatively limited when compared to that of the individuals of African ancestry. The latter group does include three individuals with overlapping $\delta^{13}C$ values with the former group, but overall the individuals of African ancestry have a greater range of isotope values that extend into more positive values. It is

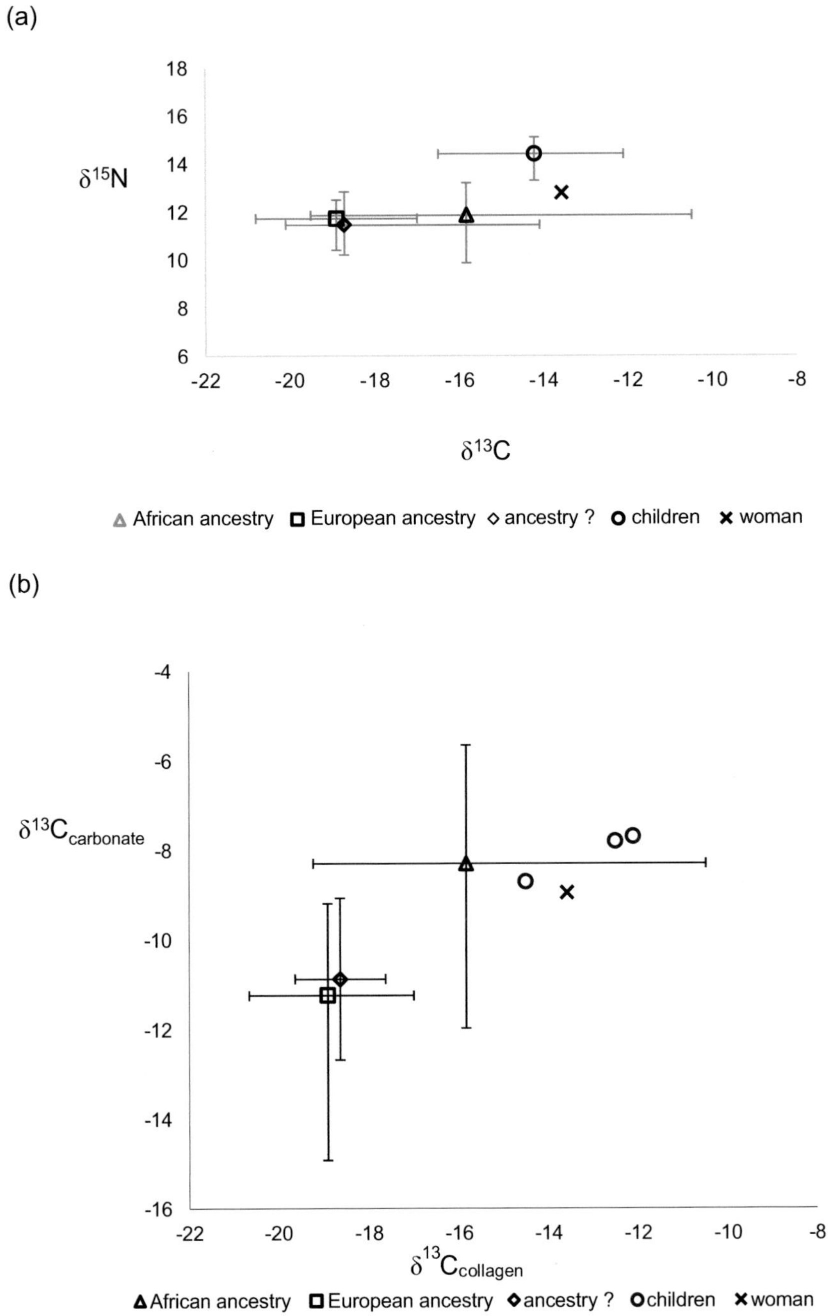

FIGURE 13.1 (a) Stable carbon and nitrogen isotope values of bone collagen for Royal Naval Hospital cemetery, Antigua (data from Varney 2011); (b) Stable carbon isotope values of bone collagen vs. bone carbonate for Royal Naval Hospital cemetery, Antigua (data from Varney 2011).

in this more positive range of values that we would expect to see if they had been consuming a diet as described in historical sources for enslaved populations. The greater diversity in the values for individuals of the African ancestry is not surprising given the wide breadth of origins and status of people potentially interred in this RN cemetery. Enslaved military personnel would have had different foodstuffs available to them through a combination of provisioning and self-procurement. Interestingly, both groups had similar $\delta^{15}N$ values, reflecting a similar source of dietary protein (see Figure 13.1a). This result was unexpected for the individuals of African ancestry since if they had been consuming the diet outlined in historical sources, their protein would have been primarily salted fish and their $\delta^{15}N$ values would have been higher. The $\delta^{15}N$ values for the African ancestry group do show more variation than the European ancestry group, but it is in their respective $\delta^{13}C$ values that show far more diversity. In summary, these results suggest that people of both ancestries were eating protein from similar sources, probably mostly from imported domestic livestock. The difference in diet lies in the source of primary starch or carbohydrate, with the individuals of African ancestry consuming greater variation in primary starches comprised of a mix of C_3 and C_4 plants.

This information was further supplemented with $\delta^{13}C$ values from bone carbonate (see Figure 13.1b). The same three individuals of African ancestry once again have isotopic values overlapping $\delta^{13}C$ values with the European ancestry individuals; however, the mean values for apatite for the two groups are more disparate than those for collagen (−8.3‰ and −11.2‰, respectively). Since the isotopic values of bone carbonate reflect the whole diet and are strongly influenced by carbohydrates and fats, the greater difference in $\delta^{13}C$ values between the two groups most likely reflects a focus on different primary starches rather than protein sources in the diet.

In other words, the African ancestry group was consuming a more diverse diet on both the group and individual levels in terms of staple carbohydrates, with some individuals eating C_3 carbohydrates similar to the European ancestry group, while others consumed more C_4 carbohydrates. In terms of protein sources, all adults regardless of ancestry were consuming a similar dietary protein source.

While this study is admittedly a small sample, at the time it was done, it was the first isotopic study of colonial-era populations in the Caribbean. Stable isotopic data for bone collagen from two slave cemeteries from former British colonies, the Harney site, Montserrat, and the Newton Plantation site, Barbados, are available for comparison in Figure 13.2 (Schroeder et al. 2009; Sparks et al. 2012). Both populations have stable C and N isotopic values that meet the expectations for the diet of enslaved populations in the British West Indian colonies, and when they are compared to the isotopic data of the bone collagen from the Naval Hospital cemetery, there are no overlapping values with the European ancestry group (Varney 2011). Stable nitrogen isotope values for bone collagen were only available for the Harney cemetery, and there is again no overlap with the RN hospital cemetery in Antigua for either ancestral group. A similar situation is seen with the $\delta^{13}C$ isotopic data of the bone carbonate, although the disparity between isotopic values for the

European ancestry group from the Naval Hospital cemetery and the enslaved individuals from the Montserrat site is even greater (see Figure 13.1b). This contrasts with the complete overlap in isotopic values from bone carbonate between the African ancestry group from the Antigua RN site and those of the Montserrat site (Varney 2011).

The African ancestry group from the Naval Hospital cemetery has a substantial range of stable C and N isotopic values indicating consumption of variable diets with much diversity in the base staple carbohydrate sources accompanied by a protein source very similar to the European ancestry group. This finding is very likely related to their variable status, including enslaved individuals owned by the RN or Army, enslaved people on loan from estate owners, and free blacks. Enslaved individuals owned by the RN would have been provisioned by the RN. The varying status of people of colour likely afforded them differential access to resources, as well as influenced their choice of which resources they preferred to eat, and which to sell or barter (Buckley 1998; Dyde 1997; Hauser 2008; Voelz 1993).

Stable C and N isotopic values were also obtained for the five children and one woman recovered from the site. The $\delta^{13}C$ and $\delta^{15}N$ of bone collagen and the $\delta^{13}C$ of bone apatite all plot closer to the values of the Montserrat and Barbados slave cemeteries and do not overlap with the values from the European ancestry group. Further comparison with the stable C and N isotope ratios (shown in Figure 13.2) for the interred at a nearby burial ground (Galleon Beach/Freeman's Bay) assumed associated with the RN Dockyard reveals very similar values with the European ancestry group of the RN Hospital cemetery (Varney et al. 2017). The former site bears signs of expedient burial, which is suggestive of a burying place for those gravely ill on arrival and dying soon before or after arrival in English Harbour (Varney et al. 2017). However, the identity of those interred at the Freeman's Bay site remains uncertain. The $\delta^{13}C$ and $\delta^{15}N$ values of the European ancestry group at the RN Hospital site are also consistent with those from other RN Hospital sites in St. Johns, Newfoundland, Canada (Munkittrick et al. 2019) and England (Roberts et al. 2012).

Lead Exposure

Another question that was addressed with archaeological bone chemistry was the question of possible lead (Pb) poisoning in the RN. In historical works on the era, excessive alcohol consumption and Pb poisoning are hypothesized as having exacerbated the poor health of the troops and sailors in the West Indies (Buckley 1998; Curtin 1998). Sources of lead were abundant in the everyday life of the RN in the West Indies. Sugar and rum production on plantations involved Pb-containing equipment and the release of toxic heavy metals, including Pb-laden effluent (Wells et al. 2017, 2018). Many other tools of daily life, such as water collection and storage equipment, paints, ceramic glazes, and eating utensils, often contained Pb. It was also commonly used as a component of cosmetics and medicine and

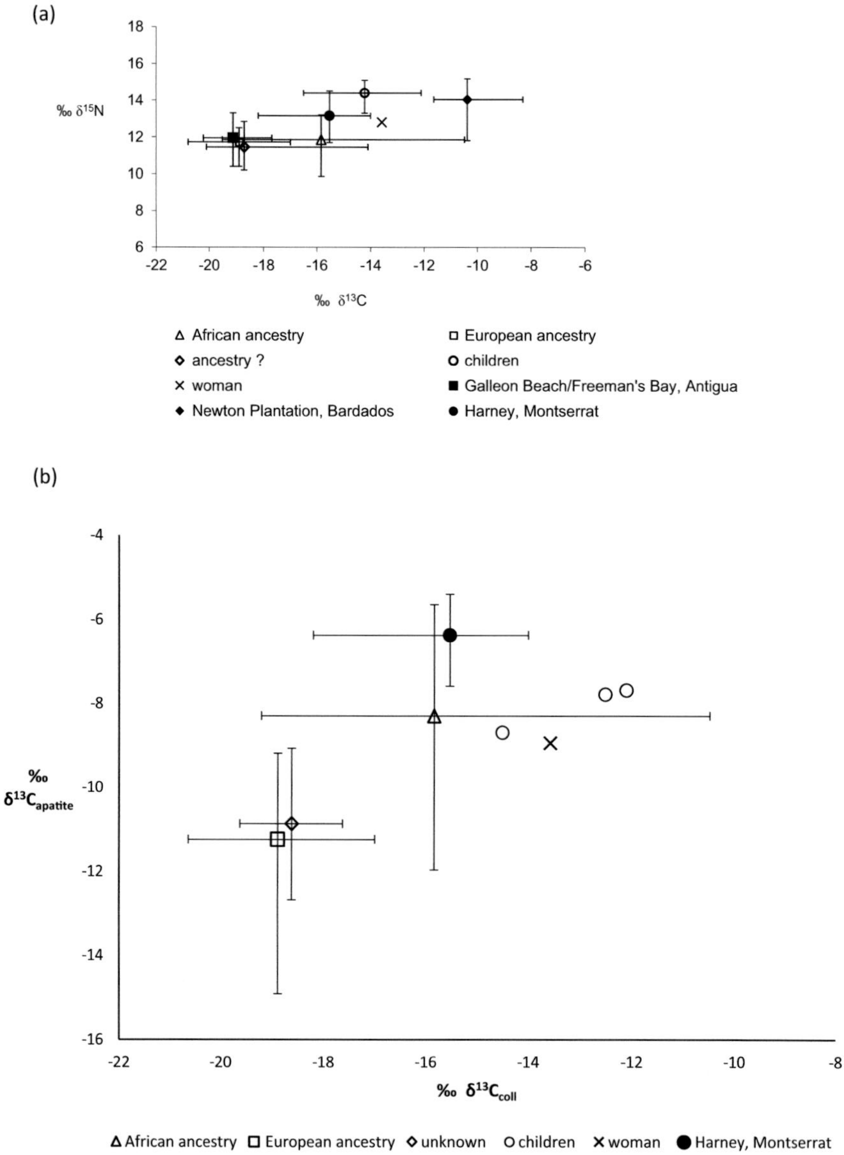

FIGURE 13.2 (a) Stable carbon and nitrogen isotope values of bone collagen for Royal Naval Hospital (Varney 2011) compared to Freeman's Bay, Antigua (FB, data from Varney et al. 2017), Newton Plantation, Barbados (BARB., data from Schroeder et al. 2009), and Harney site, Montserrat (MONT., data from Varney et al. 2012); (b) stable carbon and nitrogen values for bone collagen vs. bone carbonate for Royal Naval Hospital (Varney 2011) compared to FB (Varney et al. 2017) and MONT (Varney et al. 2012).

sometimes used to enhance food and beverage flavour (Wedeen 1984). Although the toxicity of Pb had long been recognized, its usefulness seemed to have overruled its dangers, or the dangers were unacknowledged in many circumstances (Hernberg 2000; Waldron 1973).

Pb is a chemical element that is classified as a heavy metal, and it is highly toxic to all animals, including humans. Exposure can occur through inhalation (primarily of dust or vapours), ingestion, or absorption via skin contact. It accumulates in both soft and bony tissues, and although it affects every system in the body, it mainly impacts the nervous system. Acute exposure, even at lower doses, results in symptoms that include headache, joint and muscle pain, extreme fatigue, apathy, weakness in the extremities, cognitive impairment, and distinctive severe abdominal cramping from constipation (Brodkin et al. 2007; Kosnett et al. 2007). Most concerning is that Pb negatively affects the growth and cognitive development of children. Another potential source of Pb exposure for the Antigua RN hospital cemetery population, and others, may have been medicinal treatments. Medicinal compounds of the era commonly included Pb and other toxic ingredients. Indeed, a residue containing Pb was found in a broken beaker recovered in the 1980 excavations of a midden associated with the hospital (Nicholson 1995). The usage of the RN sites mentioned in this study largely predates the RN's usage of foods preserved in cans sealed with Pb solder for large-scale provisioning (Graham 1981).

Physicians working for the colonial era British RN and army noted that Pb poisoning was likely commonplace amongst personnel (Blane 1785; Buckley 1979; Hughes 1750; Lloyd and Coulter 1961; Trapham 1679; Turnbull 1806). Lead poisoning is usually described as colic, often referred to as 'dry bellyache' (to describe abdominal cramping without dysentery), with its symptoms being likened to that of 'Devonshire cholic', which was recognized by the medical profession of the time as being caused by the 'poison of lead'. Turnbull (1806, 179), a nineteenth-century physician, listed dry bellyache as the third most common health complaint in the Navy at the time. The idea that lead poisoning was a contributor to ill health in the colonial British military (Buckley 1979; Curtin 1989) and civilian populations (Wedeen 1984) of the West Indies has been suggested but not yet tested. This RN cemetery presented a unique opportunity to address the issue.

It is known that trace elements can be incorporated into bone from the burial environment. Previous studies of Pb in past populations had been hampered by the inability to differentiate between biogenic (taken up during life) and diagenetic (due to postmortem contamination) Pb (Wittmers et al. 2008). To overcome this hurdle, a novel approach was taken by Swanston et al. (2012) and Varney et al. (2012) involving the application of synchrotron radiation. The brilliant light produced by a synchrotron increases the resolution of analyses such as the X-ray florescence imaging (XFI) used in this study. The distribution of Pb was mapped in the bone micro-structure such that it matched well with images obtained via conventional light microscopy. These results strongly suggested that Pb of biogenic origin within discrete biological structures (osteons – structures consisting of concentric rings of bone that are formed during regular bone tissue maintenance or turnover) occurred at higher levels than the surrounding bone tissue. In other words, the Pb in the bones of individuals interred in the RN cemetery was very likely deposited during life rather than being present as postmortem contamination/diagenetic origin. Furthermore, little to no detectable Pb was found in a precontact sample from the island (Swanston et al. 2012).

Giffin et al. (2017) used Inductively Coupled Plasma Mass Spectrometry (ICP-MS) to determine the bone Pb content for 17 of the older adolescent and adult individuals from the RN hospital cemetery. ICP-MS can measure trace elements at minute concentrations (in ppm to parts per million here) and represents a bulk or average measure of the burden of Pb accumulated over the lifetime of an individual. Analysis of cortical bone samples resulted in a mean value of 107.5 ppm with a substantial range of values from 13 to 336 ppm. To put these results into perspective, the 'normal' range in modern populations not occupationally exposed to Pb is in the range of 5–30 ppm (Gamblin et al. 1994; Hisanaga et al. 1988; Kowal et al. 1989). While there is no established clinical threshold, the World Health Organization [WHO] does not consider any level of Pb to be safe in the modern world (2019). If the aforementioned modern bone Pb values are taken as a point of reference, most (14/17) of the RN cemetery individuals have a bone level of 200 ppm, which could be expected to have resulted in at least mild Pb poisoning (Giffin et al. 2017). An additional nine individuals had even higher Pb levels suggestive of moderate to severe poisoning. It should be noted that lower Pb burdens can be asymptomatic throughout an individual's lifetime, and while typical symptoms may be nonspecific (e.g., headaches, fatigue, slightly lower cognitive functioning), they can still have the potential to affect quality of life. In fact, recent studies have found health concerns associated even with low levels of exposure (e.g., Lanphear et al. 2018). The wide spread of values is also interesting, especially since there is a lack of the typical positive correlation of Pb with age. Typically, such a pattern is expected since younger individuals absorb lead at a more rapid rate (Ziegler et al. 1978). However, given the vagaries of estimating age-at-death from the skeleton, the variable background of the interred, and the small sample size, perhaps it is not surprising that such a trend was not seen.

Interestingly, similar to age, no relationship was found between Pb content and ancestry. This contrasts with the aforementioned study of diet. Previous research on this assemblage focused on socially defined class based on ancestry via dietary reconstruction (Varney 2011), and it was hypothesized that Pb exposure might also follow a similar pattern.

Analysis of bone from another nearby burial ground, the Freeman's Bay site (sometimes referred to as the Galleon Beach site), also thought to likely be associated with the RN, showed a similar mean and distribution of bone Pb levels (Varney et al. 2017; see Table 13.2).

Otherwise, there have only been four other studies of Pb exposure in colonial-era West Indian populations, all using conventional means of trace element analysis,

TABLE 13.2 Bulk lead concentration values of Royal Navy Hospital and Freeman's Bay sites

Site	*Mean and Range of Lead Concentration (ppm)*
Royal Naval Hospital cemetery (n = 24)	79.6 (10.1–251.5)
Galleon Beach/Freeman's Bay (n=10)	75.1 (8.7–218.5)

on the skeletal remains of enslaved labourers from Newton Plantation on Barbados (Corruccini et al. 1987; Handler et al. 1986; Laffoon et al. 2009; Schroeder et al. 2013). Both the Pb levels and their variability are similar to the studies of the Antigua sites. However, these studies may not be the best comparators as they involve the remains of enslaved individuals who had been exposed to Pb during their labour in the sugar and rum industry, which could potentially have been a considerable portion of their lives. In contrast, many of those interred in the RN hospital cemetery, notably those of European ancestry, had likely only been in the West Indies a relatively brief time for their naval service. So other contemporaneous comparators were sought, including other eighteenth- and nineteenth-century European naval expeditions such as the Franklin Expedition (Keenleyside et al. 1996, 1997; Kowal et al. 1989), which perished entirely in the Canadian Arctic. Once again, both the range of Pb content and the pattern of distribution of Pb in the bone microstructure are similar to that seen in the Antigua RN cemeteries (Swanston et al. 2018). A bone sample from the roughly contemporaneous Southside cemetery, also thought to have been associated with a former RN hospital in Newfoundland (Munkittrick et al. 2019), also yielded a very similar distribution pattern when imaged with XFI (Choudhury et al. 2016). The only bone sample from a probable slave cemetery, the Harney site on Montserrat, that has undergone XFI also had a pattern of Pb distribution in its microstructure very similar to that of the various RN sites. So, it would seem that neither being in the RN nor living in colonial West Indies seems to have resulted in a unique exposure to the toxic elemental, at least in terms of its uptake into the skeleton. Rather, it seems that Pb exposure was a common yet variable experience of colonial-era life.

The trace element analysis of the Antigua RN hospital cemetery included measurement of multiple trace elements at once and revealed that one individual (Burial 6) had a surprisingly high level of mercury (Hg). Compared to the other individuals tested, Burial 6 had very low Pb levels but very high levels of Hg (94 ppm vs. 0–2 ppm). This burial was not distinctive in any other way, and no visible exogenous sources of Hg were apparent during excavation. This individual, a male of European ancestry who was between 25 and 29 years of age upon his death, exhibited little osseous evidence of ill health other than excessive porosity of alveolar bone, which could be consistent with Hg poisoning since increased salivation and sore, bleeding gums are two of the initial signs of the toxic effects of the once common ingredient in medicinal treatments for a wide range of maladies, particularly syphilis (Swanston et al. 2015).

Overall – what have we learned

Prior to the 1997–2002 excavations, limited documentation on the hospital site was available. Since that time, researchers travelling to the National Archives in London, UK, have contributed important information. Key documents made available in 2019 by St. Paul's Parish Church in Falmouth, Antigua, have greatly added to our knowledge as well. Both the RN hospital and the associated cemetery existed and were in use for about 50 years longer than previously known. By the time archaeological investigation of the cemetery began, all but one of the grave markers had disappeared. The one grave that had a remaining tombstone (B 2) represents a valuable resource to the project since it yielded the only individual of known identity, age-at-death, and occupation. The biological profile of the skeleton found in the associated grave was consistent with the information engraved on the stone: a male aged mid-late 20s at the time of death with no osseous evidence of disease. A brief entry in the ship's log of the Pyramus notes that Bernard died on October 16, 1821, almost exactly a month after the ship arrived at English Harbour, but does not provide any other information about his death (ADM 53/1013 Ship's Log's: Pyramus).

The age at death of those interred ranges from newborn to over 60 years. The three individuals with the oldest age at death estimates (Burials 1, 4a, 13) were of African ancestry. The men from Burials 1 and 13 were aged 40–45 years, with Burial 4a estimated to be over 60 years of age and representing the oldest individual disinterred.

The presence of teenage individuals is not surprising, as it was common for boys of such age to be in service to the military, whether army or navy (Buckley 1998, 124–127; Pope 1981, 64). Other archaeological investigations on Antigua have also found women and children interred in cemeteries associated with military facilities roughly contemporaneous with the one described here, including another nearby burial ground presumed associated with the RN (Murphy 2012, Varney et al. 2017). Permanent facilities such as the Dockyard would almost certainly have had a sizable population of affiliated women and children. Some British Army enlisted men were permitted to have their families accompany them on campaign, as attested by the presence on tombstones in army cemeteries near English Harbour (Jane 1982, 32); however, only warrant officers on larger RN ships were allowed to accompany family. Nevertheless, women were not entirely absent on Naval ships and shore facilities. The Naval authorities often simply overlooked such infringements upon official policies (Lavery 1994, 141; Pope 1981, 171–175). And although most enslaved persons working for the Navy and Army were male, women and children also worked in various capacities, including, until the mid-nineteenth century, enslaved female nurses at Naval shore hospitals, including the contractor-owned and -run hospital in English Harbour (Buckley 1998; Lloyd and Coulter 1961; ADM 1/306, Report by Admiral Frankland, 22 September 1756). That being said, the only female individual found interred at this site was the individual excavated in 2017. This skeleton was of an older adult woman who appeared to have been buried within the hospital compound rather than in the burial ground, suggesting possible special consideration in her interment. The interment of Sarah, a black enslaved woman who died at age 20 years, is recorded on January 9, 1830, in the Naval Hospital cemetery (NAAB B000293, St. Paul's Parish burial records 1824–1840, p 30, n. 235). While there is limited direct archival evidence, the parish burial records confirm the presence of women and children of enlisted military personnel interred within other military cemeteries around English Harbour (NAAB B000293, St. Paul's Parish burial records 1824–1840). This is further substantiated by engraved burial markers in several military cemeteries around English Harbour, which almost exclusively are for women and children. The lack of females in the excavated portion of the RN cemetery may very likely be due to excavation sample bias.

Since the patients treated at the hospital appear to have been diverse, it is also plausible that their dependents may have been treated and potentially buried in the cemetery, which may account for the infants and young children found buried there (B 4b; B 12b; B 15b; B 20; B 21). Nicholson (1993) cites two additional historical documents that lend credence to such a scenario: first, a baptismal record from St. Phillips Church records the baptism of Mary, the infant slave of a woman belonging to a hospital watchman's wife, by the hospital chaplain in 1815; second, a local newspaper advertisement in the Antigua Gazette, 28 January 1819, refers to the hospital at English Harbour as the General Detachment Hospital and states that Robert Hartle, Surgeon, would inoculate persons there. These public inoculations may have included a diverse range of ages, including infants. This inoculation would have to be referring to the smallpox vaccine, which was the first and only vaccine developed at the time and would represent an early public health initiative (Weiss and Esperza 2015).

Finding that the interred were of both European and African ancestry was likewise not surprising given that the population of the RN Dockyard labourers, and of Antigua in general was dominated by Africans and Afro-descendants who outnumbered Whites in Antigua, as in the other Leeward Islands, by approximately 7 to 1 by the mid-eighteenth century (Buckley 1998, 43; Waters 2020). Similarly, by 1780, the Dockyard workforce was composed of 70% African/Afro-descendant labourers many of whom were owned by the Navy (Nicholson 1991, 7). In addition, there was no doubt that there may have been some free Blacks interred at the site, albeit they were present in relatively small numbers, as well as enslaved labourers that were provided to the military by the local planters (ADM 1/306, George Thomas to Thomas Pye, Antigua, 12 July 1756; Buckley 1998; Nicholson 1991, 8). Blacks were also not uncommon on-board RN ships in the West Indies during the nineteenth century (Langley 1967). It was common practice for the RN to employ personnel from amongst the resident populations in the West Indies/overseas Dockyards so that the workforce reflected the local demography (de Foy et al. 2016). As such, the numbers of Blacks on RN ships may very well have been higher in the West Indies, where populations were predominantly of African descent. Although Foy (2016) presents evidence estimating that Blacks made up only about 6% of RN crews overall, that number was probably higher in the West Indies (Foy 2016; Langley 1967, 275) despite General Order No. 9, which dictated limits on the number of 'Negroes' aboard ships (ADM 1/309 f. 389. Admiral Young General Order No. 9 for the Leeward Islands Station, 21 November 1775).

The population of the West Indies colonies was predominately enslaved, and owners were not willing to have their property taken without compensation. Thus, the use of press gangs in the West Indies was not a feasible means of manning ships as it might have been in Britain (Brunsman 2007). However, there certainty seemed to be a perception that the RN might enlist runaway slaves into their ranks (Bolster 1998; Foy 2016; TNA CO 9/41, 24 September 1783: Antigua Assembly and Council to the Admiralty).

Initially, it was thought that all of the interred assessed as having African ancestry were most likely members of the 'King's Negros' or government/military-owned enslaved individuals (Nicholson 2002). Indeed, muster records for the RN Hospital covering much of the period from 1791 to 1816 note patients' names, ships or places received from, quality (rank), reasons for being treated at the facility, and how they were discharged (live or dead). The patients listed include men from the Naval Dockyard that are noted to be 'King's Negros', and often include their trade (e.g., mason, labourer) (ADM 120/5–11, Hospital Muster Books, Antigua, 1791–1794/1804/1806–1808/1809–1811/1812–1814/1815–1816). However, the St. Paul's Church burial records document the interment of 18 individuals between 1826 and 1834. These records confirm the diverse nature (in terms of sex, ancestry, and quality/trade/profession) of those interred in the RN cemetery (NAAB B000293, St. Paul's Burial Records 1816–1825/1826–1840) in addition to Naval personnel.

The 1980 salvage excavation of the midden also yielded clues to the diverse identity of the hospital's clientele (Nicholson 1993). Buttons from the uniforms of naval officers and army regiments from both the British and French Armies were found in the midden, suggesting that the Naval hospital also served the Army and prisoners of war. In addition, a shaker (large cap) badge bearing the insignia of the 8th West India Regiment was found, suggesting the presence of African/Afro-descended troops at the hospital (Nicholson 1993). This regiment was one of a dozen comprised mainly of African slaves that were promised their freedom in exchange for five years of military service (Buckley 1979, 26–27).

Taking all evidence into account, it seems, for at least the last seven years of its operation, the hospital served the greater community in addition to RN personnel, and more than naval personnel were interred in the cemetery. While it is not clear if everyone treated at the hospital was included among the interred, the cemetery remained in use after the RN hospital closed. Based on published accounts and St. Paul's Church burial records, it seems that class rather than gender and racial segregation existed when it came to the location of interment. Whereas officers and civilian dockyard workers were buried in the cemetery at St. Paul's, enlisted soldiers and sailors, their dependents, and enslaved Africans owned by the Crown, such as the King's Negroes or serving in one of the pioneer battalions, were buried together in military cemeteries such as the one at the Naval Hospital (NAAB B000293; St. Paul's Burial Records 1816–1825/1826–1840; for black pioneers see Chertrand 2011).

Examination of the remains revealed little osseous evidence of infectious disease in contrast to the many historical accounts of the West Indies military campaigns that cite very high rates of infectious disease morbidity and mortality (e.g., Buckley 1979, 1998; Crewe 1993; Curtin 1989; Duffy 1987). It seems that relatively few army or naval personnel on campaign in the West Indies were wounded or killed in action; rather, they suffered and died from a variety of infectious diseases, including yellow fever, malaria, dengue, dysentery, typhoid, smallpox, and other unidentifiable fevers that were exacerbated by poor nutrition and excessive alcohol consumption (Buckley 1978, 1998; Crewe 1993; Curtin 1989; Duffy 1987). Leg ulcers were another common ailment (Loudon 1981). The lack of osteological evidence is not surprising given that most diseases that people were suffering from in the West Indies were short-term ailments that would not have affected an individual long enough to induce osseous changes prior to death. Typically, a minimum of two to three weeks is required for a substantial bony response to be induced, since the generation of bone tissue is a relatively slow process (Ortner 2003). Of the ailments listed above, only leg ulcers would have been likely to affect bone, and they were a common and serious disease of the time (Loudon 1981). Many of the adult individuals had generalized striae and porosity on their long bones, particularly the lower leg bones. However, the only active bony periosteal reaction that was seen occurred on the surface of an infant's (B20) tibia. These phenomena are nonspecific indicators of stress that can be representative of many conditions of ill health, so nothing more precise can be linked to them (Weston 2008, 2012).

The child (approximately three years old) designated as Burial 12b is the only individual to show multiple signs of poor health. All of the deciduous teeth were erupted and had large carious lesions on the mid-labial and mid-lingual aspects of the anterior teeth and occlusal aspect of the molars. The lesions were most dramatic on the first deciduous molars, for which half of the crown was missing. On the remaining enamel, furrows of linear enamel hypoplasia (LEH) were evident on many of the teeth, including the second deciduous molars. LEH was often located between carious lesions, suggesting a relationship between the two lesions. This may be due to the rough, thinner, poorly mineralized enamel of the LEH lesion, which allows food and bacteria to accumulate (Alshehhi et al. 2020). Bilateral cribra orbitalia was noted, as well as porotic hyperostosis on the parietal and occipital bones, which has often been interpreted as evidence of anaemia, which may be dietary, infectious, genetic, or a combination of any of the three factors in nature (Ortner 2003; Stuart-Macadam 1989). A more recent interpretation is that cribra orbitalia is a descriptive term and is not necessarily associated with anaemia.

A variety of other causative factors has been suggested (Walker et al. 2009; Wapler et al. 2004). Despite the state of the teeth and the presence of cribra orbitalia, the remainder of the skeleton is unremarkable. The child is at the age that children were often weaned in British colonies of the North American continent (Herring et al. 1998; Katzenberg and Pfeiffer 1995; Henderson et al. 2014), which would be consistent with the location of the caries and LEH on the deciduous dentition of this child. However, it is not known if the same weaning customs apply to the islands.

Six individuals (B 1; B 3; B 4a; B 6; B 12a; B 15a; B 22) exhibited healed fractures of various bones, and one (B 13) had atrophy of the vertebral margin of the scapula, likely from a muscular injury, along with healed fractures of a fibula and rib. All of the fractures are typical of those that could be sustained in day-to-day life and are well healed without any evidence of complications. Three individuals (B 1; B 4a; B 12) had patterns of fractures that may have been suffered during falls, the most notable being seen in individual B 4a, who had healed fractures of the right clavicle, many of the right ribs, and the left hand. Some of the ribs were broken in two places, suggesting that he fell onto his side, impacting his ribs on a structure or object. The broken clavicle suggests that he may have tried to break his fall with an outstretched arm in that or another fall. It was not evident if the different injuries had been the result of one or more incidents. Although the clavicular fracture healed in badly misaligned positions, the individual shows little sign of significant disability in terms of atrophy or asymmetry of the adjacent arm bones. The case of the partially atrophied scapula (B 13) may have been caused by injury to the rhomboid muscles that serve to retract the scapula toward the spine. Again, no atrophy or other changes were noted in the adjacent arm bones, vertebrae, or ribs, indicating that these individuals were not made invalid by their injuries for any substantial time. Healed metacarpal fractures were also present in two individuals (B 4a; B 6) and a healed fracture of the nasal bones in another (B 1), leading to speculation of stereotypical brawling sailors. However, these injuries may have been due to other causes. Osteophytic lipping and porosity of joint surfaces were seen in many individuals and may have been due to age-related osteoarthritis. In some individuals, repetitive activities or accidents may have also played a role (Weiss and Jurmain 2007). Several individuals (B 3; B 6; B 14; B 15a; B 18; B 22) exhibited Schmorl's nodes on their thoracic and/or lumbar vertebrae, attributed to excessive strain on the back associated with heavy lifting and carrying, causing herniation of the intervertebral discs (Ortner 2003). One of the younger individuals (Burial 5, 16–18 yrs) also exhibited cortical defects at the site of muscle attachments that stabilize the shoulder, indicating excessive physical activity (Owsley et al. 1991). Arthritic lesions are evident in all adult individuals, but they are not excessive or patterned in such a way that they cannot be attributable to the accumulated effects of hard physical labour and/or age.

Among the many isolated skeletal elements found in the grave fill of burials was a partial femur shaft that bore distinct cutmarks indicative of a saw with no evidence of healing; this element no doubt represents an amputation. Amputations were a common medical treatment for wounds caused by severe trauma or infection (Pope 1981, 145). Unfortunately, the fragment was short and found in a disturbed context, so it is not known whether it came from a discrete grave or not. Hospitals often had a pit where amputated limbs were buried, but if the patient died very soon after the amputation, the limbs might be buried with their body (Mann et al. 1991). No other evidence of amputation was found. The topic of amputation at this site should include a mention of Dr. Ralph Cuming, a surgeon who was the first to perform an amputation of the entire shoulder girdle and arm at the Antigua RN hospital in 1808 (Keevil 1949). The patient recovered successfully, while the surgeon unfortunately passed away in one of the many yellow fever outbreaks that plagued the station (Keevil 1949).

In summary, the reconstruction of diet at the Royal Naval Hospital cemetery has allowed greater insight into the probable identities of those included among the interred at this RN Hospital and emphasized that this burial ground is not simply 'the grave of the Englishman' or the RN sailor. For this site, those individuals who could be classified into two general ancestral groups seem to have maintained different dietary habits with respect to the non-protein components of their diet. The dietary regime of those of African ancestry shows more variation, no doubt a reflection of the greater diversity of status, access to resources, and preference. These data promise to be a potentially valuable dataset for other sites where individuals are not well preserved enough for determination of ancestry, as well as providing an additional piece of information for those individuals for which such assessments were possible based on morphological criteria. Although that potential will not be completely borne out until bone collagen and carbonate isotopic values from second-generation colonists of European ancestry have been established.

The interred were also shown to have had ubiquitous yet variable exposure to the toxic element of lead. Upon examination of various other demographic and osseous variables, little stood out in terms of the social determinants of exposure. One individual also showed high levels of mercury, which, given the context of the cemetery, is not surprising given that the toxic element was commonly used in the treatment of a wide range of maladies (Swanston et al. 2015).

In terms of gross morphological evidence of health, the skeletons of all individuals over the age of 15 years were robust, as one would expect for individuals who had spent their lives doing physical labour. The health conditions most frequently indicated were the result of a life of physically demanding activity. There was little evidence of the often written about infectious diseases that plagued the island, no doubt to their swift acute nature.

In conclusion, the work presented in this chapter represents over two decades of investigation of a cemetery associated with a Royal Naval Hospital, near English Harbour, Antigua. The research has been strongly interdisciplinary, as is the nature of bioarchaeological investigation, including being the first to employ both conventional and novel techniques on colonial-era non-Indigenous populations of the Caribbean. Future directions for this research are manyfold. One goal will be to attempt to refine the possible origins and/or socioeconomic class of those interred in the RN cemetery via stable oxygen and strontium isotope analysis of dental enamel to attempt to determine probable geographic origins of the interred. In addition, extraction and analysis of DNA from the dentin of the teeth would provide more precise demographic information that would allow more in-depth interpretation of both the isotopic and trace element data. The ability to further explore demographic patterns should facilitate our elucidation of social determinants of diet, resource allocation, and acquisition, as well as lead exposure. We also plan to expand our sample to other contemporary RN, civilian, and enslaved assemblages in order to better understand life in the colonial West Indies. As the cemetery continues to be threatened, mitigation of the site is ongoing (the latest excavation took place in the summer of 2019), as is the analysis of the remains and search for additional historical documents.

Acknowledgements

We would like to give thanks to the people of the neighbourhood of the site for their support. Special thanks to Sylvan Braithwaithe and his family for their facilitating our work in such close proximity to their home. Thanks also to Faye Edwards of Falmouth Village for her assistance with the St. Paul's Church records. There are so many who have provided support over the years who have not been named – we are grateful to all of you since none of this work would have been possible without everyone involved in anyway, big or small.

Note

1 The terms Black and White are used to be consistent with the methodology used. Individuals with Black features were assumed to be of African ancestry and those with White features were assumed to be of European ancestry.

References

Historical Documents Abbreviations

ADM = Records of the Admiralty, London: National Archives.
NAAB = National Archives of Antigua and Barbuda.
TNA CO = The National Archives (Kew) Colonial Office.
Abrahams, R.D. and Szwed, J.F. (1983). *After Africa: Extracts from British Travel Accounts and Journals of the Seventeenth, Eighteenth and Nineteenth Centuries Concerning the Slaves, their Manners, and Customs in the British West Indies*. New Haven, CT: Yale University Press.
ADM 1/306 Report by Admiral Frankland 22 September 1756.
ADM 1/306 Francis Holborne to the Admiralty 18 July 1749.
ADM 1/306 George Thomas to Thomas Pye, Antigua, 12 July 1756.

ADM 1/309 f. 389. Admiral Young General Order No. 9 for the Leeward Islands Station, 21 November 1775.
ADM 51/3350 Captains' logs, including: PYRAMUS (1821 February 28–1828 September 15). PYRAMUS (1833 August 18–1833 October 23).
ADM 53/1013 Ship's Log's: Pyramus.
ADM 102/5 Hospital Muster Books, Antigua, 1791–1794.
ADM 102/6 Hospital Muster Books, Antigua, 1794.
ADM 102/7 Hospital Muster Books, Antigua, 1804.
ADM 102/8 Hospital Muster Books, Antigua, 1806–1808.
ADM 102/9 Hospital Muster Books, Antigua, 1809–1811.
ADM 102/10 Hospital Muster Books, Antigua, 1812–1814.
ADM 102/11 Hospital Muster Books, Antigua, 1815–1816.
ADM 140/1175 General plans of English Harbour.
ADM 140/1198 Sir Anderson's Report with drawings, 1853–1858.
ADM 140/1218 Elevation of hospital, 1763.
ADM 140/1220 Plan of hospital, 1779–1788.
Alshehhi, A., Al Halabi, M., Hussein, I., Salami, A., Hassan, A., and Kowash, M. (2020). Enamel defects and caries prevalence in preterm children aged 5–10 years in Dubai. *Libyan Journal of Medicine* 15(1): 1705633.
Ambrose, S.H., Butler, B.M., Hanson, D.B., Hunter-Anderson, R.L., and Krueger, H.W. 1997. Stable isotopic analysis of human diet in the Marianas Archipelago, Western Pacific. *American Journal of Physical Anthropology: The Official Publication of the American Association of Physical Anthropologists*, 104(3): 343–361.
Ambrose, S.H., and Norr, L. (1993). Experimental Evidence for the Relationship of the Carbon Isotope Ratios of Whole Diet and Dietary Protein to Those of Bone Collagen and Carbonate. In: *Prehistoric Human Bone*, edited by J.B. Lambert, and G. Grupe. Berlin, Heidelberg: Springer. https://doi.org/10.1007/978-3-662-02894-0_1
Blane, G. 1785. *Observations on the Diseases Incident to Seamen*. London: J. Cooper.
Bolster, W.J. 1998. *Black Jacks: African American Seamen in the Age of Sail*. Cambridge: Harvard University Press.
Brodkin, E., Copes, R., Mattman, A., Kennedy, J., Kling, R., and Yassi, A. 2007. Lead and mercury exposures: Interpretation and action. *Canadian Medical Association Journal* 176(1): 59–63
Brunsman, D. 2007. The Knowles Atlantic impressment riots of the 1740s. *Early American Studies* 5(2): 324–366. http://www.jstor.org/stable/23546612
Buckley, R.N. 1978. The destruction of the British army in the West Indies 1793–1815: A medical history. *Journal of the Society for Army Historical Research*, 56(226): 79–92.
Buckley, R.N. 1979. *Slaves in Red Coats: The British West India Regiments, 1795–1815*. New Haven, CT: Yale University Press.
Buckley, R.N. 1998. *The British Army in the West Indies: Society and the Military in the Revolutionary Age*. Gainsville: University Press of Florida.
Buikstra, J.E., and Ubelaker, D.H. 1994. Standards for data collection from human skeletal remains. Arkansas archaeological survey. In *Biocultural Adaptation in Prehistoric America*, edited by Robert L. Blakely, p. 144. Southern Anthropological Society Proceedings 11. Athens: University of Georgia Press.
Chartrand, R. 2011. The British Army's unknown, regular, African-West Indian Engineer and service Corps, 1783 to the 1840s. *Journal of the Society for Army Historical Research* 89(358): 117–138.

Choudhury, S., Swanston, T., Varney, T.L., Cooper, D.M.L., Gordon, R.A., George, G.N., Pickering, I.J., and Coulthard, I. 2016 Confocal x-ray fluorescence imaging Facilitates high resolution elemental mapping in fragile archaeological bone. *Archaeometry* 58(S1): 207–217.

Convertito, C. 2016. Mending the sick and wounded: The development of Naval Hospitals in the West Indies, 1740–1800. *Canadian Journal of History* 51(3): 500–533.

Corruccini, R.S., Aufderheide, A.C., Handler, J.S., Wittmers, L.E., Jr. 1987. Patterning of skeletal lead content in Barbados slaves. *Archaeometry* 29: 233–239.

Crewe, D. 1993. *Yellow Jack and the Worm: British Naval Administration in the West Indies, 1739–1748.* Liverpool Historical Studies No. 9. Liverpool: Liverpool University Press.

Curtin, P.D. 1989. *Death by Migration: Europe's Encounter with the Tropical World in the Nineteenth Century*. Cambridge: Cambridge University Press.

de Foy, B., Tong, Y., Yin, X., Zhang, W., Kang, S., Zhang, Q., Zhang, G., Wang, X., and Schauer, J.J., 2016. First field-based atmospheric observation of the reduction of reactive mercury driven by sunlight. *Atmospheric environment*, 134: 27–39.

Dietler, M. 2007. Culinary encounters: Food, identity, and Colonialism. In *The Archaeology of Food and Identity*, edited by Katheryn Twiss, pp. 218–242. Carbondale: Center for Archaeological Investigations Press, University of Southern Illinois.

Duffy, M. 1987. *Soldiers, Sugar, and Seapower: The British Expeditions to the West Indies and the War against Revolutionary France*. Oxford: Clarendon Press.

Dunn, R.S. 1972. *Sugar and Slaves: The Rise of the Planter Class in the English West Indies, 1624–1713.* Chapel Hill: The University of North Carolina Press.

Dyde, B. 1997. *The Empty Sleeve: The Story of the West India Regiments of the British Army*. Antigua, WI: Hansib Caribbean, St. John's.

Gamblin, C., Gordon, C.L., Muir, D.C., Chettle, D.R., and Webber, C.E. (1994). In vivo measurements of bone lead content in residents of southern Ontario. *Applied Radiation and Isotopes* 45(10): 1035–1038.

Giffin, K.L., Swanston, T., Coulthard, I., Murphy, A.R., Cooper, D.M.L., and Varney, T.L. (2017). Skeletal lead burden of the British Royal Navy in colonial Antigua. *International Journal of Osteoarchaeology* 27(4): 672–682.

Gill, G. W. 1986. Craniofacial criteria in forensic race identification. In *Forensic Osteology*, edited by K. Reichs, pp. 143–159. Springfield, IL: Charles C Thomas.

Graham, J.C. 1981. The French connection in the early history of canning. *Journal of the Royal Society of Medicine* 74(5): 374–381.

Handler, JS, Aufderheide, AC, Corruccini, RS, Brandon, EM, and Wittmers, L.E. 1986. Lead contact and poisoning in Barbados slaves: Historical, chemical and biological evidence. *Social Science History* 10: 399–425.

Hauser, M.W. 2008. An archaeology of black markets: Local ceramics and economies in eighteenth-century Jamaica. *African Diaspora Archaeology Newsletter* 11(4): 15.

Henderson, R.C., Lee-Thorp, J., and Loe, L. 2014. Early life histories of the London poor using δ13C and δ15N stable isotope incremental dentine sampling. *American Journal of Physical Anthropology* 154(4): 585–593.

Hernberg, S. 2000. Lead poisoning in a historical perspective. *American Journal of Industrial Medicine* 38(3): 244–254.

Herring, D.A., Saunders, S.R., and Katzenberg, M.A. 1998. Investigating the weaning process in past populations. *American Journal of Physical Anthropology* 105(4): 425–439.

Higman, B.W. 1984. *Slave Populations of the British Caribbean 1807–1834.* Baltimore, MD: John Hopkins University Press.

Hisanaga, A., Eguchi, Y., Hirata, M., and Ishinishi, N. 1988. Lead levels in ancient and contemporary Japanese bones. *Biological Trace Element Research* 16(1): 77–85.

Hughes, G. 1750. *The Natural History of Barbados: In Ten Books*. London: Printed for the author.

Jane, Charles W.E. 1982. *Shirley Heights: The Story of the Redcoats in Antigua*. English Harbour, Antigua: Nelson's Dockyard Park Foundation.

Katzenberg, M.A., and Pfeiffer, S. 1995. Nitrogen isotope evidence for weaning age in a nineteenth century Canadian skeletal sample. In *Bodies of Evidence: Reconstructing History Through Skeletal Analysis*, edited by M.A. Katzenberg and A. Grauer, pp. 221–235. Hoboken: John Wiley & Sons, Inc.

Katzenberg, M.A. and Waters-Rist, A.L. 2018. Stable isotope analysis: A tool for studying past diet, demography, and life history. In *Biological Anthropology of the Human Skeleton*, 3rd edition, edited by M.A. Katzenberg and A.L. Grauer, pp. 467–504. Oxford: Wiley-Blackwell.

Keenleyside, A., Bertulli, M., and Fricke, H.C. 1997. The final days of the Franklin expedition: New skeletal evidence. *Arctic* 50(1): 36–46.

Keenleyside, A., Song, X., Chettle, D.R., and Webber, C.E. 1996. The lead content of human bones from the 1845 Franklin expedition. *Journal of Archaeological Science* 23(3): 461–465.

Keevil, J.J. 1949. Ralph Cuming and the interscapulo-thoracic amputation in 1808. *The Journal of Bone and Joint Surgery*. British volume, 31(4): 589–595.

Knight, R., and Wilcox, M.H. 2010. *Sustaining the Fleet, 1793–1815: War, the British Navy and the Contractor State*. Woodbridge: Boydell & Brewer.

Kosnett, M.J., Wedeen, R.P., Rothenberg, S.J., Hipkins, K.L., Materna, B.L., Schwartz, B.S., H, H., and Woolf, A. 2007. Recommendations for medical management of adult lead exposure. *Environmental Health Perspectives* 115(3): 463–471.

Kowal, W.A., Krahn, P.M., and Beattie, O.B. 1989. Lead levels in human tissues from the Franklin Forensic project. *International Journal of Environmental Analytical Chemistry* 35(2): 119–126.

Laffoon, J.E., Shuler, K.A., Millard, A.R., Connelly, J.N., and Schroeder, H. 2009. Isotopic evidence for anthropogenic lead exposure on a 17th/18th century Barbadian plantation. *American Journal of Physical Anthropology* 171(3): 529–538. DOI: 10.1002/ajpa.23938

Langley, H.D. 1967. The Negro in the Navy and Merchant service--1789–1860, 1798. *The Journal of Negro History* 52(4): 273–286.

Lanphear, B.P., Rauch, S., Auinger, P., Allen, R.W., and Hornung, R.W. 2018. Low-level lead exposure and mortality in US adults: A population-based cohort study. *The Lancet Public Health* 3(4): e177–e184.

Lavery, B. 1994. *Nelson's Navy: The Ships, Men and Organization 1793–1815*. Annapolis, MD: Naval Institute Press.

Lloyd, C. and Coulter, J.L.S. 1961 *Medicine and the Navy: 1200–1900, Volume III -1714–1815.* Edinburgh and London: E and S Livingstone Ltd.

Loudon, I.S.L. 1981. Leg ulcers in the eighteenth and early nineteenth centuries. *The Journal of the Royal College of General Practitioners* 31(226): 263–273.

Lowe, R. 1968. *The Coderington Correspondance, 1743–1851.* London: Robson Lowe

Luffman, J. 1789. A brief account of the Island of Antigua, together with the customs and manners of its inhabitants as well White as Black: As also an accurate statement of the food, clothing, labor, and punishment, of slaves. In *Letters to a Friend Written in the Years 1786, 1787, 1788*, edited by J. Luffman. London.

Mann, R.W., Owsley, D.W., and Shackel, P.A. 1991. A reconstruction of 19th-century surgical techniques: Bones in Dr. Thompson's privy. *Historical Archaeology* 25(1): 106–112.

Munkittrick, T.J.A., Varney, T., Pike, K, and Grimes, V. 2019. Insights into life histories from the St. John's Royal Naval Hospital cemetery, Newfoundland using stable isotopes in bone and teeth. *Journal of Archaeological Science: Reports* 24(4): 815–828.

NAAB B00308 St. Philip Marriage Banns, 16 May 1820.

NAAB F00935 Record of Emolument, R.N. Crosse to E.J. Murray MacGregor, 14 November 1833.

NAAB B000293 St. Paul's Burial Records 1816–1825/1826–1840.

Nicholson, D.V. 1991. *The Story of English Harbour, Antigua, West Indies.* Antigua, WI: The Historical and Archaeological Society, St. John's.

Nicholson, D.V. 1993. *Mud and Blood: Artifacts from Harbour Dredging and the Naval Hospital at English Harbour, Antigua*. Antigua: Museum of Antigua and Barbuda

Nicholson, D.V. 1995. Mud and blood: Artifacts from harbour dredging and the Naval hospital at English Harbour, Antigua. In *Proceedings of the 15th Congress for Caribbean Archaeology*, Puerto Rico, 1993.

Nicholson, D.V. 2002. *The King's Negroes: The Journal of Boatswain Fox, Antigua Navy Yard, 1820–23.* Antigua: He Dockyard Museum, English Harbour.

Ortner, D.J. 2003. *Identification of Pathological Conditions in Human Skeletal Remains*. San Diego, CA: Academic Press.

Owsley, D.W., Mann, R.W., and Murphy, S.P. 1991 Injuries, surgical care and disease. In *Snake Hill: An Investigation of a Military Cemetery from the War of 1812*, edited by S. Pfeiffer and R.F. Williamson. Toronto: Dundurn Press.

Pope, D. 1981. *Life in Nelson's Navy.* London: George Allen and Unwin.

Roberts, P., Weston, S., Wild, B., Boston, C., Ditchfield, P., Shortland, A.J., and Pollard, A.M. 2012. The men of Nelson's navy: A comparative stable isotope dietary study of late 18th century and early 19th century servicemen from Royal Naval Hospital burial grounds at Plymouth and Gosport, England. *American Journal of Physical Anthropology* 148(1): 1–10.

Schoeninger, M.J. and DeNiro, M.J. 1983. Carbon isotope ratios of bone apatite and animal diet reconstruction (reply). *Nature*, *301*(5896): 178.

Schoeninger, M.J. and DeNiro, M.J. 1984. Nitrogen and carbon isotopic composition of bone collagen from marine and terrestrial animals. *Geochimica et Cosmochimica acta*, *48*(4): 625–639.

Schroeder, H., O'Connell, T.C., Evans, J.A., Shuler, K.A., and Hedges, R.E. 2009. Trans-Atlantic slavery: Isotopic evidence for forced migration to Barbados. *American Journal of Physical Anthropology* 139(4): 547–557.

Schroeder, H., Shuler, K.A., and Chenery, S.R. 2013. Childhood lead exposure in an enslaved African community in Barbados: Implications for birthplace and health status. *American Journal of Physical Anthropology* 150(2): 203–209.

Sheridan, R.B. 1985. *Doctors and Slaves: A Medical and Demographic History of Slavery in the British West Indies, 1680–1834*. Cambridge: Cambridge University Press

Sparks, H., Varney, T.L., Courtaud, P., Romon, T., and Watters, D. 2012. Reconstructing the diet of the enslaved from two Archaeological sites (Guadeloupe and Montserrat). *Caribbean Connections: A Publication of the Field Research Centre* 2(1): 1–9. Special Issue focusing in the Bioarchaeology of the Caribbean. https://fieldresearchcentre.weebly.com/index.html

Stuart-Macadam, P.L. 1989. Nutritional deficiency diseases: A survey of scurvy, rickets, and iron-deficiency anemia. In: *Reconstruction of Life from the Skeleton*, edited by M.Y. Iscan and K.A.R. Kennedy, pp. 211–220. New York: Wiley-Liss.

Swanston, T., Varney, T., Coulthard, I., Feng, R., Bewer, B., Murphy, R., Hennig, C., and Cooper, D. 2012. Element localization in archaeological bone using synchrotron radiation x-ray fluorescence: Identification of biogenic uptake. *Journal of Archaeological Science* 39: 2409–2413.

Swanston, T., Varney, T., Coulthard, I., George, G., Pickering, I., Murphy, R., and Cooper, D. 2015. Synchrotron imaging evidence of biogenic mercury identified in a burial in colonial Antigua. *Journal of Archaeological Science* 58: 26–30.

Swanston, T., Varney, T., Kozachuk, M., Choudhury, S., Bewer, B., Keenleyside, A., Coulthard, I., Nelson, A., Martin, R., Stenton, D., and Cooper, D. 2018. Franklin expedition lead exposure: New insights from high resolution confocal x-ray fluorescence imaging of skeletal microstructure. *PLoS One* 13(8): e0202983

Thomas, A. 1968. *The Newfoundland Journal of Aaron Thomas, Able Seaman in HMS Boston, 1794*, edited by J.M. Murray. Don Mills: Longmans.

TNA CO 9/41: 24 September 1783: *Antigua Assembly and Council to the Admiralty*.

Trapham, T. 1679. *A Discourse on the State of Health in the Island of Jamaica*. London: Boulter.

Twiss, K. 2012. The archaeology of food and social diversity. *Journal of Archaeological Research* 20(4): 357–395.

Varney, T.L. 2011. The Royal Naval Hospital cemetery at English Harbour, Antigua: Not just the "Grave of the Englishman". *Proceedings of the XXIII International Congress for Caribbean Archaeology, Antigua*, 2009. Print pp. 291300; digital pp. 447–459.

Varney, T., Swanston, T., Brown, M., and Murphy, R. 2017. Exposed burials on Galleon Beach, Antigua – a preliminary reporting. *Proceedings of the XXVI International Congress for Caribbean Archaeology, Sint Maarten, 2015.*

Varney, T.L., Swanston, T., Coulthard, I., Cooper, D.M.L., George, G.N., Pickering, I.J., and Murphy, A.R. 2012. A preliminary investigation of lead contamination in a Napoleonic era naval cemetery in Antigua, W.I. *Caribbean Connections: A Publication of the Field Research Centre* 2(1): Special Issue focusing in the Bioarchaeology of the Caribbean. https://fieldresearchcentre.weebly.com/index.html

Voelz, P.M. 1993. *Slave and Soldier: The Military Impact of Blacks in the Colonial Americas*. New York: Garland.

Waldron, H.A. 1973. Lead poisoning in the ancient world. *Medical History* 17(4): 391–399.

Walker, P.L., Bathurst, R.R., Richman, R., Gjerdrum, T., and Andrushko, V.A. 2009. The causes of porotic hyperostosis and cribra orbitalia: A reappraisal of the iron-deficiency anemia hypothesis. *American Journal of Physical Anthropology* 139: 109–125.

Wapler, U., Crubezy, E., and Schultz, M. 2004. Is cribra orbitalia synonymous with anemia? Analysis and interpretation of cranial pathology in Sudan. *American Journal of Physical Anthropology* 123(4): 333–339.

Ward, J.R. 1988. *British West Indian Slavery, 1750–1834: The Process of Amelioration.* Oxford: Clarendon Press.

Waters, C.K. 2020. Re-contextualizing mortality rates and population trends of enslaved Africans in Antigua. *Journal of Caribbean History* 54(1): 1–29.

Wedeen, R.P. 1984. *Poison in the Pot: The Legacy of Lead*. Carbondale: Southern Illinois University Press.

Weiss, R.A. and Esparza, J. 2015. The prevention and eradication of smallpox: A commentary on Sloane (1755) 'An account of inoculation'. *Philosophical Transactions of the Royal Society of London. Series B, Biological Sciences* 370(1666): 20140378. doi:10.1098/rstb.2014.0378

Weiss, E. and R. Jurmain. 2007. Osteoarthritis revisited: A contemporary review of aetiology. *International Journal of Osteoarchaeolology* 17: 437–450.

Wells, E.C., Waters, C.K., Tricarico, A.R., and Fox, G.L. 2017. Agroindustrial soilscapes in the Caribbean: A geochemical perspective from Betty's Hope, Antigua. *Environmental Archaeology* 22(4): 381–393.

Wells, E.C., Pratt, S.M., Fox, G.L., Siegel, P.E., Dunning, N.P., and Murphy, A.R. 2018. Plantation soilscapes: Initial and cumulative impacts of colonial agriculture in Antigua, West Indies. *Environmental Archaeology* 23(1): 23–35.

Weston, D.A. 2008. Investigating the specificity of periosteal reactions in pathology museum specimens. *American Journal of Physical Anthropology* 137: 48–59.

Weston, D.A. 2012. Non-specific infection in paleopathology: Interpreting periosteal reactions. In: *Companion to Paleopathology*, edited by A.L. Grauer, pp. 492–512. New York: Wiley-Blackwell.

Wittmers Jr, L.E., Aufderheide, A.C., Pounds, J.G., Jones, K.W., and Angel, J.L. 2008. Problems in determination of skeletal lead burden in archaeological samples: An example from the first African Baptist Church population. *American Journal of Physical Anthropology* 136(4): 379–386.

World Health Organization [WHO]. 2019. *Lead Poisoning and Health*. https://www.who.int/en/news-room/fact-sheets/detail/lead-poisoning-and-health

Ziegler, E.E., Edwards, B.B., Jensen, R.L., Mahaffey, K.R., and Fomon, S.J. 1978. Absorption and retention of lead by infants. *Pediatric Research* 12: 29–34.

Research Ethics

14

ETHICAL AND SUSTAINABLE

Toward an Archaeology of Decolonization in the Neo-Colonial World

Mirjana Roksandic

This paper is a very personal view of the ever-changing dilemmas on ethics of research in biological anthropology, bioarchaeology, and any archaeological research, including skeletal remains of past human populations, burials, and associated material culture. I have structured it as an autoethnography to avoid passing judgments on fellow archaeologists, as I firmly believe that lapses in ethical behavior stem from a lack of reflection on the issue, succumbing to cultural and academic norms and pressures, and general attitudes toward the past, rather than from any deliberate action. The questions of ethics in archaeological research as they pertain to colonial power imbalances span more than 30 years of my involvement in multidisciplinary international projects in the field, from an undergraduate team member to a principal investigator (PI). While the issues of ethical treatment of Indigenous skeletal remains in the colonial setting came to my attention only after I moved to Canada and started a PhD in archaeology at Simon Fraser University in the mid-90s, I was faced with intricacies, complexities, and power relationships in international collaborations much earlier. These personal experiences shaped much of my understanding of international collaborations and resulted in a model that I practice in my current projects. What is considered ethical is historically bound and ever changing; it is situation-specific and will, of course, vary from one locality to another. Recently, a young colleague summarized her ethics approach as "even if it means not getting to be the first one to publish on the topic, I do not want to be ashamed, 10 years down the road, of what I did in the name of expediency."

As an undergraduate student, I worked on international projects in the former Yugoslavia funded by Yugoslav government grants to local researchers (both rescue and research excavations); on projects funded through USA grants and co-directed by Yugoslav and USA researchers; on projects in France directed by French researchers on French grants; on projects in the UK funded by UK grants; and on

DOI: 10.4324/9780429444449-19

projects in France funded through collaborative grants from France and the USA. As a postdoctoral fellow, I worked closely with a Portuguese team where funding was mostly obtained through local resources (University, Township, private donations), and some were obtained through my then employer, the Max Planck Institute for Demographic Research, and later through the University of Toronto. When I started working in the Caribbean, major funding was obtained through Canadian institutions: the Social Sciences and Humanities Research Council of Canada (SSHRC) and various institutional grants. Some of it (including important logistic support) was funded through Cuban institutions. My research in the Paleolithic archaeology of Serbia was funded by regular contributions of both Canadian and Serbian grants. It is therefore reasonable to say that my personal experience has covered modalities of many different combinations of team members and funding packages, as well as the position of a team member from a non-dominant (Yugoslavia, Portugal) and dominant academic tradition (France, USA), as well as roles with decision-making capacity (Caribbean, Serbia).

These various experiences have allowed me to observe the neo-colonial attitudes stemming from funding sources, the dominant academic traditions, and power dynamics in international collaborations. It has also attuned me to the need to ponder the ethics of such endeavors and the ways in which our efforts can be stirred to cause the least harm with maximum benefit for all involved.

Archaeology as a Colonial Science

From the way the (Western) European archaeological sequence was used as a norm in defining cultures and periods in Europe and beyond, to the quest for enlarging museum collections by often unsavory practices, there is no doubt that archaeology arose from both the genuine interest in the past and colonial practices and attitudes. The European understanding of historical development, influenced/supported by nineteenth-century social thinkers like Engels and Durkheim, established a framework of unilinear evolution of societies and cultures: primitive, slavery, feudal, and capitalist. This flow from primitive to industrialized was considered universal and applied to the rest of the world, despite the very lively combination of slavery and capitalism practiced by Western Europe contradicting this neat linear progression. This general progression of society was considered an inevitable co-evolution of technology (means of production) and society (social structure). Anthropological and archaeological observations have subsequently demonstrated that this system *might* have been applicable to the known history of Western Europe, but that once transplanted into the past or into different areas of the world, the structure collapses. In archaeology, this attitude is best exemplified by the assumed correlation between pottery and agriculture. While this connection can be observed in European archaeology, it was by no means the case with other areas of the world (Japan, the Northwest Coast of North America, and the Caribbean, to name a few).

Recently, the Society of American Archaeology (SAA, 2022) has published an acknowledgment of harm that recognizes the destructive nature of archaeology when practiced within the colonial paradigm. One of the first steps in addressing the issues arising from colonial-neocolonial attitudes is recognizing that they exist. If we want to transcend our biases and allow archaeological interpretation to draw not only on one historical experience but also on the other histories and systems of knowing, archaeology must be able to recognize where it is coming from and the implications of these inherent biases on interpretation of the past, especially when it examines the past of non-dominant or colonized cultures.

Does It Have to Be Colonial?

All of us practicing archaeology in the twenty-first century agree and proclaim that the field needs to be decolonized. How exactly do we accomplish this is a matter of some thinking and soul-searching. Ethics protocols are established by most universities. They are, however, rarely relevant to archaeology as a field or to handling archaeological materials/samples. They fail to address the issues that are at the core of archaeological research, no longer limited to the question of "who curates the archaeological finds," but broadened to include important questions such as "who curates the data" and consequentially, "who gets to interpret the past?"

The positivist stance of processual archaeology, which dominated the field from the 60s to the early 80s, was characterized by an unwarranted sense that archaeologists not only **can know about the past** but that they **know better** than those whose past they are examining. This stance was amply criticized by post-processual archaeology. Science is not practiced in a vacuum. The attitude that we can do away with local ecological or historical knowledge is evident in the architecture of the university campuses from the same period where most of us still teach: no natural light and air, and no respect for environmentally sound design. While providing the necessary critique of the positivist mindset, in its more extreme form, post-processual archaeology relativized our knowledge to a degree that became counterproductive. If everything in our interpretation is based on our own biases and ideologies, why bother excavating the sites? Should we stop doing archaeology? Archaeology is inherently destructive. If we are just reinforcing our biases by uncovering more sites, there is no need to practice the discipline. Thankfully, the extreme opinions did not lead us to abandon archaeology; they made us recognize inherent fallacies and opened us to new venues of research into the behaviors, lives, and sometimes even attitudes of ancient people. As a species, we love to dwell on our past and predict the future, and we cannot leave this field to fantastical interpretations like in *Ancient Apocalypse,* the Netflix documentary series that presents our hard-won realization that there are aspects of the past that we are not likely to know, such as the failure or—worse still—the conspiracy of the discipline.

Archaeology Can Be a Force for the Better

While there is little doubt that archaeology developed as a colonial science within a Eurocentric worldview, archaeology countered the colonial narrative by demonstrating the prowess, ability, and artistic achievements of past local cultures and ancestral groups in many colonial settings. Archaeology is often one of the few (if not only) means available to the acculturated groups to reconnect with their past. If practiced with an open mind, it creates an opportunity to learn about different past adaptations about the ways humans dealt with ever-changing natural and cultural surroundings. Learning about past cultures has shifted from looking for beautiful art pieces to looking for lived experiences and the ways cultural norms interacted with the given historical and natural moment. It is no longer concerned exclusively with the elite but strives to understand the marginalized and unrecorded. In fact, other than oral histories, it is often our only connection to and means of understanding the cultures, the past, and the way of life of the marginalized, and unrecorded. In the colonial and neocolonial setting, it has the potential to give a voice to the marginalized, acknowledging the prowess of past cultures and demonstrating the connection between the past (ancestral groups) and their living descendants. It can and often does play a part in building pride and confidence in the descendant populations. In particular in the Caribbean, where most history textbooks start with European colonization, further marginalizing the Indigenous voices, archaeology can and should be practiced to uncover, document, and understand the pre-contact groups with or without recognized living descendants.

Should We Stop International Collaborations?

While no longer colonial, archaeology still suffers from the disequilibrium of power of the neo-colonial world. This disequilibrium in the power structure between the regions of archaeological interest and international teams (often, but not limited to, Western Europeans/Americans) stems from the latter commanding all or more substantial funding and command of the expertise for which no formal training or access to infrastructure exists in the country of research. The substantial funding (the source of power imbalance) comes with a set of expectations from the granting agencies, funders, and universities where international PIs are located. In addition, there is a strong expectation of immediate, significant results, as well as prompt and substantial dissemination in prominent academic journals—and, of course, a presence in both the traditional media and on social media platforms. Basically, we're peddling our results under the guise of public engagement in the hope that it will eventually lead to more funding for the fieldwork and analyses, to the benefit of both local and international researchers. The rigors of ethical scientific practice can often conflict with the demands of the granting agencies and future funding for field work. Local archaeologists could be under quite a different set of expectations from their governments, local communities, and their own academic milieu.

Even the forms of dissemination, the structure of communications, and the way data are presented could be and often are different. On top of that, the speed at which different sub-disciplines—often in the domain of the expertise of the international researchers—can present their data in publishable form favors small bits of natural science info (dates, isotopes, etc.) rather than interpretations of sites, regions, or archaeological phenomena, which need to incorporate many more lines of evidence to build a plausible (publishable) picture. However, avoiding the risk of dealing with overbearing foreigners by avoiding international teams and conducting research with only local funding and know-how comes with a greater risk: the risk of parochialism and missing out on the developments in the field.

The only viable solution is to establish an equitable relationship in which the members of the local academic community (archaeologists, historians, environmentalists, cavers, biologists, geologists, etc.), non-academic community members, Indigenous/local knowledge bearers, and international team members contribute freely and consistently. Including local knowledge bearers (academic or otherwise) in the decision-making process from the very beginning of the planning—asking relevant questions and setting up the goals—to the execution, interpretation, and dissemination of the project presupposes the ability of the whole team to communicate across different languages, systems of education, or knowledge bases. It requires an open mind and a willingness not only to accept the knowledge of other team members, but also the ability to share one's own ideas, methods, and understandings. Even if a local museum does not require researchers to publish in prominent journals on a yearly basis, communicating the requirements of the granting agencies from the outset creates a platform for equitable teamwork and decision-making in both research and dissemination. Respect for and reliance on the expertise of local researchers has an added benefit: out of 180 days that my teammates in Serbia spent in the field this year, I could—even during my research leave—only participate for 30 days. Clear communication about the goals and the roles of each team member facilitates such arrangements regardless of the sources of funding.

Sensitive Materials

In the not-so-distant past, I was attending a conference where a researcher presented results of their "work *on* the DNA of *subjects* from an Indigenous community" (my emphasis) in the context of gathering comparative data for bioarchaeological research. Informed consent notwithstanding, the idea of conducting "*objective scientific work on* a group of people" carried many negative connotations. What surprised me the most is the complete lack of awareness that this might be a problem on the part of the researcher, which only served to reinforce the stereotype of a colonial scientist in search of the archaeological El Dorado.

Thankfully, in recent years we have slowly progressed from the question of *What do you work on*? to *Who do you work with*?—emphasizing collaborative projects with descendant groups or local communities—to asking *Who do you work for*? Or, put differently, *How does your research benefit the local community*? This approach, consistent with the principles of Collective Benefit, Authority to Control, Responsibility, and Ethics (CARE) (GIDA, 2022), is developed around working with Indigenous communities and guaranties that their rights are respected when it comes to setting the goals and publishing results of research. The CARE principles seem to favor the Indigenous communities' interest over the global scientific goals only if you assume that local or Indigenous communities do not have the same curiosity toward the general archaeological questions or that they are a monolithic group of like-minded individuals. Starting your project with a local community under such an assumption is both wrong and, I would claim, unethical. There is an inherent diversity of opinions in any human group; however, the cultural norm of the community sets the standard of what is or is not appropriate and beneficial, leaving the authority of control firmly with those whose past is the core of research and avoiding the pitfalls of a colonial approach to setting the questions. Rather than limiting the science to a set of prescribed questions, this principle has a great potential to enlarge and diversify the scope of research, opening the doors to more innovative approaches and interpretations, benefiting both the Indigenous community and global science. The communication at the onset of the project also allows the researcher to present the potential of different methodologies that could provide benefits to the community that may not have been considered by the members prior to consultations.

Once the goals are set and the results start streaming in, the next important question is: *Who has the rights to the results of a joint project*? Who gets the right to publish, and when? These questions get further problematized when working on sensitive questions that could have unintended impacts on local communities: aDNA might be different from the DNA of local communities, potentially contradicting or diminishing the community's claim to the land in legal disputes with the colonial governments. How your results are communicated also matters, especially when it comes to the ancestry of an individual. When Kennewick Man was reported in *Time* magazine as having "Caucasoid traits" (Lemonick & Dorfman, 2006)—a rather nonsensical statement in the context of biological anthropology—it had strong reverberations in the lay community. Among the lay settlers, it read, "Europeans were here first, then they were chased out by First Nations peoples, then Europeans conquered them, so what is a big deal?" justifying colonial oppression in the name of history. While such a reading was far from the intention of the archaeologist who reported it, it took the aDNA to demonstrate that while the individual had Caucasoid traits that did not mean that he was of European origin—that the morphological traits observed were just that, morphological traits without relevance to the ancestry of the individual. It should have been obvious that reporting it in a way that contradicts bioanthropology's understanding of human variation but caters to popular stereotypes is unethical and would have unintended consequences with real impact on the communities. No matter where the research money comes from and what the exigencies of the funding body and needs of the researchers' careers are, when publications can have a negative impact on descendant/Indigenous communities, they should have a right to say, "We do not want this published."

When our research contradicts the received wisdom (whatever its source), an ethical approach requires us to report our findings as they are, to the best of our understanding. Our first responsibility is to report our findings to local stakeholders in a way that can be understood by non-specialists while maintaining scientific rigor and integrity, and to discuss the meaning of our results in a comprehensive way, including their possible consequences. The decision on when and what to publish should be guided by the interests of the Indigenous community, which will have a final say in the discussions of all interested parties, to prevent negative impacts of the findings on local stakeholders.

Disciplinary Imbalances

New issues in power imbalances arise from the comparative value assigned to archaeology as a social science and to the biological, geochemical, molecular, and other hard science methods used in the interpretation of archaeological data. We are all too familiar with titles like "*Genetics rewrite the history and the archaeology of* (insert region, historical period, problem)." There is no doubt in my mind, as a bio-anthropologist, that the impact of scientific methods has substantially increased the resolution at which we can examine archaeological sites. However, having learned from the critique of the positivist mindset that dominated processual archaeology, we should not repeat the mistakes of the past and uncritically accept the results of any new method as providing the absolute truth. Our information is truncated and often biased by natural and cultural factors. The inclusion of scientific methodology from excavation to analysis and interpretation benefited the field immensely. However, the source (excavation of an archaeological site) and the ultimate interpretation of all these scientific data lie in the ability of the social science of archaeology to collate, compare, and intertwine the results of many disciplines. Except in a few cataclysmic events, the accumulated data stem from a fortuitous inclusion of evidence into geological layers that need to be disentangled and interpreted by archaeologists. The interpretation of social, cultural, and bio-cultural phenomena cannot be the sole domain of hard science, especially if they lack training in social sciences. When dealing with human remains or past cultures, their interpretations, while maybe technically correct, can unintentionally promote problematic and dubious conclusions. Ultimately, archaeology is a truly interdisciplinary endeavor only if all the disciplines can cross-communicate their results.

The ethics of international collaboration in archaeological and even more so in bioarchaeological research has received a lot of attention in the last decade. Well-funded workshops with brilliant international participants debated the best ways to move forward. Our granting agencies and our academic employers insist on statements that cover the ethics of proposed research and ask us to demonstrate equity, diversity, and inclusion; the big three scientific journals (Nature, Science, PNAS) insist on an ethics statement. Being academics—good with our words—we are often able to convincingly demonstrate the adherence to the most rigorous ethical standards in our research, yet these statements have the potential to remain just that: well-written statements that are not followed by rigorous action. This is particularly relevant when new methods that require the aggregation of many samples from many different sources are required for a comprehensive research goal. The onus of obtaining samples ethically is not only on those who perform the analysis. It is the duty of the archaeologist and biological anthropologist to maintain and oversee the collection of these samples in accordance with the ethical standards of the field and the local legal procedures. Where such procedures are not available, as is the case on many Caribbean islands, we should work with local authorities to help establish the legal framework for heritage management for the benefit of the local community and future researchers.

Like other important concepts in our lives, ethics is also a product of the dialectical interplay of universal and temporal. While the universals can be summed up as practicing respect and operating within the legal framework, the particulars will depend on the local conditions and social and cultural norms of the time and the space. This interplay will provide the fine balance and the roadway in each individual situation, but some steps (that are easy to formulate) can be taken in our current practice of bioarchaeology.

From the inception of the project, work with local communities that know their landscape. Much can be learned from their insights into the area of interest, and some questions that you hadn't even dreamt of can be elucidated. Should you encounter reticence in knowledge sharing, respect it; you do not want to repeat the experiences of the extractive nature of the colonial approach. If knowledge is shared, acknowledge it from the start. Benefits to the community should be an important concern in heritage research but do not offer what you cannot deliver. In devising the best project, listen, learn, and acknowledge the expertise of all knowledge-bearers and their methodologies.

Establish an equitable relationship with local archaeologists and heritage specialists. Assume that they know their local archaeological milieu and their heritage landscape more intimately than you do until you have spent some serious time in the same area. The equitable relationship also means that you respect and are willing to share your own knowledge and perspectives. Relationships take a long time to build. While supporting the education of local specialists is one of the best ways to support local heritage institutions, avoid the role of colonial savior by truly integrating your research with the interests and perspectives of the local academic community. Work with as wide an array of sciences as you can to comprehend and incorporate into your interpretation, but do not serve as an intermediary, collecting samples for hard sciences. Archaeologists on your team should have a say in the interpretation of the results, even if the data are outside of their area of expertise. Do not duplicate samples, even if your team is in direct competition with a different lab. Collaborate; competition is often contradictory to an ethical practice of science. Do not work on the material if it is obtained unethically or without a proper legal framework. Most importantly, take your time to build a community around shared interests and goals and be open when communicating—both listening and learning and imparting your own knowledge and expertise. Local experience will vary; in order to behave ethically, respect the position of those you are working with.

References

Global Indigenous Data Alliance (GIDA). 2022. *CARE Principles for Indigenous Data Governance*. https://www.gida-global.org/care.

Lemonick, M.D., & Dorfman, A. 2006, March. Who were the first Americans? *Time*. https://content.time.com/time/subscriber/article/0,33009,1169905,00.html.

Society for American Archaeology (SAA). 2022. *Acknowledgement of Harms*. https://archaeology.blob.core.windows.net/container/docs/default-source/saa-news-and-press-statements/acknowledgment-of-harms_2022-05-05.pdf?sfvrsn=3c8c18e_4.

INDEX

Note: **Bold** page numbers refer to tables; *italic* page numbers refer to figures and page numbers followed by "n" denote endnotes.

For Product Safety Concerns and Information please contact our EU representative GPSR@taylorandfrancis.com Taylor & Francis Verlag GmbH, Kaufingerstraße 24, 80331 München, Germany

Batch number: 10733620

Printed by Printforce, the Netherlands